Lecture Notes in Artificial Intelligence   11441

Subseries of Lecture Notes in Computer Science

LNAI Series Editors

Randy Goebel
University of Alberta, Edmonton, Canada
Yuzuru Tanaka
Hokkaido University, Sapporo, Japan
Wolfgang Wahlster
DFKI and Saarland University, Saarbrücken, Germany

LNAI Founding Series Editor

Joerg Siekmann
DFKI and Saarland University, Saarbrücken, Germany

More information about this series at http://www.springer.com/series/1244

Qiang Yang · Zhi-Hua Zhou ·
Zhiguo Gong · Min-Ling Zhang ·
Sheng-Jun Huang (Eds.)

Advances in Knowledge Discovery and Data Mining

23rd Pacific-Asia Conference, PAKDD 2019
Macau, China, April 14–17, 2019
Proceedings, Part III

Springer

Editors
Qiang Yang
Hong Kong University of Science
and Technology
Hong Kong, China

Zhiguo Gong
University of Macau
Taipa, Macau, China

Sheng-Jun Huang
Nanjing University of Aeronautics
and Astronautics
Nanjing, China

Zhi-Hua Zhou
Nanjing University
Nanjing, China

Min-Ling Zhang
Southeast University
Nanjing, China

ISSN 0302-9743   ISSN 1611-3349   (electronic)
Lecture Notes in Artificial Intelligence
ISBN 978-3-030-16141-5   ISBN 978-3-030-16142-2   (eBook)
https://doi.org/10.1007/978-3-030-16142-2

Library of Congress Control Number: 2019934768

LNCS Sublibrary: SL7 – Artificial Intelligence

This Springer imprint is published by the registered company Springer Nature Switzerland AG
The registered company address is: Gewerbestrasse 11, 6330 Cham, Switzerland

PC Chairs' Preface

It is our great pleasure to introduce the proceedings of the 23rd Pacific-Asia Conference on Knowledge Discovery and Data Mining (PAKDD 2019). The conference provides an international forum for researchers and industry practitioners to share their new ideas, original research results, and practical development experiences from all KDD-related areas, including data mining, data warehousing, machine learning, artificial intelligence, databases, statistics, knowledge engineering, visualization, decision-making systems, and the emerging applications.

We received 567 submissions to PAKDD 2019 from 46 countries and regions all over the world, noticeably with submissions from North America, South America, Europe, and Africa. The large number of submissions and high diversity of submission demographics witness the significant influence and reputation of PAKDD. A rigorous double-blind reviewing procedure was ensured via the joint efforts of the entire Program Committee consisting of 55 Senior Program Committee (SPC) members and 379 Program Committee (PC) members.

The PC Co-Chairs performed an initial screening of all the submissions, among which 25 submissions were desk rejected due to the violation of submission guidelines. For submissions entering the double-blind review process, each one received at least three quality reviews from PC members or in a few cases from external reviewers (with 78.5% of them receiving four or more reviews). Furthermore, each valid submission received one meta-review from the assigned SPC member who also led the discussion with the PC members. The PC Co-Chairs then considered the recommendations and meta-reviews from SPC members, and looked into each submission as well as its reviews and PC discussions to make the final decision. For borderline papers, additional reviews were further requested and thorough discussions were conducted before final decisions.

As a result, 137 out of 567 submissions were accepted, yielding an acceptance rate of 24.1%. We aim to be strict with the acceptance rate, and all the accepted papers are presented in a total of 20 technical sessions. Each paper was allocated 15 minutes for oral presentation and 2 minutes for Q/A. The conference program also featured three keynote speeches from distinguished data mining researchers, five cutting-edge workshops, six comprehensive tutorials, and one dedicated data mining contest session.

We wish to sincerely thank all SPC members, PC members and externel reviewers for their invaluable efforts in ensuring a timely, fair, and highly effective paper review and selection procedure. We hope that readers of the proceedings will find that the PAKDD 2019 technical program was both interesting and rewarding.

February 2019

Zhiguo Gong
Min-Ling Zhang

General Chairs' Preface

On behalf of the Organizing Committee, it is our great pleasure to welcome you to Macau, China for the 23rd Pacific-Asia Conference on Knowledge Discovery and Data Mining (PAKDD 2019). Since its first edition in 1997, PAKDD has well established as one of the leading international conferences in the areas of data mining and knowledge discovery. This year, after its four previous editions in Beijing (1999), Hong Kong (2001), Nanjing (2007), and Shenzhen (2011), PAKDD was held in China for the fifth time in the fascinating city of Macau, during April 14–17, 2019.

First of all, we are very grateful to the many authors who submitted their work to the PAKDD 2019 main conference, satellite workshops, and data mining contest. We were delighted to feature three outstanding keynote speakers: Dr. Jennifer Neville from Purdue University, Professor Hui Xiong from Baidu Inc., and Professor Josep Domingo-Ferrer from Universitat Rovira i Virgili. The conference program was further enriched with six high-quality tutorials, five workshops on cutting-edge topics, and one data mining contest on AutoML for lifelong machine learning.

We would like to express our gratitude to the contributions of the SPC members, PC members, and external reviewers, led by the PC Co-Chairs, Zhiguo Gong and Min-Ling Zhang. We are also very thankful to the other Organizing Committee members: Workshop Co-Chairs, Hady W. Lauw and Leong Hou U, Tutorial Co-Chairs, Bob Durrant and Yang Yu, Contest Co-Chairs, Hugo Jair Escalante and Wei-Wei Tu, Publicity Co-Chairs, Yi Cai, Xiangnan Kong, Gang Li, and Yasuo Tabei, Proceedings Chair, Sheng-Jun Huang, and Local Arrangements Chair, Andrew Jiang. We wish to extend our special thanks to Honorary Co-Chairs, Hiroshi Motoda and Lionel M. Ni, for their enlightening support and advice throughout the conference organization.

We appreciate the hosting organization University of Macau, and our sponsors Macao Convention & Exhibition Association, Intel, Baidu, for their institutional and financial support of PAKDD 2019. We also appreciate the Fourth Paradigm Inc., ChaLearn, Microsoft, and Amazon for sponsoring the PAKDD 2019 data mining contest. We feel indebted to the PAKDD Steering Committee for its continuing guidance and sponsorship of the paper award and student travel awards.

Last but not least, our sincere thanks go to all the participants and volunteers of PAKDD 2019—there would be no conference without you. We hope you enjoy PAKDD 2019 and your time in Macau, China.

February 2019

Qiang Yang
Zhi-Hua Zhou

Organization

Organizing Committee

Honorary Co-chairs

Hiroshi Motoda   Osaka University, Japan
Lionel M. Ni   University of Macau, SAR China

General Co-chairs

Qiang Yang   Hong Kong University of Science and Technology, SAR China
Zhi-Hua Zhou   Nanjing University, China

Program Committee Co-chairs

Zhiguo Gong   University of Macau, China
Min-Ling Zhang   Southeast University, China

Workshop Co-chairs

Hady W. Lauw   Singapore Management University, Singapore
Leong Hou U   University of Macau, China

Tutorial Co-chairs

Bob Durrant   University of Waikato, New Zealand
Yang Yu   Nanjing University, China

Contest Co-chairs

Hugo Jair Escalante   INAOE, Mexico
Wei-Wei Tu   The Fourth Paradigm Inc., China

Publicity Co-chairs

Yi Cai   South China University of Technology, China
Xiangnan Kong   Worcester Polytechnic Institute, USA
Gang Li   Deakin University, Australia
Yasuo Tabei   RIKEN, Japan

Proceedings Chair

Sheng-Jun Huang   Nanjing University of Aeronautics and Astronautics, China

Local Arrangements Chair

Andrew Jiang   Macao Convention & Exhibition Association, China

Steering Committee

Co-chairs

Ee-Peng Lim   Singapore Management University, Singapore
Takashi Washio   Institute of Scientific and Industrial Research,
                                   Osaka University, Japan

Treasurer

Longbing Cao   Advanced Analytics Institute, University
                                   of Technology, Sydney, Australia

Members

Dinh Phung   Monash University, Australia (Member since 2018)
Geoff Webb   Monash University, Australia (Member since 2018)
Jae-Gil Lee   Korea Advanced Institute of Science & Technology,
                                   Korea (Member since 2018)
Longbing Cao   Advanced Analytics Institute,
                                   University of Technology, Sydney, Australia
                                   (Member since 2013, Treasurer since 2018)
Jian Pei   School of Computing Science,
                                   Simon Fraser University (Member since 2013)
Vincent S. Tseng   National Cheng Kung University,
                                   Taiwan (Member since 2014)
Gill Dobbie   University of Auckland,
                                   New Zealand (Member since 2016)
Kyuseok Shim   Seoul National University, Korea (Member since 2017)

Life Members

P. Krishna Reddy   International Institute of Information Technology,
                                   Hyderabad (IIIT-H), India (Member since 2010,
                                   Life Member since 2018)
Joshua Z. Huang   Shenzhen University, China (Member since 2011,
                                   Life Member since 2018)
Ee-Peng Lim   Singapore Management University, Singapore
                                   (Member since 2006, Life Member since 2014,
                                   Co-chair 2015–2017, Chair 2018–2020)
Hiroshi Motoda   AFOSR/AOARD and Osaka University, Japan
                                   (Member since 1997, Co-chair 2001–2003,
                                   Chair 2004–2006, Life Member since 2006)

Rao Kotagiri   University of Melbourne, Australia (Member since
                           1997, Co-chair 2006–2008, Chair 2009–2011,
                           Life Member since 2007, Co-sign since 2006)
Huan Liu   Arizona State University, USA (Member since 1998,
                           Treasurer 1998–2000, Life Member since 2012)
Ning Zhong   Maebashi Institute of Technology,
                           Japan (Member since 1999, Life Member
                           since 2008)
Masaru Kitsuregawa   Tokyo University, Japan (Member since 2000,
                           Life Member since 2008)
David Cheung   University of Hong Kong, SAR China (Member since
                           2001, Treasurer 2005–2006, Chair 2006–2008,
                           Life Member since 2009)
Graham Williams   Australian National University, Australia
                           (Member since 2001, Treasurer 2006–2017, Co-sign
                           since 2006, Co-chair 2009–2011, Chair 2012–2014,
                           Life Member since 2009)
Ming-Syan Chen   National Taiwan University, Taiwan (Member since
                           2002, Life Member since 2010)
Kyu-Young Whang   Korea Advanced Institute of Science & Technology,
                           Korea (Member since 2003, Life Member
                           since 2011)
Chengqi Zhang   University of Technology Sydney, Australia
                           (Member since 2004, Life Member since 2012)
Tu Bao Ho   Japan Advanced Institute of Science and Technology,
                           Japan (Member since 2005, Co-chair 2012–2014,
                           Chair 2015–2017, Life Member since 2013)
Zhi-Hua Zhou   Nanjing University, China (Member since 2007,
                           Life Member since 2015)
Jaideep Srivastava   University of Minnesota, USA (Member since 2006,
                           Life Member since 2015)
Takashi Washio   Institute of Scientific and Industrial Research, Osaka
                           University (Member since 2008, Life Member since
                           2016, Co-chair 2018–2020)
Thanaruk Theeramunkong   Thammasat University, Thailand (Member since 2009,
                           Life Member since 2017)

Past Members

Hongjun Lu   Hong Kong University of Science and Technology,
                           SAR China (Member 1997–2005)
Arbee L. P. Chen   National Chengchi University,
                           Taiwan (Member 2002–2009)
Takao Terano   Tokyo Institute of Technology,
                           Japan (Member 2000–2009)

Tru Hoang Cao   Ho Chi Minh City University of Technology,
                              Vietnam (Member 2015–2017)
Myra Spiliopoulou   Information Systems, Otto-von-Guericke-University
                              Magdeburg (Member 2013–2019)

Senior Program Committee

James Bailey   University of Melbourne, Australia
Albert Bifet   Telecom ParisTech, France
Longbin Cao   University of Technology Sydney, Australia
Tru Cao   Ho Chi Minh City University of Technology, Vietnam
Peter Christen   Australian National University, Australia
Peng Cui   Tsinghua University, China
Guozhu Dong   Wright State University, USA
Benjamin C. M. Fung   McGill University, Canada
Bart Goethals   University of Antwerp, Belgium
Geoff Holmes   University of Waikato, New Zealand
Qinghua Hu   Tianjin University, China
Xia Hu   Texas A&M University, USA
Sheng-Jun Huang   Nanjing University of Aeronautics and Astronautics,
                              China
Shuiwang Ji   Texas A&M University, USA
Kamalakar Karlapalem   IIIT Hyderabad, India
George Karypis   University of Minnesota, USA
Latifur Khan   University of Texas at Dallas, USA
Byung S. Lee   University of Vermont, USA
Jae-Gil Lee   KAIST, Korea
Gang Li   Deakin University, Australia
Jiuyong Li   University of South Australia, Australia
Ming Li   Nanjing University, China
Yu-Feng Li   Nanjing University, China
Shou-De Lin   National Taiwan University, Taiwan
Qi Liu   University of Science and Technology of China, China
Weiwei Liu   University of New South Wales, Australia
Nikos Mamoulis   University of Ioannina, Greece
Wee Keong Ng   Nanyang Technological University, Singapore
Sinno Pan   Nanyang Technological University, Singapore
Jian Pei   Simon Fraser University, Canada
Wen-Chih Peng   National Chiao Tung University, Taiwan
Rajeev Raman   University of Leicester, UK
Chandan K. Reddy   Virginia Tech, USA
Krishna P. Reddy   IIIT Hyderabad, India
Kyuseok Shim   Seoul National University, Korea
Myra Spiliopoulou   Otto-von-Guericke-University Magdeburg, Germany
Masashi Sugiyama   RIKEN/The University of Tokyo, Japan
Jiliang Tang   Michigan State University, USA

Kai Ming Ting   Federation University, Australia
Hanghang Tong   Arizona State University, USA
Vincent S. Tseng   National Chiao Tung University, Taiwan
Fei Wang   Cornell University, USA
Jianyong Wang   Tsinghua University, China
Jie Wang   University of Science and Technology of China, China
Wei Wang   University of California at Los Angeles, USA
Takashi Washio   Osaka University, Japan
Jia Wu   Macquarie University, Australia
Xindong Wu   Mininglamp Software Systems, China
Xintao Wu   University of Arkansas, USA
Xing Xie   Microsoft Research Asia, China
Jeffrey Xu Yu   Chinese University of Hong Kong, SAR China
Osmar R. Zaiane   University of Alberta, Canada
Zhao Zhang   Soochow University, China
Feida Zhu   Singapore Management University, Singapore
Fuzhen Zhuang   Institute of Computing Technology, CAS, China

Program Committee

Saurav Acharya   University of Vermont, USA
Swati Agarwal   BITS Pilani Goa, India
David Albrecht   Monash University, Australia
David Anastasiu   San Jose State University, USA
Luiza Antonie   University of Guelph, Canada
Xiang Ao   Institute of Computing Technology, CAS, China
Sunil Aryal   Deakin University, Australia
Elena Baralis   Politecnico di Torino, Italy
Jean Paul Barddal   Pontifícia Universidade Católica do Paraná, Brazil
Arnab Basu   Indian Institute of Management Bangalore, India
Gustavo Batista   Universidade de São Paulo, Brazil
Bettina Berendt   KU Leuven, Belgium
Raj K. Bhatnagar   University of Cincinnati, USA
Arnab Bhattacharya   Indian Institute of Technology, Kanpur, India
Kevin Bouchard   Université du Quebec a Chicoutimi, Canada
Krisztian Buza   Eotvos Lorand University, Hungary
Lei Cai   Washington State University, USA
Rui Camacho   Universidade do Porto, Portugal
K. Selcuk Candan   Arizona State University, USA
Tanmoy Chakraborty   Indraprastha Institute of Information Technology Delhi,
                                India
Shama Chakravarthy   University of Texas at Arlington, USA
Keith Chan   Hong Kong Polytechnic University, SAR China
Chia Hui Chang   National Central University, Taiwan
Bo Chen   Monash University, Australia
Chun-Hao Chen   Tamkang University, Taiwan

Lei Chen	Nanjing University of Posts and Telecommunications, China
Meng Chang Chen	Academia Sinica, Taiwan
Rui Chen	Samsung Research America, USA
Shu-Ching Chen	Florida International University, USA
Songcan Chen	Nanjing University of Aeronautics and Astronautics, China
Yi-Ping Phoebe Chen	La Trobe University, Australia
Yi-Shin Chen	National Tsing Hua University, Taiwan
Zhiyuan Chen	University of Maryland Baltimore County, USA
Jiefeng Cheng	Tencent Cloud Security Lab, China
Yiu-ming Cheung	Hong Kong Baptist University, SAR China
Silvia Chiusano	Politecnico di Torino, Italy
Jaegul Choo	Korea University, Korea
Kun-Ta Chuang	National Cheng Kung University, Taiwan
Bruno Cremilleux	Université de Caen Normandie, France
Chaoran Cui	Shandong University of Finance and Economics, China
Lin Cui	Nanjing University of Aeronautics and Astronautics, China
Boris Cule	University of Antwerp, Belgium
Bing Tian Dai	Singapore Management University, Singapore
Dao-Qing Dai	Sun Yat-Sen University, China
Wang-Zhou Dai	Nanjing University, China
Xuan-Hong Dang	IBM T.J. Watson Research Center, USA
Jeremiah Deng	University of Otago, New Zealand
Zhaohong Deng	Jiangnan University, China
Lipika Dey	Tata Consultancy Services, India
Bolin Ding	Data Analytics and Intelligence Lab, Alibaba Group, China
Steven H. H. Ding	McGill University, Canada
Trong Dinh Thac Do	University of Technology Sydney, Australia
Gillian Dobbie	University of Auckland, New Zealand
Xiangjun Dong	Qilu University of Technology, China
Dejing Dou	University of Oregon, USA
Bo Du	Wuhan University, China
Boxin Du	Arizona State University, USA
Lei Duan	Sichuan University, China
Sarah Erfani	University of Melbourne, Australia
Vladimir Estivill-Castro	Griffith University, Australia
Xuhui Fan	University of Technology Sydney, Australia
Rizal Fathony	University of Illinois at Chicago, USA
Philippe Fournier-Viger	Harbin Institute of Technology (Shenzhen), China
Yanjie Fu	Missouri University of Science and Technology, USA
Dragan Gamberger	Rudjer Boskovic Institute, Croatia
Niloy Ganguly	Indian Institute of Technology Kharagpur, India
Junbin Gao	University of Sydney, Australia

Wei Gao	Nanjing University, China
Xiaoying Gao	Victoria University of Wellington, New Zealand
Angelo Genovese	Università degli Studi di Milano, Italy
Arnaud Giacometti	University Francois Rabelais of Tours, France
Heitor M. Gomes	Telecom ParisTech, France
Chen Gong	Nanjing University of Science and Technology, China
Maciej Grzenda	Warsaw University of Technology, Poland
Lei Gu	Nanjing University of Posts and Telecommunications, China
Yong Guan	Iowa State University, USA
Himanshu Gupta	IBM Research, India
Sunil Gupta	Deakin University, Australia
Michael Hahsler	Southern Methodist University, USA
Yahong Han	Tianjin University, China
Satoshi Hara	Osaka University, Japan
Choochart Haruechaiyasak	National Electronics and Computer Technology Center, Thailand
Jingrui He	Arizona State University, USA
Shoji Hirano	Shimane University, Japan
Tuan-Anh Hoang	Leibniz University of Hanover, Germany
Jaakko Hollmén	Aalto University, Finland
Tzung-Pei Hong	National University of Kaohsiung, Taiwan
Chenping Hou	National University of Defense Technology, China
Michael E. Houle	National Institute of Informatics, Japan
Hsun-Ping Hsieh	National Cheng Kung University, Taiwan
En-Liang Hu	Yunnan Normal University, China
Juhua Hu	University of Washington Tacoma, USA
Liang Hu	University of Technology Sydney, Australia
Wenbin Hu	Wuhan University, China
Chao Huang	University of Notre Dame, USA
David Tse Jung Huang	University of Auckland, New Zealand
Jen-Wei Huang	National Cheng Kung University, Taiwan
Nam Huynh	Japan Advanced Institute of Science and Technology, Japan
Akihiro Inokuchi	Kwansei Gakuin University, Japan
Divyesh Jadav	IBM Research, USA
Sanjay Jain	National University of Singapore, Singapore
Szymon Jaroszewicz	Polish Academy of Sciences, Poland
Songlei Jian	University of Technology Sydney, Australia
Meng Jiang	University of Notre Dame, USA
Bo Jin	Dalian University of Technology, China
Toshihiro Kamishima	National Institute of Advanced Industrial Science and Technology, Japan
Wei Kang	University of South Australia, Australia
Murat Kantarcioglu	University of Texas at Dallas, USA
Hung-Yu Kao	National Cheng Kung University, Taiwan

Shanika Karunasekera	University of Melbourne, Australia
Makoto P. Kato	Kyoto University, Japan
Chulyun Kim	Sookmyung Women University, Korea
Jungeun Kim	Korea Advanced Institute of Science and Technology, Korea
Kyoung-Sook Kim	Artificial Intelligence Research Center, Japan
Yun Sing Koh	University of Auckland, New Zealand
Xiangnan Kong	Worcester Polytechnic Institute, USA
Irena Koprinska	University of Sydney, Australia
Ravi Kothari	Ashoka University, India
P. Radha Krishna	National Institute of Technology, Warangal, India
Raghu Krishnapuram	Indian Institute of Science Bangalore, India
Marzena Kryszkiewicz	Warsaw University of Technology, Poland
Chao Lan	University of Wyoming, USA
Hady Lauw	Singapore Management University, Singapore
Thuc Duy Le	University of South Australia, Australia
Ickjai J. Lee	James Cook University, Australia
Jongwuk Lee	Sungkyunkwan University, Korea
Ki Yong Lee	Sookmyung Women's University, Korea
Ki-Hoon Lee	Kwangwoon University, Korea
Sael Lee	Seoul National University, Korea
Sangkeun Lee	Korea University, Korea
Sunhwan Lee	IBM Research, USA
Vincent C. S. Lee	Monash University, Australia
Wang-Chien Lee	Pennsylvania State University, USA
Yue-Shi Lee	Ming Chuan University, Taiwan
Zhang Lei	Anhui University, China
Carson K. Leung	University of Manitoba, Canada
Bohan Li	Nanjing University of Aeronautics and Astronautics, China
Jianmin Li	Tsinghua University, China
Jianxin Li	Deakin University, Australia
Jundong Li	Arizona State University, USA
Nan Li	Alibaba, China
Peipei Li	Hefei University of Technology, China
Qian Li	University of Technology Sydney, Australia
Rong-Hua Li	Beijing Institute of Technology, China
Shao-Yuan Li	Nanjing University, China
Sheng Li	University of Georgia, USA
Wenyuan Li	University of California, Los Angeles, USA
Wu-Jun Li	Nanjing University, China
Xiaoli Li	Institute for Infocomm Research, A*STAR, Singapore
Xue Li	University of Queensland, Australia
Yidong Li	Beijing Jiaotong University, China
Zhixu Li	Soochow University, China

Defu Lian	University of Electronic Science and Technology of China, China
Sungsu Lim	Chungnam National University, Korea
Chunbin Lin	Amazon AWS, USA
Hsuan-Tien Lin	National Taiwan University, Taiwan
Jerry Chun-Wei Lin	Western Norway University of Applied Sciences, Norway
Anqi Liu	California Institute of Technology, USA
Bin Liu	IBM Research, USA
Jiajun Liu	Renmin University of China, China
Jiamou Liu	University of Auckland, New Zealand
Jie Liu	Nankai University, China
Lin Liu	University of South Australia, Australia
Liping Liu	Tufts University, USA
Shaowu Liu	University of Technology Sydney, Australia
Zheng Liu	Nanjing University of Posts and Telecommunications, China
Wenpeng Lu	Qilu University of Technology, China
Jun Luo	Machine Intelligence Lab, Lenovo Group Limited, China
Wei Luo	Deakin University, Australia
Huifang Ma	Northwest Normal University, China
Marco Maggini	University of Siena, Italy
Giuseppe Manco	ICAR-CNR, Italy
Silviu Maniu	Universite Paris-Sud, France
Naresh Manwani	International Institute of Information Technology, Hyderabad, India
Florent Masseglia	Inria, France
Tomoko Matsui	Institute of Statistical Mathematics, Japan
Michael Mayo	The University of Waikato, New Zealand
Stephen McCloskey	The University of Sydney, Australia
Ernestina Menasalvas	Universidad Politécnica de Madrid, Spain
Xiangfu Meng	Liaoning Technical University, China
Xiaofeng Meng	Renmin University of China, China
Jun-Ki Min	Korea University of Technology and Education, Korea
Nguyen Le Minh	Japan Advanced Institute of Science and Technology, Japan
Leandro Minku	The University of Birmingham, UK
Pabitra Mitra	Indian Institute of Technology Kharagpur, India
Anirban Mondal	Ashoka University, India
Taesup Moon	Sungkyunkwan University, Korea
Yang-Sae Moon	Kangwon National University, Korea
Yasuhiko Morimoto	Hiroshima University, Japan
Animesh Mukherjee	Indian Institute of Technology Kharagpur, India
Miyuki Nakano	Advanced Institute of Industrial Technology, Japan
Mirco Nanni	ISTI-CNR, Italy

Richi Nayak	Queensland University of Technology, Australia
Raymond Ng	University of British Columbia, Canada
Wilfred Ng	Hong Kong University of Science and Technology, SAR China
Cam-Tu Nguyen	Nanjing University, China
Hao Canh Nguyen	Kyoto University, Japan
Ngoc-Thanh Nguyen	Wroclaw University of Science and Technology, Poland
Quoc Viet Hung Nguyen	Griffith University, Australia
Arun Reddy Nelakurthi	Arizona State University, USA
Thanh Nguyen	Deakin University, Australia
Thin Nguyen	Deakin University, Australia
Athanasios Nikolakopoulos	University of Minnesota, USA
Tadashi Nomoto	National Institute of Japanese Literature, Japan
Eirini Ntoutsi	Leibniz University of Hanover, Germany
Kouzou Ohara	Aoyama Gakuin University, Japan
Kok-Leong Ong	La Trobe University, Australia
Shirui Pan	University of Technology Sydney, Australia
Yuangang Pan	University of Technology Sydney, Australia
Guansong Pang	University of Adelaide, Australia
Dhaval Patel	IBM T.J. Watson Research Center, USA
Francois Petitjean	Monash University, Australia
Hai Nhat Phan	New Jersey Institute of Technology, USA
Xuan-Hieu Phan	University of Engineering and Technology, VNUHN, Vietnam
Vincenzo Piuri	Università degli Studi di Milano, Italy
Vikram Pudi	International Institute of Information Technology, Hyderabad, India
Chao Qian	University of Science and Technology of China, China
Qi Qian	Alibaba Group, China
Tang Qiang	Luxembourg Institute of Science and Technology, Luxembourg
Biao Qin	Renmin University of China, China
Jie Qin	Eidgenössische Technische Hochschule Zürich, Switzerland
Tho Quan	Ho Chi Minh City University of Technology, Vietnam
Uday Kiran Rage	University of Tokyo, Japan
Chedy Raissi	Inria, France
Vaibhav Rajan	National University of Singapore, Singapore
Santu Rana	Deakin University, Australia
Thilina N. Ranbaduge	Australian National University, Australia
Patricia Riddle	University of Auckland, New Zealand
Hiroshi Sakamoto	Kyushu Institute of Technology, Japan
Yücel Saygin	Sabanci University, Turkey
Mohit Sharma	Walmart Labs, USA
Hong Shen	Adelaide University, Australia

Shoujin Wang   Macquarie University, Australia
Sibo Wang   Chinese University of Hong Kong, SAR China
Suhang Wang   Pennsylvania State University, USA
Wei Wang   University of New South Wales, Australia
Wei Wang   Nanjing University, China
Weiqing Wang   Monash University, Australia
Wendy Hui Wang   Stevens Institute of Technology, USA
Wenya Wang   Nanyang Technological University, Singapore
Xiao Wang   Beijing University of Posts and Telecommunications,
                                 China
Xiaoyang Wang   Zhejiang Gongshang University, China
Xin Wang   University of Calgary, Canada
Xiting Wang   Microsoft Research Asia, China
Yang Wang   Dalian University of Technology, China
Yue Wang   AcuSys, USA
Zhengyang Wang   Texas A&M University, USA
Zhichao Wang   University of Technology Sydney, Australia
Lijie Wen   Tsinghua University, China
Jorg Wicker   University of Auckland, New Zealand
Kishan Wimalawarne   Kyoto University, Japan
Raymond Chi-Wing Wong   Hong Kong University of Science and Technology,
                                 SAR China
Brendon J. Woodford   University of Otago, New Zealand
Fangzhao Wu   Microsoft Research Asia, China
Huifeng Wu   Hangzhou Dianzi University, China
Le Wu   Hefei University of Technology, China
Liang Wu   Arizona State University, USA
Lin Wu   University of Queensland, Australia
Ou Wu   Tianjin University, China
Qingyao Wu   South China University of Technology, China
Shu Wu   Institute of Automation, CAS, China
Yongkai Wu   University of Arkansas, USA
Yuni Xia   Indiana University—Purdue University Indianapolis
                                 (IUPUI), USA
Congfu Xu   Zhejiang University, China
Guandong Xu   University of Technology Sydney, Australia
Jingwei Xu   Nanjing University, China
Linli Xu   University of Science and Technology China, China
Miao Xu   RIKEN, Japan
Tong Xu   University of Science and Technology of China, China
Bing Xue   Victoria University of Wellington, New Zealand
Hui Xue   Southeast University, China
Shan Xue   University of Technology Sydney, Australia
Pranjul Yadav   Criteo, France
Takehisa Yairi   University of Tokyo, Japan
Takehiro Yamamoto   Kyoto University, Japan

Chun-Pai Yang	National Taiwan University, Taiwan
De-Nian Yang	Academia Sinica, Taiwan
Guolei Yang	Facebook, USA
Jingyuan Yang	George Mason University, USA
Liu Yang	Tianjin University, China
Ming Yang	Nanjing Normal University, China
Shiyu Yang	East China Normal University, China
Yiyang Yang	Guangdong University of Technology, China
Lina Yao	University of New South Wales, Australia
Yuan Yao	Nanjing University, China
Zijun Yao	IBM Research, USA
Mi-Yen Yeh	Academia Sinica, Taiwan
feng Yi	Institute of Information Engineering, CAS, China
Hongzhi Yin	University of Queensland, Australia
Jianhua Yin	Shandong University, China
Minghao Yin	Northeast Normal University, China
Tetsuya Yoshida	Nara Women's University, Japan
Guoxian Yu	Southwest University, China
Kui Yu	Hefei University of Technology, China
Yang Yu	Nanjing University, China
Long Yuan	University of New South Wales, Australia
Shuhan Yuan	University of Arkansas, USA
Xiaodong Yue	Shanghai University, China
Reza Zafarani	Syracuse University, USA
Nayyar Zaidi	Monash University, Australia
Yifeng Zeng	Teesside University, UK
De-Chuan Zhan	Nanjing University, China
Daoqiang Zhang	Nanjing University of Aeronautics and Astronautics, China
Du Zhang	California State University, Sacramento, USA
Haijun Zhang	Harbin Institute of Technology (Shenzhen), China
Jing Zhang	Nanjing University of Science and Technology, China
Lu Zhang	University of Arkansas, USA
Mengjie Zhang	Victoria University of Wellington, New Zealand
Quangui Zhang	Liaoning Technical University, China
Si Zhang	Arizona State University, USA
Wei Emma Zhang	Macquarie University, Australia
Wei Zhang	East China Normal University, China
Wenjie Zhang	University of New South Wales, Australia
Xiangliang Zhang	King Abdullah University of Science and Technology, Saudi Arabia
Xiuzhen Zhang	RMIT University, Australia
Yudong Zhang	University of Leicester, UK
Zheng Zhang	University of Queensland, Australia
Zili Zhang	Southwest University, China
Mingbo Zhao	Donghua University, China

Peixiang Zhao	Florida State University, USA
Pengpeng Zhao	Soochow University, China
Yanchang Zhao	CSIRO, Australia
Zhongying Zhao	Shandong University of Science and Technology, China
Zhou Zhao	Zhejiang University, China
Huiyu Zhou	University of Leicester, UK
Shuigeng Zhou	Fudan University, China
Xiangmin Zhou	RMIT University, Australia
Yao Zhou	Arizona State University, USA
Chengzhang Zhu	University of Technology Sydney, Australia
Huafei Zhu	Nanyang Technological University, Singapore
Pengfei Zhu	Tianjin University, China
Tianqing Zhu	University of Technology Sydney, Australia
Xingquan Zhu	Florida Atlantic University, USA
Ye Zhu	Deakin University, Australia
Yuanyuan Zhu	Wuhan University, China
Arthur Zimek	University of Southern Denmark, Denmark
Albrecht Zimmermann	Université de Caen Normandie, France

External Reviewers

Ji Feng	Zheng-Fan Wu
Xuan Huo	Yafu Xiao
Bin-Bin Jia	Yang Yang
Zhi-Yu Shen	Meimei Yang
Yanping Sun	Han-Jia Ye
Xuan Wu	Peng Zhao

Sponsoring Organizations

University of Macau

Macao Convention & Exhibition Association

Intel

Baidu Inc.

Contents – Part III

Mining Unstructured and Semi-structured Data

Behavioral Data Mining

Visual Data Mining

Knowledge Graph and Interpretable Data Mining

Representation Learning and Embedding

AAANE: Attention-Based Adversarial Autoencoder for Multi-scale Network Embedding

Lei Sang[1,2], Min Xu[2(✉)], Shengsheng Qian[3], and Xindong Wu[1]

[1] School of Computer Science and Information Technology,
Hefei University of Technology, Hefei, China
lei.sang@student.uts.edu.au, xwu@hfut.edu.cn
[2] Faculty of Engineering and IT, University of Technology Sydney, Ultimo, Australia
Min.Xu@uts.edu.au
[3] Institute of Automation, Chinese Academy of Sciences, Beijing, China
shengsheng.qian@nlpr.ia.ac.cn

Abstract. Network embedding represents nodes in a continuous vector space and preserves structure information from a network. Existing methods usually adopt a "one-size-fits-all" approach when concerning multi-scale structure information, such as first- and second-order proximity of nodes, ignoring the fact that different scales play different roles in embedding learning. In this paper, we propose an Attention-based Adversarial Autoencoder Network Embedding (AAANE) framework, which promotes the collaboration of different scales and lets them vote for robust representations. The proposed AAANE consists of two components: (1) an attention-based autoencoder that effectively capture the highly non-linear network structure, which can de-emphasize irrelevant scales during training, and (2) an adversarial regularization guides the autoencoder in learning robust representations by matching the posterior distribution of the latent embeddings to a given prior distribution. Experimental results on real-world networks show that the proposed approach outperforms strong baselines.

Keywords: Network embedding · Multi-scale · Attention · Adversarial autoencoder

1 Introduction

Network embedding (NE) methods have shown outstanding performance on many tasks including node classification [1], community detection [2,3] and link prediction [4]. These methods aim to learn latent, low-dimensional representations for network nodes while preserving network topology structure information. Networks' structures are inherently hierarchical [5]. As shown in Fig. 1, each individual is a member of several communities and can be modeled by his/her neighborhoods' structure information with different scales around him/her, which

© Springer Nature Switzerland AG 2019
Q. Yang et al. (Eds.): PAKDD 2019, LNAI 11441, pp. 3–14, 2019.
https://doi.org/10.1007/978-3-030-16142-2_1

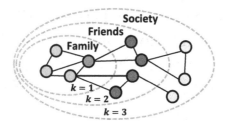

Fig. 1. Illustration of the multi-scale network with three scales.

range from short scales structure (e.g. families, friends), to long-distance scales structure (e.g. society, nation states). Every single scale is usually sparse and biased, and thus the node embedding learned by existing approaches may not be so robust. To obtain a comprehensive representation of a network node, multi-scale structure information should be considered collaboratively.

Recently, a number of methods have been proposed for learning data representations from multiple scales. For example, DeepWalk [1] models multi-scale indirectly from a random walk. Line [6] proposes primarily a breadth-first strategy, sampling nodes and optimizing the likelihood independently over short scales structure information such as 1-order and 2-order neighbors. GraRep [7] generalizes LINE to incorporate information from network neighborhoods beyond 2-order, which can embed long distance scales structure information to the node representation. More recently, some autoencoder based methods, For example, DNGR [8] learns the node embedding through stacked denoising autoencoder from the multi-scale PPMI matrix. Similarly, SDNE [9] is realized by a semi-supervised deep autoencoder model. Besides, MVE [10] aims to learn embedding from several multi-viewed networks with the same nodes but different edges, which is different from our single network setting.

Despite their strong task performance, existing methods have the following limitations: (1) *Lack of weight learning.* To learn robust and stable node embeddings, the information from multiple scales needs to be integrated. During integration, as the importance of different scales can be quite different, their weights need to be carefully decided. For example, if we consider very young kids on a social network, and they may be very tightly tied to their family and loosely tied to the society members. However, for university students, they may have relatively more ties to their friends and the society than very young kids. Existing approaches usually assign equal weights to all scales. In other words, different scales are equally treated, which is not reasonable for most multi-scale networks. (2) *Insufficient constrain for embedding distribution.* Take the autoencoder based method for example, an autoencoder is a neural network trained to attempt to copy its input to its output, which has a typical pipeline like $(x \rightarrow E \rightarrow z \rightarrow D \rightarrow x')$. Autoencoder only requires x to approach $x' = D(E(x))$, and for that purpose the decoder may simply learn to reconstruct x regardless of the distribution obtained from E. This means that $p(z)$ can be very irregular, which sometimes makes the generation of new samples difficult or even infeasible.

In this paper, we focus on the multi-scale network embedding problem and propose a novel Attention-based Adversarial Autoencoder Network Embedding (AAANE) method to jointly capture the weighted scale structure information and learn robust representation with adversarial regularization. We first introduce a set of scale-specific node vectors to preserve the proximities of nodes in different scales. The scales-specific node embeddings are then combined for voting the robust node representations. Specifically, our work has two major contributions. (1) To deal with the weights learning, we propose an attention-based autoencoder to infer the weights of scales for different nodes, and then capture the highly non-linear network structure, which is inspired by the recent progress of the attention mechanism for neural machine translation [11]. (2) To implement regularisation of the distribution for encoded data, we introduce adversarial training component [12] to the attention-based autoencoder, which can discriminatively predict whether a sample arises from the low-dimensional representations of the network or from a sampled distribution. Adversarial regularisation reduces the amount of information that may be held in the encoding, forcing the model to learn an efficient representation of the data. Through the attention-based weight learning together with the adversarial regularization, the proposed AAANE model can effectively combine the virtues of multiple scale information to complement and enhance each other.

2 Preliminaries

Network Embedding(NE): An information network is represented as $G = (V, E)$, where $V = \{v_i\}_{i=1,\cdots,N}$ consist a set of nodes, $e_{i,j} = (v_i, v_j) \in E$ is an edge indicating the relationship between two nodes. The task of NE aims to build a low-dimensional representation $x_i \in \mathbb{R}^d$ for each node $i \in V$, where d is the dimension of embedding space and expected much smaller than node number $|V|$.

We define *adjacency matrix* $\widetilde{A} \in \mathbb{R}^{|V| \times |V|}$ for a network and D is a diagonal degree matrix. To capture the transitions from one node to another, we can define the (first-order) probability transition matrix $A = D^{-1}\widetilde{A}$, where $A_{i,j}$ is the probability of a transition from node v_i to node v_j within one step. It can be observed that the matrix A is a normalized adjacency matrix where the summation of each row equals to 1.

In this paper, multi-scale structural information serves two functions: (1) the capture of long-distance relationship between two different vertices and (2) the consideration of distinct connections in terms of different transitional orders.

The (normalized) adjacency matrix A characterizes the first-order proximity which models the local pairwise proximity between vertices. As discussed earlier, we believe that the k-order (with varying k) long scale relational information from the network needs to be captured when constructing such multi-scale network embedding [7]. To compute the various scale transition probabilities, we introduce the following k-order probability proximity matrix:

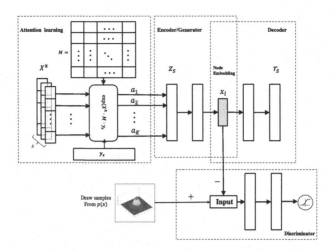

Fig. 2. The architecture of AAANE. The top row is an attention-based autoencoder that infers the weights of scales for different nodes, and then captures the highly non-linear network structure. The bottom row diagrams a discriminator trained to discriminatively predict whether a sample arises from the hidden code of the autoencoder or from a prior distribution specified by the user.

$$A^k = \underbrace{A \cdot A \cdots A}_{k} \tag{1}$$

where the entry $A^k_{i,j}$ refers to the k-order proximity between node v_i and v_j.

Multi-scale Network Embedding: Given a network $G = (V, E)$, the robust node representation $\{x_i\}_{v_i \in V} \subseteq R^d$ can be collaboratively learned from k successively network structural information representation, $A, A^2, \ldots, A^k$, where A^k captures the view of the network at scale k. Intuitively, each member of the family encodes a different view of social similarity, corresponding to shared membership in latent communities at different scales.

3   The Framework

In this section, we first give a brief overview of the proposed AAANE, and then formulate our method of multi-scale network embedding from attention based adversarial autoencoder.

3.1   An Overview of the Framework

In this work, we leverage attention-based adversarial autoencoder to help learn stable and robust node embedding. Figure 2 shows the proposed framework of Attention-based Adversarial Autoencoder for Network Embedding (**AAANE**), which mainly consists of two components, i.e., an attention-based autoencoder and an adversarial learning component.

We introduce an attention mechanism to the autoencoder for learning the weights of structure information with different scales. A standard autoencoder consists of an encoder network and a decoder network. The encoder maps the network structure information z_s into a latent code x_i, and the decoder reconstructs the input data as r_s. Then, the adversarial learning component acts as regularization for the autoencoder, by matching the aggregated posterior, which helps enhance the robustness of the representation x_i. The generator of the adversarial network is also the encoder of the autoencoder. The adversarial network and the autoencoder are trained jointly in two phases: the reconstruction phase and the regularization phase. In the reconstruction phase, the autoencoder updates the encoder and the decoder to minimize the reconstruction error of the inputs. In the regularization phase, the adversarial network first updates its discriminative network to tell apart the true samples (generated using the prior) from the generated samples (the hidden node embedding x_i computed by the autoencoder). As a result, the proposed AAANE can jointly capture the weighted scale structure information and learn robust representations.

3.2   Attention-Based Autoencoder

The Attention-based Autoencoder for network embedding (**AANE**) model uses a stacked neural network to preserve the structure information. As we discussed in Sect. 2, different k-order proximity matrices preserve network structure information in different scales. Scale vector X^k is column in each A^k, for $k = 1, 2 \ldots K$, which denotes the k-th scale structure information for the node. The length of each scale vector X^k is the same as the node size. The autoencoder component tries to capture the full range of structure information. We construct a vector representation z_s for each node as the input of the autoencoder in the first step. In general, we expect this vector representation to capture the most relevant information with regards to different scales of a node. z_s is defined as the weighted summation of every scale vector $X^k, k = 1, 2, \ldots K$, corresponding to the scale index for each node.

$$z_s = \sum_{k=1}^{K} a_k X^k \tag{2}$$

For each scale vector X^k of one node, we compute a positive weight a_k which can be interpreted as the probability that X_k is assigned by one node. Intuitively, by learning proper weights a_k for each node, our approach can obtain most informative scale information. Following the recent attention based models for neural machine translation, we define the weight of scales k for a node using a softmax unit as follows:

$$a_k = \frac{\exp(d_k)}{\sum_{j=1}^{n} \exp(d_k)}$$

$$d_k = X^{k^\top} \cdot M \cdot y_s \qquad y_s = \frac{1}{K} \sum_{k=1}^{K} X^k \tag{3}$$

where y_s is the average of different scale vector, which can capture the global context of the structure information. M is a matrix mapping between the global context embedding y_s and each structure scale vector X^k, which is learned as part of the training process. By introducing an attentive matrix M, we compute the relevance of each scale vector to the node. If X^k and y_s have a large dot product, this node believes that scale k is an informative scale, i.e., the weight of scale k for this node will be largely based on the definition.

Once we obtain the weighted node vector representation $z_s \in \mathbb{R}^{|V|}$, a stacked autoencoder is used to learn a low-dimensional node embedding. An autoencoder performs two actions, an encoding step, followed by a decoding step. In the encoding step, a function $f()$ is applied to the original vector representation z_s in the input space and send it to a new feature space. An activation function is typically involved in this process to model the non-linearities between the two vector spaces. At the decoding step, a reconstruction function $g()$ is used to reconstruct the original input vectors back from the latent representation space. The r_s is the reconstructed vector representation. After training, the bottleneck layer representations x_i can be viewed as the low dimension embedding for the input node v_i.

This attention-based autoencoder is trained to minimize the reconstruction error. We adopt the contrastive max-margin objective function, similar to previous work [13–15]. For each input node, we randomly sample m nodes from our training data as negative samples. We represent each negative sample as n_s, which is computed by averaging its scale vectors as y_s. Our objective is to make the reconstructed embedding r_s similar to the target node embedding z_s while different from those negative samples n_s. Therefore, the unregularized objective J is formulated as a hinge loss that maximizes the inner product between r_s and z_s, and minimizes the inner product between r_s and the negative samples simultaneously:

$$J(\theta) = \sum_{s \in D} \sum_{i=1}^{m} max(0, 1 - r_s z_s + r_s n_i) \qquad (4)$$

where D represents the training dataset.

3.3   Adversarial Learning

We hope to learn vector representations of the most representative scale for each node. An autoencoder consists of two models, an encoder and a decoder, each of which has its own set of learnable parameters. The encoder is used to get a latent code x_i from the input with the constraint. The dimension of the latent code should be less than the input dimension. The decoder takes in this latent code and tries to reconstruct the original input. However, we argue that training an autoencoder with contrastive max-margin objective function gives us latent codes with similar nodes being far from each other in the Euclidean space, especially when processing noisy network data. The main reason is the insufficient constrain for embedding distribution. Adversarial autoencoder (AAE) addresses these issues by imposing an Adversarial regularization to the bottleneck layer representation of autoencoder, and then the distribution of latent code may be shaped to match a desired prior distribution. Adversarial regularisation can reduce the amount of information that may be held in the encoding process, forcing the model to learn an efficient representation for the network data.

AAE typically consists of a generator $G()$ and a discriminator $D()$. Our main goal is to force output of the encoder to follow a given prior distribution $p(x)$(this can be normal, gamma .. distributions). We use the encoder as our generator, and the discriminator to tell if the samples are from a prior distribution or from the output of the encoder x_i. D and G play the following two-player minimax game with the value function V (G, D):

$$\min_{G} \max_{D} V(D, G) = \mathbb{E}_{p(x)}[\log D(x_i)] + \mathbb{E}_{q(x)}[\log(1 - D(x_i))] \qquad (5)$$

where $q(x)$ is the distributions of encoded data samples.

3.4 Training Procedure

The whole training process is done in three sequential steps: (1) The encoder and decoder are trained simultaneously to minimize the reconstruction loss of the decoder as Eq. 4. (2) The discriminator D is then trained to correctly distinguish the true input signals x from the false signals x_i, where the x is generated from target distribution, and x_i is generated by the encoder by minimizing the loss function 5. (3) The next step will be to force the encoder to fool the discriminator by minimizing another loss function: $L = -\log(D(x_i))$. More specifically, we connect the encoder output as the input to the discriminator. Then, we fix the discriminator weights and fix the target to 1 at the discriminator output. Later, we pass in a node to the encoder and find the discriminator output which is then used to find the loss.

4 Experiments

In this section, we conduct node classification on sparsely labeled networks to evaluate the performance of our proposed model.

4.1 Datasets

We employ the following three widely used datasets for node classification.

Cora. Cora is a research paper set constructed, which contains 2, 708 machine learning papers which are categorized into seven classes. The citation relationships among them are crawled form a popular social network.

Citeseer. Citeseer is another research paper set constructed, which contains 3, 312 publications and 4, 732 links among them. These papers are from 6 classes.

Wiki. Wiki contains 2, 405 web pages from 19 categories and 17, 981 links among them. Wiki is much denser than Cora and Citeseer.

4.2   Baselines and Experimental Settings

We consider a number of baselines to demonstrate the effectiveness and robustness of the proposed AAANE algorithm. For all methods and datasets, we set the embedding dimension $d = 128$.

DeepWalk [1]: DeepWalk first transforms the network into node sequences by truncated random walk, and then uses it as input to the Skip-gram model to learn representations.

LINE [6]: LINE can preserve both first-order and second-order proximities for the undirected network through modeling node co-occurrence probability and node conditional probability.

GraRep [7]: GraRep preserves node proximities by constructing different k-order transition matrices.

node2vec [16]: node2vec develops a biased random walk procedure to explore the neighborhood of a node, which can strike a balance between local properties and global properties of a network.

AIDW [17]: Adversarial Inductive DeepWalk (AIDW) is an Adversarial Network Embedding (ANE) framework, which leverages random walk to sample node sequences as the structure-preserving component.

Parameter Setting: In our experimental settings, we vary the percentage of labeled nodes from 10% to 90% by an increment of 10% for each dataset. We treat network embeddings as vertex features and feed them into a one-vs-rest logistic regression classifier implemented by LibLinear [18]. For DeepWalk, LINE, GraRep, node2vec, we directly use the implementations provided by OpenNE[1]. For our methods AAANE, the maximum matrix transition scale is set to 8, and the number of negative samples per input sample m is set to 7. For attention-based autoencoder, it has three hidden layers, with the layer structure as $512 - 128 - 512$. For the discriminator of AAANE, it is a three-layer neural network, with the layer structure as $512 - 512 - 1$. And the prior distributions are Gaussian Distribution following the original paper [19].

4.3   Multi-label Classification

Tables 1, 2 and 3 show classification accuracies with different training ratios on different datasets, where the best results are **bold-faced**. In these tables, AANE denotes our model AAANE without Adversarial component. From these tables, we have the following observations:

(1) The proposed framework, without leveraging the adversarial regularization version AANE, achieving average 2% gains over AIDW on cora and wiki when varying the training ratio from 10% to 90% in most cases, and slightly better result on Citeseer, which suggests that assigning different weights to different scales of a node may be beneficial.

---

[1] https://github.com/thunlp/OpenNE.

Table 1. Accuracy (%) of node classification on Wiki.

% Labeled nodes	10%	20%	30%	40%	50%	60%	70%	80%	90%
DeepWalk	57.2	62.98	64.03	65.78	66.74	68.69	68.36	67.85	67.22
LINE	57.09	59.98	62.47	64.38	66.5	65.8	67.31	67.15	65.15
GraRep	59.55	60.76	62.23	62.3	62.76	63.72	63.02	62.79	60.17
node2vec	58.47	61.38	63.9	63.96	66.08	66.74	67.73	67.57	66.8
AIDW	57.29	61.89	63.77	64.26	66.85	67.23	69.04	70.13	71.33
AANE	59.95	64.14	66.15	68.40	68.66	69.34	69.25	70.89	69.71
AAANE	**60.36**	**64.98**	**67.21**	**68.79**	**69.07**	**70.32**	**70.85**	**72.03**	**72.45**

Table 2. Accuracy (%) of node classification on Cora.

% Labeled nodes	10%	20%	30%	40%	50%	60%	70%	80%	90%
DeepWalk	76.37	79.6	80.85	81.42	82.35	82.1	82.9	84.32	83.39
LINE	71.08	76.19	77.32	78.4	79.25	79.06	79.95	81.92	82.29
GraRep	77.02	77.95	78.53	79.75	79.61	78.78	78.6	78.23	78.23
node2vec	75.84	78.77	79.54	80.86	80.43	80.9	80.44	79.7	77.86
AIDW	76.21	78.93	80.21	81.45	82.03	82.74	82.81	83.69	83.92
AANE	77.65	81.50	82.49	84.43	84.71	84.69	84.75	85.98	86.03
AAANE	**78.23**	**82.14**	**82.76**	**85.31**	**85.69**	**86.12**	**86.02**	**86.74**	**87.21**

Table 3. Accuracy (%) of node classification on Citeser.

% Labeled nodes	10%	20%	30%	40%	50%	60%	70%	80%	90%
DeepWalk	53.47	54.19	54.6	57.55	57	59.02	58.95	58.22	55.72
LINE	48.74	50.87	52.82	52.72	52	52.3	53.12	53.54	52.41
GraRep	53.23	54.34	53.77	54.43	54.05	54.57	54.83	55.35	55.12
node2vec	53.94	54.08	56.23	57.34	57.55	60.3	61.17	61.24	59.33
AIDW	52.17	56.23	56.87	58.26	58.45	59.27	59.34	60.38	61.3
AANE	55.02	56.15	58.65	58.76	58.52	59.93	60.97	61.39	61.23
AAANE	**55.45**	**56.73**	**59.37**	**59.81**	**60.12**	**60.58**	**61.43**	**61.72**	**62.38**

(2) After introducing Adversarial component into AANE, our Method AAANE can achieve further improvements over all baselines. It demonstrates that adversarial learning regularization can improve the robustness and discrimination of the learned representations.

(3) AAANE consistently outperforms all the other baselines on all three datasets with different training ratios. It demonstrates that attention-based weight learning together with the adversarial regularization can significantly improve the robustness and discrimination of the learned embeddings.

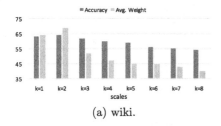
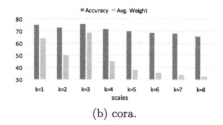

(a) wiki.   (b) cora.

Fig. 3. Comparison of performances on each individual scale and the average weights of scales. Scales with better performances usually attract more attentions from nodes.

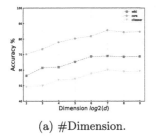
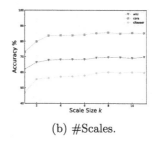

(a) #Dimension.   (b) #Scales.

Fig. 4. Parameter sensitivity of dimension d and scale size k.

4.4   Detailed Analysis of the Proposed Model

Analysis of the Learned Attentions over Scales: In our proposed AAANE model, we adopt an attention based approach to learn the weights of scales during voting, so that different nodes can focus most of their attentions on the most informative scales. The quantitative results have shown that AAANE achieves better results by learning attention over scales. In this part, we will examine the learned attention to understand why it can help improve the performances (Fig. 3).

We study which scale turn to attract more attentions from nodes. We take the Cora and Wiki datasets as examples. For each scale, we report the results of the scale-specific embedding corresponded to this scale, which achieves by taking only one scale vector A^k as an input of autoencoder. Then, we compare this scale-specific embedding with the average attention values learned by AAANE. The results are presented in Fig. 4. Overall, the performances of single scale and the average attention received by these scales are positively correlated. In other words, our approach can allow different nodes to focus on the scales with the best performances, which is quite reasonable.

Parameter Sensitivity: We discuss the parameter sensitivity in this section. Specifically, we assess how the different choices of the maximal scale size K, dimension d can affect node classification with the training ratio as 50%. Figure 4(a) shows the accuracy of AAANE over different settings of the dimension d. The accuracy shows an apparent increase at first. This is intuitive as more bits can encode more useful information in the increasing bits. However, when the number of dimensions continuously increases, the performance starts to drop slowly. The reason is that too large number of dimensions may introduce noises which will deteriorate the performance. Figure 4(b) shows the accuracy scores over different choices of K. We can observe that the setting $K = 2$ has a significant improvement over the setting $K = 1$, and $K = 3$ further outperforms $K = 2$. This confirms that different k-order can learn complementary local information. When K is large enough, learned k-order relational information becomes weak and shifts towards a steady distribution.

5   Related Work

To preserve multi-scale structure information, some random walk and matrix factorization methods [1,7] have been proposed. GraRep [7] accurately calculates k-order proximity matrix, and computes specific representation for each k using SVD based dimension reduction method, and then concatenates these embeddings. Another line of the related work is deep learning based methods. SDNE [9], DNGR [8] utilize this ability of deep autoencoder to generate an embedding model that can capture non-linearity in graphs. AIDW [17] proposes an adversarial network embedding framework, which leverages the adversarial learning principle to regularize the representation learning. However, existing approaches usually lack weight learning for different scales.

Our work is also related to the attention-based models. Rather than using all available information, attention mechanism aims to focus on the most pertinent information for a task and has been applied to various tasks, including machine translation and sentence summarization [11]. MVE [10] proposes a multi-view network embedding, which aims to infer robust node representations across different networks.

6   Conclusion

In this paper, we study learning node embedding for networks with multiple scales. We propose an effective framework to let different scales collaborate with each other and vote for the robust node representations. During voting, we propose an attention-based autoencoder to automatically learn the voting weights of scales while preserving the network structure information in a non-linear way. Besides, an Adversarial regularization is introduced to learn more stable and robust network embedding. Experiments on node classification demonstrate the superior performance of our proposed method.

References

1. Perozzi, B., Al-Rfou, R., Skiena, S.: DeepWalk: online learning of social representations. In: Proceedings of the 20th ACM SIGKDD International Conference on Knowledge Discovery and Data Mining (2014)

2. Wang, X., Cui, P., Wang, J., Pei, J., Zhu, W., Yang, S.: Community preserving network embedding. In: AAAI, pp. 203–209 (2017)
3. Sang, L., Xu, M., Qian, S., Wu, X.: Multi-modal multi-view Bayesian semantic embedding for community question answering. Neurocomputing (2018)
4. Lü, L., Zhou, T.: Link prediction in complex networks: a survey. Phys. A: Stat. Mech. Appl. **390**(6), 1150–1170 (2011)
5. Perozzi, B., Kulkarni, V., Chen, H., Skiena, S.: Don't walk, skip!: online learning of multi-scale network embeddings. In: Proceedings of the 2017 IEEE/ACM International Conference on Advances in Social Networks Analysis and Mining, pp. 258–265. ACM (2017)
6. Tang, J., Qu, M., Wang, M., Zhang, M., Yan, J., Mei, Q.: Line: large-scale information network embedding. In: Proceedings of the 24th International Conference on World Wide Web, pp. 1067–1077. International World Wide Web Conferences Steering Committee (2015)
7. Cao, S., Lu, W., Xu, Q.: Grarep: learning graph representations with global structural information. In: Proceedings of the 24th ACM International on Conference on Information and Knowledge Management, pp. 891–900. ACM (2015)
8. Cao, S., Lu, W., Xu, Q.: Deep neural networks for learning graph representations. In: AAAI, pp. 1145–1152 (2016)
9. Wang, D., Cui, P., Zhu, W.: Structural deep network embedding. In: Proceedings of the 20th ACM SIGKDD (2016)
10. Qu, M., Tang, J., Shang, J., Ren, X., Zhang, M., Han, J.: An attention-based collaboration framework for multi-view network representation learning. In: Proceedings of the 2017 ACM on Conference on Information and Knowledge Management, pp. 1767–1776. ACM (2017)
11. Luong, M.-T., Pham, H., Manning, C.D.: Effective approaches to attention-based neural machine translation. arXiv preprint arXiv:1508.04025 (2015)
12. Goodfellow, I., et al.: Generative adversarial nets. In: Advances in Neural Information Processing Systems, pp. 2672–2680 (2014)
13. Weston, J., Bengio, S., Usunier, N.: WSABIE: scaling up to large vocabulary image annotation. In: IJCAI, vol. 11, pp. 2764–2770 (2011)
14. Iyyer, M., Guha, A., Chaturvedi, S., Boyd-Graber, J., Daumé III, H.: Feuding families and former friends: unsupervised learning for dynamic fictional relationships. In: Proceedings of the 2016 Conference of the North American Chapter of the Association for Computational Linguistics: Human Language Technologies, pp. 1534–1544 (2016)
15. He, R., Lee, W.S., Ng, H.T., Dahlmeier, D.: An unsupervised neural attention model for aspect extraction. In: Proceedings of the 55th Annual Meeting of the Association for Computational Linguistics (Volume 1: Long Papers), vol. 1, pp. 388–397 (2017)
16. Grover, A., Leskovec, J.: node2vec: Scalable feature learning for networks. In: Proceedings of the 22nd ACM SIGKDD International Conference on Knowledge Discovery and Data Mining, pp. 855–864. ACM (2016)
17. Dai, Q., Li, Q., Tang, J., Wang, D.: Adversarial network embedding. In: Proceedings of AAAI (2018)
18. Fan, R.-E., Chang, K.-W., Hsieh, C.-J., Wang, X.-R., Lin, C.-J.: Liblinear: a library for large linear classification. J. Mach. Learn. Res. **9**(Aug), 1871–1874 (2008)
19. Makhzani, A., Shlens, J., Jaitly, N., Goodfellow, I., Frey, B.: Adversarial autoencoders. arXiv preprint arXiv:1511.05644 (2015)

NEAR: Normalized Network Embedding with Autoencoder for Top-K Item Recommendation

Dedong Li[1], Aimin Zhou[1(✉)], and Chuan Shi[2]

[1] Department of Computer Science and Technology, East China Normal University,
Shanghai, China
ddlecnu@gmail.com, amzhou@cs.ecnu.edu.cn
[2] School of Computer Science, Beijing University of Posts and Telecommunications,
Beijing, China
shichuan@bupt.edu.cn

Abstract. The recommendation system is an important tool both for business and individual users, aiming to generate a personalized recommended list for each user. Many studies have been devoted to improving the accuracy of recommendation, while have ignored the diversity of the results. We find that the key to addressing this problem is to fully exploit the hidden features of the heterogeneous user-item network, and consider the impact of hot items. Accordingly, we propose a personalized top-k item recommendation method that jointly considers accuracy and diversity, which is called Normalized **N**etwork **E**mbedding with **A**utoencoder for Personalized Top-K Item **R**ecommendation, namely **NEAR**. Our model fully exploits the hidden features of the heterogeneous user-item network data and generates more general low dimension embedding, resulting in more accurate and diverse recommendation sequences. We compare NEAR with some state-of-the-art algorithms on the DBLP and MovieLens1M datasets, and the experimental results show that our method is able to balance the accuracy and diversity scores.

Keywords: Network embedding · Recommendation system · Autoencoder · Heterogeneous network

1 Introduction

Recommendation systems are widely used in industry for they can recommend items that users most likely to consume. Among current methods, the network embedding algorithm attracts lots of attention. Since the network embedding algorithm can describe high-dimensional complex networks as low-dimensional dense vectors, making it possible to combine complex tasks with machine learning algorithms effectively [6,16]. In [17], it proves that the network embedding

This work is supported by the National Natural Science Foundation of China under Grant No. 61673180 and 61772082.

© Springer Nature Switzerland AG 2019
Q. Yang et al. (Eds.): PAKDD 2019, LNAI 11441, pp. 15–26, 2019.
https://doi.org/10.1007/978-3-030-16142-2_2

algorithm is actually equivalent to the matrix decomposition of a particular adjacent matrix. Simultaneously, the network embedding method is more effective than matrix decomposition in feature learning.

Recently, the deep neural network has shown a revolutionary advance in speech recognition, computer vision and natural language processing. Researchers apply deep learning methods in capturing useful information of complex networks [18]. Among these methods, autoencoder aims to map the input data to the low dimension representation while minimizing the difference between output and input, which meets the requirements of recommended systems. In [8], it shows that the autoencoder performs robustly on extracting the implicit relationships between items and users collaboratively when processing the network representation. Considering the above issues, we propose to choose a graph embedding algorithm with a deep learning method to train the model, aiming to fully exploit the hidden features of the heterogeneous user-item network.

However, on the process of dealing with diversity and novelty, we have found that the recommended list generated by network embedding methods are almost hot items, which is called the Harry Potter Problem. Experimental data has also shown that hot items always have high scores rated by users, which leads to the large value of the final vector representation. As a result, the items recommended for each user are much the same (almost hot items only), which leaves a negative impact on users permanently even if the F1 score is good. To address this problem, we resort to punishing hot items to a certain extent through the L2 normalization. Experiments show that the prediction of the hot items declines in ranking, leading to the top-k results more diverse.

In short, the contributions of this paper are listed as follows:

- We propose a Normalized Network Embedding with Autoencoder for Personalized Top-K Item Recommendation model, namely NEAR. We generate user and item embedding with autoencoder separately, combining advantages of both autoencoder and network embedding algorithms, which is a new method for the top-k recommendation.
- We choose KL-divergence to extract explicit features of ratings network and autoencoder to extract the higher-order proximity of implicit features so as to preserve the network structure more comprehensively.
- We suggest punishing the popular items through a L2 normalization in the process of network embedding, which turns out to be helpful to generate more diverse recommendations.

2   Related Work

2.1   Recommendation Systems

The task of the recommendation system includes rating prediction and ranking prediction [18]. The rating task is to predict vacant scores of the rating matrix

[5]. The ranking task generates top-k items that each user is most likely to consume. Recommendation algorithms are mainly based on collaborative filtering, content-based and hybrid recommendation systems [18]. The collaborative filtering (CF) method utilizes users' historical interactive information to recommend, including the traditional neighborhood-based algorithm (ItemKNN, UserKNN), latent factor-based algorithm (NMF, SVD) [9] and graph-based algorithm [4,13]. Many recent efforts devote to utilize deep learning models (such as CNN, RNN, RBM, and Autoencoder) to learn multi-layer representations of data to achieve better results [18,19].

2.2   Network Embedding

Network embedding maps graph data to a low dimensional space, which preserves structure information and properties of graphs [6]. Network embedding methods are mainly based on matrix eigenvector computation, neural network and matrix decomposition. Recent methods are mainly based on deep learning methods [3]. For example, the random walk [11] generates random paths on the network to make the analogy with Word2Vec [10]. LINE further mines the information of the network [14] through learning the first and second order relations of the network. Walklet generates multi-scale relationships by subsampling short random walks on the vertices of a graph [12], which further captures higher-order relationships. SDNE preserves the higher-order relations of the network based on autoencoder, aiming to preserve the overall structure of the network [15]. BiNE proposes a network embedding model for bipartite networks, which is the first work in bipartite network embedding [7].

3   The Proposed NEAR Model

3.1   Problem Definition

A top-k item recommendation system aims to recommend a sequence of items most likely to be consumed to each user in a preference decrease order.

For a recommendation system, the **input** includes: the network $G = (U, V, E)$, the weight matrix of the network R, and the embedding dimension d.

Table 1. Notations and terms of a recommendation system

Symbol	Definition	Symbol	Definition
$U = \{u_i\}_{i=1}^{|U|}$	User set	$V = \{v_j\}_{j=1}^{|V|}$	Item set
$E \in U * V$	Inter-set edges	$G = (U, V, E)$	Network
$\vec{u_i}, \vec{v_j}$	Embedding vectors	$\mathbf{U} = [\vec{u_i}], \mathbf{V} = [\vec{v_j}]$	Embedding matrix
$R = [r_{ij}]$	Rating matrix	R^U, R^V	Adjacent matrix of U, V
$X = \{\mathbf{x}_i\}_{i=1}^n, \hat{X} = \{\hat{\mathbf{x}}_i\}_{i=1}^n$	Input, reconstructed	$Y^{(l)} = \{\mathbf{y}_i^{(l)}\}_{i=1}^n$	lth layer embedding
$W^{(l)}, \hat{W}^{(l)}$	lth layer weights	$\mathbf{b}^{(l)}, \hat{\mathbf{b}}^{(l)}$	lth layer biases
$\theta = \{W^{(l)}, \hat{W}^{(l)}, \mathbf{b}^{(l)}, \hat{\mathbf{b}}^{(l)}\}$	AE parameters	d	Embedding dimension

The **output** of a system is: a predicted sequence of favorite items C_i for each user u_i. The notations and terms of a recommendation system are summarized in Table 1.

3.2   Framework

Based on the above analysis, we propose a Normalized Network Embedding with Autoencoder for Personalized Top-K Item Recommendation model (NEAR). As shown in Fig. 1, we first preprocess the rating matrix into an adjacent matrix, then we resort to learn a low dimensional representation of the adjacent matrix separately with the help of autoencoder, which is proved to preserve features of adjacent matrix effectively. To capture explicit features of the rating matrix, we propose to employ KL-divergence and L2 norm to fine-tune the embedding of the adjacent matrix. More details of the framework are given in following.

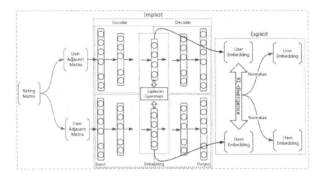

Fig. 1. NEAR Framework: the implicit part employs autoencoder to train on user adjacent matrix and item adjacent matrix separately, the explicit part fine-tunes the embedding by KL-divergence and normalization.

3.3   Explicit Relations

As shown in Fig. 2a, the edges between nodes u_i and v_j are defined as the explicit relationship in the recommendation system scenario. Inspired by the first-order proximity in LINE [14], we establish explicit relations by considering the local proximity between two connected nodes. The joint probability $P(i, j)$ between nodes u_i and v_j can be defined in (1).

Inspired by Word2Vec [10], we simulate the interaction between two entities by inner product to estimate the local proximity between two nodes in the embedding space. The interactive value is transformed into probabilistic space $\hat{P}(i, j)$ by a sigmoid function.

$$P(i, j) = \frac{r_{ij}}{\sum_{e_{ij} \in E} r_{ij}}, \quad \hat{P}(i, j) = \frac{1}{1 + exp(-\overrightarrow{\mathbf{u}_i}^T \overrightarrow{\mathbf{v}_j})} \tag{1}$$

where r_{ij} is the weight of edge e_{ij}, $\vec{u_i}$ and $\vec{v_j}$ are the embedding vectors of nodes u_i and v_j respectively.

With the empirical distribution of the co-occurrence probability between nodes and the reconstructed distribution, we can learn the embedding vector by minimizing the difference. We choose the KL-divergence as the difference measure between their distributions. The loss function is defined as follows:

$$minimize \quad O_{KL} = KL(P||\hat{P}) \propto - \sum_{e_{ij} \in E} r_{ij} log \hat{P}(i,j) \tag{2}$$

By minimizing the loss function, two connected nodes in the original network will be near in the embedding space, thus maintaining local proximity.

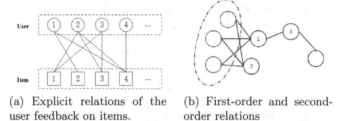

(a) Explicit relations of the user feedback on items.

(b) First-order and second-order relations

Fig. 2. Explicit and implicit relations.

3.4   Implicit Relations

We define the implicit relationship as the relationship between the same type nodes, that is, the relationship between users or between items. We can see from Fig. 2a that there is an explicit relationship between users and items in the recommendation system. However, the relationship between the users doesn't exist explicitly. Traditionally, we calculate the similarity between two users with the help of inner product. However, we find that a user with embedding [0, 2, 2, 0] will be more similar to the user with embedding [0, 5, 0, 3] than the user with [0, 2, 2, 0]. So we define a new similarity function with punishment to depress the difference as follows, for items are similar. Therefore we can get user adjacent matrix and item adjacent matrix.

$$u_{ij} = \sum_{k=1}^{n} \frac{r_{ik} * r_{jk}}{|r_{ik} - r_{jk}| + 1} \tag{3}$$

where $i = 1, 2, ..., m; j = 1, 2, ..., m$, and m, n is the shape of the rating matrix.

In order to mine local and global information of adjacent matrices, inspired by LINE [14], we model low order relations of adjacent matrices as shown in Fig. 2b. Since nodes 1 and 3 have a direct connection, they should be near in embedding space, which is called first-order relations; nodes 1 and 2 have multiple common neighbors even they don't have an explicit connection, so they should be near in embedding space, which is called second-order relations.

Second Order Relations. The idea of second-order relations is that two nodes with common neighbors tend to be closer in embedding space, so the global structure of the network can be well preserved. Inspired by SDNE [15], we can implement the autoencoder algorithm with the idea of unsupervised learning. As shown in Fig. 1, both the encoder and decoder parts contain multilayer nonlinear function, the encoder maps the raw data to the low-dimensional space, and the decoder maps the low-dimensional space representation to the reconstructed space. In this way, the network is more and more robust by minimizing the difference between the original space and the reconstructed space, so that the low-dimensional representation of each node can be obtained. For example, given $\mathbf{x}_i$, the representation of each layer of the encoder is as follows:

$$y_i^{(1)} = \sigma(W^{(1)}\mathbf{x}_i + \mathbf{b}^{(1)}) \tag{4}$$

$$y_i^{(k)} = \sigma(W^{(k)}y_i^{(k-1)} + \mathbf{b}^{(k)}), k = 2, ..., K \tag{5}$$

where $\mathbf{y}_i^{(K)}$ is a low-dimensional representation, and then decoder can perform an inverse operation to get the reconstructed $\hat{\mathbf{x}}_i$, so that the whole auto-coder can be trained by minimizing the loss functions:

$$\mathcal{L}_{2nd} = \sum_{i=1}^{n} \|\hat{\mathbf{x}}_i - \mathbf{x}_i\|_2^2 = \left\|\hat{X} - X\right\|_F^2 \tag{6}$$

First Order Relations. Inspired by laplacian eigenmaps (LE) [2], we define the loss functions $\mathcal{L}_{1st}$ as follows:

$$\mathcal{L}_{1st} = \sum_{i,j=1}^{n} s_{i,j} \left\|\mathbf{y}_i^{(K)} - \mathbf{y}_j^{(K)}\right\|_2^2 = \sum_{i,j=1}^{n} s_{i,j} \left\|\mathbf{y}_i - \mathbf{y}_j\right\|_2^2 \tag{7}$$

where s_{ij} is a penalty coefficient to make the related nodes closer.

Combining the first order relation and the second order relation, we define the joint optimal loss function as follows:

$$\mathcal{L}_{ae\_x} = \mathcal{L}_{2nd} + \alpha\mathcal{L}_{1st} + \beta\mathcal{L}_{reg} = \left\|\hat{X} - X\right\|_F^2 + \alpha \sum_{i,j=1}^{n} s_{i,j} \left\|\mathbf{y}_i - \mathbf{y}_j\right\|_2^2 + \beta\mathcal{L}_{reg} \tag{8}$$

$$\mathcal{L}_{reg} = \frac{1}{2} \sum_{k=1}^{K} \left(\left\|W^{(k)}\right\|_F^2 + \left\|\hat{W}^{(k)}\right\|_F^2\right) \tag{9}$$

3.5   Optimization

For explicit relations $e_{ij} \in E$, we use Stochastic Gradient Descent (SGD) to update the embedding vectors $\vec{\mathbf{u}}_i$ and $\vec{\mathbf{v}}_j$ so as to minimize KL-divergence O_{KL}. The updates to $\vec{\mathbf{u}}_i$ and $\vec{\mathbf{v}}_j$ are as follows:

$$\vec{\mathbf{u}}_i = \vec{\mathbf{u}}_i - \lambda\{\gamma w_{ij}[1 - \sigma(\vec{\mathbf{u}}_i^T \vec{\mathbf{v}}_j)] \cdot \vec{\mathbf{v}}_j\}, \quad \vec{\mathbf{v}}_j = \vec{\mathbf{v}}_j - \lambda\{\gamma w_{ij}[1 - \sigma(\vec{\mathbf{u}}_i^T \vec{\mathbf{v}}_j)] \cdot \vec{\mathbf{u}}_i\} \quad (10)$$

For implicit relations, we want to minimize loss function $\mathcal{L}_{ae\_user}$ and $\mathcal{L}_{ae\_item}$, take $\mathcal{L}_{ae\_user}$ for example, the patial derivative is shown below:

$$\frac{\partial \mathcal{L}_{ae\_user}}{\partial \hat{W}^{(k)}} = \frac{\partial \mathcal{L}_{2nd}}{\partial \hat{W}^{(k)}} + \frac{\partial \mathcal{L}_{reg}}{\partial \hat{W}^{(k)}}, \quad \frac{\partial \mathcal{L}_{ae\_user}}{\partial W^{(k)}} = \frac{\partial \mathcal{L}_{2nd}}{\partial W^{(k)}} + \frac{\partial \mathcal{L}_{1st}}{\partial W^{(k)}} + \frac{\partial \mathcal{L}_{reg}}{\partial W^{(k)}} \quad (11)$$

Firstly, $\frac{\partial \mathcal{L}_{2nd}}{\partial \hat{W}^{(K)}}$ and $\frac{\partial \mathcal{L}_{2nd}}{\partial \hat{X}}$ can be rephrased as follows:

$$\frac{\partial \mathcal{L}_{2nd}}{\partial \hat{W}^{(K)}} = \frac{\partial \mathcal{L}_{2nd}}{\partial \hat{X}} \cdot \frac{\partial \hat{X}}{\partial \hat{W}^{(K)}}, \quad \frac{\partial \mathcal{L}_{2nd}}{\partial \hat{X}} = 2(\hat{X} - X) \quad (12)$$

For the second term, since $\hat{X} = \sigma(\hat{Y}^{(K-1)}\hat{W}^{(K)} + \hat{b}^{(K)})$, we can get $\frac{\partial \mathcal{L}_{2nd}}{\partial \hat{W}^{(K)}}$, and based on hack-propagation, we can iteratively obtain $\frac{\partial \mathcal{L}_{2nd}}{\partial \hat{W}^{(k)}}, k = 1, ..., K-1$ and $\frac{\partial \mathcal{L}_{2nd}}{\partial W^{(k)}}, k = 1, ..., K$.

Secondly, we continue to calculate the partial derivative of $\frac{\partial \mathcal{L}_{1st}}{\partial W^{(k)}}$. The loss function of $\mathcal{L}_{1st}$ can be rephrased as follows:

$$\mathcal{L}_{1st} = \sum_{i,j=1}^{n} s_{i,j} \left\| \mathbf{y}_i - \mathbf{y}_j \right\|_2^2 = 2tr(Y^T L Y) \quad (13)$$

where $L = D - S, D \in \mathcal{R}^{n*n}$ is a diagonal matrix, $D_{i,i} = \sum_j s_{i,j}$.

Then the calculation of $\frac{\partial \mathcal{L}_{1st}}{\partial W^{(K)}}$ and $\frac{\partial \mathcal{L}_{1st}}{\partial Y}$ can be rephrased as follows:

$$\frac{\partial \mathcal{L}_{1st}}{\partial W^{(K)}} = \frac{\partial \mathcal{L}_{1st}}{\partial Y} \cdot \frac{\partial Y}{\partial W^{(K)}}, \quad \frac{\partial \mathcal{L}_{1st}}{\partial Y} = 2(L + L^T) \cdot Y \quad (14)$$

Since $Y = \sigma(Y^{(K-1)}W^{(K)} + b^{(K)})$, we can get $\frac{\partial Y}{\partial W^{(K)}}$. Similar to $\mathcal{L}_{2nd}$, we can get the calculation of partial derivative of $\mathcal{L}_{1st}$.

Since we have calculated the partial derivatives of the parameters, we can optimize our model using the stochastic gradient descent method. Empirically, we pretrain our parameters first in order to find a good region of parameter space. To sum up, the whole framework is given in Algorithm 1.

3.6   Analysis and Discussion

As we discussed before, hot items always have high scores rated by users, which leads to the large value of the final vector representation. As a result, users are recommendated too much hot items, which is called the Harry Potter Problem.

Algorithm 1. NEAR Framework

Require: Network $G = (U, V, E)$; Weight matrix of the network R; Embedding dimension d;

Ensure: Recommended list C_i for each user u_i;

  1: Construct the rating matrix R according to the network $G = (U, V, E)$;
  2: Calculate the adjacent matrix R^U and R^V according to the rating matrix R;
  3: $X = R^U (X = R^V)$;
  4: Pretrain the model to obtain the initialized parameter $\theta = \{\theta^{(1)}, ..., \theta^{(K)}\}$;
  5: **repeat**
  6:    Apply encoder to get $\hat{X}$ and $Y = Y^K$ according to X and θ;
  7:    Calculate the loss according to the loss function $\mathcal{L}_{ae\_x}(X; \theta) = \left\| \hat{X} - X \right\|_F^2 +$
     $2\alpha tr(Y^T LY) + \beta \mathcal{L}_{reg}$;
  8:    Update θ through back-propagate the entire network according to Eq. 11;
  9: **until** converge
10: Obtain the embedding matrix $\mathbf{U} = Y^{(K)}$ $(\mathbf{V} = Y^{(K)})$;
11: **for** each $edge(u_i, v_j) \in E$ **do**
12:    Update $\vec{u_i}$ and $\vec{v_j}$ using Eq. 10;
13: **end for**
14: Normalize the embedding matrix $\mathbf{U}$ and $\mathbf{V}$ with $l2$ norm;
15: Predict the ratings $\hat{R}$ according to the embedding matrix $\mathbf{U}$ and $\mathbf{V}$;
16: Obtain the top-k recommended list C_i for each user u_i according to $\hat{R}$;
17: **return** C_i;

To address this problem, we resort to punishing hot items to a certain extent through the L2 normalization.

$$\|\mathbf{x}\|_p = (\sum_{i=1}^{d} |x_i|^p)^{(1/p)} \qquad (15)$$

where p is 2 here, $\mathbf{x}$ stands for user embeding $\vec{u_i}$ or item embedding $\vec{v_j}$.

Experiments show that the prediction of the hot items declines in ranking, leading to the top-k results more diverse.

Assume that V is the total vertices, E is the total edges, I is the iteration times, K is the average degree of the network and D is the maximum dimension of the hidden layer. It is not difficult to see that the complexity of the framework is $O(I(|V|KD + |E|))$.

4 Experiments

4.1 Datasets

We choose two popular datasets MovieLens 1M[1] and DBLP[2] in our experiments. MovieLens has been widely used in the film recommendation system, where the

---

[1] https://grouplens.org/datasets/movielens/.
[2] https://dblp.uni-trier.de/xml/.

weight of an edge represents a user's rating of a movie. DBLP dataset contains the data of the author's published articles, in which the weight of a side indicates the number of papers published by an author in a venue. The statistics of the experimental dataset are summarized in Table 2.

In order to make the experimental comparison more reliable, we use the same training data and test data for all of the recommendation algorithms, and we set the scale of test data to 0.4. We set the embedding dimension of all compared methods as 64 for a fair comparison. For our model, the encoding layer dimension of auto-encoder are set [1024, 512], the embedding dimension are set 64. In practical, we need to fine-tune our hyper parameters according to different datasets to get good results.

4.2   Baseline Algorithms

We choose four contrast algorithms based on deep neural network (Walklet-Rec, BiNE-Rec) and matrix decomposition (SVD, NMF) respectively. The details are as follows:

- SVD [1]. SVD maps users and items to low-dimensional vectors on potential factors by decomposing scoring matrix.
- NMF [9]. Non-negative Matrix Factorization (NMF) is a latent factor model.
- Walklet [12] for recommendation (Walklet-Rec). Walklet is a network representation algorithm, which generates multiscale relationships by subsampling short random walks on the vertices of a graph.
- BiNE [7] for recommendation (BiNE-Rec). BiNE is a bipartite network representation learning algorithm.

Table 2. Statistics of the experimental dataset, evaluation metrics and parameters. AMD means adjacent matrix density.

Dataset	MovieLens 1M	DBLP
$|U|$	6040	6001
$|V|$	3900	1177
$|E|$	1000209	29256
Density	4.2%	0.4%
User AMD	85.4%	16.2%
Item AMD	63.1%	6.4%
Test rate	0.4	0.4
Embedding dimension	64	64
Metric	Coverage, Novelty, F1, NDCG	

4.3   Evaluation Metrics

We select four commonly used recommendation system evaluation metrics: Coverage, Novelty, F1, and NDCG. The analysis detail of four evaluation metrics are as follows.

- Coverage. Coverage score can evaluate the ability of discovering long-tailed items and the diversity of the recommended list.

$$Coverage = \frac{\left| \bigcup_{u_i \in U} C_i \right|}{|V|} \tag{16}$$

- Novelty. Novelty can be defined as follows:

$$Novelty = \frac{1}{|U|} \sum_{i=1}^{|U|} \sum_{j=1}^{k} \frac{pop_j}{k}, \ pop_j = \log_2 \frac{|U|}{d_j} \tag{17}$$

where k is the length of top-k recommended list, and d_j is the degree of item node v_j, which is the number of links between item node v_j and user set U.
- F1. F1 score is a compromise between the accuracy and the recall metric.
- NDCG. NDCG is a metric widely used in ranking learning. Normalized Discounted Cumulative Gain (NDCG) is defined as follows:

$$DCG@k = \sum_{i=1}^{k} \frac{2^{rel_i} - 1}{\log_2(i+1)}, \ NDCG@k = \frac{\sum_{u_i \in U} \frac{DCG_{u_i}@k}{IDCG_{u_i}}}{|U|} \tag{18}$$

where rel_i means the relevance of i-th item in recommended list; Ideal Discounted Cumulative Gain means the best sorted recommended list score.

Table 3. Top-K recommendation performance comparison of different algorithms on MovieLens and DBLP

Algorithm	DBLP				MovieLens 1M			
	F1@10	NDCG@10	Nov@10	Cov@10	F1@10	NDCG@10	Nov@10	Cov@10
NMF	1.16%	1.01%	8.28	4.79%	0.54%	0.41%	6.18	17.05%
SVD	2.36%	2.92%	7.19	2.46%	2.96%	2.81%	3.72	17.77%
BiNE-Rec	**11.32%**	**25.79%**	4.69	1.93%	**7.24%**	**7.08%**	3.48	2.74%
Walklet-Rec	4.64%	9.51%	8.17	19.03%	6.10%	4.89%	3.32	2.77%
NEAR	10.28%	20.92%	**8.55**	**57.63%**	6.52%	6.21%	**6.20**	**46.89%**
	F1@20	NDCG@20	Nov@20	Cov@20	F1@20	NDCG@20	Nov@20	Cov@20
NMF	1.65%	2.24%	8.53	8.13%	1.38%	1.06%	5.18	23.77%
SVD	2.54%	3.87%	7.21	4.51%	4.77%	4.23%	3.78	23.13%
BiNE-Rec	**8.04%**	**27.40%**	5.05	3.72%	**10.13%**	**9.24%**	3.98	2.88%
Walklet-Rec	4.47%	8.86%	7.62	36.62%	8.31%	7.37%	4.19	4.10%
NEAR	7.05%	21.47%	**8.68**	**67.37%**	9.30%	8.56%	**6.34**	**51.64%**

4.4   Experimental Results

From the experimental results shown in Table 3, we can see that deep neural network based algorithms (Walklet, BiNE and NEAR) perform better than matrix factorization-based methods (NMF and SVD) in terms of F1 and NDCG metric. We suppose that deep learning methods preserve sufficient hidden features of data and generate more general low dimension embedding. By comparing the results of deep neural network based methods (Walklet, BiNE) and matrix factorization-based algorithms (NMF, SVD), we can conclude that accuracy and diversity of the prediction are two mutually exclusive metrics. However, the impact of diversity on users cannot be ignored according to the previous discussion. Through the comparison of NEAR and Walklet, it can be seen that there is a significant increase in each metric, especially in terms of novelty and diversity. The main reasons might be as follows: (1) We employ a new similarity function to calculate adjacent matrices and mine the implicit relations of the network through the deep learning method based on autoencoder. (2) To further capture explicit features of the rating matrix, we propose to employ KL-divergence and L2 norm to fine-tune the embedding of vertices, we preserve sufficient hidden features of data and generate more general low dimension embedding, resulting in more accurate and diverse recommendation sequence. Among the results, we observe that BiNE algorithm performs best in accuracy while losing coverage and novelty score. On the contrary, our model NEAR has a significant increase in novelty and diversity while accuracy varies little from the optimal value.

5   Conclusion

We propose a Normalized Network Embedding with Autoencoder for Personalized Top-K Item Recommendation model, namely NEAR, to balance the accuracy and diversity scores and thus to give a more personalized recommendation. To fully exploit the hidden feature of the heterogeneous network, we design a new similarity function and employ the network embedding method based on autoencoder. To further capture explicit features of the rating matrix, we use KL-divergence and L2 norm to fine-tune the embedding of the vertex. Comparison studies on MovieLens and DBLP datasets show that our model has a good performance both on diversity and accuracy.

Since network embedding on the heterogeneous network is still in the early stages, we resort to propose more robust algorithms in our future work. Furthermore, the past decades have witnessed a great improvement in the deep neural network, we suppose to exploit more information of the recommendation dataset, such as the timestamp of each activity, the comment on each item, the information of each item and so on.

References

1. Ariyoshi, Y., Kamahara, J.: A hybrid recommendation method with double SVD reduction. In: Yoshikawa, M., Meng, X., Yumoto, T., Ma, Q., Sun, L., Watanabe, C.

(eds.) DASFAA 2010. LNCS, vol. 6193, pp. 365–373. Springer, Heidelberg (2010). https://doi.org/10.1007/978-3-642-14589-6_37

2. Belkin, M., Niyogi, P.: Laplacian eigenmaps for dimensionality reduction and data representation. Neural Comput. **15**(6), 1373–1396 (2014)

3. Cai, H.Y., Zheng, V.W., Chang, C.C.: A comprehensive survey of graph embedding: problems, techniques, and applications. IEEE Trans. Knowl. Data Eng. **30**(9), 1616–1637 (2018)

4. Cao, X., Shi, C., Zheng, Y., Ding, J., Li, X., Wu, B.: A heterogeneous information network method for entity set expansion in knowledge graph. In: Phung, D., Tseng, V.S., Webb, G.I., Ho, B., Ganji, M., Rashidi, L. (eds.) PAKDD 2018. LNCS (LNAI), vol. 10938, pp. 288–299. Springer, Cham (2018). https://doi.org/10.1007/978-3-319-93037-4_23

5. Chen, C., Zheng, X., Wang, Y., Hong, F., Lin, Z.: Context-ware collaborative topic regression with social matrix factorization for recommender systems. In: Twenty-Eighth AAAI Conference on Artificial Intelligence, pp. 9–15 (2014)

6. Cui, P., Wang, X., Pei, J., Zhu, W.: A survey on network embedding. IEEE Trans. Knowl. Data Eng. **PP**(99), 1 (2018)

7. Gao, M., Chen, L., He, X., Zhou, A.: BiNE: bipartite network embedding. In: The International ACM SIGIR Conference, pp. 715–724 (2018)

8. Li, X., She, J.: Collaborative variational autoencoder for recommender systems. In: The ACM SIGKDD International Conference, pp. 305–314 (2017)

9. Luo, X., Zhou, M., Li, S., You, Z., Xia, Y., Zhu, Q.: A nonnegative latent factor model for large-scale sparse matrices in recommender systems via alternating direction method. IEEE Trans. Neural Netw. Learn. Syst. **27**(3), 579–592 (2016)

10. Mikolov, T., Chen, K., Corrado, G., Dean, J.: Efficient estimation of word representations in vector space. In: ICLR Workshop (2013)

11. Perozzi, B., Al-Rfou, R., Skiena, S.: DeepWalk: online learning of social representations. In: ACM SIGKDD International Conference on Knowledge Discovery and Data Mining, pp. 701–710 (2014)

12. Perozzi, B., Kulkarni, V., Chen, H., Skiena, S.: Don't walk, skip! online learning of multi-scale network embeddings, pp. 258–265 (2017)

13. Shi, C., Hu, B., Zhao, X., Yu, P.: Heterogeneous information network embedding for recommendation. IEEE Trans. Knowl. Data Eng. **PP**(99), 1 (2017)

14. Tang, J., Qu, M., Wang, M., Zhang, M., Yan, J., Mei, Q.: Line: large-scale information network embedding, pp. 1067–1077 (2015)

15. Wang, D., Cui, P., Zhu, W.: Structural deep network embedding. In: ACM SIGKDD International Conference on Knowledge Discovery and Data Mining, pp. 1225–1234 (2016)

16. Wang, Z., Zhang, J., Feng, J., Chen, Z.: Knowledge graph embedding by translating on hyperplanes. In: Twenty-Eighth AAAI Conference on Artificial Intelligence, pp. 1112–1119 (2014)

17. Yang, C., Zhao, D., Zhao, D., Chang, E.Y., Chang, E.Y.: Network representation learning with rich text information. In: International Conference on Artificial Intelligence, pp. 2111–2117 (2015)

18. Zhang, S., Yao, L., Sun, A.: Deep learning based recommender system: a survey and new perspectives. CoRR abs/1707.07435 (2017)

19. Zhuang, F., Zhang, Z., Qian, M., Shi, C., Xie, X., He, Q.: Representation learning via dual-autoencoder for recommendation. Neural Netw. **90**, 83–89 (2017)

Ranking Network Embedding
via Adversarial Learning

Quanyu Dai[1(✉)], Qiang Li[2], Liang Zhang[3], and Dan Wang[1]

[1] Department of Computing, The Hong Kong Polytechnic University,
Hong Kong, China
{csqydai,csdwang}@comp.polyu.edu.hk
[2] Y-tech, Kwai, Beijing, China
leetsiang.cloud@gmail.com
[3] JD.com, Beijing, China
zhangliang16@jd.com

Abstract. Network Embedding is an effective and widely used method for extracting graph features automatically in recent years. To handle the widely existed large-scale networks, most of the existing scalable methods, e.g., DeepWalk, LINE and node2vec, resort to the negative sampling objective so as to alleviate the expensive computation. Though effective at large, this strategy can easily generate false, thus low-quality, negative samples due to the trivial noise generation process which is usually a simple variant of the unigram distribution. In this paper, we propose a Ranking Network Embedding (RNE) framework to leverage the ranking strategy to achieve scalability and quality simultaneously. RNE can explicitly encode node similarity ranking information into the embedding vectors, of which we provide two ranking strategies, vanilla and adversarial, respectively. The vanilla strategy modifies the uniform negative sampling method with a consideration of edge existance. The adversarial strategy unifies the triplet sampling phase and the learning phase of the model with the framework of Generative Adversarial Networks. Through adversarial training, the triplet sampling quality can be improved thanks to a softmax generator which constructs hard negatives for a given target. The effectiveness of our RNE framework is empirically evaluated on a variety of real-world networks with multiple network analysis tasks.

1 Introduction

Network embedding, i.e., learning low-dimensional representations for nodes in graph-structured data, can help encode meaningful semantic, relational and structural information of a graph into embedding vectors. Typically, such learning process is conducted in an unsupervised manner [1–3] due to the lack of labeled data, and thus the learned representations can be used to facilitate different kinds of downstream tasks such as network visualization, link prediction and node classification. In real-world applications, data entities with complicated relationships can be well organized with graphs. For example, paper citation networks characterize the information of innovation flow, social networks

© Springer Nature Switzerland AG 2019
Q. Yang et al. (Eds.): PAKDD 2019, LNAI 11441, pp. 27–39, 2019.
https://doi.org/10.1007/978-3-030-16142-2_3

entail complicated relationships among people, groups and organizations, and protein-protein interaction networks capture information between different proteins. Therefore, it is of great application interest to develop effective and scalable methods for unsupervised network embedding.

Network data are usually high-dimensional, very sparse and non-linear, which makes network embedding a challenging problem. Some classical methods, such as MDS [4], IsoMap [5] and LLE [6], can be used for network representation learning. However, they can neither effectively capture highly nonlinear structure of networks, nor scale to large networks. When handling large-scale networks, DeepWalk [1], LINE [2] and node2vec [7] are shown to be quite effective and efficient. These three methods preserve network structural properties in the embedding vectors through the negative sampling technique [8]. The negative sampling method is a simplified variant of negative contrastive estimation [9], which can help speed up the training process of the model. However, since the negative samples are constructed according to a unigram noise generation process, this strategy may generate false negative samples that violate pairwise relationships presented in the network structure. Here, we aim to answer two questions: (1) can we find some other ways for encoding pairwise relationships into node representations instead of the negative sampling approach? (2) how to sample better negative nodes for target-positive pairs (i.e., closely related node pairs) for training?

In this paper, we propose a Ranking Network Embedding (**RNE**) framework based on triplet ranking loss for preserving pairwise relationships of nodes in embedding vectors. Specifically, we firstly construct triplets based on network structure where each triplet consists of a target, a positive and a negative node. In the training process, the distance between embedding vectors of the target and positive node will be minimized while the distance between that of target and negative node will be maximized until they are separated by a predefined margin. Different from the negative sampling technique used in [1, 2, 7], the ranking strategy enforces a non-trivial margin between similar node pairs and dissimilar ones, thus explicitly encodes similarity ranking information among node pairs into the embedding vectors.

With the RNE framework, we propose two network embedding models by using a vanilla ranking strategy and an adversarial ranking technique respectively. In the vanilla RNE model, we utilize a simple negative node sampling method to construct triplets, which uniformly samples nodes from the node set without direct link to the target. This vanilla approach can perfectly avoid false negative samples while maintain the efficiency. Though works well to some extend, this vanilla strategy may also generate totally unrelated negative nodes for the target node, which will be of little help for the training. This phenomenon is even more common in very high-dimensional and sparse networks. To improve the vanilla RNE, we propose a generative adversarial model to unify the triplet sampling process and the learning process with the framework of Generative Adversarial Networks (GANs) [10], which leads to an adversarial RNE model. It leverages a generator for generating hard negative nodes with respect to a given target to help construct high-quality triplets, and thus achieves better node similarity rankings in the embedding space. We empirically evaluate the proposed vanilla and adversarial RNE models through several network analysis tasks, including network visualization, link prediction and node classification, on benchmark datasets. Experimental results show that both models achieve competitive performance with state-of-the-art methods.

2   Related Work

Many scalable network embedding methods, such as DeepWalk [1], LINE [2] and node2vec [7], have been proposed to learn node representations to facilitate downstream tasks. They model node conditional probability with softmax function over the whole network, which is computationally expensive. Further, the negative sampling approach [8] is usually leveraged to replace the log likelihood objective, and thus enabling a scalability to large networks. However, it can generate some negative samples violating pairwise relationships reflected by network structure because of the simple unigram noise generation process. To handle this issue, we propose to use the triplet ranking loss to learn embedding vectors and leverage an adversarial sampling method to sample negative nodes. We noticed that the triplet ranking loss is also employed by [11] in learning embeddings, but for networks with node attributes.

Recently, some methods are proposed to learn node representations through adversarial training [12,13]. In ANE [12], a prior distribution is imposed on node representations through adversarial training to achieve robustness. In [13], the authors proposed to unify the generative models and discriminative models of network embedding into the framework of GANs to help learn a stronger generator. Different from these two methods, our method aim to learn a stronger discriminator to obtain node representations.

Some knowledge graph embedding methods are also related [14–16]. TransE [14] is a translation-based knowledge graph embedding model, which learns embeddings for both data entities and relations with triplet ranking loss. KBGAN [16] is an adversarial learning framework for knowledge graph embedding. Our method is by part inspired by these works. However, this line of research has notable differences with our work. Firstly, knowledge graph is fundamentally different from the networks we study. The assumption, that two connected nodes should be similar and close in embedding space, of network embedding methods does not hold in knowledge graph. Secondly, knowledge graph embedding learns representations for both data entities (nodes) and relations (edges) simultaneously, while network embedding is designed to learn node representations only.

3   RNE: Ranking Network Embedding

3.1   Framework

The framework of Ranking Network Embedding method is shown in Fig. 1(a). It consists of two phases, i.e., the triplet construction phase and the learning phase. Firstly, we leverage some sampling methods to construct triplets based on network structure, which can help specify the similarity ranking of some pairwise relationships. Then, in the learning process, triplet ranking loss is minimized by directly updating embeddings to pull similar nodes closer in the embedding space, while pushing dissimilar nodes apart.

To help better understand our model, we first introduce some notations and describe the research problem. A network is denoted as $\mathcal{G} = (\mathcal{V}, \mathcal{E})$, with a set of nodes $\mathcal{V}$ representing data entities and a set of edges $\mathcal{E}$ each representing the relationship between two nodes. We mainly consider undirected graph in this paper. Given a graph $\mathcal{G}$, we aim to learn low-dimensional representations $\boldsymbol{u_i} \in R^d$ for each node $v_i \in \mathcal{V}$, which can capture network structural properties. We denote U as embedding matrix with $\boldsymbol{u_i}$ as its ith row.

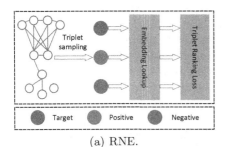
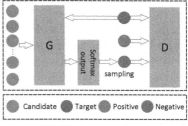

(a) RNE.   (b) Adversarial RNE.

Fig. 1. Model architecture.

3.2 Vanilla Ranking Network Embedding

The vanilla RNE model is a simple instantiation of the proposed RNE framework with uniform negative sampling method. Some detailed descriptions of its triplet sampling method and loss function are provided below.

Vanilla Triplet Sampling. Triplet sampling method plays an important role in learning good embedding vectors for downstream learning tasks. The constructed triplets directly specify pairwise relationships from network structure which will be regarded as ground-truth in learning process to be encoded into embedding vectors. We only explicitly consider first-order proximity when constructing positive pairs. The triplet set $\mathcal{T}$ is defined as follows:

$$\mathcal{T} = \{(v_t, v_p, v_n) | (v_t, v_p) \in \mathcal{E}, (v_t, v_n) \notin \mathcal{E}\}, \tag{1}$$

where (v_t, v_p, v_n) is a triplet with v_t, v_p and v_n as the target, positive and negative node, respectively. Since network is usually very sparse, for each positive pair, there can be a large number of negative nodes. To improve model efficiency, we only uniformly sample K negative nodes from the negative space for each positive pair.

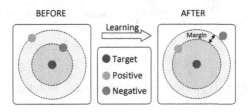

Fig. 2. The triplet ranking loss minimizes the distance between a target node and a positive node while maximizing that of the target and a negative node until they are separated by at least a margin distance. The pairwise relationships can be well preserved in embedding vectors after the learning process.

Triplet Ranking Loss. For vanilla RNE model, we seek to minimize the following loss function:

$$\mathcal{L} = \sum_{(v_t, v_p, v_n) \in \mathcal{T}} [m + D(v_t, v_p; \theta_D) - D(v_t, v_n; \theta_D)]_+, \tag{2}$$

where $[x]_+$ denotes the positive part of x, $D(v_1, v_2; \theta_D)$ is a distance function of two nodes, θ_D is the union of all node embeddings, and $m > 0$ is a margin hyperparameter separating the positive pair and the corresponding negative one. We use the squared L2 distances in the embedding space, i.e., $D(v_1, v_2; \theta_D) = \|u_1 - u_2\|^2$. The triplet ranking loss explicitly encodes similarity ranking among node pairs into the embedding vectors, and the visualization explanation can be found in Fig. 2 [17].

3.3 Adversarial Ranking Network Embedding

For the vanilla RNE model, we only use uniform negative sampling method for constructing triplets. It can easily generate totally unrelated negative nodes for the target node due to the sparsity and high-dimensionality of the network, which will be of little help for the training process. To help alleviate this problem, we propose an Adversarial Ranking Network Embedding model, which unifies the triplet sampling phase and the learning phase of the RNE method with the framework of GANs. The model architecture is presented in Fig. 1(b). It consists of a generator G and a discriminator D (we abuse the notation and directly use the distance function D to represent the discriminator). In the learning process, the discriminator tries to pull similar nodes closer in the embedding space, while pushing dissimilar nodes apart. The generator aims to generate difficult negative nodes for a given target from a set of negative candidates by optimizing its own parameters.

Discriminator. The discriminator is aimed at optimizing the following triplet ranking loss function similar to the vanilla RNE model:

$$\mathcal{L}_D = \sum_{(v_t, v_p) \in \mathcal{P}} \mathbb{E}_{v_n \sim G(\cdot | v_t; \theta_G)} [m + D(v_t, v_p; \theta_D) - D(v_t, v_n; \theta_D)]_+, \tag{3}$$

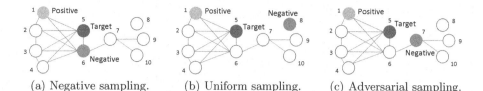

(a) Negative sampling.   (b) Uniform sampling.   (c) Adversarial sampling.

Fig. 3. For the negative sampling approach, each node is sampled according to its unigram distribution (regard each node as a word) raised to the 3/4 power, which can violate pairwise relationships reflected by network structure. For example, node 6 is very likely to be sampled as negative node for target-positive pair (5, 1), even though node 5 and 6 have strong relationship. For our triplet sampling method, such problem can be well avoided. However, simple uniform sampling method can easily generate totally unrelated nodes (node 8 in the example graph), which can be improved with adversarial sampling method.

where $\mathcal{P} = \{(v_1, v_2), (v_2, v_1) | (v_1, v_2) \in \mathcal{E}\}$ is the positive pair set in graph $\mathcal{G}$, and $G(\cdot | v_t; \theta_G)$ is the generator. Only first-order proximity is directly considered, and each edge $(v_i, v_j) \in \mathcal{E}$ corresponds to two positive pairs (v_i, v_j) and (v_j, v_i). Particularly, a softmax generator is employed to construct high-quality triplets instead of simple uniform sampling method. More detailed illustrations of this sampling method will be introduced below. Note that Eq. (3) can be directly optimized with gradient descent technique.

Generator. Softmax function is widely used in network embedding literature [1, 18] to model node conditional probability. In this paper, we also employ softmax function as the generator to sample negative nodes given a target, but it is defined over the negative node space with respect to the given positive pair according to network structure. Specifically, the generator $G(v_n | v_t; \theta_G)$ is defined as a softmax function over a set of negative candidates:

$$G(v_n | v_t; \theta_G) = \frac{\exp(\boldsymbol{u}_n^T \boldsymbol{u}_t)}{\sum_{v_{n_i} \in Neg(v_t, v_p)} \exp(\boldsymbol{u}_{n_i}^T \boldsymbol{u}_t)}, \tag{4}$$

where $Neg(v_t, v_p) = \{v_{n_1}, v_{n_2}, \cdots, v_{n_{N_c}}\}$ is a set of negative candidates with size as N_c. In implementation, $Neg(v_t, v_p)$ is a subsample of the original negative space of the positive pair to reduce the computation complexity, which is actually a common practice in network embedding literature [1,2]. For each positive pair, N_c negative nodes will be first uniformly randomly sampled from the negative space, and used as input for the generator. Then, a hard negative node will be sampled from $Neg(v_t, v_p)$ according to the probability distribution $G(v_n | v_t; \theta_G)$. Besides, in the training process, K hard negatives will be sampled for each positive pair.

The loss function of the generator is defined as follows:

$$\mathcal{L}_G = \sum_{(v_t, v_p) \in \mathcal{P}} \mathbb{E}_{v_n \sim G(\cdot | v_t; \theta_G)}[D(v_t, v_n; \theta_D)]. \tag{5}$$

It can encourage the softmax generator to generate useful negative nodes for a given positive pair instead of totally unrelated ones. The sampling process of hard negatives is discrete, which hinders the objective from directly being optimized by gradient descent method as that of the discriminator. According to [19,20], this loss can be optimized with the following policy gradient:

$$\nabla_{\theta_G} \mathcal{L}_G = \nabla_{\theta_G} \sum_{(v_t, v_p)} \mathbb{E}_{v_n \sim G(\cdot | v_t; \theta_G)}[D(v_t, v_n; \theta_D)]$$
$$= \sum_{(v_t, v_p)} \mathbb{E}_{v_n \sim G(\cdot | v_t; \theta_G)}[D(v_t, v_n; \theta_D) \nabla_{\theta_G} \log G(v_n | v_t; \theta_G)]. \quad (6)$$

The gradient of $\mathcal{L}_G$ is an expected summation of the gradient $\nabla_{\theta_G} \log G(v_n | v_t; \theta_G)$ weighted by the distance of node pair (v_t, v_n). When training the generator, the parameters will be shifted to involve high-quality negatives with high probability from softmax generator, i.e., node pairs (v_t, v_n) with small distance from discriminator will be encouraged to be generated. In practice, the expectation can be approximated with sampling in the negative space. Besides, the REINFORCE algorithm suffers from the notorious high variance, which can be alleviated by subtracting a baseline function from the reward term of the objective, i.e., adding a baseline function to the reward term in the loss [21]. Specifically, we replace $D(v_t, v_n; \theta_D)$ in the loss by its advantage function as follows:

$$D(v_t, v_n; \theta_D) + \sum_{(v_t, v_p)} \mathbb{E}_{v_n \sim G(\cdot | v_t; \theta_G)}[D(v_t, v_n; \theta_D)], \quad (7)$$

where $\sum_{(v_t, v_p)} \mathbb{E}_{v_n \sim G(\cdot | v_t; \theta_G)}[D(v_t, v_n; \theta_D)]$ is the average reward of the whole training set, and acts as the baseline function in policy gradient.

A comparison of the sampling methods is presented in Fig. 3 with toy examples. Our proposed adversarial sampling method can help select difficult negative nodes with respect to given target. With high-quality triplets, the tricky pairwise relationship rankings can be encoded into node representations through the training of the discriminator as illustrated in Fig. 2. Note that false negative nodes can still be generated by the generator due to the incompleteness and non-linearity of the real-world networks, but in a very low probability since the subsampling trick is employed for generating negative candidates among a very large negative space. So, the embedding vectors can be improved in general. This is also validated by our experiments.

Algorithm 1 presents the pseudocode of the adversarial RNE model, which employs a joint training procedure. The overall time complexity of the algorithm is linear to the number of edges, i.e., $\mathcal{O}(dKN_c|\mathcal{E}|)$ (d, K and N_c are some constants independent of the network size), which enables it scale to large networks.

Algorithm 1. The adversarial RNE algorithm

Input : $\mathcal{G}(\mathcal{V}, \mathcal{E})$, Dimension d, Margin m, Negative size K, Candidate size N_c
Output: The parameters of Discriminator θ_D
1 Initialize the Generator $G(v_n|v_t; \theta_G)$ and Discriminator $D(v_1, v_2; \theta_D)$ with pretrained embedding vectors;
2 **while** *not converge* **do**
3 | Sample a batch of positive pairs $\mathcal{B}$ from positive set $\mathcal{P}$;
4 | $\mathcal{T} = \{\}$; $\mathcal{N} = \{\}$;
5 | // Adversarial negative sampling with softmax generator
6 | **for** *each* $(v_t, v_p) \in \mathcal{B}$ **do**
7 | | **repeat**
8 | | | Sample N_c negative candidates $Neg(v_t, v_p)$ uniformly from the negative space of (v_t, v_p);
9 | | | Sample a hard negative v_n from $Neg(v_t, v_p)$ according to $G(v_n|v_t, \theta_G)$;
10 | | | $\mathcal{T} = \mathcal{T} \cup \{v_n\}$; $\mathcal{N} = \mathcal{N} \cup \{Neg(v_t, v_p)\}$;
11 | | **until** K *times*;
12 | **end**
13 | // Parameters updating
14 | update θ_D according to Eq. (3) with $\mathcal{T}$ as training batch;
15 | update θ_G according to Eq. (5) and (4) with $\mathcal{T}$ and $\mathcal{N}$ as training batch;
16 **end**

---

4 Experiments

4.1 Experiment Setup

Datasets. We conduct experiments on benchmark datasets from various real-world applications. Table 1 shows some statistics of them. Note that we regard all paper citation networks as undirected networks, and do some preprocessing on the original datasets by deleting self-loops and nodes with zero degree.

Table 1. Statistics of benchmark datasets from real-world applications

Name	Citeseer [22]	Cit-DBLP [23]	PubMed [24]	CA-GrQc [25]	CA-HepTh [25]	Wiki [26]	USA-AIR [27]
$|V|$	3,264	5,318	19,717	5,242	9,877	2,363	1,190
$|E|$	4,551	28,065	44,335	14,484	25,973	11,596	13,599
Avg. degree	1.39	5.28	2.25	2.76	2.63	4.91	11.43
#Labels	6	3	3	-	-	17	4

Baseline Models. We only consider scalable baselines in this paper. Some matrix factorization based methods such as M-NMF [3,28] are excluded from the baselines due to the $O(|V|^2)$ time complexity. The descriptions of the baseline models are as follows: Graph Factorization (GF) [29] directly factorizes the adjacency matrix to obtain the embeddings. DeepWalk (DW) [1] regards node sequence obtained from truncated random walk as word sequence, and then uses skip-gram model to learn node representations. LINE [2] preserves proximities through modeling node co-occurrence probability and node conditional probability. node2vec (n2v) [7] develops a biased random walk procedure to explore neighborhood of a node, which can strike a balance between local and global properties. We denote the vanilla RNE model as V-RNE, and the adversarial RNE model as A-RNE in the rest of the paper.

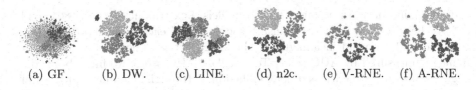

(a) GF.   (b) DW.   (c) LINE.   (d) n2c.   (e) V-RNE.   (f) A-RNE.

Fig. 4. Visualization of Cit-DBLP network.

Parameter Settings. The window size, walk length and the number of walks per node of both DeepWalk and node2vec are set to 10, 80 and 10, respectively. We use node2vec in an unsupervised manner by setting both in-out and return hyperparameters to 1.0 for fair comparison. For LINE, we follow the original paper [2] to set the parameters. For our method, the parameter settings are the margin $m = 2.5$, the negative size per edge $K = 5$, and the negative candidate size $N_c = 5$. The learning rate of V-RNE is set to 0.01, while A-RNE to 0.0001. L2-normalization is conducted on node embeddings for both the V-RNE and A-RNE model after each training epoch. Besides, the dimension of embedding vectors are set to 128 for all methods.

4.2 Network Visualization

We leverage a commonly used toolkit *t-SNE* [30] to visualize node embeddings of Cit-DBLP generated by different models. Cit-DBLP is a citation network constructed from the DBLP datsest [23], which consists of papers from publication venues including Information Sciences, ACM Transactions on Graphics and Human-Computer Interaction. These papers are naturally classified into three categories according to their publication venues, and represented with different colored nodes in the visualization.

Experimental Results. Figure 4 displays the visualization results. Papers from different publication venues are mixed together terribly for GF as shown in Fig. 4(a). In the center part of both DeepWalk and LINE, papers from different categories are mixed with each other. Visualizations from node2vec, V-RNE and A-RNE are much better as three clusters are formed with quite clear margin. Compared with V-RNE, A-RNE model has better visualization result, since the margin between different clusters are larger. The reason is that adversarial sampling method aims to generate hard negative nodes, i.e., negative nodes near the boundary, which directly contributes to producing more clear margin between different clusters. On the whole, this experiment demonstrates that ranking network embedding method can help capture intrinsic structure of original network in embedding vectors.

4.3 Link Prediction

We conduct link prediction on three benchmark datasets. For each network, we randomly and uniformly sample 20% and 50% of the edges as test labels and use the remaining network as input to the models, i.e., training ratio as 80% and 50%. When sampling edges, we ensure the degree of each node is greater than or equal to 1 to avoid meaningless embedding vectors. The prediction performance is measured by AUC score. To calculate AUC score, we first obtain the edge features from the learned node embeddings through Hadamard product of embeddings of two endpoints as many other works [7], and then train a L2-SVM classifier with under-sampling to get prediction results.

Table 2. AUC score for link prediction

Training ratio	80%			50%		
Dataset	Wiki	CA-GrQc	CA-HepTh	Wiki	CA-GrQc	CA-HepTh
GF	0.583 ± 0.008	0.593 ± 0.003	0.554 ± 0.001	0.566 ± 0.002	0.572 ± 0.003	0.531 ± 0.001
DeepWalk	0.656 ± 0.001	0.694 ± 0.001	0.683 ± 0.001	0.639 ± 0.001	0.657 ± 0.002	0.630 ± 0.001
LINE	0.649 ± 0.007	0.638 ± 0.005	0.630 ± 0.001	0.627 ± 0.014	0.600 ± 0.003	0.561 ± 0.002
node2vec	0.634 ± 0.016	0.690 ± 0.007	0.668 ± 0.003	0.621 ± 0.010	0.667 ± 0.010	0.624 ± 0.007
V-RNE	0.647 ± 0.008	0.691 ± 0.005	0.657 ± 0.005	0.627 ± 0.007	0.655 ± 0.004	0.606 ± 0.004
A-RNE	$\mathbf{0.670 \pm 0.005}$	$\mathbf{0.708 \pm 0.004}$	$\mathbf{0.688 \pm 0.004}$	$\mathbf{0.655 \pm 0.006}$	$\mathbf{0.673 \pm 0.004}$	$\mathbf{0.639 \pm 0.004}$

Experimental Results. The link prediction results are the average of 10 different runs, which are shown in Table 2. The AUC scores of A-RNE model consistently outperform those of the V-RNE model. It validates that A-RNE can help achieve better node similarity rankings in embedding space, since link prediction task can be considered as similarity ranking among node pairs. The performance of the proposed RNE method is competitive with the baselines, which shows that using ranking strategy for learning node representations is a good practice. In particular, the AUC scores of A-RNE model are superior to all the baselines in all test datasets when the training ratios are 80% and 50%.

Table 3. Accuracy (%) of multi-class classification on USA-AIR and PubMed

Dataset	USA-AIR					Pubmed				
Ratio	10%	30%	50%	70%	90%	10%	30%	50%	70%	90%
GF	41.10	42.21	42.27	41.12	41.60	35.63	36.69	37.56	37.74	38.08
DeepWalk	43.43	51.79	53.41	55.74	56.05	69.43	71.33	71.74	71.82	72.37
LINE	48.80	53.95	56.35	56.72	58.91	67.23	69.20	69.84	69.97	70.48
node2vec	42.76	47.07	48.62	49.86	50.76	79.66	80.89	81.09	81.07	81.27
V-RNE	55.20	58.96	60.05	61.29	61.09	77.56	79.08	79.39	79.46	79.73
A-RNE	**56.94**	**61.96**	**62.79**	**65.71**	**64.12**	**80.48**	**81.20**	**81.58**	**81.56**	**81.64**

4.4   Node Classification

Node classification can be conducted to dig out missing information. In this section, we carry out experiments on the air-traffic network USA-AIR and paper citation network PubMed. The learned embedding vectors are used as feature input for the classification model. We randomly sample a portion of nodes as training data ranging from 10% to 90%, and the rest for testing. For both datasets, multi-class classification is conducted, and accuracy score is used for performance comparison. All experiments are conducted with support vector classifier in Liblinear package[1] [31] with default settings.

Experimental Results. The experimental results are presented in Table 3. Both V-RNE and A-RNE perform competitively with baseline models for these two datasets while varying the train-test split from 10% to 90%. It shows the effectiveness of the proposed Ranking Network Embedding models for learning discriminative embedding vectors for classification. Specifically, both V-RNE and A-RNE achieve better performance in USA-AIR, and A-RNE obtains the best results in these two datasets across all training ratios. In particular, A-RNE gives us 13.32% gain on average over the best baseline, i.e., LINE on USA-AIR. Besides, A-RNE consistently achieves more excellent performance than V-RNE as shown in the tables, which demonstrates that adversarial sampling method contributes to learning more discriminative node representations.

5 Conclusion

This paper presented a novel scalable Ranking Network Embedding method, which can explicitly encode node similarity ranking information into the embedding vectors. Firstly, a vanilla RNE model was proposed with uniform negative sampling method. Then, we improved the vanilla RNE model by unifying the triplet sampling phase and the learning phase with the framework of GANs which leads to an adversarial RNE model. The adversarial RNE model utilizes a softmax generator to generate hard negatives for a given a target, which can help strengthen the discriminator. Empirical evaluations prove the effectiveness of the proposed method on several real-world networks with a variety of network analysis tasks.

References

1. Perozzi, B., Al-Rfou, R., Skiena, S.: DeepWalk: online learning of social representations. In: KDD, pp. 701–710 (2014)
2. Tang, J., Qu, M., Wang, M., Zhang, M., Yan, J., Mei, Q.: LINE: large-scale information network embedding. In: WWW, pp. 1067–1077 (2015)
3. Cao, S., Lu, W., Xu, Q.: Grarep: learning graph representations with global structural information. In: CIKM, pp. 891–900 (2015)
4. Cox, T.F., Cox, M.A. (eds.): Multidimensional Scaling. CRC Press, Boca Raton (2000)
5. Tenenbaum, J.B., de Silva, V., Langford, J.C.: A global geometric framework for nonlinear dimensionality reduction. Science **290**, 2319–2323 (2000)

---

[1] https://www.csie.ntu.edu.tw/~cjlin/liblinear/.

6. Roweis, S.T., Saul, L.K.: Nonlinear dimensionality reduction by locally linear embedding. Science **290**, 2323–2326 (2000)
7. Grover, A., Leskovec, J.: node2vec: Scalable feature learning for networks. In: KDD, pp. 855–864 (2016)
8. Mikolov, T., Sutskever, I., Chen, K., Corrado, G.S., Dean, J.: Distributed representations of words and phrases and their compositionality. In: NIPS, pp. 3111–3119 (2013)
9. Gutmann, M., Hyvärinen, A.: Noise-contrastive estimation of unnormalized statistical models, with applications to natural image statistics. J. Mach. Learn. Res. **13**, 307–361 (2012)
10. Goodfellow, I.J., et al.: Generative adversarial nets. In: NIPS, pp. 2672–2680 (2014)
11. Duran, A.G., Niepert, M.: Learning graph representations with embedding propagation. In: NIPS, pp. 5125–5136 (2017)
12. Dai, Q., Li, Q., Tang, J., Wang, D.: Adversarial network embedding. In: AAAI (2018)
13. Wang, H., et al.: Graph representation learning with generative adversarial nets. In: AAAI, Graphgan (2018)
14. Bordes, A., Usunier, N., Garcia-Duran, A., Weston, J., Yakhnenko, O.: Translating embeddings for modeling multi-relational data. In: NIPS (2013)
15. Wang, P., Li, S., Pan, R.: Incorporating GAN for negative sampling in knowledge representation learning. In: AAAI (2018)
16. Cai, L., Wang, W.Y.: KBGAN: adversarial learning for knowledge graph embeddings. CoRR, abs/1711.04071 (2017)
17. Schroff, F., Kalenichenko, D., Philbin, J.: FaceNet: a unified embedding for face recognition and clustering. In: CVPR, pp. 815–823 (2015)
18. Yang, Z., Cohen, W.W., Salakhutdinov, R.: Revisiting semi-supervised learning with graph embeddings. In: ICML, pp. 40–48 (2016)
19. Schulman, J., Heess, N., Weber, T., Abbeel, P.: Gradient estimation using stochastic computation graphs. In: NIPS, pp. 3528–3536 (2015)
20. Yu, L., Zhang, W., Wang, J., Yu, Y.: SeqGAN: sequence generative adversarial nets with policy gradient. In: AAAI, pp. 2852–2858 (2017)
21. Sutton, R.S., Mcallester, D., Singh, S., Mansour, Y.: Policy gradient methods for reinforcement learning with function approximation. In: NIPS, pp. 1057–1063. MIT Press (2000)
22. McCallum, A., Nigam, K., Rennie, J., Seymore, K.: Automating the construction of internet portals with machine learning. Inf. Retr. **3**(2), 127–163 (2000)
23. Tang, J., Zhang, J., Yao, L., Li, J., Zhang, L., Su, Z.: ArnetMiner: extraction and mining of academic social networks. In: KDD, pp. 990–998 (2008)
24. Nandanwar, S., Narasimha Murty, M.: Structural neighborhood based classification of nodes in a network. In: KDD, pp. 1085–1094 (2016)
25. Leskovec, J., Kleinberg, J.M., Faloutsos, C.: Graph evolution: densification and shrinking diameters. TKDD **1**(1), 2 (2007)
26. Sen, P., Namata, G., Bilgic, M., Getoor, L., Galligher, B., Eliassi-Rad, T.: Collective classification in network data. AI Mag. **29**(3), 93–106 (2008)
27. Ribeiro, L.F.R., Saverese, P.H.P., Figueiredo, D.R.: struc2vec: Learning node representations from structural identity. In: KDD, pp. 385–394 (2017)
28. Wang, X., Cui, P., Wang, J., Pei, J., Zhu, W., Yang, S.: Community preserving network embedding. In: AAAI, pp. 203–209 (2017)

29. Ahmed, A., Shervashidze, N., Narayanamurthy, S.M., Josifovski, V., Smola, A.J.: Distributed large-scale natural graph factorization. In: WWW, pp. 37–48 (2013)
30. van der Maaten, L., Hinton, G.: Visualizing data using t-SNE. JMLR **9**, 2579–2605 (2008)
31. Fan, R.-E., Chang, K.-W., Hsieh, C.-J., Wang, X.-R., Lin, C.-J.: LIBLINEAR: a library for large linear classification. JMLR **9**, 1871–1874 (2008)

Selective Training: A Strategy for Fast Backpropagation on Sentence Embeddings

Jan Neerbek[1,2]([✉]), Peter Dolog[3], and Ira Assent[1]

[1] Department of Computer Science, DIGIT Center, Aarhus University,
Aarhus, Denmark
{jan.neerbek,ira}@cs.au.dk
[2] Alexandra Institute, Aarhus, Denmark
[3] Department of Computer Science, Aalborg University, Aalborg, Denmark
dolog@cs.aau.dk

Abstract. Representation or embedding based machine learning models, such as language models or convolutional neural networks have shown great potential for improved performance. However, for complex models on large datasets training time can be extensive, approaching weeks, which is often infeasible in practice. In this work, we present a method to reduce training time substantially by selecting training instances that provide relevant information for training. Selection is based on the similarity of the learned representations over input instances, thus allowing for learning a non-trivial weighting scheme from multi-dimensional representations. We demonstrate the efficiency and effectivity of our approach in several text classification tasks using recursive neural networks. Our experiments show that by removing approximately one fifth of the training data the objective function converges up to six times faster without sacrificing accuracy.

Keywords: Selective training · Machine learning · Neural network · Recursive models

1 Introduction

Recent years have seen substantial performance improvements in machine learning for *deep models* in a variety of application domains [10]. However, training times for deep models can easily be in the order of days [27] or even weeks [3]. Being able to efficiently train and evaluate new models is important in order to preserve our ability to investigate and develop better machine learning models. Thus, training effort may be a critical factor in the deployment and advancement of more powerful, expressive machine learning models. This is certainly true for deep neural network models where the quest for stronger and better neural models drives doubling of models sizes (number of neurons) approximately every 2.4 years [10].

© Springer Nature Switzerland AG 2019
Q. Yang et al. (Eds.): PAKDD 2019, LNAI 11441, pp. 40–53, 2019.
https://doi.org/10.1007/978-3-030-16142-2_4

In this work, we present *Selective Training*, an effective training strategy for artificial neural network models. In a nutshell, by focusing on instances with relevant information for training, our approach requires fewer training iterations to converge to a stable and effective model. Selective Training adjusts training based on multi-dimensional representations of what the network has learned. It can be used with different training methodologies such as standard backpropagation or adaptive training approaches like Adam where the learning rate is adjusted depending on the loss gradient [14].

In this paper we focus on the deep structured gradient backpropagation training approach, *backpropagation-through-structure (BPTS)* [9,12] for text classification tasks with high training times, where we observe substantial improvements in training time. Still, our results are not limited to this application, but generalize to classifier models which generate distributed instance representations. We demonstrate in our empirical study on several document datasets that the gains in training time do not come at the cost of accuracy, but may even bring a slight improved accuracy score[1].

Our method identifies obsolete training samples through clustering in representation space. The cluster approach makes it possible to select those parts of the training data that matter for training, and to focus on these in order to reduce training time. This is in contrast to approaches such as instance selection [25] or active learning [28] where the goal is to find the minimum representative instances (or equivalently instance lookups). This difference in goals leads to two major methodological differences; (1) We remove entire clusters rather than instances and (2) we work on embedded representations rather than on instance features. A major challenge for instance selection as reported in [25] is the need for comparing new instances to all previously selected instances. This costly comparison is a challenge for scale-up, that our method does not suffer from. Clustering into relatively few clusters is sufficient and efficient, and can even be further scaled up using sub-sampling or hierarchical approaches [2].

Our contributions include a selection strategy for training with substantial speed up while still maintaining high model accuracy, an empirical study on four real world datasets that demonstrates the effectiveness and efficiency of our approach, robustness with respect to parameterization, as well as a detailed error analysis.

2 Background

In the following, we study efficient training for complex deep models for text classification, as a concrete instance of costly training problems in deep learning. In the text classification, distributed word embeddings [1] have been immensely successful for a wide range of tasks including sentiment analysis [31], POS-tagging [4] and text classification [13]. Many approaches learn unsupervised word embeddings on large document sets such as word2vec [23] and GloVe [26]. VecAvg

---

[1] code   https://bitbucket.alexandra.dk/projects/TAB,   data   https://dataverse.harvard.edu/dataverse/enron-w-trees.

proposed to define sentence embeddings as the average of all word embeddings in a given sentence [20]. VecAvg has since been superseded by more expressive models such as recurrent neural networks (RNN) using gated memory cells such as the LSTM [11].

A generalization of the recurrent model has been proposed in [30] as *recursive neural networks (RecNN)*[2]. RecNN models can incorporate semantic knowledge about the sentence in a tree or graph like structure. To evaluate a RecNN a walk is required from node to node through the entire tree. This may negatively impact performance, and the walk may be hard to parallelize, therefore various restricted versions of the RecNN have been proposed such as Hierarchical ConvNet [5] and Graph Convolutional Networks (GCN) [15] where the number of steps in the graph is restricted to a fixed constant number. In our experimental study, we consider embeddings generated by the full complexity of the RecNN model over the constituency parse trees as proposed in [29]. As we only assume embeddings, our approach can generally be applied to any text or sentence embedding generating approach as well.

3   Problem Definition

Given a dataset D of n text documents $D = \{d_1, d_2, \ldots, d_n\}$, a ground-truth labeling $L : D \to \{1, 2, \ldots C\}$ with C classes, and label $L(x)$ for $x \in D$, the goal is to train a given model m using as few training cycles as possible while maintaining m's accuracy. Model m is parametrized by a set of parameters $\theta \in \Theta$, where Θ denotes all possible model parametrizations. The approach used to find a good model is referred to as the learning approach, which we denote T_m. For example, T_m could be the application of backpropagation on model m. Given (mini-)batches of b texts per cycle the learning approach updates the set of parameters $T_m : (\Theta, D^b) \to \Theta$, i.e., given a parametrized model m_θ and a subset $D' \subseteq D$ of b texts the training function returns a new set of parameters θ': $\theta' = T_m(\theta, D')$.

Efficiency of the training approach is then the expected number of random batches of training text documents that needs to be processed by T_m before the performance of the model m_θ converges. That is, further batches do not bring m_θ closer to L, as indicated by an error measure, such as the squared error function $(L(x) - m_\theta(x))^2$.

We wish to minimize the objective function $obj(m_\theta) = \frac{1}{|D|} \sum_{x \in D} (L(x) - m_\theta(x))^2$. Given a threshold ϵ, model $m_{\theta'}$ as the model after training m_θ on additional δ batches, then we say that the model m has converged iff $obj(m_\theta) - obj(m_{\theta'}) < \epsilon$, i.e., further batches do not improve m_θ. In this paper, we use the expected number of batches to produce a converged model on the training dataset as a measure of the training time, denoted $t(T_m)$ or just t_m for short. We define the training time optimization problem for a dataset D with labeling

---

[2] In this work we refer to recursive neural networks as *RecNN* to avoid name clash with RNNs.

function L, a model architecture m and training method T_m as the minimization of $t(T_m)$ under the constraint that model accuracy be preserved.

For embedding based approaches, a distributed representation of the input is learned in order to predict the correct label. Thus, our model m_θ can be split into a predictive part m_θ^p and an embedding generating part m_θ^e as follows: $m_\theta = m_\theta^p m_\theta^e$. In complex models, the embedding layer in fact typically consists of several layers $m_\theta^e = m_\theta^L \ldots m_\theta^2 m_\theta^1$. For our method, we only use the most informative embedding which is the final embedding produced just before a prediction is made. Evaluating model m_θ on training instance x yields $m_\theta(x) = m_\theta^p m_\theta^e(x)$. We refer to the output of the embedding layer as a *representation* of x: $repr(x) = m_\theta^e(x)$.

The key idea in this work is to exploit this representation as a means to identify a subset of the training instances that does not contain information for training. Excluding it from further training reduces training time $t(T_m)$ while not hurting accuracy.

4   Our Approach

To minimize the number of training cycles $t(T_m)$, our goal is to select the most informative training samples. Given a model m_θ and input x, we extract representations $repr(x)$ of what the model has learned about the input x. In a deep neural network, this generally is the second last layer before the final prediction layer.

If the model has learned to differentiate between two inputs x and x' with different labels $L(x) \neq L(x')$, then their representations should differ as well. If the representations were similar then it would be challenging for the model prediction layer to distinguish between the two representations. We observe that the structure of the representation space allows us to select training instances of interest, namely groups of instances with different labels in close vicinity in representation space, as those have not yet been properly differentiated with respect to their labels. Clearly, training on instances with similar representations and identical label provides less information for learning to distinguish between classes.

We therefore pose the problem of finding subsets of interesting training samples as a clustering problem. While any hard clustering method could be used, we use the well established K-means [2,7,21] as a simple and efficient choice. In brief, K-means assigns representations into k clusters such that the total sum (TS) of $L2$ distances between cluster centroid and cluster members is minimized: $TS = \sum_{x_i \in D} \|repr(x_i) - c_j(x_i)\|^2$, where $c_j(x_i)$ is the centroid of the cluster that input x_i is assigned to.

Let $C = \{C_1, C_2, \ldots, C_k\}$ be the resulting clusters, grouped exclusively based on the similarity in representation space. We now analyze them with respect to label purity to identify sets of samples in representation space that have already been learned and can be omitted from further training. We formalize this analysis as the ratio of the *most frequently occurring (MFO)* class label ratio in a cluster,

MFO_i, as $MFO_i = \max_{\ell \in L} \frac{\sum_{x \in C_i : L(x) = \ell} 1}{|C_i|}$ i.e., the ratio of the most frequently occurring class label in a cluster C_i is the maximum (over possible labels) of the ratio between the number of instances with that label and the cardinality of the cluster. Clusters with low MFO_i are valuable for training, whereas those where MFO_i is close to 1 are uninteresting as little more can be learned. A strong model has MFO_i for all clusters close to 1 (otherwise accuracy is low, see above). If there are only two classes $\{0, 1\}$, we simplify using the ratio for only one class $f_i = \frac{\sum_{x \in C_i : L(x) = 1} 1}{|C_i|}$ and obtain $MFO_i = \max(f_i, 1 - f_i)$.

Our goal is to separate interesting from uninteresting clusters by finding a suitable threshold for the MFO ratio to filter uninteresting training instances away and focus training on interesting ones only. More formally we wish to choose the lowest possible MFO threshold such that models trained with our approach m_θ^s satisfy the optimization goal, i.e., that the objective function value over the model trained using our selective strategy is at least as good as the objective value obtained with full training over all data samples over all training cycles.

Algorithm 1. Proposed fast training approach

1: **procedure**
2:     $D \leftarrow$ corpus of labeled documents
3:     $k \leftarrow$ Number of clusters to generate
4:     $MFO_{cut} \leftarrow$ cutoff for filtering
5:     **while** *pretraining* **do**
6:         $\Delta Acc \leftarrow$ rate of acc. improv.
7:         **if** ΔAcc starts dropping **then**
8:             break pretraining
9:     Cluster using K-means
10:     $MFO_i \leftarrow MFO$ for cluster i
11:     $D' \leftarrow$ clusters with $MFO_i \leq MFO_{cut}$
12:     **while** *main-training* **do**
13:         train on reduced dataset D'
14:         **if** convergence is achieved **then**
15:             break main-training

This can be seen as a balance between two forces: (1) high MFO cutoff means filtering only data where we are sure of the label, and do not mistakenly dismiss information and in turn decrease model accuracy. (2) low MFO cutoff means reducing training to fewer instances, thus fewer minibatches and finally lower training time. Clearly, the best filtering cutoff is a trade-off. We propose to study the decrease in MFO in the log-scale, and define $\Delta MFO = -\log_{10}(1 - MFO)$. Our empirical study suggests values in the range $2 \geq \Delta MFO \geq 1.5$. At prediction time we match new data to clusters. For data in removed clusters we use its dominating class label. For other clusters, we use the model trained on the reduced set. Algorithm 1 outlines our selective training strategy. In the experiments, we demonstrate that our approach indeed converges faster, i.e., uses fewer training epochs and converges to an accuracy that is at least as high as the one for full time training on all data.

5   Evaluation

5.1   Data and Experimental Setup

We use the large-scale, open access Enron data [18], comprised of more than $500,000$ documents that vary greatly in style, language and length and thus provide excellent insight into performance on varied text. All documents are split into sentences using the Punkt sentence boundary detection approach [16]. Constituency parse trees (splitting sentences into phrases) are generated from a probabilistic context-free grammar [17], trained over the Stanford Penn Treebank [32]. We train a recursive neural network model over these parse trees, which allows us to learn an embedding for each phrase. We make this data available online (see footnote 1).

We use labels by domain experts from the TREC competition [6,33], where topics were labeled by at least 3 human annotators (using majority where different). We evaluate on 4 binary topics where a sentence is *true* if it belongs to the topic and *false* otherwise: $FCAST$: 267366 sentences regarding Enron's financial state. We use 40000 sentences for validation, 40000 for testing, the rest for training. The percentage of positive (i.e., true) sentences is 31%. FAS: 178266 sentences where Enron claims compliance with Financial Accounting Standards[3]. We use 27000 sentences for validation, 27000 for training; 59% positive sentences. $PPAY$: 134256 sentences about financial *prepay transactions*. We use 15000 sentences for validation, 15000 for testing; 13% positive sentences. $EDENCE$: 167913 sentences discussing tampering with evidence. We use 25000 sentences for validation, 25000 for testing; positive sentences 23%. For further details see [6,24,33].

We test Selective Training on an RecNN where the intrinsic model is applied recursively over parse-trees. For long sentences the number of layers in the final recursive network ("unfolded" in time and structure) reaches several hundreds layers. We report findings where the intrinsic model has a single hidden layer and where the hidden layer has 100 neurons. We have tested on a smaller development set and found these hyperparameters to yield robust performance. Standard full training takes in the order of 1 day per $1,000,000$ minibatches, whereas clustering takes few minutes. Minibatch count is thus an appropriate measure for training time of full and selective training.

---

[3] http://www.fasb.org/jsp/FASB/Document_C/DocumentPage?cid=1218220124871.

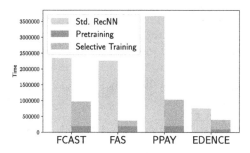

Fig. 1. Training time (in mini-batches) on 4 datasets

5.2   Empirical Study and Discussion

Figure 1 reports training times on 4 different datasets of varying complexities. Standard backpropagation through structure (denoted Std. RecNN in the figure) converges on the $EDENCE$ dataset in $758,000$ minibatches, and on the $PPAY$ dataset in $3,674,000$ minibatches, even though the datasets are of comparable size. Our approach reduces the training time by a factor of approximately 2 to 6, depending on the dataset under study. $FCAST$ and FAS have approximately the same runtime on the full dataset, which our approach reduces for $FCAST$ by factor 2.42, while for FAS we obtain an impressive factor of 6.1. It seems that FAS can be learned from fewer training instances, which our method picks up on.

It is interesting to note that our approach indeed selects the most relevant sets of samples for training, even being able to slightly improve accuracy. Comparing standard full training and our selective training strategy (in parentheses), we have $FCAST$ 83.24% (83.41%), FAS 96.05% (96.40%), $PPAY$ 95.93% (96.09%) and $EDENCE$ 89.02% (89.12%).

Stopping pre-training, i.e., when to cluster and remove instances, can be determined from the graph over accuracy as a function of training time (see Fig. 2 for accuracy on $PPAY$ using backpropagation; other datasets show similar behaviour; omitted here due to space limitations).

Vertical cuts on the training graph show pretraining stopping points and the number of additional mini-batch visits required after filtering. The minimum (*Total* column) is at $200,000$ mini-batches. Comparing with cut-lines in Fig. 2, we note that the best stopping point for pretraining is where the curve "bends", i.e., where the rate of improvement starts to plateau. At this point we have gained most (further gains are more expensive) and thus have the best potential for out-performing full training. At this point, large-scale statistical properties of the data are encoded (Table 1).

To generate the curve in Fig. 2, we use the "pocket" algorithm where the best performing model seen so far is used to output the accuracy on a small validation set. This process yields a nice monotonically increasing curve and allows for a simple pretraining cutoff criteria. In our experiments, we stopped training manually based on the curve, but this could be automated based on the shape and slope of the curve.

Table 1. Pretraining cutoff on $PPAY$

Pretraining	Training	Total
0	3,674,000	3,674,000
50,000	2,128,000	2,178,000
100,000	2,365,000	2,465,000
150,000	913,000	1,063,000
200,000	**830,000**	**1,030,000**
250,000	1,036,000	1,286,000
500,000	1,818,000	2,318,000

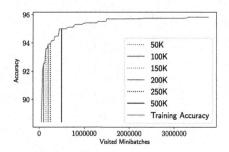

Fig. 2. Pre-train cutoffs, $PPAY$ training curve

Table 2. Filter percentage, $PPAY$, $200K$ minibatches

MFO	ΔMFO	Count	Percentage	Total training
-	(0)	-	-	3,674,000
0.9970	2.52	1	3.6%	3,362,000
0.9945	2.26	3	11.9%	2,574,000
0.9940	2.22	4	19.3%	1,194,000
0.9867	1.88	5	23.0%	1,030,000
0.9863	1.85	6	26.9%	891,000
0.9858	1.85	8	34.3%	1,480,000
0.9800	1.70	12	50.6%	1,739,000
0.9720	1.55	15	59.9%	>5,000,000

To determine how many clusters should be filtered out, we study the model after 200,000 minibatches and its sentence representations. These representations are clustered using K-means and we filter out all clusters with a higher MFO ratio than a filter cutoff. Figure 3 shows clusters sorted by the ratio of sentences with class 1 (using the simplified approach for 2-class problems as described above). Clusters to be filtered (i.e., high MFO ratios) are at the far left and right and clusters to keep are at the center. Note that the MFO ratio for the far left clusters is much higher than the MFO ratio for the far right clusters. For dataset $PPAY$ it seems easier finding pure clusters of label 0, whereas clusters with a high ratio of label 1 tend to be mixed with many examples of label 0 occurrences. Thus, here we only filter clusters on the far left.

We show converge times for different filtering cutoffs from 1 cluster (3.6% of data filtered) to 15 clusters (59.9%) in Table 2. We observe marked improvements in the range 19.3% to 26.9%. Cutoffs are shown as vertical lines against class 1 ratio in Fig. 3.

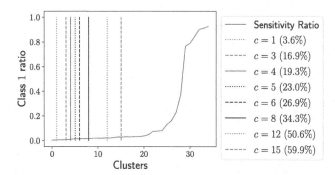

Fig. 3. Filter cutoffs (Table 2) over class 1 ratio.

Table 2 shows a large drop in ΔMFO from 2.22 to 1.88, where we thus should place our cut to obtain a runtime of $1,194,000$, which is significantly less than the standard full training runtime of $3,674,000$.

Table 3 shows robustness to hyperparameter k. We test values between 15 and 70 and compare runtime of 3 different filtering cuts. Each cut is a row in Table 3 ("Small (S)", "Medium (M)", "Large (L)"). "Medium (M)" corresponds to the optimal size cut and the others are smaller and larger, respectively. We set the MFO cutoff such that we filter approximately the same amount of examples across different k values (ratio of filtered samples given in column "Size"). Please note that the actual ratio of samples filtered varies only slightly, due to filtering an integral number of clusters, but still allowing a comparison of runtimes across different values of k.

Across k values we experiment with filtering a "Small (S)" amount ($\approx 12\%$) of data, a "Medium (M)" amount ($\approx 27\%$) and a "Large (L)" amount ($\approx 33\%$). Table 3 shows that runtimes only vary slightly, i.e., an average difference from mean of 6.32%, 6.19%, 11.21% for Small, Medium and Large, respectively. Thus, k does not impact reduction in training time significantly.

Table 3. Time (measured as number of minibatches) over k and filtering cutoffs (Small (S), Medium (M) and Large (L)). Size is filtered data ratio

	$k = 15$		$k = 35$		$k = 70$		Avg. Diff
	Size	Time	Size	Time	Size	Time	
S	11.92%	$1,486K$	11.91%	$1,683K$	11.87%	$1,756K$	6.32%
M	27.84%	$924K$	26.92%	$891K$	26.92%	$1,040K$	6.19%
L	33.66%	$1,551K$	34.32%	$1,480K$	33.07%	$1,933K$	11.21%

6   Detailed Error Analysis

We study whether selective training still learns models that generalize as well as full training models do. We report prediction accuracy in $PPAY$ (76067 sentences with embeddings from both selected and full training set) in the interest of space; findings are similar for the other datasets.

Table 4 shows that the standard full training model misclassifies 1926 instances, whereas our approach only misclassifies 1565. Out of these 1565, 869 are made by the full training approach as well, i.e., our approach makes 18.74% fewer errors and out of the remaining, 55% are identical to those of full training. We conclude that our speed up does not jeopardize accuracy, but even leads to slightly improved accuracy. We analyze errors using K-means clustering of embeddings (Table 5). Here Id refers to cluster id. $Size$ is the number of elements (errors) in a cluster and $Accuracy$ the overall accuracy when we cluster all data elements (not just errors) into these 10 clusters and calculate the accuracy of our approach per cluster. The errors are fairly uniformly distributed; we observe that except for group 7 with 310 elements, all other clusters have size around 100–200. This uniform size distribution (especially compared to Fig. 3) suggests that the errors are not concentrated in any particular part of the embedding space. Our approach thus is expected to generalize well.

Table 4. Shared embeddings in $PPAY$.

Total number of shared embeddings	76067
Errors of the standard training	1926
Accuracy of the standard training	97.47%
Errors of our training method	1565
Accuracy of our training method	97.94%

Table 5. Error statistics. See text for details.

Id	Size	Accuracy
1	97	100.0%
2	113	100.0%
3	112	99.1071%
4	188	98.9362%
5	134	89.5522%
6	202	78.7229%
7	310	74.8387%
8	78	71.0526%
9	185	48.6486%
10	148	41.8919%

A concern for the clustering step could be that K-Means fails to do meaningful clustering for very high representation sizes. Such investigation falls outside the work done here. However we observe that this degradation of K-means has been investigated in [2] where the authors find that hierarchical K-means may provide strong clustering even for high dimensionalities.

Before studying actual examples, we note that the first four groups in Table 5 contain 510 out of 1565 of our errors (32.59%) with high MFO, all for label 1, which makes some of the errors in these groups challenging to resolve. Groups 6, 7, 8 have a medium MFO score, with group 5 showing a slightly higher score. Finally, groups {9, 10} show lower MFO. From Table 6 with error examples from each group, we observe the following types of errors: **Soft errors.** Clusters 2, 3, 4 contain examples of prepay transactions (label 1) where wording and structure seem to be good indicators of the class. These clusters also contain challenging samples. Cluster 7 seems similar, but shows greater diversity in label distribution, which is even more challenging to resolve. **Poor filtering.** Cluster 8 contains examples of sentences which should have been filtered out in pre-processing, and can therefore be used to inform the pre-processing step. **Short emails and headers.** Clusters 6, 9 contain examples of short email sentences. Again, this can be used to inform pre-processing e.g. combining them with email subject, to/from or the like. **Hard errors.** Clusters 1, 5, 10 contain headlines, locations and sentences with little information, which we consider unlikely targets for improvement based on sentences alone.

Table 6. Samples from each of the 10 error groups with description of type of error

Group Id	Sentence	Type of error
1	Section 12.1 Duration	Document headlines
2	I spoke to Tim today regarding the prepayment transactions with terminated counterparties	Generic sentence
3	Let me know, but as I understand it, the Chase agreement should be a good format for this master agreement	Stating facts
4	A similar swap was entered into between the American Public Energy Agency and Chase on the same date	Prepay technicalities
5	Maybe this happens in a region of not prepay and that is why... New York, New York 10043	Specific location
6	If there is anything else you need please feel free to call me on 0207 783 5404	Short standard email
7	Please let me know the nature of the transaction with National Steel	General question wrt. entity
8	[...] – – – – – - Enron-6.11.00.ppt	Bad text/filtering
9	RE: Swap Transaction; Deal No M180816	Preambles, email header
10	The PSCO Project will be sold into TurboPark on 1/19/01	Abbreviation and numbers

In summary, some groups can be used directly to improve performance further in pre-processing, others offer potential for future work, and few contain too little information to be correctly predicted. Models trained using our faster selective training strategy seem to generalize just as well as full training, providing even slightly better accuracy.

7   Related Work

There is a large body of research on different training strategies, often with the aim of improving the accuracy or stability of a classifier. The classical approach is AdaBoost [8] which adapts training depending on the error observed by weighing "difficult" examples higher. At an abstract level, boosting also refocuses training. However, there are two core differences between our approach and boosting: first, we aim to reduce training time (i.e., speeding up convergence of the model parameters) by removing samples that are not expected to benefit. And secondly, we base the selection of samples on their multi-dimensional representation rather than on the one-dimensional difference in ground truth label and predicted label. For an extensive survey of bagging and boosting in classification we refer the interested reader to [19].

Adaptive learning methods, such as Adam [14], automatically adjust the learning rate based on adaptive estimates of lower-order moments in the loss function. [14] observe faster convergence of the model during training to a particular accuracy (cost) value compared to other popular learning optimizers (AdaGrad, RMSProp, SGDNesterov, AdaDelta). Adaptive learning methods thus essentially adjust the learning rate, whereas our approach takes a fundamentally different approach that adapts the training data being used to learn accurate models much more efficiently. [22] estimates which samples provide largest decrease in loss function based on estimates of previous decreases incurring an $N \log N$ sorting/rank penalty over all samples. By contrast, we use the rich information embedded in the representations, using clustering for easy selection. Zhao et al. [34] use K-means clustering to differentiate local from global word context for improved text embeddings, but not for training time reduction.

8   Conclusion and Future Work

We present an efficient selective strategy for reducing training time of complex models, focusing on recurrent neural networks over phrase trees on large text datasets. We propose studying groups in representation space to identify where learning from training data seems to be completed, and where more training is expected to improve model accuracy. Discarding clusters with pure label distribution, we refocus training to those samples that lead to high accuracy models with less training time. We show how to easily infer the parameters for selecting clusters using rate of improvement on training graphs and our proposed ΔMFO measure. In thorough experiments, we demonstrate up to 6 times faster training without loss of accuracy on a number of datasets.

Our method generalizes to similar training problems of complex models that generate distributed instance representations. We intend to study representation based models, such as Recurrent Neural Networks, Long Short Term Memory Networks and Convolutional Neural Networks.

Acknowledgments. This project has received funding from the European Union's Horizon 2020 research and innovation programme under grant agreement No. 732240 (Synchronicity Project). The authors would like to thank the anonymous reviewers for valuable comments and suggestions.

References

1. Bengio, Y., Ducharme, R., Vincent, P., Jauvin, C.: A neural probabilistic language model. JMLR **3**, 1137–1155 (2003)
2. Coates, A., Ng, A.Y.: Learning feature representations with K-means. In: Montavon, G., Orr, G.B., Müller, K.-R. (eds.) Neural Networks: Tricks of the Trade. LNCS, vol. 7700, pp. 561–580. Springer, Heidelberg (2012). https://doi.org/10.1007/978-3-642-35289-8_30
3. Collobert, R., Weston, J.: A unified architecture for natural language processing: deep neural networks with multitask learning. In: ICML, pp. 160–167 (2008)
4. Collobert, R., Weston, J., Bottou, L., Karlen, M., Kavukcuoglu, K., Kuksa, P.: Natural language processing (almost) from scratch. JMLR **12**, 2493–2537 (2011)
5. Conneau, A., Kiela, D., Schwenk, H., Barrault, L., Bordes, A.: Supervised learning of universal sentence representations from natural language inference data. In: EMNLP, pp. 670–680 (2017)
6. Cormack, G.V., Grossman, M.R., Hedin, B., Oard, D.W.: Overview of the TREC 2010 legal track. In: TREC (2010)
7. Forgy, E.W.: Cluster analysis of multivariate data: efficiency versus interpretability of classification. Biometrics **21**(3), 768–769 (1965)
8. Freund, Y., Schapire, R.E.: Experiments with a new boosting algorithm. In: ICML, pp. 148–156 (1996)
9. Goller, C., Kuchler, A.: Learning task-dependent distributed representations by backpropagation through structure. In: IEEE ICNN, pp. 347–352 (1996)
10. Goodfellow, I., Bengio, Y., Courville, A.: Deep Learning. MIT Press, Cambridge (2016)
11. Hochreiter, S., Schmidhuber, J.: Long short-term memory. Neural Comput. **9**(8), 1735–1780 (1997)
12. İrsoy, O., Cardie, C.: Deep recursive neural networks for compositionality in language. In: NIPS, pp. 2096–2104 (2014)
13. Kim, Y.: Convolutional neural networks for sentence classification. In: EMNLP, pp. 1746–1751 (2014)
14. Kingma, D.P., Ba, J.: Adam: a method for stochastic optimization. In: ICLR (2015)
15. Kipf, T.N., Welling, M.: Semi-supervised classification with graph convolutional networks. In: ICLR (2017)
16. Kiss, T., Strunk, J.: Unsupervised multilingual sentence boundary detection. Comput. Linguist. **32**(4), 485–525 (2006)
17. Klein, D., Manning, C.D.: Accurate unlexicalized parsing. In: ACL, pp. 423–430 (2003)

18. Klimt, B., Yang, Y.: The enron corpus: a new dataset for email classification research. In: Boulicaut, J.-F., Esposito, F., Giannotti, F., Pedreschi, D. (eds.) ECML 2004. LNCS (LNAI), vol. 3201, pp. 217–226. Springer, Heidelberg (2004). https://doi.org/10.1007/978-3-540-30115-8_22

19. Kotsiantis, S.B.: Bagging and boosting variants for handling classifications problems: a survey. Knowl. Eng. Rev. **29**(1), 78–100 (2014)

20. Le, Q.V., Mikolov, T.: Distributed representations of sentences and documents. In: ICML, pp. 1188–1196 (2014)

21. Lloyd, S.: Least squares quantization in PCM. IEEE TIT **28**(2), 129–137 (1982)

22. Loshchilov, I., Hutter, F.: Online batch selection for faster training of neural networks. In: ICLR Workshop (2016)

23. Mikolov, T., Sutskever, I., Chen, K., Corrado, G.S., Dean, J.: Distributed representations of words and phrases and their compositionality. In: NIPS, pp. 3111–3119 (2013)

24. Neerbek, J., Assent, I., Dolog, P.: Detecting complex sensitive information via phrase structure in recursive neural networks. In: Phung, D., Tseng, V.S., Webb, G.I., Ho, B., Ganji, M., Rashidi, L. (eds.) PAKDD 2018. LNCS (LNAI), vol. 10939, pp. 373–385. Springer, Cham (2018). https://doi.org/10.1007/978-3-319-93040-4_30

25. Olvera-López, J.A., Carrasco-Ochoa, J.A., Martínez-Trinidad, J.F., Kittler, J.: A review of instance selection methods. AI Rev. **34**(2), 133–143 (2010)

26. Pennington, J., Socher, R., Manning, C.D.: Glove: global vectors for word representation. In: EMNLP, pp. 1532–1543 (2014)

27. Rush, A.M., Chopra, S., Weston, J.: A neural attention model for abstractive sentence summarization. In: EMNLP (2015)

28. Settles, B.: Active learning. Synth. Lect. Artif. Intell. Mach. Learn. **6**(1), 1–114 (2012)

29. Socher, R., Huang, E.H., Pennin, J., Manning, C.D., Ng, A.Y.: Dynamic pooling and unfolding recursive autoencoders for paraphrase detection. In: NIPS, pp. 801–809 (2011)

30. Socher, R., Manning, C.D., Ng, A.Y.: Learning continuous phrase representations and syntactic parsing with recursive neural networks. In: NIPS Deep Learning and Unsupervised Feature Learning Workshop, pp. 1–9 (2010)

31. Socher, R., et al.: Recursive deep models for semantic compositionality over a sentiment treebank. In: EMNLP, pp. 1631–1642 (2013)

32. Taylor, A., Marcus, M., Santorini, B.: The penn treebank: an overview. In: Abeillé, A. (ed.) Treebanks. TLTB, pp. 5–22. Springer, Heidelberg (2003). https://doi.org/10.1007/978-94-010-0201-1_1

33. Tomlinson, S.: Learning task experiments in the TREC 2010 legal track. In: TREC (2010)

34. Zhao, Z., Liu, T., Li, B., Du, X.: Cluster-driven model for improved word and text embedding. In: ECAI, pp. 99–106 (2016)

Extracting Keyphrases from Research Papers Using Word Embeddings

Wei Fan[1], Huan Liu[1], Suge Wang[2], Yuxiang Zhang[1(✉)], and Yaocheng Chang[1]

[1] School of Computer Science and Technology, Civil Aviation University of China,
Tianjin, China
{wfan,yxzhang}@cauc.edu.cn, hliu_h@outlook.com, chang.yc@outlook.com
[2] School of Computer and Information Technology, Shanxi University,
Taiyuan, Shanxi, China
wsg@sxu.edu.cn

Abstract. Unsupervised random-walk keyphrase extraction models mainly rely on global structural information of the word graph, with nodes representing candidate words and edges capturing the co-occurrence information between candidate words. However, integrating different types of useful information into the representation learning process to help better extract keyphrases is relatively unexplored. In this paper, we propose a random-walk method to extract keyphrases using word embeddings. Specifically, we first design a new word embedding learning model to integrate local context information of the word graph (*i.e.*, the local word collocation patterns) with some crucial features of candidate words and edges. Then, a novel random-walk ranking model is designed to extract keyphrases by leveraging such word embeddings. Experimental results show that our approach outperforms 8 state-of-the-art unsupervised methods on two real datasets consistently for keyphrase extraction.

Keywords: Keyphrase extraction · Word embeddings ·
Ranking model

1 Introduction

Automatic keyphrase extraction extracts a set of *representative phrases* that are related to the main *topics* discussed in a document [10]. Since keyphrases can provide a high-level topic description of a document, they are very useful for a wide range of natural language processing (NLP) tasks, such as text summarization, information retrieval and question answering. However, the performance of existing methods is still far from being satisfactory [8]. The main reason is it is very challenging to determine if a phrase or set of phrases accurately capture main topics presented in a document.

Existing methods for keyphrase extraction can be broadly divided into supervised and unsupervised methods. Specifically, supervised methods usually treat

© Springer Nature Switzerland AG 2019
Q. Yang et al. (Eds.): PAKDD 2019, LNAI 11441, pp. 54–67, 2019.
https://doi.org/10.1007/978-3-030-16142-2_5

the keyphrase extraction as a binary classification task, in which a classifier is trained on the features of labeled keyphrases to determine whether a candidate phrase is a keyphrase [8]. By way of contrast, unsupervised approaches directly treat keyphrase extraction as a ranking problem, scoring each word using various measures such as *tf-idf* (term frequency-inverse document frequency), graph-based ranking scores (*e.g.*, *PageRank*) [4,12].

The random-walk models (*i.e.*, PageRank-based models) are widely used in the unsupervised scenario and considered as the state-of-the-arts [8]. These models first build a *word graph* in which each node denotes a candidate word and each edge/link represents a co-occurrence relation between words within a document. Random-walk techniques are subsequently employed on the word graph to rank words. Since Text-Rank [12] firstly computed PageRank scores on the word graph, many PageRank-based extensions have been proposed, aiming at integrating various types of information into ranking model to improve the performance of keyphrase extraction.

Although many efforts have been made on PageRank-based models for keyphrase extraction, how to effectively integrate various types of information is still not well studied. The PageRank-based models, by their nature, make integrating *local* context information with *features* of words and links of the word graph difficult. More specifically, a PageRank-based model is a way of deciding on the importance of a node in the word graph, only by taking into account *global* information recursively computed from the entire word graph, and ignoring the *local* context information. In addition, other types of information are incorporated into a PageRank-based model only by directly modifying its transition probability or reset probability. Thus, existing PageRank-based models are not effective to fuse deeply various types of information to achieve better performance on keyphrase extraction.

In this paper, we first design a heterogeneous text graph representation learning approach to deeply incorporate the *local context information* of the word graph, *topical information* expressed in the document and *co-occurrence information* between words, which are important for keyphrase extraction task. Although many word embedding methods, such as Skip-gram [13], PTE [16] and TWE [9], have been proposed, they are mainly designed for general NLP tasks such as text classification or document visualization, rather than for keyphrase extraction. Secondly, we propose a novel PageRank-based ranking model to leverage the learned word embeddings and global structure information of the word graph to rank candidate words in the word co-occurrence graph. Finally, we conduct comprehensive experiments over two publicly-available datasets (SIGKDD and SIGIR) in Computer Science area. Our experimental results show that our approach outperforms 8 state-of-the-art unsupervised methods.

2   Related Work

Many supervised and unsupervised approaches to keyphrase extraction have been proposed. Supervised approaches use a set of features of labeled keyphrases to train a classifier for identifying keyphrases from a document. In the unsupervised approaches, PageRank-based approaches have been widely used for keyphrase extraction and proved to be effective in this task. Our work is closely related to these approaches. TextRank [12] is the first to use PageRank algorithm [14], which provides a way to explore the *global structural information* in the ranking procedure, to rank each candidate words by iteratively computing the entire word graph.

Many PageRank-based modifications have been proposed as extensions to the Text-Rank. These studies usually use different types of background knowledge to enhance the accuracy of the word graph. ExpandRank [18] uses a small number of nearest neighbor documents to compute more accurate co-occurrences in the word graph. Topical PageRank (abbreviated as TPR) [10] first incorporates topical information into Text-Rank model which increases the weight of important topics generated by LDA model. Single-TPR [15] and Salience-Rank [17] only run PageRank once instead of L times (L is the number of topics used in LDA model) as in the TPR. CiteTextRank [7] incorporates evidence from a citation context to enhance co-occurrences in the word graph. PositionRank [6] integrates position information from all positions of a word's occurrences into a biased PageRank. Additionally, both link-associated and word-associated information are simultaneously integrated by PageRank optimization models, such as SEAFARER [21] and MIKE [22]. However, these methods didn't consider the *local context information* of the word graph.

Although a few researches have integrated some background knowledge into PageRank model using word embedding techniques, the embeddings used in these studies are learned by the typical representation learning models such as C&W model on Wikipedia [19] and Skip-gram on domain-specific corpus [20]. In contrast, the word representation learning model proposed in our work is *designed especially* for the key-phrase extraction task, rather than for other general NLP tasks.

3 Proposed Model: WeRank

Definition 1 (Keyphrase Extraction). Let $D = \{d_1, d_2, ..., d_m\}$ be a set of m text documents. Each document $d_i \in D$ includes a set of candidate words $W_i = \{w_{i1}, w_{i2}, ..., w_{in_i}\}$. The goal of a keyphrase extraction model is to find a function to map each $w_{ij} \in W_i$ into a score, and then extract a ranked list of phrases (which consist of consecutive words) that best represents d_i.

The basic framework of our proposed method involves four key steps: (1) constructing a heterogeneous text graph (including word-word graph, word-topic graph and topic-topic graph), (2) learning word embeddings from this graph in which some crucial piece of information (*e.g.*, local word collocation patterns and topical information) are preserved, (3) ranking candidate words using the modified random-walk ranking model with the learned word embeddings, and (4) scoring the phrase using the sum scores of individual words that comprise it. To summarize, the core technical contribution of our method is to learn the **Word**

Embedding and then incorporates it into the random-walk **Rank**ing model for keyphrase extraction, *i.e.*, WeRank.

3.1 Word-Word Graph Embedding

Word-word graph, denoted as $G_{ww} = (W, E_{ww}, \omega_{ww})$, captures the word co-occurrence information between words of a given text corpus. W is a vocabulary of words, E_{ww} is a set of co-occurrence edges between words, and ω_{ww} is a set of edge weights.

Given a word-word graph, besides its global link structure which is used by existing random-walk methods, two types of information related to the word co-occurrences are crucial for keyphrase extraction task. The first type of information is the local contexts (*i.e.*, *local collocation patterns* between words). It is difficult to integrate directly this information into the random-walk model. Fortunately, recent advances on word representation learning [1] have shed light on this problem, which makes it possible to integrate this local context information into the process of keyphrase extraction through the word representation techniques.

The second type of information is the weight of co-occurrence between two words, which is used to reflect the degree of cooperation between two words. Specifically, this edge weight is not only related to the co-occurrence frequency between two words, but also associated with the distance between two words within a given window size. Thus, the edge weight $\omega_{ij} \in \omega$ between word w_i and word w_j is defined as follows: $\omega_{ij} = \sum_{k=1}^{fr(w_i,w_j)} \frac{1}{l_k(w_i,w_j)}$, where $fr(w_i, w_j)$ is the co-occurrence frequency, and $l_k(w_i, w_j)$ is the number of words between w_i and w_j in original text of k-th co-occurrence.

The word-word graph embedding model aims to learn a low-dimensional vector $\overrightarrow{w} \in \mathbb{R}^d$ for each word, in which the aforementioned two types of information are preserved. In order to achieve this goal, we can make the joint distribution $p(w_i, w_j)$ of two words be close to its empirical distribution $\hat{p}(w_i, w_j)$, which can be achieved by minimizing the KL-divergence between such two distributions, given as:

$$\mathcal{L}(W) = \sum_{(i,j)\in E_{ww}} \hat{p}(w_i, w_j) \log \frac{\hat{p}(w_i, w_j)}{p(w_i, w_j)}, \tag{1}$$

where the empirical distribution is defined as $\hat{p}(w_i, w_j) = \frac{\omega_{ij}}{\sum_{(i,j)\in E_{ww}} \omega_{ij}}$, and the joint distribution is computed as $p(w_i, w_j) = 1/(1 + exp(-\overrightarrow{w_i}^{\mathsf{T}} \cdot \overrightarrow{w_j}))$.

Note that most of existing word embedding methods model the relationship between a target word w_i and its context w_j using the conditional probability $p(w_i|w_j)$, which is used to predict the target word w_i according to its context w_j. In the process of extracting keyphrases, we treat two words occurring in the same co-occurrence as equally important. Thus, we use the joint probability $p(w_i, w_j)$, where the (w_i, w_j) pairs are trained to obtain higher scores.

Besides the local contexts and weight of the co-occurrence between two words, the topical information, which is integrated into the random-walk model by modifying its rest probability in almost all existing graph-based approaches [10, 15], is also very important for keyphrase extraction, and can be used to improve the word representations.

3.2   Word-Topic Graph Embedding

Word-topic graph, represented as $G_{wz} = (W, Z, E_{wz}, \phi_{wz}, \psi_w^z)$, is a bipartite network where W is a set of words and Z is a set of topics. E_{wz} is the set of edges between words and topics. The weight $\phi_{ik} \in \phi_{wz}$ of the edge between word $w_i \in W$ and topic $z_k \in Z$ is defined as the empirical conditional probability $\hat{p}(z_k|w_i)$ calculated by LDA [2].

In addition, the weight $\psi_i^z \in \psi_w^z$ is an importance weighting of w_i contributing to all the topics, and defined as the *topical specificity* of word w_i, which was proposed in the related work [17]. It describes how informative the specific word w_i is for determining the generating topic, versus a randomly-selected word, given as:

$$\psi_i^z = \sum_{k \in Z} \hat{p}(z_k|w_i) \log \frac{\hat{p}(z_k|w_i)}{\hat{p}(z_k)}, \tag{2}$$

where $\hat{p}(z_k)$ is the likelihood that any randomly-selected word is generated by topic z_k.

To preserve the local word-topic collocation patterns and topical influence on each candidate word, the objective is to minimize the KL-divergence of two probability distributions:

$$\mathcal{L}(W, Z) = \sum_{(i,k) \in E_{wz}} \psi_i^z \, \hat{p}(z_k|w_i) \, \log \frac{\hat{p}(z_k|w_i)}{p(z_k|w_i)}, \tag{3}$$

where $p(z_k|w_i)$, which is defined as the probability of w_i generates z_k when compared with how w_i generates other topics, is estimated using the softmax function:

$$p(z_k|w_i) = \frac{exp(\vec{z_k}^\mathsf{T} \cdot \vec{w_i})}{\sum_{k' \in Z} exp(\vec{z_{k'}}^\mathsf{T} \cdot \vec{w_i})}, \tag{4}$$

where $\vec{w_i} \in \mathbb{R}^d$ and $\vec{z_k} \in \mathbb{R}^d$ are the embedding of word w_i and topic z_j, respectively, and d is the dimension of embeddings. In this process of embedding, the conditional probability is used due to the assumption that two words with similar distributions over the topics are similar to each other.

In addition to the aforementioned information, the interaction between topics is important for keyphrase extraction, and can also be used to improve the word embeddings.

3.3   Topic-Topic Graph Embedding

Topic-topic graph, denoted as $G_{zz} = (Z, E_{zz}, \varphi_{zz})$, captures the interaction between topics, where Z is a set of topics, and E_{zz} is a set of edges. Given any two topics z_i and z_k, there exists an edge $(i, k) \in E_{zz}$ if z_i and z_k link with a same word. The weight $\varphi_{ik} \in \varphi_{zz}$ of this edge is defined as $\varphi_{ik} = \sum_w \phi_{w,z_i} * \phi_{w,z_k}$, where $\phi_{w,z}$ is set by the conditional probability $\hat{p}(z|w)$ calculated by LDA model.

Similar to the word-word graph embedding, the objective is to minimize the KL-divergence of two probability distributions:

$$\mathcal{L}(Z) = \sum_{(i,k) \in E_{zz}} \hat{p}(z_i, z_k) \, \log \frac{\hat{p}(z_i, z_k)}{p(z_i, z_k)}, \tag{5}$$

where the empirical distribution is defined as $\hat{p}(z_i, z_k) = \frac{\varphi_{ik}}{\sum_{(i,k) \in E_{zz}} \varphi_{ij}}$, and $p(z_i, z_k)$ is the joint distribution.

3.4   Joint Embedding

In order to capture the mutual interaction among the aforementioned three graphs, an intuitive approach is to collectively embed the three graphs, which can be achieved by minimizing the following objective function:

$$\mathcal{L} = \mathcal{L}(W) + \mathcal{L}(W, Z) + \mathcal{L}(Z). \tag{6}$$

The objective function in Eq. (6) is optimized in the pre-training and fine-tuning strategies. Specifically, in our experiments, we learn the word embeddings and topic embeddings with the word-word graph and topic-topic graph, respectively, and then fine-tune the embeddings with the word-topic graph.

3.5   Model Optimization

We train our model using the stochastic gradient descent, which is suitable for large-scale data processing. However, optimizing directly the objective function in Eq. (6) is problematic. Firstly, for the objective function $\mathcal{L}(W)$ in Eq. (1) or $\mathcal{L}(Z)$ in Eq. (5), there exists a trivial solution: $\overrightarrow{w_{i,d}}$ or $\overrightarrow{z_{i,d}} = \infty$, for all i and all d. Secondly, computing the gradients of the conditional probability $p(z_k|w_i)$ in Eq. (3) the costly summation over all inner product with every node in the word-topic graph. To address these problems, we adopt negative sampling approach [13]. The equivalent counterparts of $\log p(z_k|w_i)$ can be derived, given as follows:

$$\log p(z_k|w_i) \propto \log \sigma(\overrightarrow{z_k}^\top \cdot \overrightarrow{w_i}) + \sum_{k=1}^{K} E_{z_{k'} \sim p_{k'}(z)} \log \sigma(-\overrightarrow{z_{k'}}^\top \cdot \overrightarrow{w_i}), \tag{7}$$

where $\sigma(x) = 1/(1 + exp(-x))$ is the sigmoid function. K is the number of negative samples $z_{k'}$ sampled from the "noisy distribution" of $p_{k'}(z) = d_z^{3/4}$ as proposed in [13], and d_z is the output degree. Similarly, we can easily obtain the equivalent counterparts of $\log p(w_i, w_j)$ by just changing $\overrightarrow{z_k}$ to $\overrightarrow{w_j}$ in $\log p(z_k|w_i)$, and of $\log p(z_i, z_k)$ by changing $\overrightarrow{w}$ to $\overrightarrow{z}$ in $\log p(w_i, w_j)$. Thus, the gradients of the

Algorithm 1. Learning Embeddings Algorithm

Input: *(1) Graphs:* G_{ww}, G_{wz}, G_{zz}; *(2) Embedding dimension d; (3) Learning rate η;*
  (4) Number of negative samples K; (5) Number of edge samples #samples.
Output: *word embedding vector* $\overrightarrow{w_i}$
 1: **Initialize:** $\overrightarrow{w_i}$, $\overrightarrow{w_j}$, $\forall w \in W$; $\overrightarrow{z_k}$, $\forall z \in Z$
 2: **while** *iter* $\leq$ *#samples* **do**
 3:     Sample one edge (w_i, w_j) from G_{ww} and update $\overrightarrow{w_i}$, $\overrightarrow{w_j}$ based on Eq. (8) with
     η;
 4:     **for** $k = 0$; $k < K$; $k = k + 1$ **do**
 5:         Sample a negative word $w_{j'}$ for w_i;
 6:         Sample a negative word $w_{i'}$ for w_j;
 7:         Update $\overrightarrow{w_i}$, $\overrightarrow{w_{j'}}$, $\overrightarrow{w_j}$ and $\overrightarrow{w_{i'}}$ based on Eqs. (8), (10) with η;
 8:     **end for**
 9: **end while**
10: pre-training $\overrightarrow{z_k}$ for all topics in G_{zz} by procedures similar to Lines 2-9;
11: **while** *iter* $\leq$ *#samples* **do**
12:     Sample one edge (w_i, z_k) from G_{wz} and update $\overrightarrow{w_i}$, $\overrightarrow{z_k}$ based on Eqs. (8), (9)
     with η;
13:     **for** $k = 0$; $k < K$; $k = k + 1$ **do**
14:         Sample a negative topic $z_{k'}$ and update $\overrightarrow{w_i}$, $\overrightarrow{z_{k'}}$ based on Eqs. (8), (11) with
     η;
15:     **end for**
16: **end while**
17: **return** word embedding vector $\overrightarrow{w_i}$

---

objective function $\mathcal{L}$ with respect to $\overrightarrow{w_i}$, $\overrightarrow{z_k}$, $\overrightarrow{w_{j'}}$ and $\overrightarrow{z_{k'}}$ can be formulated as follows:

$$\frac{\partial \mathcal{L}}{\partial \overrightarrow{w_i}} = \theta_w \left(\sigma(\overrightarrow{w_i}^\top \cdot \overrightarrow{w_{j'}}) \overrightarrow{w_{j'}} - \sigma(-\overrightarrow{w_i}^\top \cdot \overrightarrow{w_j}) \overrightarrow{w_j} \right) + \\ \theta_w^z \left(\sigma(\overrightarrow{z_{k'}}^\top \cdot \overrightarrow{w_i}) \overrightarrow{z_{k'}} - \sigma(-\overrightarrow{z_k}^\top \cdot \overrightarrow{w_i}) \overrightarrow{z_k} \right) \tag{8}$$

$$\frac{\partial \mathcal{L}}{\partial \overrightarrow{z_k}} = \theta_z \left(\sigma(\overrightarrow{z_{i'}}^\top \cdot \overrightarrow{z_k}) \overrightarrow{z_{i'}} - \sigma(-\overrightarrow{z_i}^\top \cdot \overrightarrow{z_k}) \overrightarrow{z_i} \right) - \theta_w^z \, \sigma(-\overrightarrow{z_k}^\top \cdot \overrightarrow{w_i}) \overrightarrow{w_i} \tag{9}$$

$$\frac{\partial \mathcal{L}}{\partial \overrightarrow{w_{j'}}} = \theta_w \, \sigma(\overrightarrow{w_i}^\top \cdot \overrightarrow{w_{j'}}) \overrightarrow{w_i} \tag{10}$$

$$\frac{\partial \mathcal{L}}{\partial \overrightarrow{z_{k'}}} = \theta_z \, \sigma(\overrightarrow{z_{k'}}^\top \cdot \overrightarrow{z_i}) \overrightarrow{z_i} + \theta_w^z \, \sigma(\overrightarrow{z_{k'}}^\top \cdot \overrightarrow{w_i}) \overrightarrow{w_i} \tag{11}$$

where θ_w, θ_w^z and θ_z are defined as $\theta_w = \hat{p}(w_i, w_j)$, $\theta_w^z = \psi_i^z \hat{p}(z_k|w_i)$ and $\theta_z = \hat{p}(z_i, z_k)$, respectively. Due to space limitation, we omit the gradients of $\overrightarrow{w_j}$, $\overrightarrow{z_i}$, $\overrightarrow{w_{i'}}$ and $\overrightarrow{z_{i'}}$ which are easily derived.

Note that the magnitudes of these gradients vary considerably, which is caused by θ_w, θ_w^z and θ_z, so it is very difficult to find a good learning rate during optimization. A more reasonable solution is to alternatively sample edges using the technique of edge sampling proposed in [16], which a binary edge is sampled with the probability proportional to its empirical value θ_w, θ_w^z or θ_z.

The number of edge samples is set to several times of the number of edges of respective graph, which will be discussed in detail in Subsect. 4.2.

We summarize the detailed training algorithm in Algorithm 1. The overall time complexity of our model is $O(dK|E|)$, where d is the dimension of embeddings, K is the number of negative samples and $|E|$ is the number of edges.

3.6 Word Ranking Using Embedding

Almost all PageRank-based approaches score each candidate words in the word co-occurrence graph using a unified random-walk framework. That is, the PageRank score $R(w_i)$ for a candidate word w_i is computed recursively from the entire word graph, given as follows:

$$R(w_i) = \alpha \sum_{j:w_j \to w_i} \frac{e(w_j, w_i)}{out(w_j)} R(w_j) + (1-\alpha), \tag{12}$$

where α is a damping factor typically set to 0.85, $e(w_j, w_i)$ is a weight of edge (w_j, w_i), $out(w_j) = \sum_{w_k : w_j \to w_k} e(w_j, w_k)$ is a out-degree of word w_j.

In our modified random-walk ranking model, the weight $e(w_j, w_i)$ is computed as the product of *position relationship strength* $prs(w_j, w_i)$ which reflects the degree of position correlation over semantic similarity between two words, and dice coefficient $dice(w_j, w_i)$ [5] which measures the probability of two words co-occurring in a phrase, given as:

$$e(w_j, w_i) = \underbrace{\frac{pos(w_j) \cdot pos(w_i)}{\|\vec{w_j} - \vec{w_i}\|_2}}_{prs(w_j, w_i)} \cdot \underbrace{\frac{2 \cdot fr(w_j, w_i)}{tf(w_j) + tf(w_i)}}_{dice(w_j, w_i)}, \tag{13}$$

where $tf(w_i)$ is a term frequency of the word w_i, and $pos(w_i) = \sum_{k=1}^{tf(w_i)} \frac{1}{p_k(w_i)}$ is an importance weighting of the position of word w_i with its frequency occurring in a document. $p_k(w_i)$ is the k-th position where word w_i occurs in the document. Note that the position information is shown to be a very effective feature in supervised keyphrase extraction. In this work, we first incorporate the position information fused semantic similarity into the PageRank-based framework by modifying the transition probability.

4 Experiments

4.1 Experimental Datasets and Settings

Benchmark Datasets. The first dataset provided by [7] consists of research papers from ACM Conference on Knowledge Discovery and Data Mining (KDD). The other dataset is a new dataset collected by us, and is made up of research papers from ACM Conference on Research and Development in Information Retrieval (SIGIR). Some statistics of the two datasets are summarized in Table 1, including the total number of abstracts and keyphrases in the original dataset (#Abs/#KPs(All)), the number of abstracts for which at least one author-labeled keyphrase could be located and the total number of keyphrases located (#Abs/#KPs(Locatable)), percentage of keyphrases not present in the abstracts (MissingKPs), average number of keyphrases per paper (#AvgKPs), and the number of keyphrases with one, two, three and more than three tokens found in these abstracts.

Table 1. Statistics of the two benchmark datasets.

Dataset	#Abs/#KPs(All)	#Abs/#KPs(Loc.)	MissingKPs	AvgKPs	#uni.	#bi.	#tri.	#>trigrams
KDD	365/1471	315/719	51.12%	4.03	363	853	189	66
SIGIR	560/2137	560/1257	41.18%	3.81	503	1283	303	48

For data preprocessing, we first use Python and Natural Language Toolkit (NLTK) package[1] to tokenize the raw text, and then assign parts of speech (POS) to each word. Finally, we retain only nouns and adjectives as candidate words by POS filtering.

Evaluation Metrics. We have employed 4 widely used evaluation metrics for keyphrase extraction, including precision, recall, F1-score and Mean Reciprocal Rank (MRR) to evaluate various methods [11]. Note for each metric, the top k predicted keyphrases are examined in evaluation. MRR is used to evaluate how the first correct keyphrase for each document is ranked. Specifically, for a document d, MRR is defined as MRR $= \frac{1}{|D|} \sum_{d \in D} \frac{1}{rank_d}$ where D is the set of target documents and $rank_d$ is the rank of the first correct keyphrase from all our extracted keyphrases.

Comparative Methods. We have compared our proposed WeRank with 8 state-of-the-arts, including 5 keyphrase extraction methods, 3 representation learning models. Specifically, the 5 keyphrase extraction methods are: (1) *TF-IDF*, which calculates the ranking score of candidate words based on words' *tf-idf* values in the document; (2) *TextRank* [12], which is the first method to score candidate words by directly applying Page-Rank on a word graph built from adjacent words within a document only; (3) *Single-TPR* [15], which integrates the full topical information into the reset probability of PageRank; (4) *PositionRank* [6], which incorporates the position information into the reset probability of PageRank; (5) *WordAttractionRank* [19], which first uses word embeddings pre-trained over Wikipedia to enhance word co-occurrence relations. In our repeated experiments, we also use the publicly-available word embeddings trained by fastText[2] over Wikipedia [3]. The 3 representation learning models are summarized as follows: (1) *Skip-gram* [13], which is capable of accurately modeling the context (i.e., surrounding words) of the target word within a given

---

[1] http://www.nltk.org/.
[2] https://github.com/facebookresearch/fastText.

corpus; (2) *TWE* [9], which first assigns different topics obtained by LDA model for each target word in the corpus, and then learns different topical word embeddings for each word-topic combination; (3) *fastText* [3], which is a newly proposed approach based on the Skip-gram model, where each word is represented as the sum of the n-gram embeddings.

4.2 Parameters and Influences

Some parameters of WeRank are empirically set as follows: (1) the size of co-occurrence window is set as $window\_size = 2$, which is used to construct word graph in process of ranking candidate words mentioned in Subsect. 3.6; (2) damping factor α is set to 0.85, same with many existing PageRank-based methods; (3) similar to some of existing embedding studies, the learning rate is set as $\eta_i = \eta_0(1-i/S)$, in which S is the total number of mini-batches or edge samples and $\eta_0 = 0.025$; (4) the number of negative samples is set as $K = 5$; (5) all the word embeddings are finally normalized by setting $\|\vec{w}\|_2 = 1$; (6) a phrase is scored by using the sum scores of individual words that comprise the phrase, computed by $R(p) = \gamma_p \sum_{w \in p} R(w)$, where $R(w)$ represents the ranking score of candidate word w, and γ_p is a weight of p according to the length of phrase p. γ_p is set as follows: $\gamma_p = 1$, if $|p| = 1$; $\gamma_p = 0.62$, if $|p| = 2$; $\gamma_p = 0.3$, if $|p| \geq 3$.

Besides empirical parameters mentioned above, we firstly study how the *window size*, which is used to construct the word graph G_{ww} for learning word embeddings, impacts the performance of our WeRank. In our experiments, to show the influence of the *window size*, we test values of this parameter in the

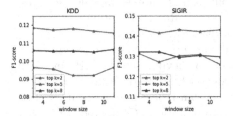

Fig. 1. Influence of the window size.   **Fig. 2.** Influence of the number of topics.

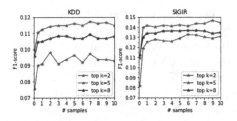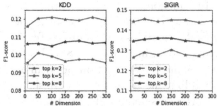

Fig. 3. Influence of the number of samples.   **Fig. 4.** Influence of dimension of embeddings.

range of 3 to 11 and plot the results in Fig. 1. As can be seen from it, the performance of WeRank do not change significantly as the *window size* varies on both KDD and SIGIR datasets. The best-performing setting is *window size* = 3 on two datasets, which is finally used in the comparison experiments.

Secondly, to illustrate the influence of the number of topics in our WeRank, we test values of this parameter *#topics* in the range of 0 to 500 and plot the results in Fig. 2. Note that the *#topics* = 0 means that the topical information has not been added to our model. We observe that the performance of WeRank is influenced by the changes on the number of topics. Generally, the performance increases and then slowly decreases on both KDD and SIGIR datasets. The best-performing setting is *#topics* = 50 on two datasets, which is finally used in the comparison experiments.

Thirdly, to study the influence of the number of edge samples *#samples* in our WeRank, we test values of this parameter *#samples* in the range of 0.1 to 10 times of the number of edges in respective graphs and plot the results in Fig. 3. We can see that the performance of WeRank increases rapidly and then slowly converges as *#samples* varies. *#samples* is set to 7 times of the number of edges on both datasets in our comparison experiments due to time efficiency.

Finally, as can be seen from Fig. 4, the dimension of embeddings of words or topics has little impact on the performance of WeRank and is set to 100 in the experiments.

4.3   Results and Analysis

In our experiments, the predicted keyphrases have to be exactly the same as the author-labeled keyphrases when computing the number of true positives. Table 2 shows the comparison of results of our WeRank with other methods at top $k = 2, 5, 8$ predicted keyphrases on two datasets.

We first discuss the comparison of results at top $k = 5$ predicted keyphrases. This value is close to the average numbers of keyphrases AvgKPs in given research papers (AvgKPs = 4.03 on KDD and 3.81 on SIGIR dataset), as shown in Table 1. The benefit is that the experiment can reflect real application environment. All results are shown in Rows 10–18 of Table 2. Compared with other 5 unsupervised approaches, our WeRank gets the best results in terms of all performance measures on both KDD and SIGIR datasets at top $k = 5$, as presented in Rows 10–15 of Table 2.

In addition, we further conduct experiments to compare the *joint embedding* method used in our WeRank with other three word representation learning models (TWE-1, Skip-gram and fastText) at top $k = 5$ predicted keyphrases. As the results shown in Rows 15–18 of Table 2, WeRank with our embeddings gets the best results in terms of all performance measures. This indicates that our joint embedding method, which is designed especially for the keyphrase extraction and preserves some different types of useful information, substantially outperforms other three recent word embedding models. Furthermore, the fastText trained over Wikipedia gives the worst performance. This illustrates that word

Table 2. Comparison of WeRank with other approaches at top k predicted keyphrases.

k		Unsupervised method	KDD				SIGIR			
			Precision	Recall	F1-score	MRR	Precision	Recall	F1-score	MRR
	1	TF-IDF	0.1055	0.0523	0.0700	0.1616	0.1304	0.0683	0.0896	0.1857
	2	TextRank	0.0712	0.0354	0.0473	0.1014	0.1098	0.0575	0.0755	0.1518
	3	Single-TPR	0.1082	0.0537	0.0718	0.1575	0.1152	0.0603	0.0792	0.1554
	4	WordAttractionRank	0.1082	0.0537	0.0718	0.1685	0.1420	0.0744	0.0976	0.2098
	5	PositionRank	0.1342	0.0666	0.0891	0.2137	0.1723	0.0903	0.1185	0.2527
2	6	**WeRank**	**0.1452**	**0.0721**	**0.0963**	**0.2178**	**0.1830**	**0.0959**	**0.1258**	**0.2813**
	7	WeRank(TWE-1)	0.1425	0.0707	0.0945	0.2027	0.1786	0.0935	0.1228	0.2768
	8	WeRank(Skip-gram)	0.1397	0.0693	0.0927	0.2096	0.1830	0.0959	0.1258	0.2759
	9	WeRank(fastText)	0.1342	0.0666	0.0891	0.1904	0.1598	0.0837	0.1099	0.2402
	10	TF-IDF	0.0927	0.1149	0.1026	0.2150	0.1050	0.1375	0.1191	0.2430
	11	TextRank	0.0765	0.0945	0.0845	0.1576	0.0896	0.1174	0.1017	0.2050
	12	Single-TPR	0.0898	0.1108	0.0992	0.2080	0.0975	0.1277	0.1106	0.2144
	13	WordAttractionRank	0.0881	0.1088	0.0973	0.2148	0.1021	0.1338	0.1158	0.2584
	14	PositionRank	0.0994	0.1224	0.1097	0.2588	0.1218	0.1595	0.1381	0.308
5	15	**WeRank**	**0.1097**	**0.1346**	**0.1209**	**0.2682**	**0.1286**	**0.1684**	**0.1458**	**0.3354**
	16	WeRank(TWE-1)	0.0990	0.1217	0.1092	0.2465	0.1246	0.1632	0.1414	0.3308
	17	WeRank(Skip-gram)	0.1023	0.1258	0.1128	0.2581	0.1211	0.1586	0.1373	0.3239
	18	WeRank(fastText)	0.0962	0.1183	0.1061	0.2350	0.1171	0.1534	0.1328	0.2956
	19	TF-IDF	0.0719	0.1414	0.0954	0.2271	0.0837	0.1754	0.1133	0.2574
	20	TextRank	0.0642	0.1244	0.0847	0.1697	0.0791	0.1656	0.1070	0.2208
	21	Single-TPR	0.0731	0.1414	0.0963	0.2212	0.0786	0.1646	0.1064	0.2279
	22	WordAttractionRank	0.0712	0.1380	0.0939	0.2277	0.0833	0.1745	0.1128	0.2725
	23	PositionRank	0.0799	0.1543	0.1053	0.2706	0.0936	0.1960	0.1267	0.3202
8	24	**WeRank**	**0.0803**	**0.1550**	**0.1058**	**0.2759**	**0.0996**	**0.2086**	**0.1348**	**0.3490**
	25	WeRank(TWE-1)	0.0788	0.1523	0.1038	0.2576	0.0974	0.2039	0.1318	0.3444
	26	WeRank(Skip-gram)	0.0761	0.1468	0.1002	0.2661	0.0956	0.2002	0.1294	0.3380
	27	WeRank(fastText)	0.0778	0.1502	0.1025	0.2477	0.0915	0.1918	0.1239	0.3086

embeddings learned from the documents similar to the target document are more conducive to keyphrase extraction.

Finally, we discuss the comparison of results at top $k = 2, 8$ predicted keyphrases. All the results are presented in Rows 1–9 (top $k = 2$) and 19–27 (top $k = 8$) of Table 2. As the results show, our WeRank gets the best results in terms of all performance measures, indicating that our method indeed outperforms the other approaches on two datasets.

5   Conclusions

We studied the problem of extracting keyphrases from scientific research papers. A novel representation learning model with the objective to learn the word embedding is first proposed, which not only deeply integrates some different types of crucial information, especially for integrating *local context information* of the word graph, but also has a strong predictive power for the keyphrase extraction task. Secondly, a novel PageRank-based model which ranks the candidate words is proposed to incorporate the embedded information, especially *global structural information* of the word graph. Our extensive experiments conducted on two benchmark datasets demonstrate that our proposed method outperforms 8 state-of-the-art methods consistently.

Acknowledgements. This work was partially supported by grants from the National Natural Science Foundation of China (No. 61632011, 61573231, U1633110, U1533104, U1333109) and Open Project Foundation of Intelligent Information Processing Key Laboratory of Shanxi Province (No. CICIP2018004).

References

1. Bengio, Y., Courville, A., Vincent, P.: Representation learning: a review and new perspectives. IEEE Trans. PAMI **35**(8), 1798–1828 (2013)
2. Blei, D.M., Ng, A.Y., Jordan, M.I.: Latent Dirichlet allocation. J. Mach. Learn. Res. **3**(1), 993–1022 (2003)
3. Bojanowski, P., Grave, E., Joulin, A., Mikolov, T.: Enriching word vectors with subword information. Trans. ACL **5**(1), 135–146 (2017)
4. Caragea, C., Bulgarov, F., Godea, A., Gollapalli, S.D.: Citation-enhanced keyphrase extraction from research papers: a supervised approach. In: Proceedings of EMNLP, pp. 1435–1446 (2014)
5. Dice, L.R.: Measures of the amount of ecologic association between species. Ecology **26**(3), 297–302 (1945)
6. Florescu, C., Caragea, C.: Positionrank: an unsupervised approach to keyphrase extraction from scholarly documents. In: Proceedings of ACL, pp. 1105–1115 (2017)
7. Gollapalli, S.D., Caragea, C.: Extracting keyphrases from research papers using citation networks. In: Proceedings of AAAI, pp. 1629–1635 (2014)
8. Hasan, K.S., Ng, V.: Automatic keyphrase extraction: a survey of the state of the art. In: Proceedings of ACL, pp. 1262–1273 (2014)
9. Liu, Y., Liu, Z., Chua, T.S., Sun, M.: Topical word embeddings. In: Proceedings of AAAI, pp. 2418–2424 (2015)
10. Liu, Z., Huang, W., Zheng, Y., Sun, M.: Automatic keyphrase extraction via topic decomposition. In: Proceedings of EMNLP, pp. 366–376 (2010)
11. Manning, C.D., Raghavan, P., Schütze, H.: Introduction to Information Retrieval. Cambridge University Press, Cambridge (2008)
12. Mihalcea, R., Tarau, P.: Textrank: bringing order into text. In: Proceedings of EMNLP, pp. 404–411 (2004)
13. Mikolov, T., Sutskever, I., Chen, K., Corrado, G.S., Dean, J.: Distributed representations of words and phrases and their compositionality. In: Proceedings of NIPS, pp. 3111–3119 (2013)
14. Page, L., Brin, S., Motwani, R., Winograd, T.: The pagerank citation ranking: bringing order to the web. Technical report, Stanford InfoLab (1999)
15. Sterckx, L., Demeester, T., Deleu, J., Develder, C.: Topical word importance for fast keyphrase extraction. In: Proceedings of WWW, pp. 121–122 (2015)
16. Tang, J., Qu, M., Wang, M., Zhang, M., Yan, J., Mei, Q.: Line: large-scale information network embedding. In: Proceedings of WWW, pp. 1067–1077 (2015)

17. Teneva, N., Cheng, W.: Salience rank: efficient keyphrase extraction with topic modeling. In: Proceedings of ACL, pp. 530–535 (2017)
18. Wan, X., Xiao, J.: Single document keyphrase extraction using neighborhood knowledge. In: Proceedings of AAAI, pp. 855–860 (2008)
19. Wang, R., Liu, W., McDonald, C.: Corpus-independent generic keyphrase extraction using word embedding vectors. In: Proceedings of DL-WSDM, pp. 39–46 (2015)
20. Wang, Y., Jin, Y., Zhu, X., Goutte, C.: Extracting discriminative keyphrases with learned semantic hierarchies. In: Proceedings of COLING, pp. 932–942 (2016)
21. Zhang, W., Feng, W., Wang, J.: Integrating semantic relatedness and words' intrinsic features for keyword extraction. In: Proceedings of IJCAI, pp. 139–160 (2013)
22. Zhang, Y., Chang, Y., Liu, X., Gollapalli, S.D., Li, X., Xiao, C.: Mike: keyphrase extraction by integrating multidimensional information. In: Proceedings of CIKM, pp. 1349–1358 (2017)

Sequential Embedding Induced Text Clustering, a Non-parametric Bayesian Approach

Tiehang Duan[1(✉)], Qi Lou[2], Sargur N. Srihari[1], and Xiaohui Xie[2]

[1] Department of Computer Science and Engineering,
State University of New York at Buffalo, Buffalo, NY 14260, USA
tiehangd@buffalo.edu, srihari@cedar.buffalo.edu
[2] Department of Computer Science, University of California, Irvine,
Irvine, CA 92617, USA
{qlou,xhx}@ics.uci.edu

Abstract. Current state-of-the-art nonparametric Bayesian text clustering methods model documents through multinomial distribution on bags of words. Although these methods can effectively utilize the word burstiness representation of documents and achieve decent performance, they do not explore the sequential information of text and relationships among synonyms. In this paper, the documents are modeled as the joint of bags of words, sequential features and word embeddings. We proposed **S**equential **E**mbedding induced **D**irichlet **P**rocess **M**ixture **M**odel (SiDPMM) to effectively exploit this joint document representation in text clustering. The sequential features are extracted by the encoder-decoder component. Word embeddings produced by the continuous-bag-of-words (CBOW) model are introduced to handle synonyms. Experimental results demonstrate the benefits of our model in two major aspects: (1) improved performance across multiple diverse text datasets in terms of the normalized mutual information (NMI); (2) more accurate inference of ground truth cluster numbers with regularization effect on tiny outlier clusters.

1 Introduction

The goal of text clustering is to group documents based on the content and topics. It has wide applications in news classification and summarization, document organization, trend analysis and content recommendation on social websites [13, 17]. While text clustering shares the challenges of general clustering problems including high dimensionality of data, scalability to large datasets and prior estimation of cluster number [1], it also bears its own uniqueness: (1) text data is inherently sequential and the order of words matters in the interpretation of document meaning. For example, the sentence "people eating vegetables" has a totally different meaning from the sentence "vegetables eating people", although two sentences share the same bag-of-words representation. (2) Many English

© Springer Nature Switzerland AG 2019
Q. Yang et al. (Eds.): PAKDD 2019, LNAI 11441, pp. 68–80, 2019.
https://doi.org/10.1007/978-3-030-16142-2_6

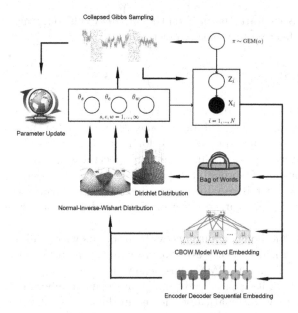

Fig. 1. Illustration of the proposed sequential embedding induced Dirichlet process mixture model (SiDPMM).

words have synonyms. Clustering methods taking synonyms into account will possibly be more effective to identify documents with similar meanings.

Pioneering works in text clustering have been done to address the general challenges of clustering. Among them nonparametric Bayesian text clustering utilizes Dirichlet process to model the mixture distribution of text clusters and eliminate the need of pre-specifying the number of clusters. Current methods use bag of words for document modeling. In this work, as shown in Fig. 1, the Bayesian nonparametric model is extended to utilize knowledge extracted from an encoder-decoder model and word2vec embedding, and documents are jointly modeled by bag of words, sequential features and word embeddings. We derive an efficient collapsed Gibbs sampling algorithm for performing inference under the new model.

Our Contributions. (1) The proposed SiDPMM is able to incorporate rich feature representations. To the best of our knowledge, this is the first work that utilizes sequential features in nonparametric Bayesian text clustering. The features are extracted through an encoder-decoder model. It also takes synonyms into account by including CBOW word embeddings as text features, considering that documents formed with synonym words are more likely to be clustered together. (2) We derive a collapsed Gibbs sampling algorithm for the proposed model, which enables efficient inference. (3) Experimental results show that our model outperforms current state-of-the-art methods across multiple datasets, and have a more accurate inference on the number of clusters due to its desirable regularization effect on tiny outlier clusters.

2   Related Work

Traditional clustering algorithms such as K-means, Hierarchical Clustering, Singular Value Decomposition, Affinity Propagation have been successfully applied in the field of text clustering (see [23] for a comparison of these methods on short text clustering). Algorithms utilizing spectral graph analysis [4], sparse matrix factorization [25], probabilistic models [24] were proposed for performance improvement. As text is usually represented as a huge sparse vector, previous works have shown that feature selection [7,14] and dimension reduction [9] are also crucial.

Most classic methods require access to prior knowledge about the number of clusters, which is not always available in many real-world scenarios. Dirichlet Process Mixture Model (DPMM) has achieved state-of-the-art performance in text clustering with its capability to model arbitrary number of clusters [27,29]; number of clusters is automatically selected in the process of posterior inference. Variational inference [2] and Gibbs sampling [6,21] can be applied to infer cluster assignments in these models.

A closely related field of text clustering is topic modeling. Instead of clustering the documents, topic modeling aims to discover latent topics in document collections [3]. Recent works showed performance of topic modeling can be significantly improved by integrating word embeddings in the model [16,26,30].

The encoder-decoder model was recently introduced in natural language processing and computer vision to model sequential data such as phrases [10,11] and videos [12]. It has shown great performance on a number of tasks including machine translation [5], question answering [22] and video description [12]. Its strength of extracting sequential features is revealed in these applications.

3   Description of SiDPMM

Our text clustering model is based on the Dirichlet process mixture model (DPMM), the limit form of the Dirichlet mixture model (DMM). When DPMM is applied to clustering, the size of clusters are characterized by the stick-breaking process, and prior of cluster assignment for each sample is characterized by the Chinese restaurant process. The Dirichlet process can model arbitrary number of clusters which is typically inferred via collapsed Gibbs sampling or variational inference. We refer readers to [2,21] for more details about DPMM.

We tailor DPMM to our task by learning clusters with multiple distinct information sources for documents, i.e., bag-of-words representations, word embeddings and sequential embeddings, which requires specifically designed likelihood, priors, and inference mechanism.

Table 1. Notations

Notation	Meaning	Notation	Meaning
d_i	the i-th document	$u_{k,\neg i}^t$	occurrence of word t in cluster k excluding d_i
$\mathbf{d}_{k,\neg i}$	documents belonging to cluster k excluding d_i	w_i	the set of bag of words in d_i
K	total number of clusters	s_i	sequential information embedding of d_i
c_i	cluster assignment of d_i	e_i	word embedding of d_i
$\mathbf{c}_{k,\neg i}$	cluster assignments of cluster k excluding document i	V	vocabulary size
		Θ_s	set of hyper-parameters $\{\mu_s, \lambda_s, \nu_s, \Sigma_s\}$
θ_k	parameters of cluster k	Θ_e	set of hyper-parameters $\{\mu_e, \lambda_e, \nu_e, \Sigma_e\}$
r_k	number of documents in cluster k	α	parameter of Chinese restaurant process
u_i	number of words in document i	β	hyper-parameter for multinomial modeling of bag of words
u_i^t	occurrence of word t in document i	ϵ	dimensionality of sequential embedding vector
$u_{k,\neg i}$	number of words in cluster k excluding d_i	δ_k	parameter of multinomial distribution for the k-th cluster

To start with, we first introduce the likelihood function $F(d_i|\theta_k)$ over documents:

$$F(d_i|\theta_k) = \text{Mult}(w_i|\delta_k)\mathcal{N}(e_i|\mu_e^k, \Sigma_e^k)\mathcal{N}(s_i|\mu_s^k, \Sigma_s^k) \qquad (1)$$

where $\theta_k = (\mu_e^k, \Sigma_e^k, \mu_s^k, \Sigma_s^k, \delta_k)$, with $\delta_k = (\delta_k^1, \ldots, \delta_k^V)$ and $\sum_{j=1}^{V} \delta_k^j = 1$. e_i is the word embedding and s_i is the encoded sequential vector. The multinomial component $\text{Mult}(w_i|\delta_k)$ captures the distribution of bag of words; the Normal components $\mathcal{N}(e_i|\mu_e^k, \Sigma_e^k)$, $\mathcal{N}(s_i|\mu_s^k, \Sigma_s^k)$ measure similarities of word and sequential embeddings. This model is general enough to model the characteristic of any text and also specific enough to capture the key information of each document including word embeddings and sequential embeddings.

The prior is set to be conjugate with the likelihood for integrating out the cluster parameters during the inference phase (Table 1). As Dirichlet distribution is the conjugate prior of multinomial distribution and Normal-inverse-Wishart (NiW) is the conjugate prior of normal distribution, we used the composition of Dirichlet distribution and NiW distribution to serve as the conjugate prior $\mathbb{G}_0$, which is defined as:

$$\mathbb{G}_0(\theta_k) = \text{Diri}(\delta_k|\beta)\text{NiW}(\mu_s^k, \Sigma_s^k|\Theta_s)\text{NiW}(\mu_e^k, \Sigma_e^k|\Theta_e) \qquad (2)$$

where Diri denotes the Dirichlet distribution and NiW denotes the Normal-inverse-Wishart distribution. Θ_s denotes hyper-parameters $\{\mu_{s0}, \lambda_{s0}, \nu_{s0}, \Sigma_{s0}\}$ for the encoder-decoder component and Θ_e denotes hyper-parameters $\{\mu_{e0}, \lambda_{e0}, \nu_{e0}, \Sigma_{e0}\}$ for CBOW word embedding component.

4   Inference via Collapsed Gibbs Sampling

We adopt collapsed Gibbs sampling for inference due to its efficiency. It reduces the dimensionality of the sampling space by integrating out cluster parameters, which leads to faster convergence.

The cluster assignment k for document i is decided based on the posterior distribution $p(c_i = k|\boldsymbol{c}_{\neg i}, \boldsymbol{d}, \theta)$. It can be represented as product of cluster prior and document likelihood.

$$p(c_i|\boldsymbol{c}_{\neg i}, \boldsymbol{d}, \theta) = \frac{p(c_i, \boldsymbol{c}_{\neg i}, \boldsymbol{d}|\theta)}{p(\boldsymbol{c}_{\neg i}, \boldsymbol{d}|\theta)} \propto \frac{p(\boldsymbol{c}, \boldsymbol{d}|\theta)}{p(\boldsymbol{c}_{\neg i}, \boldsymbol{d}_{\neg i}|\theta)} = \frac{p(\boldsymbol{c}|\theta)}{p(\boldsymbol{c}_{\neg i}|\theta)} \frac{p(\boldsymbol{d}|\boldsymbol{c}, \theta)}{p(\boldsymbol{d}_{\neg i}|\boldsymbol{c}, \theta)}$$

$$= p(c_i|\boldsymbol{c}_{\neg i}, \theta)p(d_i|\boldsymbol{d}_{\neg i}, \boldsymbol{c}, \theta) \tag{3}$$

Based on the Chinese restaurant process depiction of DPMM, we have

$$p(c_i|\boldsymbol{c}_{\neg i}, \theta) = p(c_i|\boldsymbol{c}_{\neg i}, \alpha)$$

$$= \begin{cases} \frac{r_{k,\neg i}}{D-1+\alpha} & \text{choose an existing cluster } k \\ \frac{\alpha}{D-1+\alpha} & \text{create a new cluster} \end{cases} \tag{4}$$

$(D-1)$ is the total number of documents in the corpus excluding current document i.

Given the number of variables introduced in the model, direct sampling from the joint distribution is not practical. Thus, we assume conditional independence on the variables by allowing the factorization of the second term in (3) as:

$$p(d_i|\boldsymbol{d}_{\neg i}, \boldsymbol{c}, \theta) \propto p(w_i|\boldsymbol{d}_{\neg i}, \boldsymbol{c}, \theta)p(e_i|\boldsymbol{d}_{\neg i}, \boldsymbol{c}, \theta)p(s_i|\boldsymbol{d}_{\neg i}, \boldsymbol{c}, \theta) \tag{5}$$

The calculation for each component $p(w_i|\boldsymbol{d}_{\neg i}, \boldsymbol{c}, \theta)$, $p(e_i|\boldsymbol{d}_{\neg i}, \boldsymbol{c}, \theta)$ and $p(s_i|\boldsymbol{d}_{\neg i}, \boldsymbol{c}, \theta)$ is derived below:

$$p(w_i|\boldsymbol{d}_{\neg i}, \boldsymbol{c}, \theta) = p(w_i|c_i = k, \boldsymbol{d}_{k,\neg i}, \beta) = \int p(w_i|\delta_k)p(\delta_k|\boldsymbol{d}_{k,\neg i}, \beta)d\delta_k \tag{6}$$

where the first term in the above integral is

$$p(w_i|\delta_k) = \prod_{t \in w_i} \text{Mult}(t|\delta_k) = \prod_{t=1}^{V} \delta_{k,t}^{u_i^t} \tag{7}$$

$\delta_{k,t}$ is the probability of term t bursting in cluster k and u_i^t is the count of term t in document i. The second term in (6) is

$$p(\delta_k|\boldsymbol{d}_{k,\neg i}, \beta) = \frac{p(\delta_k|\beta)p(\boldsymbol{d}_{k,\neg i}|\delta_k)}{\int_k p(\delta_k|\beta)p(\boldsymbol{d}_{k,\neg i}|\delta_k)d\delta_k} \tag{8}$$

By defining $\Delta(\boldsymbol{\beta}) = \frac{\prod_{k=1}^{K}\Gamma(\beta)}{\Gamma(\sum_{k=1}^{K}\beta)}$ similar to [28], we have

$$p(\delta_k|\boldsymbol{d}_{k,\neg i}, \beta) = \frac{\frac{1}{\Delta(\beta)} \prod_{t=1}^{V} \delta_{k,t}^{\beta-1} \prod_{t=1}^{V} \delta_{k,t}^{u_{k,\neg i}^t}}{\int_k \frac{1}{\Delta(\beta)} \prod_{t=1}^{V} \delta_{k,t}^{\beta-1} \prod_{t=1}^{V} \delta_{k,t}^{u_{k,\neg i}^t} d\delta_k}$$

$$= \frac{1}{\Delta(\boldsymbol{u}_{k,\neg i} + \beta)} \prod_{t=1}^{V} \delta_{k,t}^{u_{k,\neg i}^t + \beta - 1} \tag{9}$$

Based on (7) and (9), (6) becomes

$$p(w_i|\boldsymbol{d}_{\neg i}, \boldsymbol{c}, \theta) = \int_k \frac{1}{\Delta(\boldsymbol{u}_{k,\neg i} + \beta)} \prod_{t=1}^{V} \delta_{k,t}^{u_{k,\neg i}^t + \beta - 1} \prod_{t=1}^{V} \delta_{k,t}^{u_i^t} d\delta_k$$

$$= \frac{\prod_{t=1}^{V} \prod_{j=1}^{u_i^t} (u_{k,\neg i}^t + \beta + j - 1)}{\prod_{j=1}^{u_i} (u_{k,\neg i} + V\beta + j - 1)}$$

(10)

As we see from (10), the high dimensionality challenge of text clustering is naturally circumvented by multiplying one dimension of the vector space at a time. $p(e_i|\boldsymbol{d}_{\neg i}, \boldsymbol{c}, \theta)$ and $p(s_i|\boldsymbol{d}_{\neg i}, \boldsymbol{c}, \theta)$ in (5) are derived based on properties of NiW distribution:

$$p(s_i|\boldsymbol{d}_{\neg i}, \boldsymbol{c}, \theta) = p(s_i|c_i = k, \boldsymbol{d}_{k,\neg i}, \theta)$$

$$= \int_{\mu_k} \int_{\Sigma_k} p(s_i|\mu_k, \Sigma_k) p(\mu_k, \Sigma_k|c_i = k, \boldsymbol{d}_{k,\neg i}, \theta) d\mu_k d\Sigma_k$$

$$= \int_{\mu_k} \int_{\Sigma_k} \mathcal{N}(s_i|\mu_k, \Sigma_k) \text{NiW}(\mu_k, \Sigma_k|\Theta_s^{k,\neg i}) d\mu_k d\Sigma_k$$

(11)

where μ and Σ are the mean and variance of the sequential embedding, $\Theta_s^{k,\neg i}$ includes $\{\mu_s^{k,\neg i}, \lambda_s^{k,\neg i}, \nu_s^{k,\neg i}, \Sigma_s^{k,\neg i}\}$ which is the hyper-parameter in the NiW distribution of cluster k.

We define the normalization constant $Z(\epsilon, \lambda, \nu, \Sigma)$ of NiW distribution as

$$Z(\epsilon, \lambda, \nu, \Sigma) = 2^{\frac{(\nu+1)\epsilon}{2}} \pi^{\frac{\epsilon(\epsilon+1)}{4}} \lambda^{\frac{-\epsilon}{2}} |\Sigma|^{\frac{-\nu}{2}} \prod_{i=1}^{\epsilon} \Gamma(\frac{\nu+1-i}{2})$$

(12)

where ϵ is the dimensionality of sequential embedding vector. Therefore

$$p(s_i|\boldsymbol{d}_{\neg i}, \boldsymbol{c}, \theta)$$

$$= \int_{\mu_k} \int_{\Sigma_k} \mathcal{N}(s_i|\mu_k, \Sigma_k) \text{NiW}(\mu_k, \Sigma_k|\Theta_s^{k,\neg i}) d\mu_k d\Sigma_k$$

$$= (\pi)^{\frac{-\epsilon}{2}} (\frac{\lambda_s^k}{\lambda_s^{k,\neg i}})^{\frac{-\epsilon}{2}} \frac{|\Sigma_s^k|^{\frac{-\nu_s^k}{2}}}{|\Sigma_s^{k,\neg i}|^{\frac{-\nu_s^{k,\neg i}}{2}}} \prod_{j=1}^{\epsilon} \frac{\Gamma(\frac{\nu_s^k+1-j}{2})}{\Gamma(\frac{\nu_s^{k,\neg i}+1-j}{2})}$$

(13)

As $\nu_s^k = \nu_s^{k,\neg i} + 1$, we have

$$p(s_i|\boldsymbol{d}_{\neg i}, \boldsymbol{c}, \theta) = (\pi)^{\frac{-\epsilon}{2}} (\frac{\lambda_s^k}{\lambda_s^{k,\neg i}})^{\frac{-\epsilon}{2}} \frac{|\Sigma_s^k|^{\frac{-\nu_s^k}{2}}}{|\Sigma_s^{k,\neg i}|^{\frac{-\nu_s^{k,\neg i}}{2}}} \frac{\Gamma(\frac{\nu_s^k}{2})}{\Gamma(\frac{\nu_s^k-\epsilon}{2})}$$

(14)

The derivation of $p(e_i|\boldsymbol{d}_{\neg i}, \boldsymbol{c}, \theta)$ is analogous to that of $p(s_i|\boldsymbol{d}_{\neg i}, \boldsymbol{c}, \theta)$ as they are following the same form of distribution, thus,

$$p(e_i|\boldsymbol{d}_{\neg i}, \boldsymbol{c}, \theta) = (\pi)^{\frac{-\epsilon}{2}} (\frac{\lambda_e^k}{\lambda_e^{k,\neg i}})^{\frac{-\epsilon}{2}} \frac{|\Sigma_e^k|^{\frac{-\nu_e^k}{2}}}{|\Sigma_e^{k,\neg i}|^{\frac{-\nu_e^{k,\neg i}}{2}}} \frac{\Gamma(\frac{\nu_e^k}{2})}{\Gamma(\frac{\nu_e^k-\epsilon}{2})}$$

(15)

Algorithm 1 presents the complete inference procedure.

Algorithm 1. Inference of SiDPMM Model

Data   : For each document i, the bag of words $\boldsymbol{w}_i$, word embedding $\boldsymbol{e}_i$, sequential
              embedding $\boldsymbol{s}_i$
Result: Number of clusters K, cluster assignments for each document $\boldsymbol{c}$
/* Initialization   */
1 K=0
2 **for** *each document i* **do**
3 | compute cluster prior $p(c_i|\boldsymbol{c}_{\neg i}, \alpha) \triangleright$ (4)
4 | calculate $p(w_i|\boldsymbol{d}_{k,\neg i}, c_i = k, \theta) \triangleright$ (10)
5 | calculate $p(s_i|\boldsymbol{d}_{k,\neg i}, c_i = k, \theta) \triangleright$ (14)
6 | calculate $p(e_i|\boldsymbol{d}_{k,\neg i}, c_i = k, \theta) \triangleright$ (15)
7 | calculate $p(d_i|\boldsymbol{d}_{k,\neg i}, c_i = k, \theta) \triangleright$ (5)
8 | sample cluster $c_i \sim p(c_i = k|\boldsymbol{c}_{\neg i}, \boldsymbol{d}, \theta) \triangleright$ (3)
9 | **if** $c_i = K + 1$ **then**
10 | | K=K+1
11 | **end**
12 | update parameters of cluster c_i
13 **end**
   /* Collapsed Gibbs Sampling, N iterations   */
14 **for** *Iter= 1 to N* **do**
15 | **for** *each document i* **do**
16 | | delete document i from cluster c_i, update parameters of cluster c_i
17 | | **if** *cluster c_i is empty* **then**
18 | | | K=K-1
19 | | **end**
20 | | repeat line 3 to line 7
21 | | sample a new cluster c_i for document $i \triangleright$ (3)
22 | | **if** $c_i = K + 1$ **then**
23 | | | K=K+1
24 | | **end**
25 | | update parameters of cluster c_i
26 | **end**
27 **end**

---

5 Extraction of Sequential Feature and Synonyms Embedding

In this section, we describe how to extract sequential embeddings with an encoder-decoder component and synonyms embeddings with the CBOW model.

The encoder-decoder component is formed with two LSTM stacks [8], one is for mapping the sequential input data to a fixed length vector, the other is for decoding the vector to a sequential output. To learn embeddings, we set the input sequence and output sequence to be the same. An illustration of the encoder-decoder mechanism is shown in Fig. 2a. The last output of the encoder LSTM stack contains information of the whole phrase. In machine translation, researchers have found the information is rich enough for the original phrase to be decoded into translations of another language [18].

Current state-of-the-art text clustering methods adopt one-hot encoding for word representation. It neglects semantic relationship between similar words.

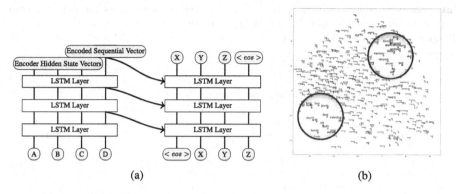

Fig. 2. (a) The Encoder-Decoder Component. It is formed by two LSTM stacks, one is for mapping a sequential input data to a fixed-length vector, the other is for decoding the vector to a sequential output. (b) Word embedding of Google News Title Set. Words describing the same topic have similar embeddings and are clustered together

Recently, researchers have shown multiple degrees of similarity can be revealed among words with word embedding techniques [20]. Utilizing such embeddings means we can cluster the documents based on *meaning of words* instead of the word itself. As shown in Fig. 2b, words describing the same topic have similar embeddings and are clustered together. The CBOW model is used to learn word embeddings by predicting each word based on word context (weighted nearby surrounding words). The embedding vector e_i is the average of word embeddings in d_i. Readers are referred to [19] for details about the CBOW model.

6   Experiments

In this section, we will demonstrate the effectiveness of our approach through a series of experiments. The detailed experimental settings are as follows:

Datasets. We run experiments on four diverse datasets including 20 News Group (20NG)[1], Tweet Set[2], and two datasets from [27]: Google News Title Set (T-Set) and Google News Snippet Set (S-Set). The 20NG dataset contains long documents with an average length of 138 while the documents in T-Set and Tweet Set are short with average length less than 10. Phrase structures are sparse in T-Set, while rich in 20NG and S-Set. The Tweet Set contains moderate phrase structures.

Baselines. We compare SiDPMM against two classic clustering methods, K-means and latent Dirichlet allocation (LDA), and two recent methods GSDMM [28] and GSDPMM [27] that are state-of-the-art in nonparametric Bayesian text clustering.

---

[1] http://qwone.com/~jason/20Newsgroups/.

[2] http://trec.nist.gov/data/microblog.html.

Table 2. NMI scores on various dataset-parameter settings. K is the prior number of clusters for K-means, LDA and GSDMM, set to be four different values including the ground truth for each dataset. K is not used for SiDPMM and GSDPMM. 20 independent runs for each setting.

	K	SiDPMM	SiDPMM-sf[a]	SiDPMM-we[b]	K-means	LDA	GSDMM	GSDPMM
20NG	10	.689 ± .006	.686 ± .005	.680 ± .006	.235 ± .008	.585 ± .013	.613 ± .007	.667 ± .004
	20	.689 ± .006	.686 ± .005	.680 ± .006	.321 ± .006	.602 ± .012	.642 ± .004	.667 ± .004
	30	.689 ± .006	.686 ± .005	.680 ± .006	.336 ± .005	.611 ± .012	.649 ± .005	.667 ± .004
	50	.689 ± .006	.686 ± .005	.680 ± .006	.348 ± .006	.617 ± .013	.656 ± .002	.667 ± .004
T-Set	100	.878 ± .003	.872 ± .003	.877 ± .005	.687 ± .005	.769 ± .012	.830 ± .004	.873 ± .002
	150	.878 ± .003	.872 ± .003	.877 ± .005	.721 ± .009	.784 ± .015	.852 ± .009	.873 ± .002
	152	.878 ± .003	.872 ± .003	.877 ± .005	.720 ± .007	.786 ± .014	.853 ± .009	.873 ± .002
	200	.878 ± .003	.872 ± .003	.877 ± .005	.730 ± .008	.806 ± .013	.868 ± .006	.873 ± .002
S-Set	100	.916 ± .004	.910 ± .005	.902 ± .003	.739 ± .006	.848 ± .005	.854 ± .004	.891 ± .004
	150	.916 ± .004	.910 ± .005	.902 ± .003	.756 ± .006	.850 ± .006	.867 ± .008	.891 ± .004
	152	.916 ± .004	.910 ± .005	.902 ± .003	.757 ± .007	.852 ± .005	.867 ± .009	.891 ± .004
	200	.916 ± .004	.910 ± .005	.902 ± .003	.768 ± .007	.862 ± .004	.885 ± .005	.891 ± .004
Tweet	50	.894 ± .007	.887 ± .006	.884 ± .005	.696 ± .008	.775 ± .012	.844 ± .006	.875 ± .005
	90	.894 ± .007	.887 ± .006	.884 ± .005	.725 ± .007	.797 ± .011	.862 ± .008	.875 ± .005
	110	.894 ± .007	.887 ± .006	.884 ± .005	.732 ± .006	.806 ± .010	.867 ± .006	.875 ± .005
	150	.894 ± .007	.887 ± .006	.884 ± .005	.742 ± .006	.811 ± .012	.871 ± .004	.875 ± .005

[a] SiDPMM model only integrating sequential features.
[b] SiDPMM model only integrating word embeddings.

Metrics. We take the normalized mutual information (NMI) as the major evaluation metric in our experiments since NMI is widely used in this field. NMI scores range from 0 to 1. Perfect labeling is scored to 1 while random assignments tend to achieve scores close to 0.

Encoder-Decoder Component. We truncate the sequence length to be 48 for Tweet Set and Google News dataset and 240 for 20NG dataset. The document with characters length shorter than this sequence length is padded with zeros. The encoder-decoder model is trained for 10 iterations. The length of hidden vectors is set to be 40, and length of input vector is 67 (number of different characters). Weights in the LSTM stack are uniformly initialized to be 0.01. Adam [15] optimizer is used to optimize the network with its learning rate set to 0.01.

Word Embedding Component. The vocabulary size is set to 100,000 which is enough to accommodate most of the words present in the dataset. We set the embedding vector length to be 40. To facilitate training with small datasets such as the Tweet Set, we augment each dataset with a well-known large-scaled text dataset[3] during training. Window size is set to be 1, meaning we only consider the words that are neighbors of the target word as its word context. We apply stochastic gradient descent for optimization with a total of 100,000 descent steps.

---

[3] http://mattmahoney.net/dc/text8.zip.

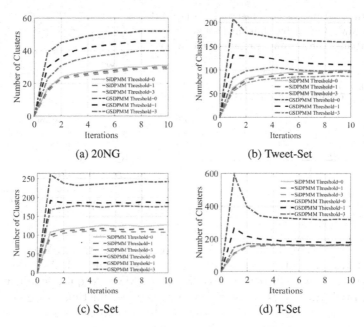

Fig. 3. Number of clusters with size above a given threshold found in each iteration by SiDPMM and GSDPMM. A cluster with size smaller than the given threshold does not count. Plots (a)–(d) are for the datasets 20NG, Tweet-Set, S-Set and T-Set respectively.

Priors. Hyper-parameter α of the Dirichlet process is set to be $0.1 \times |\boldsymbol{d}|$, where $|\boldsymbol{d}|$ is number of documents in the dataset. Hyper-parameter β for the Multinomial modeling of bag of words is $0.002 \times V$, and parameters for the prior NiW distribution of word embedding and sequential embedding are $\{\mu_0 = \mathbf{0}, \lambda_0 = 1, \nu_0 = \epsilon, \Sigma_0 = I\}$.

6.1   Empirical Results

Table 2 reports the mean and standard deviation of the NMI scores across various settings. From Table 2, we observe that SiDPMM outperforms K-means, LDA and GSDMM across all the settings by significant margins. GSDPMM has comparative performance with SiDPMM on T-Set, while SiDPMM performs better in other three datasets. We noted the average length of T-Set is short and phrase structures are scarce in its documents. To unveil the influence of each of the component on the model performance, we included implementation of SiDPMM model only integrating sequential features (denoted as SiDPMM-sf) and SiDPMM model only integrating word embeddings (denoted as SiDPMM-we) into the comparison. We noted the contribution from sequential embedding is significant in 20NG, S-Set and moderate in Tweet-Set.

SiDPMM and GSDPMM can automatically determine the number of clusters. Table 3 shows that number of clusters inferred by SiDPMM are much more accu-

Table 3. Inferred number of clusters by SiDPMM and GSDPMM. Other baseline methods are not included because they require pre-specified number of clusters.

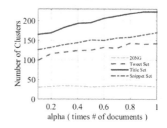

	Number of clusters			Diff. ratio	
	Ground truth	GSDPMM	SiDPMM	GSDPMM	SiDPMM
20NG	20	52	31	160%	55%
T-Set	152	323	171	113%	13%
S-Set	152	246	126	62%	17%
Tweet	110	161	99	46%	10%

Fig. 4. Number of clusters found by SiDPMM with different α values, revealing the relative strength of prior (compared to likelihood) in determining posterior distribution

rate compared to those from GSDPMM across all the datasets. We can observe that GSDPMM tends to create more clusters than SiDPMM. As illustrated in Fig. 3, many of those clusters created by GSDPMM are quite small; while in constrast, SiDPMM tends to suppress tiny clusters and thus are more robust to outliers. The sequential and word embedding components in SiDPMM are responsible for this regularization effect on number of clusters.

The hyper-parameter α in the Dirichlet process determines the prior probability of creating a new cluster (see Eq. (4)). We explore the influence of different α values on our model. Fig. 4 shows that the number of clusters typically grows with α; as observed for Tweet Set, T-Set and S-Set, but not the case for the 20NG dataset. This reveals the relative strength of prior (compared to likelihood) in determining posterior cluster distribution. The documents in 20NG have large average length (137.5 words per document). In the sampling process, the likelihood dominates the posterior distribution and the small difference caused by different α in the prior distribution is negligible, while for documents with small average length, the difference in likelihood is not significant and thus prior affects more of the posterior distribution.

7   Conclusion

In this paper, we propose a nonparametric Bayesian text clustering method (SiDPMM) which models documents as the joint of bag of words, word embeddings and sequential features. The approach is based on the observation that sequential information plays a key role in the interpretation of phrases and word embedding is very effective for measuring similarity between synonyms. The sequential features are extracted with an encoder-decoder component and word embeddings are extracted with the CBOW model. A detailed collapsed Gibbs sampling algorithm is derived for the posterior inference. Experimental results show our approach outperforms current state-of-the-art methods, and is more accurate in inferring the number of clusters with the desirable regularization effect on tiny scattered clusters.

References

1. Berkhin, P.: A survey of clustering data mining techniques. In: Kogan, J., Nicholas, C., Teboulle, M. (eds.) Grouping Multidimensional Data, pp. 25–71. Springer, Heidelberg (2006). https://doi.org/10.1007/3-540-28349-8_2
2. Blei, D.M., Jordan, M.I.: Variational inference for Dirichlet process mixtures. Bayesian Anal. 1(1), 121–143 (2006). https://doi.org/10.1214/06-BA104
3. Blei, D.M., Ng, A.Y., Jordan, M.I.: Latent Dirichlet allocation. J. Mach. Learn. Res. 3, 993–1022 (2003)
4. Cai, D., He, X., Han, J.: SRDA: an efficient algorithm for large-scale discriminant analysis. IEEE Trans. Knowl. Data Eng. 20(1), 1–12 (2008)
5. Cho, K., et al.: Learning phrase representations using RNN encoder-decoder for statistical machine translation. In: EMNLP 2014, pp. 1724–1734. Association for Computational Linguistics, Doha, Qatar, October 2014. http://www.aclweb.org/anthology/D14-1179
6. Duan, T., Pinto, J.P., Xie, X.: Parallel clustering of single cell transcriptomic data with split-merge sampling on Dirichlet process mixtures. Bioinformatics p. bty702 (2018). https://doi.org/10.1093/bioinformatics/bty702
7. Duan, T., Srihari, S.N.: Pseudo boosted deep belief network. In: Villa, A.E.P., Masulli, P., Pons Rivero, A.J. (eds.) ICANN 2016. LNCS, vol. 9887, pp. 105–112. Springer, Cham (2016). https://doi.org/10.1007/978-3-319-44781-0_13
8. Duan, T., Srihari, S.N.: Layerwise interweaving convolutional LSTM. In: Mouhoub, M., Langlais, P. (eds.) AI 2017. LNCS, vol. 10233, pp. 272–277. Springer, Heidelberg (2017). https://doi.org/10.1007/978-3-319-57351-9_31
9. Gomez, J.C., Moens, M.F.: PCA document reconstruction for email classification. Comput. Stat. Data Anal. 56(3), 741–751 (2012)
10. Gu, Y., Chen, S., Marsic, I.: Deep multimodal learning for emotion recognition in spoken language. CoRR abs/1802.08332 (2018)
11. Gu, Y., Li, X., Chen, S., Zhang, J., Marsic, I.: Speech intention classification with multimodal deep learning. In: Mouhoub, M., Langlais, P. (eds.) AI 2017. LNCS (LNAI), vol. 10233, pp. 260–271. Springer, Cham (2017). https://doi.org/10.1007/978-3-319-57351-9_30
12. Hori, C., Hori, T., Lee, T., Sumi, K., Hershey, J.R., Marks, T.K.: Attention-based multimodal fusion for video description. CoRR abs/1701.03126 (2017)
13. Hotho, A., Staab, S., Maedche, A.: Ontology-based text clustering. In: Proceedings of the IJCAI 2001 Workshop Text Learning: Beyond Supervision (2001)
14. Huang, R., Yu, G., Wang, Z.: Dirichlet process mixture model for document clustering with feature partition. IEEE Trans. Knowl. Data Eng. 25(8), 1748–1759 (2013)
15. Kingma, D.P., Ba, J.: Adam: a method for stochastic optimization. CoRR abs/1412.6980 (2014). http://arxiv.org/abs/1412.6980
16. Li, Y., et al.: Towards differentially private truth discovery for crowd sensing systems. CoRR abs/1810.04760 (2018)
17. Liu, M., Chen, L., Liu, B., Wang, X.: VRCA: a clustering algorithm for massive amount of texts. In: IJCAI 2015, pp. 2355–2361. AAAI Press (2015). http://dl.acm.org/citation.cfm?id=2832415.2832576
18. Luong, M., Pham, H., Manning, C.D.: Effective approaches to attention-based neural machine translation. CoRR abs/1508.04025 (2015)
19. Mikolov, T., et al.: Distributed representations of words and phrases and their compositionality. In: Burges, C.J.C., Bottou, L., Welling, M., Ghahramani, Z., Weinberger, K.Q. (eds.) NIPS, pp. 3111–3119. Curran Associates, Inc. (2013)

20. Mikolov, T., Yih, W., Zweig, G.: Linguistic regularities in continuous space word representations. In: HLT-NAACL, pp. 746–751 (2013)
21. Neal, R.M.: Markov chain sampling methods for Dirichlet process mixture models. J. Comput. Graph. Stat. **9**(2), 249–265 (2000)
22. Nie, Y., Han, Y., Huang, J., Jiao, B., Li, A.: Attention-based encoder-decoder model for answer selection in question answering. Front. Inf. Technol. Electron. Eng. **18**(4), 535–544 (2017)
23. Rangrej, A., Kulkarni, S., Tendulkar, A.V.: Comparative study of clustering techniques for short text documents. In: Proceedings of the 20th International Conference Companion on World Wide Web, WWW 2011, pp. 111–112. ACM, New York (2011)
24. Shafiei, M.M., Milios, E.E.: Latent Dirichlet co-clustering. In: Sixth International Conference on Data Mining (ICDM 2006), pp. 542–551, December 2006
25. Wang, F., Zhang, C., Li, T.: Regularized clustering for documents. In: SIGIR 2007, pp. 95–102. ACM, New York (2007)
26. Xun, G., Li, Y., Zhao, W.X., Gao, J., Zhang, A.: A correlated topic model using word embeddings. In: IJCAI 2017, pp. 4207–4213 (2017)
27. Yin, J., Wang, J.: A model-based approach for text clustering with outlier detection. In: 2016 IEEE 32nd International Conference on Data Engineering (ICDE), pp. 625–636, May 2016
28. Yin, J., Wang, J.: A Dirichlet multinomial mixture model-based approach for short text clustering. In: KDD 2014, pp. 233–242. ACM, New York (2014)
29. Yu, G., Huang, R., Wang, Z.: Document clustering via Dirichlet process mixture model with feature selection. In: KDD 2010, pp. 763–772. ACM, New York (2010)
30. Zhang, H., Li, Y., Ma, F., Gao, J., Su, L.: Texttruth: an unsupervised approach to discover trustworthy information from multi-sourced text data. In: KDD 2018, pp. 2729–2737. ACM, New York (2018). https://doi.org/10.1145/3219819.3219977

SSNE: Status Signed Network Embedding

Chunyu Lu[1], Pengfei Jiao[2(✉)], Hongtao Liu[1], Yaping Wang[1], Hongyan Xu[1],
and Wenjun Wang[1]

[1] College of Intelligence and Computing, Tianjin University, Tianjin, China
{tjulcy,htliu,yapingwang,hongyanxu,wjwang}@tju.edu.cn
[2] Center for Biosafety Research and Strategy, Tianjin University,Tianjin, China
pjiao@tju.edu.cn

Abstract. This work studies the problem of signed network embedding, which aims to obtain low-dimensional vectors for nodes in signed networks. Existing works mostly focus on learning representations via characterizing the social structural balance theory in signed networks. However, structural balance theory could not well satisfy some of the fundamental phenomena in real-world signed networks such as the direction of links. As a result, in this paper we integrate another theory **Status Theory** into signed network embedding since status theory can better explain the social mechanisms of signed networks. To be specific, we characterize the status of nodes in the semantic vector space and well design different ranking objectives for positive and negative links respectively. Besides, we utilize graph attention to assemble the information of neighborhoods. We conduct extensive experiments on three real-world datasets and the results show that our model can achieve a significant improvement compared with baselines.

Keywords: Signed network embedding · Attention · Status theory

1 Introduction

The rapid growth of social media has greatly promoted the development of social network analysis. Recently, network embedding(NE), an effective tool to analyze large-scale network, has attracted quite a lot of attention and aims to obtain low-dimensional representations for nodes in networks [3,10,12,17]. The representation of nodes learned in network embedding preserves the structure information and can be applied in downstream tasks of network analysis, such as node classification [11], link prediction [15], community detection [16].

Most existing works on network embedding are based on unsigned networks (i.e., there are only positive links in networks). However, signed networks are becoming quite ubiquitous in real-world data. In the signed network, the relations between two nodes are represented with positive and negative links. Positive links usually indicate that two nodes are of similarity while negative links are opposite or different. For example, in Slashdot[1], a famous online discuss sites, users may

---

[1] https://slashdot.org/.

© Springer Nature Switzerland AG 2019
Q. Yang et al. (Eds.): PAKDD 2019, LNAI 11441, pp. 81–93, 2019.
https://doi.org/10.1007/978-3-030-16142-2_7

tag other users as "friends" or "foes", which indicate positive and negative links respectively. The heterogeneity of links leads that the traditional methods on unsigned network embedding are insufficient for signed networks. In this paper we will study the problem of signed network embedding.

There are several existing works on signed network embedding [6,14,18,20]. For example, SNE [18] designs the log-bilinear model and learns node representation along a given path. SiNE [14] introduces social theories and adopts a deep learning framework. SIGNet [6] proposes a scalable node embedding method which preserves social theories in higher order neighborhoods.

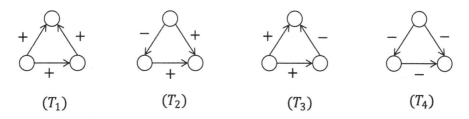

Fig. 1. The illustration of 4 types of triads. For triad T_1, both structural balance theory and status theory are satisfied, however, structural balance theory can not explain triads T_2, T_3 and T_4, while all of them satisfy status theory

However, these works mostly learn signed network embedding under the guidance of the structural balance theory [5], which is popular for "the friend of my friend is my friend; the foes of my foes is my friend" and has been widely used in signed network analysis. Nevertheless, structural balance theory is naturally defined for undirected networks and it could not work well in a directed network [9]. As Fig. 1 shows, the cases T_2, T_3, T_4 cannot satisfy balance theory. However, these cases of signed network exist in many real-world data such as Slashdot, Wikipedia signed network and etc. To address the direction in signed networks, another theory named Status Theory was proposed [4,9]. Status Theory defines **status** for nodes to describes a ranking of nodes and suggests that node u has a higher **status** than node v if there is a positive link from v to u, or a negative link from u to v. All cases in Fig. 1 can be well satisfied under the status theory. Status theory provides a better understanding of the real-world signed networks and plays a role in many lines of work in the social sciences [9]. However, to the best of our knowledge, status theory has never been used in previous signed network embedding work yet.

To address the issue above, in this paper we propose a novel signed network embedding model named SSNE (Status Signed Network Embedding), incorporating the status theory into network embedding process. We try to quantify the status of nodes as the transformation in the embedding space. Specifically inspired by the knowledge graph embedding model TransE [1], we design different translation strategies for positive and negative links to characterize the ranking of status. Furthermore, we utilize the information of neighborhoods via graph attention mechanism to obtain a robust node representation.

The major contributions of the paper are as follows:

- We are the first to consider the status theory to learn signed network embedding.
- We define positive and negative links in signed networks as translations in embedding space and design an energy-based ranking objective function to learn embedding of nodes and links.
- We conduct extensive experiments on three real-world signed networks. Experimental results show that our model SSNE is more effective than the baselines.

2   Related Work

Different from traditional network analysis methods, network embedding aims to learn a low-dimension representation of nodes or edges. It has received extensive attention in recent years and a large number of methods are proposed [3,10,12,13]. For example, DeepWalk [10] leverages skip-gram model to learn node embedding with truncated random walks. LINE [12] proposes an objective function to preserve first-order and second-order proximity. However, there are few works on signed network embedding, because the negative links in signed networks which usually denote distrust or foe relation make the theories different from unsigned network [9]. These works mostly learn signed network embedding under the guidance of the structural balance theory [5]. For example, SNE [18] designs a log-bilinear model which considers the sign of edges along a given path. SiNE designs an objective function guided by social balance theory [2] and proposes a deep learning framework to learn node representations. SIDE [7] learns representation by leveraging balance theory and socio-psychological theories along random walks. SIGNet [6] proposes a scalable node embedding method which can preserve social theories in higher order neighborhoods.

On the other hand, social theories, which are a powerful tool for network analysis, major include structural balance theory [2,5] and status theory [9] in signed network. Structural balance theory which is introduced in Heider [5], can be summarized as four rules: "A friend of my friend is my friend," "A friend of my foe is my foe," "A foe of my friend is my foe," and "A foe of my foe is my friend." Then Cygan et al. [2] extended the aforementioned theory to users should be able to have their "friends" closer than their "foes". However, structural balance theory is naturally defined for undirected networks and it could not work well in a directed network [9]. Status theory [4,9] which is naturally defined for directed signed network describes a ranking of nodes. Guha et al. [4] observed that a link in signed network from u to v can have multiple possible scenarios. Depending on the intention of u in creating the link, they also develop a framework of trust propagation schemes in cyclic triad. Then Leskovec et al. [9] extended in acyclic triad and developed a new theory called Status Theory which suggests that node u has a higher **status** than node v if there is a positive link from v to

u, or a negative link from u to v. The status levels can be passed along the link to produce a multi-step relation, which usually results in predictions that are different from the structural balance theory. In this paper, we propose a novel network embedding method called SSNE that characterizes the status of nodes in the semantic vector space and has a well-designed ranking loss function for positive and negative links respectively to learn node representations.

3 Methodology

3.1 Problem Formulation

Definition 1 (Signed directed network). A signed network can be formalized as $G = (V, E)$, where $V = \{v_1, v_2, \ldots, v_n\}$ is the set of vertices and $E \subset V \times V$ represents the relations of the nodes. In this paper, for any $e_{ij} \in E$, we define $e_{ij} = 1$ represents positive edge from v_i to v_j and $e_{ij} = -1$ represents negative edge from v_i to v_j.

Definition 2 (Signed network embedding). Given a signed network $G = (V, E)$, signed network embedding aims to learn a function $f : V \to \mathbb{R}^d$, where $d \ll |V|$ and V is the set of vertices.

Definition 3 (Status triplet). a status triplet can be formalized as (h, ℓ, t), where $h \in V$ denotes the header vertices of link, $t \in V$ denotes the tail vertices of link and ℓ denotes the link from h to t. Particularly we define ℓ_+ to represent positive link, ℓ_- to represent negative link and ℓ_0 indicates that there is no edge between h and t. In particular, (h, ℓ_+, t) denotes the status of t is higher than h and (h, ℓ_-, t) denotes the status of h is higher than t in the perspective of status.

3.2 The Model

As mentioned above, status theory suggests that v_i has a higher status than v_j if v_j has a positive link to v_i or v_i has a negative link to v_j, which can be translated that the ranking of node embedding $\boldsymbol{v_j}$ higher than $\boldsymbol{v_i}$ for the positive links or $\boldsymbol{v_i}$ higher than $\boldsymbol{v_j}$ for the negative links (note that the bold letters used in mathematics that appear in this paper represent vector representations of nodes or relations). Specifically, for the positive link which can be converted to the triplet (h, ℓ_+, t), we want to maintain a ranking relation $t > h$ from the perspective of status theory, which can be translated into a vector relation: $\boldsymbol{h} + \boldsymbol{\ell_+} \approx \boldsymbol{t}$. On the other hand, for the negative link which can be converted to the triplet (h, ℓ_-, t), we want to maintain a ranking relation $h > t$ from the perspective of status theory, which can be translated into a vector relation: $\boldsymbol{h} - \boldsymbol{\ell_-} \approx \boldsymbol{t}$. In order to satisfy the vector relations above, we define the energy-based ranking function d where d is a distance measure and we adopt L_1 or L_2-norm for d. Then we propose the following constraints to make $\boldsymbol{h} + \boldsymbol{\ell_+}$ closer to $\boldsymbol{t}$ for the positive links:

$$d(\boldsymbol{h} + \boldsymbol{\ell_+}, \boldsymbol{t}) < d(\boldsymbol{h} + \boldsymbol{\ell_+}, \boldsymbol{t'}),$$

where t' is selected from V with negative sampling and the details can be found in the next subsection. In a similar way, we propose the following constraints to make $h - \ell_-$ closer to t for the negative links:

$$d(h - \ell_-, t) < d(h - \ell_-, t').$$

The illustration of our model is shown in Fig. 2.

Given a set of triplets (h, ℓ, t) as training set S, we minimize an energy-based ranking loss function as follows:

$$\mathcal{L} = \sum_{(h, \ell, t) \in S} \sum_{(h, \ell, t') \in S'_{(h, \ell, t)}} \max\left(0, \gamma + d(h + s_\ell \ell, t) - d(h + s_\ell \ell, t')\right),$$

where h, t is the node embedding and ℓ is the relation embedding, s_ℓ is the sign of ℓ ($s_\ell = 1$ if $\ell_{h,t} = \ell+$ and $s_\ell = -1$ if $\ell_{h,t} = \ell_-$), γ is a margin hyperparameter and $S'_{(h, \ell, t)}$ is the set of triplets by negative sampling. Minimizing the loss function naturally makes $h + \ell_+$ closer to t than t' for the positive links, which depicts the implicit relation: the **status** of h is higher than t. The case of the negative link is similar to the positive link. Therefore, the vector after training is satisfied by status theory.

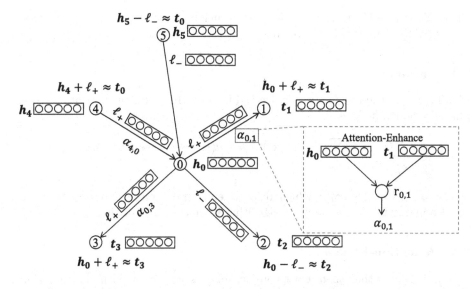

Fig. 2. The illustration of SSNE

3.3 Optimization via Negative Sampling

For the signed networks, it has been demonstrated that the negative links often contain addition information which plays an import role for downstream machine learning tasks [14] and nodes should be able to have their "friends" closer than their "foes" [2]. Inspired by this, we design a more reasonable negative sampling method: given a triplet (h, ℓ_+, t), we first sample node t' from the negative neighbors $N_{h-} = \{t \in V | (h, \ell_-, t) \in S\}$. Once N_{h-} is $\emptyset$, we then random sample t' from the set of V while h has no link to t'. On the other hand, given a triplet (h, ℓ_-, t), we first sample node t' from the positive neighbors $N_{h+} = \{t \in V | (h, \ell_+, t) \in S\}$. Once N_{h+} is $\emptyset$, we then random sample t' from the set of V while h has no link to t'. In this way, negative sampling sufficiently preserves the information of negative links and enlarges the difference between positive and negative link.

In order to distinguish the node t' obtained by negative sampling, we define a new relation between h and t', called ℓ_0, which means that there is no edge between h and t', so the triplet obtained by negative sampling becomes (h, ℓ_0, t') when h has no link to t'.

Therefore, $S'_{(h, \ell, t)}$ is improved as the following:

$$
S'_{(h, \ell, t)} = \begin{cases} \{(h, \ell', t') | t' \in N_{h-}, \ell' = \ell\} \\ \cup \{(h, \ell', t') | t' \in V \setminus N_h, \ell' = \ell_0\} & \text{if } \ell = \ell_+ \\ \{(h, \ell', t') | t' \in N_{h+}, \ell' = \ell\} \\ \cup \{(h, \ell', t') | t' \in V \setminus N_h, \ell' = \ell_0\} & \text{if } \ell = \ell_- \end{cases},
$$

where $N_h = \{t \in V | (h, \ell, t) \in S\}$ denotes the neighbors of h and $V \setminus N_h$ represents the nodes that h has no links to them.

3.4 Training

Finally, we get the loss function corrected by a carefully designed negative sampling method as follows:

$$
\mathcal{L} = \sum_{(h, \ell, t) \in S} \sum_{(h, \ell', t') \in S'_{(h, \ell, t)}} \max \left(0, \gamma_{\ell'} + d(\boldsymbol{h} + s_\ell \boldsymbol{\ell}, \boldsymbol{t}) - d(\boldsymbol{h} + s_\ell \boldsymbol{\ell'}, \boldsymbol{t'})\right),
$$

where $\gamma_{\ell'} \in \{\gamma_{\ell_+}, \gamma_{\ell_-}, \gamma_{\ell_0}\}$ which respectively represent the hyper-parameters corresponding to ℓ'. The detailed algorithm is shown in Algorithm 1.

3.5 Attention-Enhance

It is well known that there exist complex interactions between nodes in social networks, so we add the "context information" of the nodes to enhance the representation and adopt the attention mechanism to distinguish different interactions between different nodes. Cygan et al. [2] have demonstrated that users should be closer to their "friends" (positive links) than their "enemies" (negative links) in signed network, therefore, we only consider the positive link interaction by concatenating an attention-enhanced neighbor embedding to its own representation. Inspired by attention models in Graphs [8,17], Given the embedding

---

Algorithm 1. Training algorithm for SSNE

Input: signed network $G = (V, E)$, margin $\gamma_{\ell_+}, \gamma_{\ell_-}, \gamma_{\ell_0}$, embedding dimensional k.
Output: node embedding h, relation embedding ℓ.
1: Initialize $h \leftarrow uniform(-\frac{6}{\sqrt{k}}, \frac{6}{\sqrt{k}}), \ell \leftarrow uniform(-\frac{6}{\sqrt{k}}, \frac{6}{\sqrt{k}})$
2: $S \leftarrow$ convert link to status triplets and split training set
3: **for** $(h, \ell, t) \in S$ **do**
4:     $S'_{(h,\ell,t)} \leftarrow$ negative sample $S_{(h,\ell,t)}$
5:     $\mathcal{L} \leftarrow \sum_{(h,\ell,t) \in S} \sum_{(h,\ell',t') \in S'_{(h,\ell,t)}} \max(0, \gamma_{\ell'} + d(h + s_\ell \ell, t) - d(h + s_\ell \ell', t'))$
6:     Update embedding h, ℓ with Adadelta
7: **end for**
8: **return** node embedding h, relation embedding ℓ.

---

of nodes, for any node v_i and its neighborhood node $v_j \in N_{i+}$, attention weights defined as follows:

$$\alpha_{i,j} = \frac{\exp(r_{i,j})}{\sum_{k \in N_{i+}} \exp(r_{i,k})}$$

where $r_{i,j}$ denotes the relevance from v_i to v_j. Particularly we use the inner product of the embedding of v_i and v_j to define $r_{i,j}$. Finally, the embedding of node v_i is improved to the following:

$$v'_i = v_i \oplus \sum_{k \in N_{i+}} \alpha_{i,k} v_k$$

where v'_i is the new embedding of v_i and v_i is the original embedding gained by Algorithm 1.

3.6 Computational Complexity Analysis

Since one edge corresponds to one status triplet, there are a total of $|E|$ triples, so the algorithm takes $O((n+1)|E|)$ time to negative sample where n is the number of negative samples. Considering that the dimension k of each embedding, the complexity of our approach is $O(k(n+1)|E|)$.

4 Experiments

4.1 Datasets

We conduct experiments on three real-world signed directed networks: Epinions, Slashdot, and Wikipedia, which are provided by the Stanford Network Analysis Project (SNAP)[2]. Epinions is a product review site that allows users to mark trust (positive) and distrust (negative) relations with others. Slashdot is a technology news website where users can build friends (positive) and foes (negative) relations with other users. The wiki dataset is derived from Wikipedia's voting

---

[2] http://snap.stanford.edu/.

records, where users vote for (positive) or against (negative) others who apply for administrator privileges. In particular, the examinations on all three signed networks conducted by Leskovec et al. [9] demonstrate that more than 90% of triads satisfy status theory. The statistical results of the datasets are shown in Table 1.

Table 1. Statistic of the datasets.

	Epinions	Slashdot	Wiki
# of nodes	131828	82144	7118
# of edges	841372	549202	103747
% of positive edges	85.3%	77.4%	78.8%
% of negative edges	14.7%	22.6%	21.2%

4.2 Experimental Settings

We conduct link sign prediction on SSNE and several state-of-art unsigned or signed network embedding models to evaluate performance.

Baselines

- Node2vec [3]: This method designs a biased random walk strategy and leverages neural language model to learn node representation for the unsigned network.
- SNE [18]: This method designs a log-bilinear model and learns node representation along a given path and SNE does not consider any social theory for the signed network.
- SiNE [14]: This method introduces structural balance theory in signed network embedding and designs a deep learning framework to learn the representation.
- SIGNet [6]: This method proposes a scalable node embedding method which preserves social theories in higher order neighborhoods.

It should be noted that in order to evaluate the performance of Node2vec which is designed for unsigned network, we convert the signed networks to unsigned networks by ignoring the sign of the edges, while the downstream task for Link Sign Prediction is the same as other signed network embedding methods.

Parameter Setting

By default, we set the number of the negative sampling as 2 for all datasets and set the dimension k of node representation and relation representation as 140. For hyper-parameters $\{\gamma_{\ell_+}, \gamma_{\ell_-}, \gamma_{\ell_0}\}$, we empirically set $\gamma_{\ell_+} = 4$, $\gamma_{\ell_-} = 2$, $\gamma_{\ell_0} = 4$, respectively. The gradient is calculated using back-propagation and optimized using Adadelta [19] algorithm. For all other parameters in the baselines, we use the settings mentioned in their own papers, which can achieve the best results.

4.3 Link Sign Prediction

Link sign prediction is an important downstream machine learning task for signed network embedding. In this subsection, we discuss the performance of SSNE for link sign prediction. Similar to SIGNet [6], for all datasets, we randomly split 50% of edges as training set and the remaining 50% of edges as the test set. Then we learn the node embedding from the training set and generate edge embedding by combining the source node embedding and target node embedding, the combination methods include element-wise product, concatenation and average, the experimental results shows that generating edge embedding by concatenation always achieves the best results. Finally, using the embedding of the edges as features and the sign of the edges as labels, we train a simple logistic regression classifier on the training set and evaluate the performance on the test set. In real-world signed networks, the ratio of the negative links is much smaller than the positive links, so we adopt F1-micro to better evaluate the performance. We repeat the above experiment five times and use the average as the final result.

Table 2. The average F1-micro score on link sign prediction.

	Epinions	Slashdot	Wiki
Node2vec	0.831	0.776	0.749
SNE	0.854	0.778	0.751
SiNE	0.856	0.779	0.752
SIGNet	0.920	0.832	0.845
SSNE	**0.928**	**0.839**	**0.866**

The detailed experimental results are shown in Table 2. We have the observations as follows:

1. SSNE performs much better than node2vec which is a network embedding method for the unsigned network, this is because node2vec does not consider the difference between the negative and positive links at all, but the negative links between two nodes indicates that they are not similar.
2. The performance of SSNE is better than SNE. This is due to SNE only considers the difference between positive and negative edges but ignores the social theories, which are effective for signed network analysis.
3. SSNE, naturally, has significantly improved over SiNE and SIGNet. This is because the status theory utilized by the SSNE performs much better than the structure balance theory utilized by SiNE and SIGNet for the social networks mentioned in this paper.

In summary, the experiments show that SSNE is more effective than state-of-the-art methods. Particularly, SSNE has improved performance by 15.3% and 2.5% over SNE and SIGNet on Wiki dataset.

Table 3. F1-micro score on link sign prediction on effect of status and attention

t%	10	20	30	40	50	60	70	80	90
SSNE-R	0.795	0.837	0.852	0.858	0.860	0.862	0.864	0.865	0.869
SSNE-A	**0.808**	0.839	0.849	0.853	0.858	0.860	0.864	0.864	0.866
SSNE	0.804	**0.842**	**0.853**	**0.859**	**0.866**	**0.866**	**0.871**	**0.870**	**0.879**

Table 4. F1-macro score on link sign prediction on effect of status and attention

t%	10	20	30	40	50	60	70	80	90
SSNE-R	**0.689**	0.743	0.761	0.768	0.77	0.772	0.775	0.774	0.786
SSNE-A	0.669	0.727	0.748	0.753	0.765	0.766	0.771	0.773	0.777
SSNE	**0.689**	**0.746**	**0.764**	**0.774**	**0.783**	**0.786**	**0.791**	**0.790**	**0.806**

4.4 Effect of Status and Attention

To demonstrate whether the improved part of the model affects performance, we conduct a contrast experiment on SSNE and its variants. In order to investigate the effectiveness of introducing the relation ℓ_0, we propose the model called SSNE-R, which is similar to SSNE except that ℓ' is removed in negative sample triplet. On the other hand, whether attention-enhance is effective can be found by comparing SSNE with the model called SSNE-A which only removes the attention-enhance. In detail, we conduct link sign prediction on the wiki dataset and report the F1-micro and F1-macro scores, we randomly sample $t\%$ from 10% to 90% of the edges as the training set and other parameters are set to default values. From Tables 3 and 4, as the training set ratio increases, SSNE performs better than SSNE-R due to introducing the relation ℓ_0 and also perform better than SSNE-A due to the attention-enhance, so the relation ℓ_0 and attention mechanism work for our model.

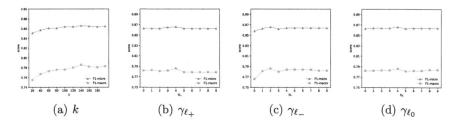

| (a) k | (b) γ_{ℓ_+} | (c) γ_{ℓ_-} | (d) γ_{ℓ_0} |

Fig. 3. Parameter w.r.t. dimension k, margin loss hyper-parameters γ_{l_+}, γ_{l_-} and γ_{l_0}

4.5   Parameter Sensitivity

In this subsection, we investigate whether and how the node embedding dimension k, margin loss hyper-parameters $\gamma_{\ell'} \in \{\gamma_{\ell_+}, \gamma_{\ell_-}, \gamma_{\ell_0}\}$ affect the performance of SSNE. We report F1-micro score and F1-macro score of link sign prediction on the Wiki dataset.

Figure 3(a) reports the performance of our model w.r.t the dimension k. Notice that when we study k, other parameters are set to default values. We find that as the dimension of k increases, the initial performance will increase significantly. This is because we embed the status into the vector space, so increasing k can encode more information, which is favorable for depicting the "status levels" between nodes. However, As k continues to grow since the dimensions of embedding are large enough to contain most of the node information, the performance of the embedding is stable.

The parameter sensitivity analysis for $\gamma_{\ell_+}, \gamma_{\ell_-}, \gamma_{\ell_0}$ are respectively shown in Fig. 3(b), (c) and (d). $\gamma_{\ell_+}, \gamma_{\ell_-}, \gamma_{\ell_0}$ are used to balance the impact of different margin loss of different relations, so the larger $\gamma_{\ell'} \in \{\gamma_{\ell_+}, \gamma_{\ell_-}, \gamma_{\ell_0}\}$ is, the more significant the effect of the loss term corresponding to $\gamma_{\ell'}$. From Fig. 3(b), we can see that the performance of $\gamma_{\ell_+} = 2$ and $\gamma_{\ell_+} = 4$ are better than the performance of $\gamma_{\ell_+} = 0$, which demonstrates that γ_{ℓ_+} is indispensable for our model. The similar conclusions which are observed in Fig. 3(b), can be found in Fig. 3(c) and (d).

5   Conclusion

In this paper, we propose a novel signed network embedding method called SSNE, which leverages the status theory and characterizes the status of nodes in the semantic vector, then we well design different ranking objectives and negative sampling method for positive and negative links respectively. The experiments on three signed networks demonstrate that our model is effective. However, our model only considers the information of network structure. In the future, we plan to investigate the status relations of node attributes and try to combine the status relations for node structure and the status relations for node attributes to learn the representations.

Acknowledgments. This work was supported by the National Social Science Foundation Project (15BTQ056), the National Key R&D Program of China (2018YFC0809800, 2016QY15Z2502-02, 2018YFC0831000), the National Natural Science Foundation of China (91746205, 91746107, 51438009), and the Applied Basic Research Project of Qinghai Province (No: 2018-ZJ-707).

References

1. Bordes, A., Usunier, N., Garcia-Duran, A., Weston, J., Yakhnenko, O.: Translating embeddings for modeling multi-relational data. In: Advances in Neural Information Processing Systems, pp. 2787–2795 (2013)
2. Cygan, M., Pilipczuk, M., Pilipczuk, M., Wojtaszczyk, J.O.: Sitting closer to friends than enemies, revisited. In: Rovan, B., Sassone, V., Widmayer, P. (eds.) MFCS 2012. LNCS, vol. 7464, pp. 296–307. Springer, Heidelberg (2012). https://doi.org/10.1007/978-3-642-32589-2_28
3. Grover, A., Leskovec, J.: node2vec: scalable feature learning for networks. In: Proceedings of the 22nd ACM SIGKDD International Conference on Knowledge Discovery and Data Mining, pp. 855–864. ACM (2016)
4. Guha, R., Kumar, R., Raghavan, P., Tomkins, A.: Propagation of trust and distrust. In: Proceedings of the 13th International Conference on World Wide Web, pp. 403–412. ACM (2004)
5. Heider, F.: Attitudes and cognitive organization. J. Psychol. **21**(1), 107–112 (1946)
6. Islam, M.R., Aditya Prakash, B., Ramakrishnan, N.: SIGNet: scalable embeddings for signed networks. In: Phung, D., Tseng, V.S., Webb, G.I., Ho, B., Ganji, M., Rashidi, L. (eds.) PAKDD 2018. LNCS (LNAI), vol. 10938, pp. 157–169. Springer, Cham (2018). https://doi.org/10.1007/978-3-319-93037-4_13
7. Kim, J., Park, H., Lee, J.E., Kang, U.: Side: representation learning in signed directed networks. In: Proceedings of the 2018 World Wide Web Conference on World Wide Web, pp. 509–518. International World Wide Web Conferences Steering Committee (2018)
8. Lee, J.B., Rossi, R.A., Kim, S., Ahmed, N.K., Koh, E.: Attention models in graphs: A survey. arXiv preprint arXiv:1807.07984 (2018)
9. Leskovec, J., Huttenlocher, D., Kleinberg, J.: Signed networks in social media. In: Proceedings of the SIGCHI Conference on Human Factors in Computing Systems, pp. 1361–1370. ACM (2010)
10. Perozzi, B., Al-Rfou, R., Skiena, S.: Deepwalk: online learning of social representations. In: Proceedings of the 20th ACM SIGKDD International Conference on Knowledge Discovery and Data Mining, pp. 701–710. ACM (2014)
11. Sen, P., Namata, G., Bilgic, M., Getoor, L., Galligher, B., Eliassi-Rad, T.: Collective classification in network data. AI Mag. **29**(3), 93 (2008)
12. Tang, J., Qu, M., Wang, M., Zhang, M., Yan, J., Mei, Q.: Line: large-scale information network embedding. In: Proceedings of the 24th International Conference on World Wide Web, pp. 1067–1077. International World Wide Web Conferences Steering Committee (2015)
13. Wang, D., Cui, P., Zhu, W.: Structural deep network embedding. In: Proceedings of the 22nd ACM SIGKDD International Conference on Knowledge Discovery and Data Mining, pp. 1225–1234. ACM (2016)
14. Wang, S., Tang, J., Aggarwal, C., Chang, Y., Liu, H.: Signed network embedding in social media. In: Proceedings of the 2017 SIAM International Conference on Data Mining, pp. 327–335. SIAM (2017)
15. Wang, S., Tang, J., Aggarwal, C., Liu, H.: Linked document embedding for classification. In: Proceedings of the 25th ACM International on Conference on Information and Knowledge Management, pp. 115–124. ACM (2016)
16. Wang, X., Cui, P., Wang, J., Pei, J., Zhu, W., Yang, S.: Community preserving network embedding. In: AAAI, pp. 203–209 (2017)

17. Xu, H., Liu, H., Wang, W., Sun, Y., Jiao, P.: NE-FLGC: network embedding based on fusing local (first-order) and global (second-order) network structure with node content. In: Phung, D., Tseng, V.S., Webb, G.I., Ho, B., Ganji, M., Rashidi, L. (eds.) PAKDD 2018. LNCS (LNAI), vol. 10938, pp. 260–271. Springer, Cham (2018). https://doi.org/10.1007/978-3-319-93037-4_21

18. Yuan, S., Wu, X., Xiang, Y.: SNE: signed network embedding. In: Kim, J., Shim, K., Cao, L., Lee, J.-G., Lin, X., Moon, Y.-S. (eds.) PAKDD 2017. LNCS (LNAI), vol. 10235, pp. 183–195. Springer, Cham (2017). https://doi.org/10.1007/978-3-319-57529-2_15

19. Zeiler, M.D.: Adadelta: an adaptive learning rate method. arXiv preprint arXiv:1212.5701 (2012)

20. Zheng, Q., Skillicorn, D.B.: Spectral embedding of signed networks. In: Proceedings of the 2015 SIAM International Conference on Data Mining, pp. 55–63. SIAM (2015)

On the Network Embedding in Sparse Signed Networks

Ayan Kumar Bhowmick$^{(\boxtimes)}$, Koushik Meneni, and Bivas Mitra

Department of CSE, Indian Institute of Technology Kharagpur, Kharagpur, India
ayankb@iitkgp.ac.in, koushik190498@gmail.com, bivas@cse.iitkgp.ernet.in

Abstract. Network embedding, that learns low-dimensional node representations in a graph such that the network structure is preserved, has gained significant attention in recent years. Most state-of-the-art embedding methods have mainly designed algorithms for representing nodes in unsigned social networks. Moreover, recent embedding approaches designed for the sparse real-world signed networks have several limitations, especially in the presence of a vast majority of disconnected node pairs with opposite polarities towards their common neighbors. In this paper, we propose *sign2vec*, a deep learning based embedding model designed to represent nodes in a sparse signed network. *sign2vec* leverages on signed random walks to capture the higher-order neighborhood relationships between node pairs, irrespective of their connectivity. We design a suitable objective function to optimize the learned node embeddings such that the link forming behavior of individual nodes is captured. Experiments on empirical signed network datasets demonstrate the effectiveness of embeddings learned by *sign2vec* for several downstream applications while outperforming state-of-the-art baseline algorithms.

Keywords: Signed network embedding · Autoencoders ·
Conflicting node pairs

1 Introduction

Signed social networks such as *Epinion, Slashdot, Wikipedia* [6] where links may have positive or negative sign depicting the polarity of relationships between nodes connected by them, have gained huge popularity recently [6]. Performing network embedding on signed networks involves mapping of nodes in such networks to a low-dimensional feature space, that preserves the sign information of links besides the network structure. A vast majority of the state-of-the-art network embedding approaches, that have been developed for learning node embeddings in unsigned social networks [1,8–10] containing only positive links, cannot be simply extended to signed networks. First, unsigned network embedding methods aim to capture different measures of proximity (first-order, second-order etc.) between users connected by only positive links reflected by their network structure. However, it is non-trivial to preserve these proximity

© Springer Nature Switzerland AG 2019
Q. Yang et al. (Eds.): PAKDD 2019, LNAI 11441, pp. 94–106, 2019.
https://doi.org/10.1007/978-3-030-16142-2_8

measures for signed networks due to the presence of positive & negative link polarities. Additionally, negative links have distinct topological properties compared to positive links that simply violate the assumptions of unsigned network embedding approaches.

Recently, few attempts have been made in the gamut of signed network embedding [11,12,15], where the objective is to place positively linked nodes close to each other in the embedding space than the negatively linked nodes. Moreover, [3,5,14] leverage on random walks for embedding nodes in signed directed networks considering both polarity and direction of links into account. However, the major limitation of the aforementioned endeavours is that computation of embeddings are mostly dependent on the presence of links connecting node pairs, their respective direction & polarities and the constituent connected triplets (say for structural balance theory [6]). Hence, such methods fail to quantify the extent of similarity & dissimilarity between the vast majority of *disconnected* node pairs in real-world signed networks.

Close inspection reveals that most real-world signed social networks are sparse [2]; hence, there exists a significant volume of node pairs that are disconnected. Further, there exist *conflicting pairs* of nodes in real-world signed networks, which are essentially disconnected node pairs having common neighbors linked with opposite (conflicting) polarities; hence, it is non-trivial to capture the similarity between such node pairs in the embedding space based on higher-order proximities. The presence of *conflicting node pairs* severely compromise the embedding quality of state-of-the-art signed network embedding methods due to conflicting polarities and absence of a link between them. Notably, the neighborhood structure of individual nodes can reveal their inherent link forming behavior (positive or negative) towards neighbors of different polarities. Subsequently, one can leverage on random walks in a signed network [4] to capture higher-order neighborhood relationships between node pairs, whether connected or disconnected, and compute the extent of similarity between such pairs in terms of their link forming behavior. This can help to determine the relative positions of the vast majority of disconnected node pairs in the embedding space, including *conflicting pairs*. Our paper takes an important step in this direction to learn node representations in a sparse signed network.

The major contribution of this paper is to develop a network embedding framework for sparse signed networks, which ensures that nodes exhibiting similar link forming behavior are placed closer in the latent embedding space, irrespective of their connectivity. First, we formulate the problem of signed network embedding and present the challenges in the light of *conflicting node pairs* (Sect. 2). Subsequently, we describe the methodology for computing the extent of similarity in link forming behavior between any node pair. We develop *sign2vec*, a deep autoencoder framework and design suitable objective functions that are optimized to capture the similarity in link forming behavior reflected by higher-order neighborhood relationships between node pairs in the learned embeddings (Sect. 3). We evaluate the performance of *sign2vec* by conducting experiments on empirical signed network datasets. We first analyze the correctness of embedding vectors learned using *sign2vec* through extensive statistical significance tests. Finally, our evaluation shows effectiveness of *sign2vec* embeddings towards achieving discriminative visualizations, good accuracy (62%) for node clustering, efficiency in performing multi-class classification (86%) and signed link prediction (93%) with high accuracy, outperforming state-of-the-art methods (Sect. 4).

2   Problem Definition

In this section, we first formulate the problem of network embedding in a signed network. Next, we present the challenges involved in learning such embeddings.

2.1   Problem Statement

Let $\mathcal{G} = (\mathcal{U}, \mathcal{E}_p, \mathcal{E}_n)$ be a signed directed network consisting of a set of N nodes $\mathcal{U} = \{u_1, u_2, \ldots, u_N\}$. Here $\mathcal{E}_p$ and $\mathcal{E}_n$ denote the set of directed positive and negative links respectively in $\mathcal{G}$ such that $\mathcal{E}_p = \{(u_i, u_j, +) : u_i, u_j \in \mathcal{U}\}$ and $\mathcal{E}_n = \{(u_i, u_j, -) : u_i, u_j \in \mathcal{U}\}$. $A \in \mathbb{R}^{N \times N}$ represents the adjacency matrix of $\mathcal{G}$ where (i, j)-th entry $a_{ij} = 1$ for positive link and $a_{ij} = -1$ for negative link from node u_i to u_j. $a_{ij} = 0$ means u_i and u_j are disconnected. In this paper, our aim is to learn a function $f(.)$ that maps each node u_i in $\mathcal{G}$ to a low-dimensional embedding vector $\mathbf{y_i} \in \mathbb{R}^d$ where $d \ll N$ is the embedding dimension. Mathematically, $f : u_i \rightarrow \mathbf{y_i}, \forall u_i \in \mathcal{U}$. Thus, our objective is to find a function $f(.)$ that places nodes with similar behavior towards forming links of a given polarity (positive or negative) close to each other in embedding space.

2.2   Challenges

In order to demonstrate the challenges, we first introduce *conflicting pairs* of nodes covering a significant fraction (40%) of node pairs in real signed networks.

***Conflicting Pairs* of Nodes:** Given the signed network $\mathcal{G}$ and node $u_i \in \mathcal{U}$, let $\mathcal{N}_{u_i}^+$ and $\mathcal{N}_{u_i}^-$ denote the positively and negatively connected neighbors of u_i respectively where the neighbor set $\mathcal{N}_{u_i} = \mathcal{N}_{u_i}^+ \cup \mathcal{N}_{u_i}^-$. We define a pair of nodes u_i and u_j ($u_i, u_j \in \mathcal{U}$) as *conflicting* if (a) u_i and u_j are disconnected in $\mathcal{G}$ and (b) there exists atleast one common neighbor $u \in \mathcal{U}$ of u_i and u_j ($u \in \mathcal{N}_{u_i} \cap \mathcal{N}_{u_j}$) that is linked with u_i and u_j via opposite polarities. For instance, for the signed social network shown in Fig. 1, nodes x and w form a *conflicting pair* while nodes w and t denote a *non-conflicting pair*.

For embedding of a node in a signed network, the state-of-the-art attempts [12,14] *only* rely on the presence of positive and negative links with its neighbors. For instance, structural balance theory [6] provides a guidance to develop an objective function such that nodes should be placed closer to their 'friends' (positively connected neighbors) than their 'foes' (negatively connected neighbors) in the embedding space. However, the presence of *conflicting node pairs* in real-world signed networks make the state-of-the-art node embedding techniques highly erroneous due to following reasons: (a) Absence of direct links between *conflicting pairs* results in difficulty in measuring the similarity between such node pairs; it fails to capture the extent of similarity or dissimilarity between those pairs. (b) In case of *conflicting node pairs* (say x, w in Fig. 1), balance theory infers opposite polarities (positive and negative), depending on the triplets being considered ($\{x, y, w\}$ and $\{x, z, w\}$ respectively). These conflicting polarities affect the correctness of learned embedding vectors while optimizing the respective objective function [12].

3   *sign2vec*: A Model for Signed Network Embedding

In this section, we present the proposed *sign2vec* model by leveraging on relative trustworthiness scores between node pairs based on signed random walks, in addition to the presence or absence of connectivity between them to learn the node embeddings. First, we describe the methodology used for computing the relative trustworthiness score matrix P. Next, we develop the deep autoencoder framework for learning the node representations in a signed network.

3.1   Characterizing Similarity in Link Forming Behavior

We introduce the relative trustworthiness score matrix $P \in \mathbb{R}^{N \times N}$ where each entry p_{ij} represents the relative trustworthiness of a node u_i with respect to node u_j in $\mathcal{G}$. A high value of relative trustworthiness score p_{ij} denotes that nodes u_i and u_j exhibit similar link forming behavior with their respective neighbors. Essentially, for the pair (u_i, u_j), p_{ij} captures the complex higher-order neighborhood relationships between this node pair in the presence of link polarities, even if u_i and u_j are disconnected, conflicting and several hops away in $\mathcal{G}$.

We leverage on Signed Random Walk with Restart (SRWR) model proposed in [4] to compute P for signed network $\mathcal{G}$. In a signed random walk, a sign is introduced into a random surfer in order to consider the link polarity in the walk. The surfer starts with a positive sign from u_i and flips her sign if she encounters a negative link in the walk, else the surfer retains her sign. We consider a surfer as trustful if the surfer, starting at u_i (with polarity positive or negative), retains the same polarity after reaching u_j. On the other hand, if the polarity gets flipped after landing at u_j (from positive to negative or vice versa), we designate the surfer as distrustful. Finally, the SRWR model assigns a score p_{ij} for the node pair (u_i, u_j) as the difference in probabilities of the trustful surfer and the distrustful surfer, initiated at u_i, visiting u_j. A high positive value of p_{ij} indicates that the trustful surfer, initiated at u_i, visits u_j in majority of random walks, whereas a high negative score indicates that the distrustful surfer visits u_j more often. Finally, we construct the matrix P by computing this score for every node pair, such that the link forming behavior is captured in learned node embeddings.

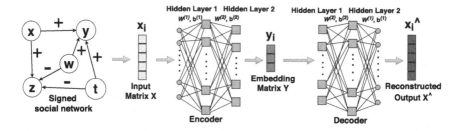

Fig. 1. Framework of *sign2vec*

3.2 Proposed Model

Leveraging on the computation of the matrix P, we develop the *sign2vec* model to learn the node embeddings.

Model Overview. Our proposed *sign2vec* model is based on deep autoencoders comprising of a neural network architecture. Autoencoders are unsupervised learning algorithms that applies backpropagation to learn the representation of high-dimensional data, typically used for dimensionality reduction. It consists of two symmetrical deep-belief networks: encoder and decoder. In our model, we use a deep autoencoder with two sets of hidden layers (see Fig. 1). The first set of hidden layers form the encoder block while the second set forms the decoder block of the autoencoder. Given the adjacency matrix A (introduced in Sect. 2) and the relative trustworthiness score matrix P (obtained in Sect. 3.1), we form the augmented matrix $X = A|P$, where the ith row of X denoted by $\mathbf{x_i}$ is obtained by concatenating the corresponding rows of A and P respectively. Here $X \in \mathbb{R}^{N \times 2N}$ is the input matrix. Let $Y \in \mathbb{R}^{N \times d}$ denote the low-dimensional embedding matrix where d is the embedding dimension. The ith row of Y is denoted by $\mathbf{y_i}$ representing the latent embedding vector for the node u_i. The encoder block encodes the input matrix X to learn the latent representation Y by reducing the dimensionality. The output of the last encoder layer essentially represents the generated embedding Y which is fed as input to the symmetric decoder block. The embeddings are then decoded to $\hat{X}$ to finally reconstruct the input matrix at the last layer of decoder. We apply backpropagation to minimize the loss between the output $\hat{X}$ and the input X matrices.

Model Construction. We construct the input matrix X from the matrices A and P such that the ith row $\mathbf{x_i} = \mathbf{a_i}|\mathbf{p_i}$, denoting that $\mathbf{p_i}$ is concatenated to the end of $\mathbf{a_i}$. The matrix X is then fed as input to the first hidden layer of the encoder network. The output of each layer of the encoder is:

$$Y^{(1)} = \sigma(W^{(1)}X + \mathbf{b^{(1)}}), \ Y^{(2)} = \sigma(W^{(2)}Y^{(1)} + \mathbf{b^{(2)}}) \tag{1}$$

where *sigmoid* activation function σ is used between the hidden layers of the encoder. $Y^{(2)}$ is the encoder output which is fed as input to the decoder. The output of corresponding layers of the decoder, that basically reverses the computations of the encoder is:

$$Y^{\hat{(1)}} = \sigma(W^{(2)}Y^{(2)} + \mathbf{b^{(2)}}), \ \hat{X} = \sigma(W^{(1)}Y^{\hat{(1)}} + \mathbf{b^{(1)}}) \tag{2}$$

The decoder output $\hat{X}$ denotes the reconstructed output of the autoencoder. The weight matrices $W^{(1)}, W^{(2)}$ are initialized to random values from a normal distribution while elements of bias vectors $\mathbf{b^{(1)}}$, $\mathbf{b^{(2)}}$ are initialized to ones at the corresponding layers. The model learns the weights and the biases in each iteration through backpropagation of the decoder loss to the input matrix X.

Loss Computation. The objective of the autoencoder is to minimize the loss through backpropagation between the reconstructed output $\hat{X}$ and the input matrix X. The objective (loss) function $||\hat{X} - X||_2^2$ is designed in a way that makes the reconstructed output very close to the original input matrix. In order to ensure that nodes connected by a link with positive polarity are placed very close in the embedding space, we penalize the reconstruction loss by upweighting the corresponding terms in the objective function

$$\mathcal{L}_{rec} = ||(\hat{X} - X) \odot B||_2^2 \tag{3}$$

where $\odot$ is the Hadamard product operation and the (i, j)-th entry of matrix $B \in \mathbb{R}^{N \times 2N}$ is denoted by b_{ij}. We define $b_{ij} = \rho$ if $a_{ij} = 1$ and $b_{ij} = 1/\rho$ if $a_{ij} = -1$ where $\rho > 1$. We keep $b_{ij} = 1$ for all other entries of B. This choice of the reconstruction loss $\mathcal{L}_{rec}$ in Eq. 3 ensures that we impose high penalty (via ρ) to the reconstruction error of the elements in X which depicts positively connected node pairs $a_{ij} = 1$ in the signed network $\mathcal{G}$, so that such elements are reconstructed with higher precision. Thus, this reconstruction criterion guarantees that positively linked nodes ($a_{ij} = 1$) are mapped close to each other, while placing negatively linked nodes ($a_{ij} = -1$) farther away from each other in the embedding space.

Besides preserving the link polarities of the underlying signed network in the learned embeddings, we also seek to preserve the similarity in link forming behavior between a pair of nodes. For this purpose, we design another loss function at the *encoder output* that minimizes the embedding loss by penalizing the error between embeddings of similar nodes with large value of relative trustworthiness score denoted by p_{ij}:

$$\mathcal{L}_{emb} = \sum_{i,j=1}^{N} p_{ij}||(\mathbf{y_i^{(2)}} - \mathbf{y_j^{(2)}})||_2^2 \tag{4}$$

The embedding loss $\mathcal{L}_{emb}$ in Eq. 4 simply ensures that we highly penalize the loss when nodes with similar link forming behavior (say u_i & u_j with embeddings $\mathbf{y_i}$ & $\mathbf{y_j}$ respectively) are placed far away in the embedding space. Finally, we combine the losses corresponding to $\mathcal{L}_{rec}$ (Eq. 3) and $\mathcal{L}_{emb}$ (Eq. 4) to jointly minimize the combined objective function

Table 1. Dataset statistics

Dataset	#Nodes	#Edges	#Positive edges	# Negative edges	Density
Wiki-elec	7118	107080	83962	23118	2.1×10^{-3}
Slashdot	82140	549202	417943	131259	8.1×10^{-5}

$$\mathcal{L} = ||(\hat{X} - X) \odot B||_2^2 + \lambda \sum_{i,j=1}^{N} p_{ij}||(\mathbf{y_i^{(2)}} - \mathbf{y_j^{(2)}})||_2^2 \qquad (5)$$

where λ is the regularizer for the embedding loss to prevent overfitting. The combined loss (Eq. 5) is minimized by performing optimization with respect to the parameter set $\theta = \{W^{(1)}, W^{(2)}, \mathbf{b^{(1)}}, \mathbf{b^{(2)}}\}$. The error is backpropagated to update θ using an optimizer at a learning rate of α after every training epoch. Finally, after a fixed number of training epochs, when the loss $\mathcal{L}$ converges, we obtain the embedding matrix $Y = Y^{(2)}$ as the output of *sign2vec*.

4 Experiments and Evaluation

In this section, we first describe our experimental setup. Then we evaluate the effectiveness of the *sign2vec* embeddings through downstream prediction tasks.

4.1 Experimental Setup

Dataset: We conduct our experiments on the following two real-world signed directed networks, summarized in Table 1.

Wiki-elec: This dataset consists of all administrator elections and vote history data extracted from Wikipedia page edit history from January 2008[1]. A Wikipedia member may cast a vote accompanied by a textual comment towards another member. This induces a signed directed network in which nodes represent Wikipedia members while signed directed links represent polarity of votes cast by members (supporting $(+1)$, opposing (-1) or neutral). In this dataset, all members are eligible to cast and receive votes.

Slashdot: Slashdot is a technology-related news website that features user-submitted and editor-evaluated current technology oriented news. The website, known for its specific user community, allows users to annotate other users as friends or foes. This induces a signed directed network containing friend/foe links between the users of Slashdot. The network was obtained in February 2009[2].

---

[1] https://snap.stanford.edu/data/wiki-Elec.html.
[2] https://snap.stanford.edu/data/soc-sign-Slashdot090221.html.

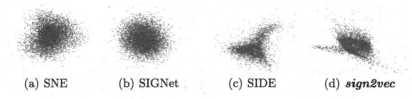

(a) SNE   (b) SIGNet   (c) SIDE   (d) *sign2vec*

Fig. 2. Visualization of network embeddings on *Wiki-elec* dataset. Color indicates the ground truth user label. Red: 'positive', blue: 'negative'. (Color figure online)

Table 2. Analysis of *sign2vec* embeddings based on average Euclidean distance between different types of node pairs

Datasets	Connected pairs			Conflicting pairs		
	$\mu^{\mathcal{E}_P}$	$\mu^{\mathcal{E}_n}$	p-value	μ^{E_I}	μ^{E_O}	p-value
Wiki-elec	0.01 ± 0.004	0.08 ± 0.138	$3.448E - 08$	0.03 ± 0.005	0.07 ± 0.04	$1.396E - 06$
Slashdot	0.03 ± 0.010	0.06 ± 0.098	$1.506E - 07$	0.04 ± 0.004	0.06 ± 0.05	$6.118E - 03$

Baseline Algorithms. We introduce the following state-of-the-art baseline algorithms to learn network embeddings on a signed network:

(a) **SNE:** This method adopts the log-bilinear model to combine edge sign information and node representations of all nodes along a given path in the learned node embeddings [14].

(b) **SiNE:** This method uses the extended structural balance theory to embed nodes in a signed undirected network using a deep learning framework [12].

(c) **SIGNet:** This relies on targeted node sampling for random walks, leveraging upon structural balance theory to learn interpretable representations [3].

(d) **SIDE:** This method is based on truncated random walks for learning embeddings in a signed directed network [5].

Ground Truth Node Labeling. We extract the ground truth labels of users in the signed network datasets based on the proportion of incoming positive or negative links for a user in the underlying network [13]. We designate each user to one of the three classes: 'positive' (denoting a trustful user), 'negative' (denoting a distrustful user) and 'neutral'. If the majority (>0.6) of incoming links associated with a user u_i has positive (or negative) polarity, we label u_i as 'positive' (or 'negative'); all the remaining users are labeled as 'neutral'. We validate the ground truth node labels for users in the *Wiki-elec* dataset through sentiment analysis of the textual comments posted for a user. Using the sentiment analyzer module in the *NLTK* toolkit[3] available in Python, for each user u_i, we calculate the positive (p^{u_i}) and negative polarity (n^{u_i}) scores for the entire corpus of textual comments posted for u_i. We find that $p^{u_i} > n^{u_i}$ for 92% of

---

[3] https://github.com/nltk/nltk.

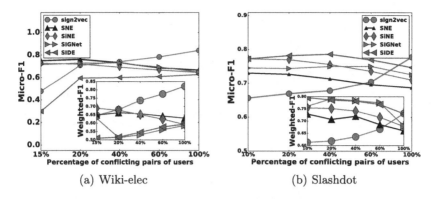

(a) Wiki-elec   (b) Slashdot

Fig. 3. Performance of multi-class classification in presence of conflicting node pairs

Table 3. Node clustering results. Here ACC = Accuracy, NMI = Normalized Mutual Information, WF = Weighted-F1.

Dataset	SNE			SiNE			SIGNet			SIDE			sign2vec		
	ACC	NMI	WF	ACC	NMI	WF	ACC	NMI	WF	ACC	NMI	WF	ACC	NMI	WF
Wiki-elec	0.286	0.001	0.383	0.3	0.0005	0.226	0.249	0.0005	0.288	0.205	0.001	0.292	0.265	**0.127**	**0.418**
Slashdot	0.402	0.009	0.453	0.446	0.010	0.367	0.298	0.007	0.405	0.268	0.007	0.333	**0.615**	**0.023**	**0.526**

users labeled as 'positive' in *Wiki-elec* while $p^{u_i} < n^{u_i}$ for 96% of users labeled as 'negative' in *Wiki-elec*. This validates the ground truth node labels for the *Wiki-elec* dataset. We do not validate the ground truth node labels for *Slashdot* in the absence of any secondary textual information.

Parameter Settings. To generate the node embeddings using our proposed model *sign2vec*, we set the number of hidden layers $H = 2$ for both encoder and decoder. We take the size of the first and second hidden layers of the encoder to be 1024 and 128 respectively. This ensures that the dimension of the generated embedding vectors is $d = 128$ at the encoder output. We use the *sigmoid* activation function between consecutive hidden layers of the autoencoder. We employ parameter sharing to ensure that the same weight matrices are used at corresponding hidden layers of the encoder and decoder. The error is computed using Eq. 5 in terms of the *binary cross-entropy loss* where the values of the hyperparameters ρ and λ have been set to 10 and 50 respectively. To minimize this loss, we use the *RMSProp* optimizer with learning rate $\alpha = 0.001$ and set the number of training epochs to 100.

4.2   Evaluation of Embedding Quality

In this section, we investigate whether the learned embeddings generated by our model can indeed capture the link forming behavior of individual users and place similar nodes close to each other while placing dissimilar nodes far away in the

Table 4. Multi-class classification results

Algorithm	SNE		SiNE		SIGNet		SIDE		sign2vec	
Model	Micro-F1	Weighted-F1	Micro-F1	Weighted-F1	Micro-F1	Weighted-F1	Micro-F1	Weighted-F1	Micro-F1	Weighted-F1
Wiki-elec	0.650	0.613	0.626	0.576	0.626	0.572	0.642	0.604	**0.856**	**0.850**
Slashdot	0.685	0.661	0.693	0.659	0.708	0.681	0.736	0.731	**0.821**	**0.777**

Table 5. Performance of signed link prediction

Algorithm	SNE		SiNE		SIGNet		SIDE		sign2vec	
Model	Accuracy	F1-score	Accuracy	F1-score	Accuracy	F1-score	Accuracy	F1-score	Accuracy	F1-score
Wiki-elec	0.784	0.882	0.812	0.882	0.833	0.891	0.903	0.918	**0.934**	**0.957**
Slashdot	0.792	0.874	0.822	0.887	0.838	0.882	0.897	0.911	**0.926**	**0.937**

embedding space. For this purpose, we consider the set of node pairs that are connected by positive links $\mathcal{E}_p$ and compute the Euclidean distance between their respective embedding vectors and obtain the average Euclidean distance over all node pairs in $\mathcal{E}_p$ as $\mu^{\mathcal{E}_p}$. Similarly, we compute the average Euclidean distance between respective embeddings of all node pairs in $\mathcal{E}_n$ connected by negative links as $\mu^{\mathcal{E}_n}$. We report the average and standard deviation of the Euclidean distances for all datasets in Table 2. We then conduct two sample t-tests on the distribution of Euclidean distance values for positively and negatively connected node pairs in each dataset to test the null hypothesis $H_0 : \mu^{\mathcal{E}_p} \geq \mu^{\mathcal{E}_n}$. The alternative hypothesis is $H_1 : \mu^{\mathcal{E}_p} < \mu^{\mathcal{E}_n}$. We reject the null hypothesis at the significance level of $\alpha = 0.01$ with p-values shown in Table 2. This shows that our learned embedding vectors can effectively preserve the network structure as well as the link polarities in a signed network.

Next, we compare the average Euclidean distance between respective embedding vectors of conflicting node pairs having *majority* of common neighbors with *identical* polarities denoted by μ^{E_I} to those having *majority* of common neighbors with *opposite* polarities denoted by μ^{E_O} as shown in Table 2. Here E_I and E_O denote the population of Euclidean distance values for conflicting pairs having *majority* of common neighbors with *identical* and *opposite* polarities respectively. We conduct two sample t-tests on these distributions for all datasets to test the null hypothesis $H_0 : \mu^{E_I} \geq \mu^{E_O}$. The alternative hypothesis is $H_1 : \mu^{E_I} < \mu^{E_O}$. We reject the null hypothesis at the significance level of $\alpha = 0.01$ with p-values as shown in Table 2. This gives a strong evidence on difference of both means and distributions between the two sets of conflicting node pairs, demonstrating the effectiveness of *sign2vec* embeddings.

4.3   Evaluation on Downstream Tasks

We evaluate the quality of embeddings generated by *sign2vec* through various graph mining applications on signed networks such as visualization, node clustering, multi-class classification and signed link prediction.

Visualization. In Fig. 2, we visualize the learned embeddings obtained from *sign2vec* as well as the baseline algorithms in a 3-D space using t-SNE approach [7] only for *Wiki-elec* dataset due to brevity. Each point represents a user in the 3-D space with a color denoting the corresponding ground truth label. *sign2vec* performs the best as visualizations show that users with same ground truth label (same color) are placed close to each other in the embedding space for *sign2vec*. On the contrary, there is high overlap between points of different colors for the baseline methods.

Node Clustering. We apply k-means clustering (taking $k = 3$, as we have three classes in ground truth) on the embedding vectors obtained from *sign2vec* as well as baseline algorithms to cluster the users. We evaluate the quality of the detected clusters using standard evaluation metrics such as *Normalized Mutual Information (NMI)*, *Accuracy (ACC)* and *Weighted-F1 (WF)* that compare the detected cluster labels with ground truth labels. In Table 3, we observe that *sign2vec* outperforms all the baseline algorithms in terms of clustering performance across all datasets. Since baseline embeddings fail to encode similarities (or dissimilarities) between disconnected node pairs, often such node pairs with identical ground truth labels get placed in different clusters in case of baselines.

Multi-class Classification. Next, we evaluate the quality of embeddings generated by different algorithms on multi-class classification task. In a supervised learning framework, we apply the generated embedding vectors as features to classify a user into one of the three ground truth labels. We consider different machine learning algorithms to train the classifier; we randomly sample 80% of the labeled users as the training set and the remaining 20% as test set. We predict the label of a user in the test set and report the average *Micro-F1* and *Weighted-F1* scores over 100 iterations using *Random Forest* classifier in Table 4, since it gives best performance among all classifiers. From the table, we observe that *sign2vec* consistently outperforms all baselines for all datasets with significant performance improvement for multi-class classification. In this paper, we claim that the novelty of *sign2vec* is to suitably handle the conflicting user pairs, which is the major weakness of baseline embeddings. In Fig. 3, we show the classification performance as we increase the proportion of *conflicting pairs* present in the network. Interestingly, baselines achieve superior performance when the network contains only a small fraction of *conflicting pairs*; performance of *sign2vec* mostly suffers due to lack of data points in terms of *conflicting pairs*. However, as we increase the fraction of *conflicting pairs*, performance of *sign2vec* gradually increases and finally outperforms all baselines.

Signed Link Prediction. Finally, we evaluate the performance of *sign2vec* on signed link prediction task using the learned embedding vectors. For this purpose, we combine the two d-dimensional node representations for a node pair connected by a link (positive or negative) to obtain a single d-dimensional link representation, using element-wise Hadamard product as the operator (as used in node2vec [1]). We then apply these link representations as features to train a 3-way classifier to predict the label of the corresponding link as positive, negative or no link (in case of disconnected node pairs). We use 10-fold cross validation for training and testing the link representations using *one-vs-rest logistic regression* classifier. First, we sample a number of disconnected node pairs equal to the number of negative links in the signed network; we further perform undersampling on the set of positive links to balance the dataset. Accordingly, we report the accuracy and F1-score for signed link prediction on both the datasets in Table 5 across different algorithms. This table shows the superiority of *sign2vec* over the baselines in performing accurate signed link prediction. The high accuracy in link prediction for *sign2vec* stems from the fact that trustworthiness score p_{ij} in P is able to correctly discriminate between node pairs u_i & u_j connected by either positive link or negative link from the disconnected node pairs.

5 Conclusion

In this paper, we have proposed *sign2vec*, a novel signed network embedding model to represent sparse signed networks. In case of sparse signed networks, quality of state-of-the-art embedding methods gets severely compromised due to the absence of connecting links. This problem further gets compounded in the presence of *conflicting node pairs* in real-world signed networks. We have relied on the higher-order neighborhood relationships by leveraging on signed random walks to quantify the relative trustworthiness between node pairs, irrespective of their connectivity, to determine their relative positions in the embedding space. The learned embedding vectors capture the link forming behavior of individual nodes in a signed network. Leveraging on the relative trustworthiness between node pairs, we have developed the deep autoencoder based *sign2vec* framework that effectively preserves both network structure and link polarities of a signed network in the latent embedding space, confirmed through extensive statistical analysis of the learned embedding vectors. Experimental results on empirical signed networks have demonstrated the effectiveness of *sign2vec* over state-of-the-art baseline algorithms on various downstream prediction tasks such as visualization, node clustering, multi-class classification and signed link prediction. Notably, the performance improvement is substantial in the presence of large number of conflicting node pairs in sparse signed networks.

References

1. Grover, A., Leskovec, J.: node2vec: scalable feature learning for networks. In: KDD, pp. 855–864 (2016)
2. Hsieh, C.-J., Chiang, K.-Y., Dhillon, I.S.: Low rank modeling of signed networks. In: KDD, pp. 507–515 (2012)

3. Islam, M.R., Aditya Prakash, B., Ramakrishnan, N.: SIGNet: scalable embeddings for signed networks. In: Phung, D., Tseng, V.S., Webb, G.I., Ho, B., Ganji, M., Rashidi, L. (eds.) PAKDD 2018. LNCS (LNAI), vol. 10938, pp. 157–169. Springer, Cham (2018). https://doi.org/10.1007/978-3-319-93037-4_13

4. Jung, J., Jin, W., Sael, L., Kang, U.: Personalized ranking in signed networks using signed random walk with restart. In: ICDM, pp. 973–978 (2016)

5. Kim, J., Park, H., Lee, J.-E., Kang, U.: Side: representation learning in signed directed networks. In: WWW, pp. 509–518 (2018)

6. Leskovec, J., Huttenlocher, D., Kleinberg, J.: Signed networks in social media. In: SIGCHI, pp. 1361–1370 (2010)

7. Van Der Maaten, L., Hinton, G.: Visualizing data using t-SNE. JMLR 9, 2579–2605 (2008)

8. Perozzi, B., Al-Rfou, R., Skiena, S.: Deepwalk: online learning of social representations. In: KDD, pp. 701–710 (2014)

9. Tang, J., Qu, M., Wang, M., Zhang, M., Yan, J., Mei, Q.: Line: large-scale information network embedding. In: WWW, pp. 1067–1077 (2015)

10. Wang, D., Cui, P., Zhu, W.: Structural deep network embedding. In: KDD, pp. 1225–1234 (2016)

11. Wang, S., Aggarwal, C., Tang, J., Liu, H.: Attributed signed network embedding. In: CIKM, pp. 137–146 (2017)

12. Wang, S., Tang, J., Aggarwal, C., Chang, Y., Liu, H.: Signed network embedding in social media. In: SIAM, pp. 327–335 (2017)

13. Wu, Z., Aggarwal, C.C., Sun, J.: The troll-trust model for ranking in signed networks. In: WSDM, pp. 447–456 (2016)

14. Yuan, S., Wu, X., Xiang, Y.: SNE: signed network embedding. In: Kim, J., Shim, K., Cao, L., Lee, J.-G., Lin, X., Moon, Y.-S. (eds.) PAKDD 2017. LNCS (LNAI), vol. 10235, pp. 183–195. Springer, Cham (2017). https://doi.org/10.1007/978-3-319-57529-2_15

15. Zheng, Q., Skillicorn, D.B.: Spectral embedding of signed networks. In: SIAM, pp. 55–63 (2015)

MSNE: A Novel Markov Chain Sampling Strategy for Network Embedding

Ran Wang, Yang Song, and Xin-yu Dai$^{(\boxtimes)}$

National Key Laboratory for Novel Software Technology,
Nanjing University, Nanjing 210023, China
{wangr,songy}@nlp.nju.edu.cn, daixinyu@nju.edu.cn

Abstract. Network embedding methods have obtained great progresses on many tasks, such as node classification and link prediction. Sampling strategy is very important in network embedding. It is still a challenge for sampling in a network with complicated topology structure. In this paper, we propose a high-order **M**arkov chain **S**ampling strategy for **N**etwork **E**mbedding (MSNE). MSNE selects the next sampled node based on a distance metric between nodes. Due to high-order sampling, it can exploit the whole sampled path to capture network properties and generate expressive node sequences which are beneficial for downstream tasks. We conduct the experiments on several benchmark datasets. The results show that our model can achieve substantial improvements in two tasks of node classification and link prediction. (Datasets and code are available at https://github.com/SongY123/MSNE.)

Keywords: Network embedding · Random walk · Sampling strategy

1 Introduction

Network embedding provides an effective and efficient way for network representation learning. It has been widely applied in many areas including including biology [19], social sciences [7] and linguistics [2]. Network embedding converts nodes and edges into vectors in a low dimensional space, in which the network structure is preserved.

Many network embedding methods learn node representations based on random walk statistics. When the network is too large to measure entirely, random walk is an useful way to approximate many properties in the network, such as node centrality [14] and similarity [6]. In another word, random walk samples the context for nodes to capture the network properties, instead of observing the whole network.

With the sampled context from random walk, DeepWalk [15] adopts a neural language model (Skip-gram [12]) for network embedding. The objective for Skip-gram is to learn node representations by maximizing the co-occurrence

This work is supported by the National Science Foundation of China (61472183).

Q. Yang et al. (Eds.): PAKDD 2019, LNAI 11441, pp. 107–118, 2019.
https://doi.org/10.1007/978-3-030-16142-2_9

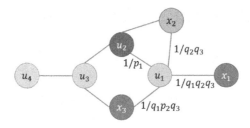

Fig. 1. An illustration of the random walk procedure in network. The walk path starts from u_4 to u_1, i.e., (u_4, u_3, u_2, u_1) and is now evaluating the next step among the neighbors of node u_1. Edge labels indicate unnormalized sampling probability. p and q are parameters of model.

probability among the "words" that appear within a window. Similar to Deep-Walk [15], Node2vec [9] also converts the networks into embedding based on random walk. The key characteristic of Node2vec is that it employs biased random walk that provides a trade-off between breadth-first (BFS) and depth-first (DFS) sampling strategies. And it produces higher-quality and more informative embedding than DeepWalk. There have also been a number of further extensional algorithms based on random walk. For example, Walklets [16] extends the Deep-Walk to learn representations that "skip" or "hop" over multiple nodes at each step. Metapath2vec [5] and HINE [10] design meta-path-based methods to cope with heterogeneous networks.

Those network embedding methods have obtained great progresses on many applications and tasks. However, the random walk policies in them are rigid. They cannot flexibly capture the diversity of connectivity patterns. We can present an example as follows to show their deficiency. As illustrated in Fig. 1, there is a simple network containing seven nodes. The current partial walk contains (u_4, u_3, u_2, u_1) and is now evaluating its next step among the node u_1's neighbors, which all have different characteristics: u_2 is previously sampled node; x_2 is a part of triangle [4] containing u_1, u_2, x_2; x_3 forms a quadrilateral [4] with u_1, u_2, u_3; x_1 is a outlier which is just connected to u_1. DeepWalk [15] just uses unbiased *1-order sampling strategy* to approximate the structure of networks, and it treats all four candidates as equal. Thus it cannot distinguish the difference among them. Node2vec [9] employs *2-order sampling strategy* of balancing the exploration-exploitation trade-off. After transitioning to node u_1 from u_2, its return parameter p and in-out parameter q control the probability of a walk staying inward revisiting nodes (u_2), staying close to the preceding nodes (x_2), or moving outward farther away (x_1, x_3). But unfortunately, Node2vec still cannot distinguish subtle difference between x_1 and x_3. To make matters worse, real-world networks often consist of numerous nodes formed in more complicated patterns.

To address above problem, we propose *a high-order **M**arkov chain **S**ampling strategy* for **N**etwork **E**mbedding (named MSNE). Based on the distance between candidates and previous sampled nodes, MSNE can efficiently explore diverse neighborhoods and select the next step. Actually, MSNE generalizes Markov chain sampling from 2-order to n-order. And due to n-order sampling strategy, it can exploit the whole sampled path to capture network properties and generate expressive node sequences which are beneficial for network embedding and downstream tasks.

Empirically, we conduct the experiments on five real-world network datasets. The effectiveness of the node embeddings is evaluated on two common prediction tasks: node classification and link prediction. The experiment results show that MSNE outperforms other competitive baselines.

Contributions: (1) We provide a high-order Markov chain sampling strategy, called MSNE. It can exploit the whole sampled path to capture the diversity of the next sample candidates and generate expressive node sequences which are beneficial for network embedding and downstream tasks. (2) We extensively evaluate our generated representation on node classification and link prediction on several real-world datasets and demonstrate the effectiveness of the MSNE.

2 Preliminary

A network is denoted as $G = (V, E)$, where $V = \{v_1, \cdots, v_n\}$ represents n nodes, and $E \subseteq (V \times V)$ represents edges. Network embedding framework aims to build a powerful function $f : V \to \mathbb{R}^d$, which converts each node $v \in V$ into a low-dimensional vector $x \in \mathbb{R}^d$. Here, the parameter, $d \ll |V|$, specifies the dimension of representation space.

For learning network embedding, firstly, random walk is used over complicated networks to generate a number of node sequences, which preserve highly non-linear network structures. Then, following the Skip-gram model [12], a mapping function is learned by maximizing the log-probability of the sampled node sequence, which is the context of a source node.

Specifically, let $W(u) = (v_{u1}, \cdots, v_{uk})$ denote the sampled node sequence. As same as Skip-gram model [12], the objective of our network embedding framework is to maximize the log-probability as follow:

$$\max_f \sum_{u \in V} \log \Pr \left(W(u) | f(u) \right). \tag{1}$$

We can obtain the network representations that capture network properties by solving the optimization problem. But the context $W(u)$ is a node sequence making Eq. 1 intractable. Hence, under the assumption that the likelihood of predicting a context node is independent of any other, we approximate the conditional probability as follow:

$$\Pr(W(u)|f(u)) = \prod_{v \in W(u)} \Pr(v|f(u)). \tag{2}$$

Moreover, we consider a source node u and a network sampling context node $v \in W(u)$ as a symmetric role over each other in feature space, which is also utilized in [9,15,20]. Furthermore, $\Pr(W(u)|f(u))$ is modeled as a softmax unit between every source-context node representation pair:

$$\Pr(v|f(u)) = \frac{\exp(f(v) \cdot f(u))}{\sum_{n \in V} \exp(f(n) \cdot f(u))}, \tag{3}$$

here, $v \in W(u)$. Hence, the objective of our model is simplified to

$$\max_f \sum_{n \in V} \left[-\log Z_n + \sum_{v \in W(u)} \exp(f(v) \cdot f(u)) \right] \tag{4}$$

It is worth noting that the denominator $Z_n = \sum_{n \in V} \exp(f(n) \cdot f(u))$ in Eq. 4 needs to consider every node pair existing in networks, which requires expensive overhead and becomes impractical in real-world networks. To speed up the training process, there are two technologies: hierarchical softmax [13] and negative sampling [13]. In this work, we use Skip-gram with negative sampling (SGNS) to approximate it in the interest of model performance.

Overall, our network embedding framework consists of two main components: a sampling strategy, which transforms highly non-linear networks into numerous linear sampled paths, and an embedding architecture, i.e. SGNS [13]. SGNS has been originally developed to handle with linear text, where the notion of a neighborhood can be naturally defined using a sliding window over consecutive words.

It is obvious that the quality of network representation produced by Skip-gram depends largely on that of the node sequences generated by the sampling strategy. As we mention before, DeepWalk employs unbiased 1-order sampling strategy which treats different candidates as equal and ignores their different characteristics. To capture the diversity of connectivity patterns observed in network, Node2vec balances the exploration-exploitation trade-off by 2-order sampling strategy. Illustrated in Fig. 1, Node2vec can differentiate candidates into three groups: "revisiting" nodes (u_2), "inward" nodes (x_2) and "outward" nodes (x_1, x_3). However, just 2-order sampling strategy fails to distinguish subtle difference among nodes in same group (x_1 is a outlier while x_3 are a part of quadrilateral, i.e., $\{u_1, u_2, u_3, x_3\}$). As we show, this is a major shortcoming of prior work which fails to offer enough flexibility in sampling nodes from a network. To overcome this limitation, we provide a *high-order* sampling strategy with more parameters to tune the explored search space.

3   MSNE: A Novel Markov Chain Sampling Strategy for Network Embedding

In this section, we present the details of our *high-order* **M**arkov chain **S**ampling strategy for **N**etwork **E**mbedding (named MSNE), which generalizes the 2-order Markov chain sampling strategy in Node2vec to n-order. For convenience, we list some of the terms and notations in Table 1 which will be used later in advance.

Table 1. Terms and notations.

Term	Definition
G	Network
V	Nodes
E	Edges
n	Order of MSNE
$(u_n, \cdots, u_i, \cdots, u_1)$	Previous sampled nodes
$d(u, v)$	Length of the shortest path from node u to v
π	Sampling strategy based on a distance metric
$\pi_i(x, u_i)$	Sampling strategy based on $d(x, u_i)$
p_i, q_i	Parameters p_i and q_i of strategy π_i
$D_k(v)$	$D_k(v) = \{u \mid u, v \in V \wedge d(u, v) = k\}, \forall k \geq 0$

3.1 n-Order Markov Chain Random Walk

Formally, given a partial random walk path $walk = (u_n, \cdots, u_i, \cdots, u_1)$, we are now evaluating its next step x among the neighbors of node u_1 based on the following probability distribution.

$$\Pr(x|u_1, \cdots, u_n) = \begin{cases} \pi(x, u_2, \cdots, u_n)/Z, & \text{if } (u_1, x) \in E; \\ 0, & \text{otherwise.} \end{cases} \tag{5}$$

where $\pi(x, u_2, \cdots, u_n)$ is the unnormalized probability of sampling node x, which is biased by the sampled node sequence $(u_n, \cdots, u_2)$, and Z is the normalizing factor.

In order to make $\pi(x, u_2, \cdots, u_n)$ tractable, we simplify it as follow:

$$\pi(x, u_2, \cdots, u_n) = \prod_{i=2}^{n} \pi_i(x, u_i) \tag{6}$$

where $\pi_i(x, u_i)$ denotes unnormalized probability biased by node u_i and x.

The sampling strategy $\pi_i(x, u_i)$ hopefully can differentiate network structures and help to evaluate the next step. Back to the Fig. 1, we have a sampled sequence starting from u_4 to u_1, i.e., (u_4, u_3, u_2, u_1), and now residing at node u_1. We observe that a distance metric of shortest length between candidates and previous sampled nodes, i.e., $d(x, u_i)$, can help to distinguish the subtle diversity among candidates. Specifically, when considering *1-order* information, $d(x, u_1)$, we treat four candidates $\{x_1, x_2, x_3, u_2\}$ as equal. Fortunately, when taking account of $d(x, u_2)$, the *2-order* information helps us to differentiate candidates into three groups: "revisiting" nodes (u_2), "inward" nodes (x_2) and "outward" nodes (x_1, x_3). Furthermore, when we consider *3-order* information, i.e., $d(x, u_3)$, it is easy to distinguish between x_1 and x_3. The shortest length from x_1 to u_3 is 3, while it is 1 from x_3 to u_3. Overall, the higher order information and the distance metric can help the sampling strategy $\pi_i(x, u_i)$ to distinguish the subtle difference of network structures, and generate expressive node sequences for network embedding.

However, in Eq. 6, sampling strategy π still suffers from the scope of each distance $d(x, u_i)$ and combinatorial explosion. Thus, we simplify the Eq. 6 as follow.

$$\pi(x, u_2, \cdots, u_n) = \pi_2(x, u_2) \triangleright \cdots \triangleright \pi_{i-1}(x, u_{i-1}) \triangleright \pi_i(x, u_i) \cdots \triangleright \pi_n(x, u_n), \quad (7)$$

where $\triangleright$ means that those sampling strategies $\pi_2(x, u_2), \cdots, \pi_n(x, u_n)$ are applied *sequentially* instead of *simultaneously*.

Due to the sequential nature of strategy π, we define each $\pi_i(x, u_i)$ with consideration of $d(u_{i-1}, x)$ as follow.

$$\pi_i(x, u_i | d(u_{i-1}, x) = k) = \begin{cases} 1/p_i, & \text{if } d(u_i, x) = k - 1; \\ 1, & \text{if } d(u_i, x) = k; \\ 1/q_i, & \text{if } d(u_i, x) = k + 1. \end{cases} \quad (8)$$

We will interpret the Eq. 8 from the following three aspects in detail.

1. The distance from candidate x to u_{i-1} is not arbitrary. Let k denote the value of $d(u_{i-1}, x)$, and k must be less or equal to $i-1$. x is the neighbor node of u_1, so $d(u_1, x) = 1$. And there is a path from u_{i-1} to u_1, so $d(u_{i-1}, u_1) \leq i - 2$. Thus, it is easy to prove that $d(u_{i-1}, x) \leq i - 1$, i.e., $k \leq i - 1$.
2. From above proof, the scope of $d(u_i, x)$ is from 0 to i. Furthermore, we can prove that $d(u_i, x)$ can be limited to $\{k - 1, k, k + 1\}$, given $d(u_{i-1}, x) = k$. We can prove it shortly as follows. Given $d(u_{i-1}, x) = k$, and $d(u_i, u_{i-1}) = 1$, there is a trivial path of length $k + 1$ connecting node u_i and x. Meanwhile, there is no path of length less than $k - 1$ connecting u_i to x, otherwise it will contradict $d(x, u_{i-1}) = k$. Therefore, given $d(x, u_{i-1}) = k$, the distance $d(x, u_i)$ must be one of $\{k-1, k, k+1\}$. This is the reason why two parameters are necessary and sufficient to guide the sampling strategies in Eq. 8.
3. Furthermore, parameters p_i and q_i in Eq. 8 can lead the sample strategy π_i to select different candidates.[1] For each strategy $\pi_i(x, u_i)$, a high value of parameter p_i $(> \max(q_i, 1))$ ensures that we are less likely to sample an already-visited node in the next steps. On the other hand, if p_i is low $(< \min(q_i, 1))$, it would lead the walk to backtrack a step and this would keep the walk close to the starting node. Meanwhile, another parameter q_i of each strategy π_i allows the search to differentiate between "inward" and "outward" candidates. If $q_i > 1$, the sampling is biased towards nodes close to preceding node ("inward" nodes). In contrast, if $q_i < 1$, the walk is more inclined to visit nodes which are further away from the preceding node ("outward" nodes). Overall, the parameters provide us a way to tune the explored search space.

---

[1] In fact, the parameters p_i, q_i can be different for strategies $\pi_i(x, u_i)$ with different distances $d(u_{i-1}, x)$. But in this work, we share the parameters across strategy π_i in order to reduce the burden of parameter searching.

In summary, our high-order MSNE provides an n-order Markov chain sampling strategy. It can help generate the next step during sampling based on the distance (shortest path) between the candidates and the previous sampled nodes. We also present a tractable and efficient way to inference the probability, as shown in Eq. 8.

3.2   The MSNE Algorithm

Algorithm 1. MSNE-Walk

Inputs: Network $G = (V, E)$; Order n;
   Walk Length l; Walks pre node r;
   Parameters $\{p_2, q_2\}, \cdots, \{p_n, q_n\}$
1: Initialize Walks $walks = []$
2: Distance Sets $D = $ PreprocessData(G)
3: **for** iter to r **do**
4:    **for all** node $u \in V$ **do**
5:       Initialize Walk $walk = [u]$
6:       Choose one node from $D_1(u)$ and
         append to $walk$
7:       **while** $|walks| < l$ **do**
8:          Compute $step$ by
            MSNE-Step($n, D, walk, \pi$)
9:          Append $step$ to $walk$
10:      **end while**
11:      Append $walk$ to $walks$
12:   **end for**
13: **end for**
14: **return** $walks$

Algorithm 2. MSNE-Step

Inputs: Order n;
   Data $D = \{D_0, D_1, \cdots, D_n\}$;
   Partial Walk $walk = [\cdots, u_n, \cdots, u_1]$;
   Parameters $\{p_2, q_2\}, \cdots, \{p_n, q_n\}$
1: Initialize Candidate $Cand = D_1(u_1)$
2: Initialize $i = 2$
3: Update $n = \min\{|walk|, n\}$
4: **while** $|Cand| > 1$ and $i \leq n$ **do**
5:    Update candidate $Cand$ according
      to Eq. 8 specified by parameters
      $\{p_i, q_i\}$
6:    $i = i + 1$
7: **end while**
8: Choose one node $step$ from candidate
   $Cand$ at random.
9: **return** $step$

The pseudo-code for sampling nodes in n-order MSNE is given in Algorithms 1 and 2. From line 3 to 13 of *MSNE-Walk*, we repeat r rounds sampling which starts from each node in the network to learn representations. Given the starting node, MSNE generates the next sampling node according to Eq. 8 specified by parameters $\{p_i, q_i\}$ until the length of sampled node sequences reaches l, as shown in line 5 of *MSNE-Step*. It is worth noting that, in the interest of high sampling efficiency we compute the distance sets in advance, as *PreprocessData* shown in line 2 of *MSNE-Walk*. Instead of computing expensive all-pairs shortest paths algorithms, we only need to calculate a small amount of distance information between nodes, i.e. $D = \{D_0, D_1, \cdots, D_n\}$, where $D_i = \{D_i(v)|v \in V\}$. By doing it, sampling of nodes while simulating the random walk can be done efficiently in $O(1)$ time when applying n-order MSNE.

4   Experiments

With flexible exploration in networks offered by MSNE, we can learn expressive representations to a wide variety of network. Our experiments mainly evaluate representations learned by our model on two tasks: node classification and link prediction.

4.1 Experimental Setup

We compare MSNE with other models on several datasets. Furthermore, we choose several representative models listed below as the baseline:

- **DeepWalk** [20]: DeepWalk adopts truncated random walk to transforms graph structures into linear sequences. Besides, it uses Skip-gram with hierarchical softmax as the loss function.
- **Node2vec** [9]: Node2vec is an extension of DeepWalk which introduces a biased random walk procedure to efficiently explore diverse neighborhoods.
- **LINE** [17]: LINE can preserve both first- and second-order proximities through modeling node co-occurrence probabilities.
- **GraRep** [3]: GraRep adopts singular value decomposition [8] in k-step probability transition matrices to obtain low-dimensional representation of nodes.

For all experiments, the dimension of each node is 128, and the length of generated node sequence is fixed as 80. We repeat sampling for 10 times. In addition, the window size is 10 for Skip-gram and the k-step of GraRep is 4.

Parameters Search. In the experiment, we search for best parameters in a greedy way for convenience. First of all, we find the best parameters $\{p_2, q_2\}$ for 2-order MSNE. Then, we use parameters $\{p_2, q_2\}$ of 2-order MSNE and find the best parameters $\{p_3, q_3\}$ for 3-order MSNE. And so on, we use parameters $\{p_2, q_2\}, \cdots, \{p_{n-1}, q_{n-1}\}$ of $n-1$-order MSNE and find the best parameters for n-order MSNE. What calls for special attention is that for each level of parameters, the best parameters are decided by the results of the development dataset. We stop at 4-order during parameter searching to reduce the number of experiments.

4.2 Node Classification

In the node classification, each node is assigned one or more labels. The task is to predict the labels of some nodes in the network. We evaluate models on the following datasets:

- BlogCatalog [18]: This is a social network of bloggers on BlogCatalog website. There are 10,312 nodes, 333,983 edges and *39 labels*.
- Cora [11]: This is a citation network of scientific publications. There are 2,708 nodes classified with 5429 edges into one of *7 classes*.

We evaluate our method using the same experimental procedure outlined in DeepWalk [15]. Logistic Regression Classifier with L2 regularization is adopted as the classifier. We randomly sample T_f (10% to 90%) fraction of the labeled nodes and use them as training data with the rest as a test data set. This process is repeated 10 times, after which we report the mean Micro-F1 and Macro-F1 scores.

Experimental Results. Experimental setup is described as mentioned earlier. In case of BlogCatalog, the best strategy of Node2vec is $\{p = 0.25, q = 0.25\}$,

Table 2. Micro-F1 and Macro-F1 scores of our MSNE and competing methods on BlogCatalog and Cora. MSNE-3rd and MSNE-4th denote 3-order and 4-order MSNE, respectively. Paired t-test results are shown for our method compared to Node2vec (statistical significance is indicated with $**(p < 0.001)$).

	ratio	Blogcatalog						Cora					
		Deep Walk	Node2vec	LINE	GraRep	MSNE -3rd	MSNE -4th	Deep Walk	Node2vec	LINE	GraRep	MSNE -3rd	MSNE -4th
Micro -F1	0.1	0.340	0.338	0.329	**0.363**	0.339	0.340	0.771	0.774	0.651	0.760	**0.778**	0.764
	0.2	0.367	0.373	0.350	**0.381**	0.373	0.371	0.801	**0.807**	0.728	0.776	0.794	0.791
	0.3	0.382	**0.389**	0.364	0.388	0.383	0.387	**0.818**	0.807	0.741	0.784	0.807	0.802
	0.4	0.387	**0.400**	0.373	0.393	0.394	0.397	**0.826**	0.819	0.758	0.791	0.824	0.814
	0.5	0.397	0.398	0.377	0.393	0.403	**0.405**	**0.826**	0.817	0.750	0.787	0.825	0.808
	0.6	0.405	0.406	0.379	0.399	0.411	**0.412***	0.817	0.817	0.755	0.786	**0.827***	0.816
	0.7	0.410	0.408	0.388	0.403	0.411	**0.418***	0.814	0.813	0.766	0.787	**0.839***	0.822
	0.8	0.416	0.418	0.398	0.408	0.421	**0.422***	0.825	0.821	0.766	0.799	**0.828***	0.819
	0.9	0.421	0.417	0.404	0.417	0.430	**0.437***	0.823	0.841	0.801	0.808	**0.849***	0.849
Macro -F1	0.1	0.189	0.192	0.171	0.194	0.200	**0.202**	0.756	0.753	0.630	0.745	**0.761**	0.750
	0.2	0.222	0.233	0.195	0.218	0.234	**0.238**	0.789	**0.793**	0.721	0.759	0.782	0.779
	0.3	0.238	0.247	0.211	0.225	0.251	**0.253**	**0.807**	0.795	0.729	0.771	0.796	0.789
	0.4	0.245	0.265	0.217	0.229	0.266	**0.268**	**0.817**	0.809	0.737	0.779	0.813	0.807
	0.5	0.255	0.271	0.223	0.230	0.272	**0.282***	0.808	0.804	0.737	0.770	**0.809***	0.800
	0.6	0.258	0.274	0.227	0.236	0.286	**0.287***	0.805	0.810	0.748	0.772	**0.813***	0.807
	0.7	0.266	0.276	0.245	0.243	0.286	**0.295***	0.802	0.803	0.744	0.768	**0.821***	0.803
	0.8	0.279	0.292	0.254	0.248	0.294	**0.302***	0.815	0.812	0.755	0.774	**0.821***	0.795
	0.9	0.286	0.291	0.269	0.266	0.315	**0.316***	0.790	0.810	0.767	0.783	**0.823***	0.813

while the best strategy of MSNE is $\{p_2 = 2, q_2 = 0.25; p_3 = 4, q_3 = 1; p_4 = 1, q_4 = 1\}$. In case of Cora dataset, the best strategy of Node2vec is $\{p = 2, q = 0.25\}$, and the best strategy of MSNE is $\{p_2 = 4, q_2 = 0.5; p_3 = 0.25, q_3 = 4\}$.

The results of our experiments in BlogCatalog are shown in Table 2. When the training ratio gets higher, the performance of MSNE gets better. In case of BlogCatalog, when the training ratio is 0.9, MSNE gives us 5.0% gain over Node2vec in Micro-F1 score and 8.7% gain in Macro-F1 score. In terms of Cora, when the training ratio is 0.9, MSNE gives us 0.95% gain over Node2vec in Micro-F1 score and 1.6% gain in Macro-F1 score.

From the experiment, we find that when training ratio is high, MSNE shows greater improvements over other models. That means with the help of higher order Markov chain used in MSNE, we can dig out more information of its structural information to make the classifier better. However, when the training ratio is low, MSNE does not acquire too much improvements due to the lack of label information of its neighborhood.[2]

4.3 Link Prediction

In link prediction, part of edges in network are removed. The task is to predict these missing edges. We preprocess the network data to generate negative samples (i.e. edges which do not exist in the original network) and positive samples. Specifically, we remove 10% of edges randomly from the network while ensuring the residual network connected, and combine positive samples with negative samples as dataset. We use 50% to train and the left to test. Meanwhile, we use Hadamard Operator, i.e., element-wise multiplication, to learn edge features.

---

[2] In the supplementary material (at https://github.com/SongY123/MSNE), we discuss why MSNE with higher order sometimes does not get better results.

We evaluate models on the following datasets:

- arXiv-AstroPh[3]: arXiv-AstroPh is a co-authorship network extracted from arXiv of Astro Physics category, containing 18,772 nodes and 198,110 edges.
- arXiv-GrQc[4]: arXiv-GrQc is also a co-authorship network extracted from arXiv of General Relativity and Quantum Cosmology category. The network consists of 5,242 nodes and 14,496 edges.
- Protein-Protein Interactions (PPI) [1]: In PPI network, the nodes are proteins, and edges represent the interactions between two proteins. This network contains 3,890 nodes and 76,584 edges.

Experimental Results. In order to avoid searching for threshold, we use Area Under Curve (AUC) score for link prediction. Besides, the best parameters (p, q) of Node2vec for arXiv-AstroPh, arXiv-GrQc and PPI are $(0.25, 4)$, $(0.25, 4)$ and $(4, 4)$. The best parameters of MSNE for arXiv-AstroPh, arXiv-GrQc and PPI are $\{p_2 = 0.25, q_2 = 4; p_3 = 0.5, q_3 = 1; p_4 = 1, q_4 = 1\}$, $\{p_2 = 0.25, q_2 = 4; p_3 = 1, q_3 = 4\}$ and $\{p_2 = 4, q_2 = 4; p_3 = 0.5, q_3 = 0.25; p_4 = 0.25, q_4 = 0.5\}$.

Table 3. Area Under Curve (AUC) scores of our MSNE and competing methods on arXiv-AstroPh, arXiv-GrQc and PPI. Paired t-test results are shown for our method compared to Node2vec (statistical significance is indicated with *($p < 0.001$)).

Datasets	Methods					
	DeepWalk	Node2vec	LINE	GraRep	MSNE-3rd	MSNE-4th
arXiv-AstroPh	0.887	0.939	0.918	0.927	0.946*	**0.949***
arXiv-GrQc	0.877	0.937	0.927	0.878	**0.953***	0.947*
PPI	0.661	0.677	0.659	0.671	0.704*	**0.708***

As is illustrated in Table 3, our model achieves best results on all datasets over other four models. In general, MSNE improves AUC score on arXiv-AstroPh, arXiv-GrQc and PPI by 1.1% to 7.0%, 1.7% to 8.7% and 4.6% to 7.4%. Specially, in spite of high AUC score achieved by Node2vec in arXiv-AstroPh, our model can also get an increase of 1.1% over Node2vec.

5   Conclusions

In this paper, we propose a novel high-order Markov chain sampling strategy for network embedding, named MSNE, based on the distance between the candidates and the previous sampled nodes. In the experiment, we show that the MSNE outperforms state-of-art network representation learning techniques. Especially in node classification, when the training ratio is high, MSNE

---

[3] https://snap.stanford.edu/data/ca-AstroPh.html.
[4] https://snap.stanford.edu/data/ca-GrQc.html.

shows significant advantages over other baselines. Not surprisingly, MSNE can be applied to different kinds of network to achieve good performance. In the future, we plan to design several other methods to search parameter in MSNE and introduce the supervised information of downstream tasks into the sampling strategy.

References

1. Breitkreutz, B.-J., et al.: The BioGRID interaction database: 2008 update. Nucleic Acids Res **36**(Database–Issue), 637–640 (2008)
2. Ferrer, R.I.C., Solé, R.V.: The small world of human language. Proc. Royal Soc. Lond. B Biol. Sci. **268**(1482), 2261–2265 (2001)
3. Cao, S., Lu, W., Xu, Q.: GraRep: learning graph representations with global structural information. In: Proceedings of the 24th ACM CIKM International Conference on Information and Knowledge Management, pp. 891–900 (2015)
4. Diestel, R.: Graph Theory. GTM, vol. 173. Springer, Heidelberg (2017). https://doi.org/10.1007/978-3-662-53622-3
5. Dong, Y., Chawla, N.V., Swami, A.: metapath2vec: scalable representation learning for heterogeneous networks. In: Proceedings of the 23rd ACM SIGKDD International Conference on Knowledge Discovery and Data Mining, pp. 135–144 (2017)
6. Fouss, F., Pirotte, A., Renders, J.-M., Saerens, M.: Random-walk computation of similarities between nodes of a graph with application to collaborative recommendation. IEEE Trans. Knowl. Data Eng. **19**(3), 355–369 (2007)
7. Linton, C.: Freeman, visualizing social networks. J. Soc. Struct. **1**, (2000)
8. Golub, G.H., Reinsch, C.: Singular value decomposition and least squares solutions. Numer. Math. **14**(5), 403–420 (1970)
9. Grover, A., Leskovec, J.: node2vec: scalable feature learning for networks. In: Proceedings of the 22nd ACM SIGKDD International Conference on Knowledge Discovery and Data Mining, pp. 855–864 (2016)
10. Huang, Z., Mamoulis, N.: Heterogeneous information network embedding for meta path based proximity, CoRR abs/1701.05291 (2017)
11. McCallum, A., Nigam, K., Rennie, J., Seymore, K.: Automating the construction of internet portals with machine learning. Inf. Retr. **3**(2), 127–163 (2000)
12. Mikolov, T., Chen, K., Corrado, G., Dean, J.: Efficient estimation of word representations in vector space, CoRR abs/1301.3781 (2013)
13. Mikolov, T., Sutskever, I., Chen, K., Corrado, G.S., Dean, J.: Distributed representations of words and phrases and their compositionality. In: Advances in Neural Information Processing Systems, pp. 3111–3119 (2013)
14. Newman, M.E.J.: A measure of betweenness centrality based on random walks. Soc. Netw. **27**(1), 39–54 (2005)
15. Perozzi, B., Al-Rfou, R., Skiena, S.: DeepWalk: online learning of social representations. In: 20th ACM SIGKDD International Conference on Knowledge Discovery and Data Mining, pp. 701–710 (2014)
16. Perozzi, B., Kulkarni, V., Skiena, S.: Walklets: multiscale graph embeddings for interpretable network classification, CoRR abs/1605.02115 (2016)
17. Tang, J., Qu, M., Wang, M., Zhang, M., Yan, J., Mei, Q.: LINE: large-scale information network embedding. In: Proceedings of the 24th International Conference on World Wide Web, pp. 1067–1077 (2015)

18. Tang, L., Liu, H.: Relational learning via latent social dimensions. In: Proceedings of the 15th ACM SIGKDD International Conference on Knowledge Discovery and Data Mining, pp. 817–826 (2009)
19. Theocharidis, A., Van Dongen, S., Enright, A.J., Freeman, T.C.: Network visualization and analysis of gene expression data using BioLayout express(3d). Nat. Protoc. **4**(10), 1535–1550 (2009)
20. Tu, C., Zhang, W., Liu, Z., Sun, M.: Max-margin deepwalk: discriminative learning of network representation. In: Proceedings of the Twenty-Fifth International Joint Conference on Artificial Intelligence, pp. 3889–3895 (2016)

Auto-encoder Based Co-training Multi-view Representation Learning

Run-kun Lu, Jian-wei Liu$^{(\boxtimes)}$, Yuan-fang Wang, Hao-jie Xie, and Xin Zuo

Department of Automation, China University of Petroleum, Beijing 102249, China
zsylrk@gmail.com, {liujw,zuox}@cup.edu.cn

Abstract. Multi-view learning is a learning problem that utilizes the various representations of an object to mine valuable knowledge and improve the performance of learning algorithm, and one of the significant directions of multi-view learning is sub-space learning. As we known, auto-encoder is a method of deep learning, which can learn the latent feature of raw data by reconstructing the input, and based on this, we propose a novel algorithm called Auto-encoder based Co-training Multi-View Learning (ACMVL), which utilizes both complementarity and consistency and finds a joint latent feature representation of multiple views. The algorithm has two stages, the first is to train auto-encoder of each view, and the second stage is to train a supervised network. Interestingly, the two stages share the weights partly and assist each other by co-training process. According to the experimental result, we can learn a well performed latent feature representation, and auto-encoder of each view has more powerful reconstruction ability than traditional auto-encoder.

Keywords: Multi-view · Auto-encoder · Co-training

1 Introduction

In real word applications, multi-view learning problems are widespread and they often exist in two ways. The first one is that multiple views exist naturally in data, such as we can easily obtain three views from web pages of Facebook, they include the content of the web page, the text of any web pages linking to this web page, and the link structure of all linked pages. The second one is that the raw data is not multi-view data and we need to construct multiple views for data, which include random approaches [1, 3, 4] reshape or decompose approaches [13], and the methods that perform feature set partitioning automatically [6]. Once we get multiple views of raw data, we can utilize the advantages of multi-view learning to improve the performance of learning tasks like regression, classification and clustering, where multi-view learning methods can be classified into

This work was supported by the National Key R&D Program of China (No. 2016YFC0303703).

Q. Yang et al. (Eds.): PAKDD 2019, LNAI 11441, pp. 119–130, 2019.
https://doi.org/10.1007/978-3-030-16142-2_10

three categories: co-training, multiple kernel learning, and subspace learning [14]. In this paper, we focus on subspace learning and propose a novel multi-view learning algorithm called Auto-encoder based Co-training Multi-View Learning (ACMVL) which utilizes both complementarity and consistency and finds a joint latent feature representation of multiple views. Note that "co-training" in our proposed algorithm's name is a training strategy instead of co-training multi-view learning method.

Multiple views of raw data have two wonderful properties, which are consistency and complementarity. Consistency represents the common information of multiple views, and complementarity represents the special information of each view. Only consistency and complementarity of multiple views can be utilized to improve the performance of learning tasks [2,5,10,12], however, both consistency and complementarity are significant that it is a waste of information if we ignore one of them. In [11], they find a joint latent representation which include both common and special features of multiple views, and followed this work, [16] made some improvements. In their works, they compute the special feature of each view as well as the common feature of all views according to matrix factorization, and then concatenate them together in to a joint latent feature. However, this kind of method has two constraints:

(1) it is not reasonable to define all views share a common feature, maybe they share a common space and each view has its own instantiation in this space;

(2) the optimization algorithm is hard to adapt large scale data set, because the algorithm requires to feed all the training data instead of a batch at one time.

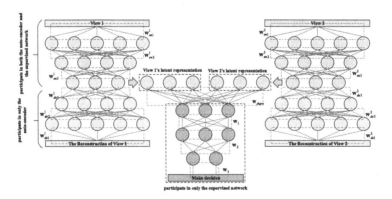

Fig. 1. Diagram of ACMVL: we illustrate a two-views' problem in this figure, therefore we have three networks. The left and right ones are view 1 and view 2's auto-encoder, and the middle one is the supervised network. Note that the yellow layers not only participate in the training process of auto-encoder but also supervised network, and our joint latent representation is the first purple layer of supervised network. (Color figure online)

To solve such two problems, our proposed algorithm ACMVL builds a framework including multiple auto-encoders and a supervised network (a multi-layer perceptron which predicts labels of instances and minimize the cross-entropy loss between prediction and true label). Our proposed approach can easily solve the second question by running mini-batch gradient descent on large scale training sets. As for the first question, we first compute each view's special feature and then map each special feature into a same space by weight sharing and add them together, furthermore, with a nonlinear activation function we can get a joint latent representation. Besides, alternating co-training is another salient characteristic of ACMVL, which is a training strategy that lets our auto-encoders and supervised network partially share model parameters, and also let supervised network help auto-encoders meliorate their encoders' model parameters. And surprisingly, we find that by using this strategy, we can not only accelerate the convergence of the algorithm but also improve the learning performance of each auto-encoder significantly, which means each view will obtain a much better special feature that can help with the construction of joint latent feature. As a result, we will make a summarize of contributions we made as follows:

(1) We propose a novel multi-view learning algorithm ACMVL, which utilizes both consistency and complementarity to build a joint latent representation of multiple views, where multiple views' auto-encoders consider the consistency, and the weight sharing method considers the complementarity;
(2) Compared with the algorithms proposed by [11,16], ACMVL is neural network-based algorithm which is easy to use mini-batch to adapt large scale data set.
(3) We propose an alternating co-training strategy which let our auto-encoders and supervised network partially shared model parameters and also let supervised network helps each auto-encoder to meliorate their encoder's weight. And this strategy can accelerate the training process and improve the learning performance of each auto-encoder significantly.

2 Framework

2.1 Notations

In this paper, bold uppercase characters are used to denote matrices, bold lowercase characters are used to denote vectors, and other characters which are not bold are all used to denote scalars. Supposed that $(\mathbf{X}^v, \mathbf{Y})$ is the sample of view v, where $v = 1, \cdots, V$. Among of them, $\mathbf{X}^v \in \Re^{M^v \times N}$ is the set of input instances of view v, $\mathbf{Y} \in \Re^N$ is the label, where N is the number of instances, M^v is the feature number of each instance of view v. More specific, we have V version of raw data, each version can be expressed as $\mathbf{X}^v$, and $\mathbf{X}^v = [\mathbf{x}_1^v, \cdots, \mathbf{x}_N^v]$, $\mathbf{x}_i^v \in \Re^{M^v}$. Note that all the views share the label $\mathbf{Y}$ because they are the various representation of raw input date, and $\mathbf{Y} = [\mathbf{y}_1, \cdots, \mathbf{y}_N]$, $\mathbf{y}_i \in \Re$.

2.2 Core Concept of Framework

As we known, auto-encoder is an algorithm that can compute the latent feature of raw data, and we can compute each view's latent feature by using auto-encoder. However, the V views' latent features we obtained only consider the complementarity of different views, and we cannot guarantee that all of the auto-encoders can generate good latent features. Therefore, we aim to find a joint latent representation by combining the V views' latent features we obtained according to some rules. In this paper, we build a simple multi-layer perceptron with its input of multiple views' latent features to supervise the process of the generation of joint latent representation. Furthermore, we adopt a novel training strategy to train multiple auto-encoders in each view as well as supervised network, we will give a description in detail in next subsection.

2.3 Description of Framework

Our proposed method ACMVL has a co-training process, which has two stages, one is the stage of learning latent feature that we need to train auto-encoders of multiple views, and the other is the stage of meliorating feature and learning joint latent feature that we need to train a supervised network. Next, we will explain the two scenarios separately.

Latent feature learning: as shown in Fig. 1, we illustrate an example with two views. In Fig. 1, there are two auto-encoders because we need to compute the latent feature of each view. Therefore, we need to train the model parameters to minimize the reconstruction error of each view, and we can formulate this problem as:

$$\min \frac{1}{V} \sum\nolimits_{v=1}^{V} \left\| \mathbf{X}^v - \hat{\mathbf{X}}^v \right\|_F \tag{1}$$

where $\hat{\mathbf{X}}^v$ is the reconstruction of view v's input. In each auto-encoder, the activation function is ReLu except for the last layer because the last layer of each auto-encoder is the reconstruction of raw data, and the optimizer algorithm we used is AdaDelta [15]. After training two auto-encoders, we should save the model parameters $\theta_{en}^v = \{\mathbf{w}_{en1}^v, \mathbf{w}_{en2}^v, \mathbf{w}_{en3}^v\}$, and $\theta_{de}^v = \{\mathbf{w}_{de1}^v, \mathbf{w}_{de2}^v, \mathbf{w}_{de3}^v\}$, note that each view's auto-encoder has its own parameters.

Meliorate feature and joint feature learning: when finish the training process of auto-encoders, we take out the third layer of each auto-encoder as the input of the supervised network as shown in the middle of Fig. 1. By mapping each latent feature representation of each view into a same subspace and add them together, we can easily find a joint latent representation by using a nonlinear mapping, which can be formulated as follows:

$$g \left(\sum\nolimits_{v=1}^{V} \mathbf{w}_{share} \mathbf{h}^v \right) \tag{2}$$

where $g\left(\cdot\right)$ is a nonlinear activation function, and we use ReLu in this paper. Note that we share the transform matrix $\mathbf{w}_{share}$ which helps us to find the consistent

First Epoch: $\theta_{en}^v \leftarrow \theta_{en,0}^v, \theta_{de}^v \leftarrow \theta_{de,0}^v, \theta_{sup} \leftarrow \theta_{sup,0}$ according to Xavier

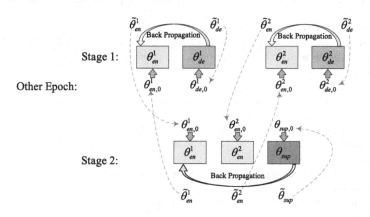

Fig. 2. Parameter updating process

factors in various views. Furthermore, in this network, the objective function of supervised network is to minimize the cross-entropy loss between prediction and true label, which can be formulated as follows:

$$\min \frac{1}{N} \sum_{i=1}^{N} \left(\mathbf{y}_i \log(\hat{\mathbf{y}}_i) + (1 - \mathbf{y}_i) \log(1 - \hat{\mathbf{y}}_i) \right) \tag{3}$$

where $\hat{\mathbf{y}}_i$ is the prediction of i-th instance. As for the choice of activation function and optimizer algorithm, expect that last layer uses Softmax as activation function, others choose ReLu, and AdaDelta is selected to optimize the objective function. It is remarkable that not only we should train the parameters $\theta_{sup} = \{\mathbf{w}_{share}, \mathbf{w}_1, \mathbf{w}_2, \mathbf{w}_3\}$ as shown in Fig. 1, but also need to update the value of $\theta_{en}^v = \{\mathbf{w}_{en1}^v, \mathbf{w}_{en2}^v, \mathbf{w}_{en3}^v\}$, which is inherited from last stage, where $v = 1, \cdots, V$. Same as last stage, we need to save parameter θ_{sup} as well as the updated parameter θ_{en}^v. The two stages we illustrate above is only just one epoch of training procedure, and we will design a co-training process. We define in one epoch, auto-encoder will be trained for R_1 rounds, and the supervised network will be trained for R_2 rounds. In first epoch, we need to initialize θ_{en}^v, θ_{de}^v, and θ_{sup} as $\theta_{en,0}^v$, $\theta_{de,0}^v$, and $\theta_{sup,0}$ according to the method of Glorot [7], and with these initial value we can conduct the first stage and obtain the best model parameters $\tilde{\theta}_{en}^v$ and $\tilde{\theta}_{de}^v$. Next, in stage 2, we use $\tilde{\theta}_{en}^v$ and $\theta_{sup,0}$ to initialize the supervised network and obtain the best value after R_2 rounds' training. Similarly, in other epochs, the training strategy only has minor difference that we do not initialize parameters according to Xavier. Specifically, in first stage, we initialize θ_{en}^v as $\tilde{\theta}_{en}^v$ from stage 2 of last epoch, and initialize θ_{de}^v as $\tilde{\theta}_{de}^v$ from stage 1 of last epoch; in second stage, we initialize θ_{en}^v as $\tilde{\theta}_{en}^v$ from stage 1 of this epoch, and initialize θ_{sup} as $\tilde{\theta}_{sup}^v$ from stage 2 of last epoch. To illustrate this process more clearly, we summarized the whole algorithm of ACMVL in Algorithm 1 and illustrate the parameter updating process of co-training in Fig. 2.

Algorithm 1. ACMVL

Initialize θ_{en}^v, θ_{de}^v and θ_{sup} as $\theta_{en,0}^v$, $\theta_{de,0}^v$, and $\theta_{sup,0}$ according to the method of Xavier;

For each epoch **do**:

  Stage 1:

  For each view $v = 1, \cdots, V$ **do**:

    If not first epoch:

      Initialize θ_{en}^v and θ_{de}^v:

      $\theta_{en}^v : \theta_{en,0}^v \leftarrow \tilde{\theta}_{en}^v$, where $\tilde{\theta}_{en}^v$ comes from last epoch of stage 2;

      $\theta_{de}^v : \theta_{de,0}^v \leftarrow \tilde{\theta}_{de}^v$, where $\tilde{\theta}_{de}^v$ comes from last epoch of stage 1;

    For each R_1 **do**:

      Update θ_{en}^v and θ_{de}^v using AdaDelta and set the best one as $\tilde{\theta}_{en}^v$ and $\tilde{\theta}_{de}^v$ (select the weight of the round with the least reconstruction error);

    End For

  End For

  Stage 2:

  If not first epoch:

    Initialize θ_{en}^v and θ_{sup}^v:

    $\theta_{en}^v : \theta_{en,0}^v \leftarrow \tilde{\theta}_{en}^v$, where $\tilde{\theta}_{en}^v$ comes from this epoch of stage 1;

    $\theta_{sup}^v : \theta_{sup,0}^v \leftarrow \tilde{\theta}_{sup}^v$, where $\tilde{\theta}_{sup}^v$ comes from last epoch of stage 2;

  For each R_2 **do**:

    Update θ_{en}^v and θ_{sup}^v using AdaDelta and set the best one as $\tilde{\theta}_{en}^v$ and $\tilde{\theta}_{sup}^v$ (select the weight of the round with the least reconstruction error);

  End For

End For

2.4 Tricks

As we known, training a neural network needs to determine many hyperparameters and also needs to adopt some tricks. However, this network is not hard to train, when selecting the node number, we only need to remember that the number is decreasing layer by layer for encoder. For example, if our input is a 500-dimensional data, then for encoder like view 1's auto-encoder in Fig. 1, the numbers of node is [256, 64, 32], and for decoder is [64, 256], where 32 is the middle-hidden layer's node number and usually we define these layer's node numbers are the same for multiple views even each view's input dimension is different. As for learning rate of optimizer algorithm, for each view's auto-encoder, learning rate usually sets to 0.5 or 0.3, and for supervised network, learning rate usually equals to 0.9. Note that suitable learning rates will let each auto-encoder and supervised network help with each other to accelerate the convergence speed. Additionally, we further emphasize that we will obtain the joint latent representation by mapping each view's latent feature into a same space with a same transformation matrix $\mathbf{w}_{share}$, because a same transformation matrix can project different data into a same subspace and that is also the reason why we define all views' latent feature as the same dimension. Lastly, one of the most significant tricks is early stopping, because the training loss of a neural network cannot always decrease, and after a period of time, training loss will not decrease any more and even increase. Therefore, early stopping is necessary that it can stable and accelerate the training process. Such as in a training epoch, we set $R_1 = R_2 = 1000$, may in round 400, the loss is lowest but we save the model parameters of round 1000, and we miss the best model parameters and will train more rounds which is a waste of time (because our early stopping rule is that if $R_1, R_2 \geq 200$, and the loss no longer decrease for 200 rounds, we will break the loop and save the parameters belong to the round corresponds to the best loss).

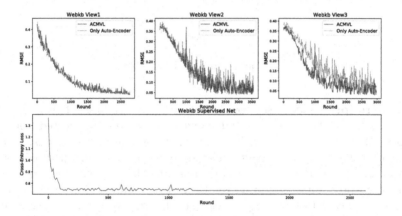

Fig. 3. Convergence analysis of WebKb

3 Experiment

3.1 Data Set Partition

In this subsection, we give a short description of the data sets and introduce the method to divide the data set into different views.

WebKb: The WebKb data set contains web information from computer science departments of four different universities, obviously, we actually have four data sets, but we compute the average value of four data sets. There are three views in each data set: the words in the main text in each web page of one of the universities is a kind of view; the clickable words in the hyperlinks pointing to other web pages of one of the universities is another view; and the words in the titles of each web page is also a view. On the other hand, there are seven categories in this data set, where we choose four most representative categories in this experiment. In general, we have 3 views in this data set.

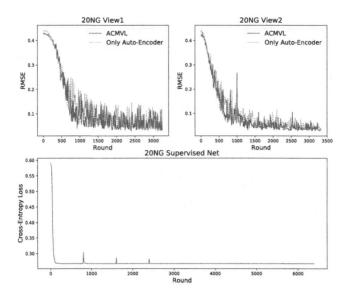

Fig. 4. Convergence analysis of 20NG

20NewsGroup: The data set consists of 20 News group, that is to say, this data set contains 20 categories. 200 documents are randomly selected from each category. As a result, we define 20 tasks corresponding to the classes, and the documents belong to the category related to the task are defined as positive instances, and from other different categories are defined as negative ones. Next, we take the words appearing in all the tasks as a common view, and the words only existing in each task as a special view. In this way, we get 21 views, however, there are only two views in each task, 19 views are missed in each task [8]. Now this is a multi-task and multi-view data, but we can conduct the experiment on each task and compute the average value of them. In general, we have 2 views in this data set.

Leaves: The leaves data set includes leaves from one hundred plant species that are divided into 32 different genera, and 16 samples of leaves for each plant species are presented. 3 geniuses that have 3 or more plant species are selected to form the data set, and the aim of the problem is to discriminate different species in a genus. And in this data set, three views of features are available, including shape descriptor, fine scale margin and texture histogram, and each view has 64 features. In general, we have 3 views in this data set.

3.2 Convergence Analysis of Training Process

In Figs. 3, 4, and 5, the first row of figures shows each views' reconstruction error of auto-encoder, where, red line shows the original auto-encoder's curve, and the blue one is our proposed method's curve. It's not hard to see that in Figs. 2 and 4 let each auto-encoder converges faster than the original auto-encoder, ACML has

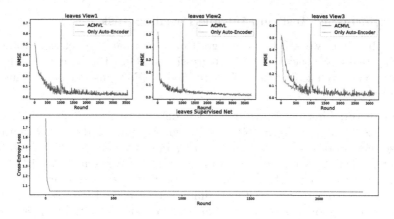

Fig. 5. Convergence analysis of leaves

less fluctuation in training process than original ones. However, we find in Fig. 5 original auto-encoder performs better because Leaves is a small data set that only original auto-encoder can fit it soon, but our proposed method will train the model alternately. Recall that after R_1 rounds in one epoch of training each auto-encoder, we will need to train the supervised network for R_2 rounds and modify each auto-encoder's parameters of encoder, and then in the next epoch, we should retrain each auto-encoder. In such way, when facing small data, the speed of convergence of ACMVL may be slower than the original auto-encoder's. However, when dealing with bigger data set, ACMVL can accelerate convergence speed of each auto-encoder as shown in Figs. 3 and 4.

On the other hand, with the help of each auto-encoder, supervised network converges fast as shown in the second row of Figs. 3, 4, and 5. As a result, in ACMVL's framework, each view's auto-encoder and supervised network help with each other according to the co-training rules.

Table 1. Classification task

Classification		LR		LR-AE		LR-AE-ACMVL		LR-ACMVL	
		ACC	F1	ACC	F1	ACC	F1	ACC	F1
WebKb	View 1	0.8230	0.7281	0.7876	0.6217	0.8142	0.7330	**0.9115**	**0.8703**
	View 2	0.8673	0.7866	0.8142	0.6991	0.7434	0.6667		
	View 3	0.7080	0.6477	0.7168	0.6031	0.7256	0.5839		
20NG	View 1	0.6267	0.6256	0.5333	0.5326	0.5200	0.5169	**0.9733**	**0.9732**
	View 2	0.9666	0.9667	0.8600	0.8592	0.9200	0.9200		
Leaves	View 1	1.000	1.000	1.000	1.000	1.000	1.000	**1.000**	**1.000**
	View 2	0.7708	0.7481	0.7917	0.7680	0.7708	0.7514		
	View 3	0.9167	0.9132	0.8750	0.8713	0.9375	0.9325		

3.3   Performance Test on Multiple Learning Task

In this subsection, we will test the algorithm performance on classification and clustering tasks. We select Logistic Regression (LR) as the baseline method of classification as well as Gaussian Mixture Model (GMM) as the baseline method of clustering.

Classification: First of all, we divide the data of each view into training set and testing set at a ratio of 50%. And then, we first conduct an experiment only use these data with the classifier of LR, and the result is list in the column "LR" of Table 1; Second, we only train each view's auto-encoder without the help of supervised network, and then we use the feature computed by auto-encoder which corresponds to the training set to train LR classifier and use the feature corresponding to the testing set to test the result. The result is list in the column "LR-AE" of Table 1; Thirdly, we will do the same thing as the second experiment but let supervised network helps each auto-encoder's training process, and the result is list in the column "LR-AE-ACMVL" of Table 1. Note that this experiment use feature computed by each view's auto-encoder to test each view's performance; Lastly, we conduct the third experiment again, but we use the joint latent feature computed by supervised network to test the performance, and therefore, this experiment only has one view. The result is list in the column "LR-ACMVL" of Table 1.

In classification experiment, we select Accuracy (ACC) and F1 score (F1) as the metrics. We can easily find that when using ACMVL, most of auto-encoders' performance get better and the joint latent feature's performance is much better than each view's, which verify ACMVL is an effective approach to compute joint latent feature of multiple views.

Table 2. Clustering task

Clustering		GMM		GMM-AE		GMM-AE-ACMVL		GMM-ACMVL	
		NMI	JC	NMI	JC	NMI	JC	NMI	JC
WebKb	View 1	0.0935	0.1947	0.1539	0.1460	0.3004	0.2088	**0.5809**	**0.0310**
	View 2	0.4739	0.1593	0.1221	0.0752	0.3752	0.0885		
	View 3	0.0843	0.1947	0.0433	0.1858	0.0752	0.1549		
20NG	View 1	0.0195	0.5167	0.0057	0.3567	0.0110	**0.3500**	0.3171	0.5100
	View 2	0.1413	0.5133	0.1968	0.4533	**0.4021**	0.4133		
Leaves	View 1	0.8567	0.2917	0.8461	0.3333	0.8642	0.1250	**0.9781**	0.1667
	View 2	0.7215	0.3021	0.3240	0.0938	0.7999	**0.0104**		
	View 3	0.7620	0.1042	0.8555	0.2083	0.8024	0.0208		

Clustering: We will conduct some semi-clustering experiment, first of all, we divide the data of each view into training set and testing set at a ratio of 50%. And then, we first conduct an experiment only use testing data with the clustering algorithm GMM, and the result is list in the column "GMM" of Table 2;

Second, we only use training data to train each view's auto-encoder without the help of supervised network, and then we use the model to compute testing data's feature representation and conduct clustering task on these feature representations. The result is list in the column "GMM-AE" of Table 2; Thirdly, we will do the same thing as the second experiment but let supervised network helps each auto-encoder's training process, and the result is list in the column "GMM-AE-ACMVL" of Table 2. Note that this experiment use feature computed by each view's auto-encoder to test each view's performance; Lastly, we conduct the third experiment again, but we use the joint latent feature computed by supervised network to test the performance, and therefore, this experiment only has one view. The result is list in the column "GMM-ACMVL" of Table 2.

In clustering experiment, we select Normalized Mutual Information (NMI) and Jaccard Coefficient (JC, the smaller of JC, the performance of clustering is better) as the metrics. We can easily find that when using ACMVL, most of auto-encoders' performance get better and the joint latent feature's performance is much better than each view's, which verify ACMVL is an effective approach to compute joint latent feature of multiple views.

4   Conclusion

In this paper, we propose a novel multi-view learning algorithm called Auto-Encoder based Co-Training Multi-View Representation Learning (ACMVL), which is aimed to subspace learning and model training strategy. We utilize the latent feature learning ability of auto-encoder to grasp the complementarity of multiple views, and at the same time, by using weight sharing we can map each view's latent representation in to a same space and learn the consistency of multiple views. Besides, we adopt co-training strategy to accelerate the training procedure of each view's auto-encoder by co-training and model parameters partially shared. And according to experimental results, we find that our proposed method can learn a suitable joint latent representation which is competent to classification and clustering learning tasks.

Our proposed method in this paper is a deterministic model which cannot measure the uncertainty of latent space. Therefore, in the future, a main target is to find a generative method to obtain the distribution of the joint latent space instead of an instantiation of the space. To our knowledge, variational auto-encoder [9] may be a good choice to solve this problem. Generally, multi-view subspace learning is a great research direction which is hard but deserved to pay more attention on it, and we will make more attempts and explorations in this field.

References

1. Bickel, S., Scheffer, T.: Multi-view clustering. In: Proceedings of the 4th IEEE International Conference on Data Mining (ICDM 2004), Brighton, UK, 1–4 November 2004, pp. 19–26 (2004)

2. Blum, A., Mitchell, T.M.: Combining labeled and unlabeled data with co-training. In: Proceedings of the Eleventh Annual Conference on Computational Learning Theory, COLT 1998, Madison, Wisconsin, USA, 24–26 July 1998, pp. 92–100 (1998)
3. Brefeld, U., Büscher, C., Scheffer, T.: Multi-view discriminative sequential learning. In: Gama, J., Camacho, R., Brazdil, P.B., Jorge, A.M., Torgo, L. (eds.) ECML 2005. LNCS (LNAI), vol. 3720, pp. 60–71. Springer, Heidelberg (2005). https://doi.org/10.1007/11564096_11
4. Brefeld, U., Scheffer, T.: Co-EM support vector learning. In: Proceedings of the Twenty-First International Conference on Machine Learning, (ICML 2004), Banff, Alberta, Canada, 4–8 July 2004 (2004)
5. Chaudhuri, K., Kakade, S.M., Livescu, K., Sridharan, K.: Multi-view clustering via canonical correlation analysis. In: Proceedings of the 26th Annual International Conference on Machine Learning, ICML 2009, Montreal, Quebec, Canada, 14–18 June 2009, pp. 129–136 (2009)
6. Chen, M., Weinberger, K.Q., Chen, Y.: Automatic feature decomposition for single view co-training. In: Proceedings of the 28th International Conference on Machine Learning, ICML 2011, Bellevue, Washington, USA, 28 June–2 July 2011, pp. 953–960 (2011)
7. Glorot, X., Bengio, Y.: Understanding the difficulty of training deep feedforward neural networks. In: Proceedings of the Thirteenth International Conference on Artificial Intelligence and Statistics, AISTATS 2010, Chia Laguna Resort, Sardinia, Italy, 13–15 May 2010, pp. 249–256 (2010)
8. Jin, X., Zhuang, F., Wang, S., He, Q., Shi, Z.: Shared structure learning for multiple tasks with multiple views. In: Blockeel, H., Kersting, K., Nijssen, S., Železný, F. (eds.) ECML PKDD 2013. LNCS (LNAI), vol. 8189, pp. 353–368. Springer, Heidelberg (2013). https://doi.org/10.1007/978-3-642-40991-2_23
9. Kingma, D.P., Welling, M.: Auto-encoding variational bayes. CoRR abs/1312.6114 (2013)
10. Kursun, O., Alpaydin, E.: Canonical correlation analysis for multiview semisupervised feature extraction. In: Rutkowski, L., Scherer, R., Tadeusiewicz, R., Zadeh, L.A., Zurada, J.M. (eds.) ICAISC 2010. LNCS (LNAI), vol. 6113, pp. 430–436. Springer, Heidelberg (2010). https://doi.org/10.1007/978-3-642-13208-7_54
11. Liu, J., Jiang, Y., Li, Z., Zhou, Z., Lu, H.: Partially shared latent factor learning with multiview data. IEEE Trans. Neural Netw. Learn. Syst. **26**(6), 1233–1246 (2015)
12. Ou, W., Long, F., Tan, Y., Yu, S., Wang, P.: Co-regularized multiview nonnegative matrix factorization with correlation constraint for representation learning. Multimed. Tools Appl. **77**(10), 12955–12978 (2018)
13. Wang, Z., Chen, S., Gao, D.: A novel multi-view learning developed from single-view patterns. Pattern Recogn. **44**(10–11), 2395–2413 (2011)
14. Xu, C., Tao, D., Xu, C.: A survey on multi-view learning. CoRR abs/1304.5634 (2013)
15. Zeiler, M.D.: ADADELTA: an adaptive learning rate method. CoRR abs/1212.5701 (2012)
16. Zhang, Z., Qin, Z., Li, P., Yang, Q., Shao, J.: Multi-view discriminative learning via joint non-negative matrix factorization. In: Pei, J., Manolopoulos, Y., Sadiq, S., Li, J. (eds.) DASFAA 2018. LNCS, vol. 10828, pp. 542–557. Springer, Cham (2018). https://doi.org/10.1007/978-3-319-91458-9_33

Robust Semi-supervised Representation Learning for Graph-Structured Data

Lan-Zhe Guo, Tao Han, and Yu-Feng Li[(✉)]

National Key Laboratory for Novel Software Technology, Nanjing University,
Nanjing 210023, China
{guolz,hant,liyf}@lamda.nju.edu.cn

Abstract. The success of machine learning algorithms generally depends on data representation and recently many representation learning methods have been proposed. However, learning a good representation may not always benefit the classification tasks. It sometimes even hurt the performance as the learned representation maybe not related to the ultimate tasks, especially when the labeled examples are few to afford a reliable model selection. In this paper, we propose a novel robust semi-supervised graph representation learning method based on graph convolutional network. To make the learned representation more related to the ultimate classification task, we propose to extend label information based on the smooth assumption and obtain pseudo-labels for unlabeled nodes. Moreover, to make the model robust with noise in the pseudo-label, we propose to apply a large margin classifier to the learned representation. Influenced by the pseudo-label and the large-margin principle, the learned representation can not only exploit the label information encoded in the graph-structure sufficiently but also can produce a more rigorous decision boundary. Experiments demonstrate the superior performance of the proposal over many related methods.

Keywords: Robust · Representation learning ·
Semi-supervised learning · Graph convolutional network

1 Introduction

The performance of machine learning methods is heavily dependent on the choice of data representation (or features). Representation learning, i.e., learning representations of the data that needed for the learning classifiers, has already become an important field in machine learning [2].

One challenge of representation learning is that it faces a paradox between preserving as much information about the input as possible, and attaining nice properties for the output learning task [5]. Recently, there are many researches pointed out that, the representation learning may fail to improve the performance of classification task [2,5]. The main reason is that the learned representation may be far from the ultimate learning task. For example, representation

© Springer Nature Switzerland AG 2019
Q. Yang et al. (Eds.): PAKDD 2019, LNAI 11441, pp. 131–143, 2019.
https://doi.org/10.1007/978-3-030-16142-2_11

learning typically pursuits the whole factors of the raw data, while the ultimate task may only be related to a small subset of these factors. For this reason, it is essential to learn a robust representation, especially when the labeled examples are few to afford a reliable model selection.

In this paper, we focus on learning a robust representation for semi-supervised graph-structured data. It is widely accepted that graph-structured data occurs in numerous application domains, such as social networks [14], citation networks [9] and many others [7]. Learning an appropriate vector representation of nodes in graphs has proved extremely useful for a wide variety of predictive and graph analysis tasks [6,14,16]. Figure 1 illustrates a visualization of a classical graph-structured data and the corresponding node embeddings. A number of graph representation learning methods such as DeepWalk [14], LINE [16], have been proposed recently. However, these methods require a multi-step pipeline where the representation learning model and the classifier are trained separately. In other words, the learned representation may be far from the classification task, and thus hurt the performance.

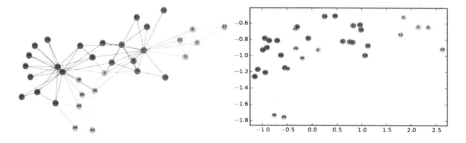

Fig. 1. Graph structure of the Zachary Karate Club social network (left) and the two dimensional visualization of node embeddings (right).

Most recently, Graph Convolutional Network (GCN) [9] is proposed to fill the gap. Unlike previous studies where the learned representation and the trained classifier are conducted separately, GCN jointly optimizes the representation learning model and the ultimate classifier. Nevertheless, most ultimate classifiers in GCN work under the labeled data, which is insufficient to learn a robust representation in semi-supervised learning.

In this paper we propose to obtain high-confidence pseudo-labels for unlabeled nodes from the well-known label propagation strategy to enhance the label capacity. Our basic idea is that given graph-structured data, many label information are encoded in the graph structure based on the smooth assumption [22], i.e., connected nodes are likely to share similar labels. We further propose a large-margin classifier to overcome the noise pseudo-labels induced from label propagation. Figure 2 shows the pipeline of the proposal.

In conclusion, we make several noteworthy contributions as follows:

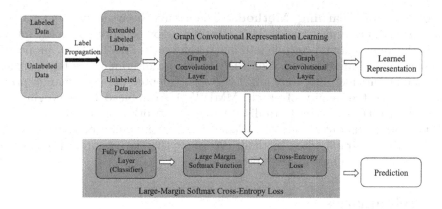

Fig. 2. The structure of the proposed RoGraph.

- We propose a robust representation learning method RoGraph for semi-supervised graph-structured data, with the idea of the classical label propagation and large margin principle, which is very easy to implement.
- Experiments on real-world network datasets are conducted. The experimental results demonstrate that RoGraph achieves clearly better results than many related methods.

The rest of this paper is organized as follows. We first introduce related works and present preliminaries with an introduction to GCN. Next, we present the proposed RoGraph, and then show the experimental results and discuss why the large-margin principle can benefit graph representation learning. Finally, we conclude this paper.

2   Related Work

The proposed algorithm is conceptually related to semi-supervised graph representation learning and large-margin learning methods.

Semi-supervised Graph Representation Learning. Inspired by the Skip-Gram [13], many semi-supervised learning methods for graph-structured data have been proposed in recent years. DeepWalk [14] learns embeddings via the prediction of the local neighborhood of nodes, sampled from random walks on the graph. There are also many works based on DeepWalk, such as LINE [16] extends DeepWalk with more sophisticated random walk and node2vec [6] extends Deep-Walk with breadth-first search schemes. For all these methods, however, a multi-step pipeline including random walk generation and semi-supervised training is required where each step has to be optimized separately, so the learned representation may not the best representation for the classification task. Planetoid [20] alleviated this by injecting label information in the process of learning embeddings. GCN [9] generalize traditional convolutional networks to graph-structured data and can learn representations end-to-end.

Large-Margin Learning Methods. There have been many large-margin learning methods in many fields. Max-margin markov network [17] firstly introduced the large-margin principle into markov networks. MedLDA [21] proposed a maximum entropy discrimination LDA to learn a discriminative topic model (e.g., latent Dirichlet allocation [3]). LEAD [10] adopted the large-margin principle to judge the quality of graph. MMDW [18] combined large-margin loss function with DeepWalk to learn discriminative network representations.

It is notable that to our best knowledge, the large-margin principle has rarely been applied to graph-based methods. We show that large-margin principle is indeed helpful for robust graph representation.

3   Preliminaries

In this section, we introduce some notations used in our method and give a brief introduction to the idea of graph convolutional networks.

3.1   Notations

We consider the problem of representation learning on a graph $\mathcal{G} = (\mathcal{V}, \mathcal{E})$, where $\mathcal{V}$ is the node set and $\mathcal{E}$ is the edge set. The given information includes a feature matrix $\mathbf{X} \in \mathbb{R}^{N \times M}$ which x_i is a feature description for every node i, N is the number of nodes and M is the dimension of input features; an adjacency matrix $\mathbf{A} = A_{ij} \in \mathbb{R}^{N \times N}$, where $A_{ij} = A_{ji} = 1$ if node i and node j has a link, otherwise $A_{ij} = A_{ji} = 0$; a labeling matrix $\mathbf{Y} \in \mathbb{R}^{N \times K}$ with K being the number of classes. In the setting of semi-supervised learning, we have set $\mathcal{Y}_L$ which includes all labeled nodes and set $\mathcal{Y}_U$ which includes all unlabeled nodes. The size of $\mathcal{Y}_L$ is much smaller than the size of $\mathcal{Y}_U$. The learned representation is matrix $\bar{\mathbf{X}} \in \mathbb{R}^{N \times F}$, where F is the dimension of output feature per node. The prediction is matrix $\mathbf{Z} \in \mathbb{R}^{N \times K}$, where Z_{ij} indicates the probability that node i belongs to class j.

3.2   Graph Convolutional Networks

Graph convolutional network [9] generalizes the convolutional network into graph-structured data and proposes an efficient layer-wise propagation rule. The traditional GCN model contains the following components:

(1) Renormalization: Adding an self-loop to each node, which results in a new adjacency matrix $\tilde{\mathbf{A}} = \mathbf{A} + \mathbf{I}$ where $\mathbf{I}$ is the identity matrix and the new degree matrix $\tilde{\mathbf{D}}$ with $\tilde{D}_{ii} = \sum_j \tilde{A}_{ij}$. After that, symmetrically normalize $\tilde{\mathbf{A}}$ and obtain $\tilde{\mathbf{A}}_s = \tilde{\mathbf{D}}^{-\frac{1}{2}} \tilde{\mathbf{A}} \tilde{\mathbf{D}}^{-\frac{1}{2}}$.

(2) Graph Convolutional layer: The graph convolutional layer uses the propagation rule:

$$\mathbf{H}^{(l+1)} = \sigma(\tilde{\mathbf{A}}_s \mathbf{H}^{(l)} \mathbf{W}^{(l)}) \tag{1}$$

where $\mathbf{H}^{(l)}$ is the matrix of activations in the l-th layer and $\mathbf{H}^{(0)} = \mathbf{X}$, $\mathbf{H}^{(L)} = \bar{\mathbf{X}}$ with L is the number of layers in the network, $\mathbf{W}^{(l)}$ is a layer specific trainable weight matrix in layer l, and $\sigma(\cdot)$ denotes an activation function, such as the ReLU$(\cdot) = \max(0, \cdot)$.

(3) Softmax cross-entropy loss: Applying a fully connected layer as the classifier to the learned representation $\bar{\mathbf{X}}$: $\mathbf{Z} = \bar{\mathbf{X}}\mathbf{W}^L$, where $\mathbf{W}^L \in \mathbb{R}^{F \times K}$ is a trainable weight matrix of the fully connected layer. Then evaluates the softmax cross-entropy loss over labeled nodes. The loss can be written as:

$$\mathcal{L} = \sum_{i \in \mathcal{Y}_L} -\ln\left(\frac{e^{Z_{iy_i}}}{\sum_j e^{Z_{ij}}}\right) \tag{2}$$

where $\mathcal{Y}_L$ is the set of labeled nodes and y_i is the label of the i-th node.

4 Our Proposed Method

In semi-supervised learning, the number of labeled nodes is usually limit to provide a reliable model selection, thus, the learned representation using only the labeled data may be not robust for the ultimate classification task.

Observed that in graph-structured data connected nodes are likely to share the same label, we propose to assign a high-confidence pseudo label for some unlabeled nodes according to this property so that we can exploit more label information encoded in the graph structure and enhance labeled data. Moreover, the large-margin principle is often used to train a robust classifier, thus, we propose to adopt the large-margin softmax cross-entropy loss function [12] instead of the original softmax function to help decrease the impact of noise in the pseudo-label and produce even more robust representation.

4.1 Enhance Labeled Data

The underlying assumption in graph-based semi-supervised learning is the smooth assumption, i.e., connected nodes likely to share the same label. With this assumption, we can exploit more label information using the graph structure information and produce a classification task related representation.

A simple method to mine the label information encoded in the graph structure for unlabeled nodes is label propagation [22]. The label propagation method only takes the graph matrix $\mathbf{A}$ and the labeling matrix $\mathbf{Y}$ as input and the objective is to find a prediction matrix $\hat{\mathbf{Y}} \in \mathbb{R}^{N \times K}$ of the same size as the labeling matrix $\mathbf{Y}$ by minimizing both fitting error and smooth regularization:

$$\hat{\mathbf{Y}} = \arg\min_{\hat{\mathbf{Y}}} \mathcal{C}(\hat{\mathbf{Y}}) \tag{3}$$

$$= \arg\min_{\hat{\mathbf{Y}}} \{ \underbrace{||\hat{\mathbf{Y}} - \mathbf{Y}||_2^2}_{\text{fitting error}} + \underbrace{\alpha \operatorname{tr}(\hat{\mathbf{Y}}^\top \mathbf{L}\hat{\mathbf{Y}})}_{\text{regularization}} \}$$

$$= \arg\min_{\hat{\mathbf{Y}}} \{ \sum_{i \in \mathcal{Y}_L} (\hat{\mathbf{Y}}_i - \mathbf{Y}_i)^2 + \alpha \sum_{i,j}^{n} A_{ij}(\hat{\mathbf{Y}}_i - \hat{\mathbf{Y}}_j)^2 \}$$

where $\mathbf{L}$ is the graph laplacian matrix.

In Eq. (3), the fitting error term enforces the prediction matrix $\hat{\mathbf{Y}}$ to agree with the label matrix $\mathbf{Y}$, while the smooth regularization term enforces each column of $\hat{\mathbf{Y}}$ to be smooth along the edges. The scalar α is a balancing parameter.

A closed-form solution of the unconstrained quadratic optimization problem can be obtained by setting the derivative of the objective function to zero:

$$\hat{\mathbf{Y}} = (\mathbf{I} + \alpha\mathbf{L})^{-1}\mathbf{Y} \tag{4}$$

For small-scale data, we can simply use Eq. (4) to get the prediction of unlabeled nodes. However, for large-scale data, Eq. (4) is time consuming because it needs to compute the inverse of the matrix $\mathbf{I} + \alpha\mathbf{L}$. To address this problem, we use Stochastic Gradient Descent (SGD) to solve Eq. (3).

Let

$$\mathcal{C}_i(\hat{\mathbf{Y}}) = \alpha \sum_{j=1}^{n} A_{ij}(\hat{\mathbf{Y}}_i - \hat{\mathbf{Y}}_j)^2 + \mathbb{I}(i) \cdot (\hat{\mathbf{Y}}_i - \mathbf{Y}_i)^2 \tag{5}$$

where $\mathbb{I}(i) = 1$ if $i \in \mathcal{Y}_L$, otherwise, $\mathbb{I}(i) = 0$, $i \in \{1, 2, \cdots, N\}$ represents the index of the chosen instance. It is easy to verify that $\mathbb{E}[\nabla\mathcal{C}_i(\hat{\mathbf{Y}})] = \frac{1}{n}\nabla\mathcal{C}(\hat{\mathbf{Y}})$.

Therefore $\nabla\mathcal{C}_i(\hat{\mathbf{Y}})$ is an unbiased estimator of $\frac{1}{n}\nabla\mathcal{C}(\hat{\mathbf{Y}})$ where $\frac{1}{n}$ is a constant given a graph. Hence, we can adopt SGD strategy to solve $\hat{\mathbf{Y}}$ by updating:

$$\hat{\mathbf{Y}}^{(t+1)} = \hat{\mathbf{Y}}^{(t)} - \eta\nabla\mathcal{C}_i(\hat{\mathbf{Y}})$$

where the gradient $\nabla\mathcal{C}_i(\hat{\mathbf{Y}}) = \alpha\mathbf{A}_i^\top(\hat{\mathbf{Y}}_i - \hat{\mathbf{Y}}_j) + \mathbb{I}(i) \cdot (\hat{\mathbf{Y}}_i - \mathbf{Y}_i)$. We adopt the interesting stochastic label propagation method [11] to derive the gradient efficiently. The element $\hat{Y}_{ij}$ in the learned prediction matrix $\hat{\mathbf{Y}}$ indicates the probability of node i belongs to class j. For an unlabeled node i, if the maximum Y_{ij} in $\hat{\mathbf{Y}}_i$ is greater than a threshold, we think the node i has a high confidence in class j and add node i to the labeled node set $\mathcal{Y}_L$. After this process, we derive a larger labeled node set $\tilde{\mathcal{Y}}_L$ to learn a better representation.

4.2   Large-Margin Cross-Entropy Loss

Obviously the pseudo-label derived by the label propagation may consist of noise. To make the learned representation robust, it is necessary to overcome the affect caused by noise. Intuitively, if the decision boundary has a large margin to the nearest training data point, the model turns out to be a robust classifier according to margin theory. An additional benefit is that the large margin principle works as a regularizer and thus help avoid overfitting issues, which is particularly useful when labeled data is limited. Thus, to learn a robust representation [12], we use a generalization of the original softmax cross-entropy loss function termed Large-Margin Cross-Entropy Loss.

Observed in GCN that $Z_{ij} = \bar{\mathbf{X}}_i \mathbf{W}_j^L$. This can be reformulated as: $Z_{ij} = \|\mathbf{W}_j^L\| \, \|\bar{\mathbf{X}}_i\| \cos(\theta_j)$ where θ_j is the angle between the vector $\mathbf{W}_j^L$ and $\bar{\mathbf{X}}_i$. Thus the original softmax cross-entropy loss function can be rewritten as:

$$L_i = -\ln\left(\frac{e^{\|\mathbf{W}_{y_i}^L\| \, \|\bar{\mathbf{X}}_i\| \cos(\theta_{y_i})}}{\sum_j e^{\|\mathbf{W}_j^L\| \, \|\bar{\mathbf{X}}_i\| \cos(\theta_j)}}\right) \tag{6}$$

In ROGRAPH, we propose to use large-margin softmax instead of the original softmax. For example, once an instance x with the label $+1$, the original softmax is to force $\mathbf{W}_1^\top \mathbf{x} > \mathbf{W}_2^\top \mathbf{x}$, i.e., $\|\mathbf{W}_1\| \, \|\mathbf{x}\| \cos(\theta_1) > \|\mathbf{W}_2\| \, \|\mathbf{x}\| \cos(\theta_2)$, in order to classify x correctly. In contrast, the large-margin softmax want to make the classification more rigorous in order to produce a large-margin decision boundary. Thus, the large-margin softmax requires $\|\mathbf{W}_1\| \, \|\mathbf{x}\| \cos(m\theta_1) > \|\mathbf{W}_2\| \, \|\mathbf{x}\| \cos(\theta_2)(0 \le \theta_1 \le \frac{\pi}{m})$ where m is a positive integer.

Observed that the following inequality always holds:

$$\|\mathbf{W}_1\| \, \|\mathbf{x}\| \cos(\theta_1) \ge \|\mathbf{W}_1\| \, \|\mathbf{x}\| \cos(m\theta_1) > \|\mathbf{W}_2\| \, \|\mathbf{x}\| \cos(\theta_2) \tag{7}$$

Therefore, we have $\|\mathbf{W}_1\| \, \|\mathbf{x}\| \cos(\theta_1) > \|\mathbf{W}_2\| \, \|\mathbf{x}\| \cos(\theta_2)$. Therefore, the new classification criteria correctly classifies x, and produces a more rigorous decision boundary. Figure 3 illustrates a geometric interpretation for the advantage of the large-margin softmax function [12].

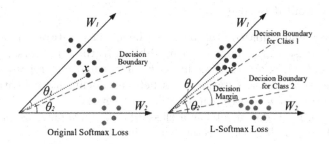

Fig. 3. Illustrative geometric interpretation. The left (right) presents the original softmax (large-margin softmax).

Formally, the large-margin softmax cross-entropy loss is defined as:

$$L_i = -\ln\left(\frac{e^{\|\mathbf{W}_{y_i}^L\| \, \|\bar{\mathbf{X}}_i\| \, \phi(\theta_{y_i})}}{e^{\|\mathbf{W}_{y_i}^L\| \, \|\bar{\mathbf{X}}_i\| \, \phi(\theta_{y_i})} + \sum_{j \ne y_i} e^{\|\mathbf{W}_j^L\| \, \|\bar{\mathbf{X}}_i\| \, \phi(\theta_j)}}\right) \tag{8}$$

where

$$\phi(\theta) = \begin{cases} \cos(m\theta), & 0 \le \theta \le \frac{\pi}{m} \\ \mathcal{D}(\theta), & \frac{\pi}{m} \le \theta \le \pi \end{cases}$$

and $\mathcal{D}(\theta)$ is a monotonically decreasing function, $\mathcal{D}(\frac{\pi}{m}) = \cos(\frac{\pi}{m})$.

According to the work in [12], we let $\phi(\theta) = (-1)^k \cos(m\theta) - 2k, \theta \in [\frac{k\pi}{m}, \frac{(k+1)\pi}{m}]$ where $k \in [0, m-1]$. The m is related to the classification margin. The larger the value m is, the larger the classification margin becomes. Then, we have the final loss function

$$\text{L-Softmax}(\bar{\mathbf{X}}, \mathbf{W}^L) = \sum_{i \in \tilde{\mathcal{Y}}_L} L_i \tag{9}$$

where $\tilde{\mathcal{Y}}_L$ is the set of extended labeled nodes after label propagation.

In short summary, the proposed RoGraph includes these steps: we first propagate the label information from the labeled nodes to unlabeled nodes using the graph structure. Then, we renormalize the adjacency matrix and use graph convolutional layers to produce the representation of each node. Finally, we add a fully connected layer to the learned representation as the classifier and adopt the large-margin cross-entropy loss to derive the final representation.

5   Experiments

In this section, we evaluate the proposed RoGraph in benchmark network datasets and show the effectiveness of our proposal. Besides, we give the loss vs. epoch in both training set and validation set.

5.1   Experimental Setup

Cora, CiteSeer and PubMed [15] are three benchmark network datasets. The statistics of datasets are summarized in Table 1. In these networks, nodes are documents and edges are citation links. Each document is represented by a sparse 0/1 feature vector. Citation links between documents constitute a 0/1 undirected graph. If v_i cites v_j or vice versa, then $A_{ij} = A_{ji} = 1$, otherwise $A_{ij} = A_{ji} = 0$. Each document has a class label. For training, we only use 20 labels per class and all feature vectors for each dataset.

Table 1. The statistics of experimental network datasets.

Dataset	Nodes	Edges	Classes	Features
CiteSeer	3,327	4,732	6	3,703
Cora	2,708	5,429	7	1,433
PubMed	19,717	44,338	3	500

For the used datasets, we trained RoGraph with two graph convolutional layers and a fully connected layer and evaluate the prediction accuracy on a test set of 1,000 labeled examples. We train our models on all three datasets for a maximum of 200 epochs using Adam [8] with a learning rate 0.01, 0.5 dropout rate, 5×10^{-4} weight decay rate and early stopping with a window size of 10. We use a hidden layer of 16 units and we initialize weights using the Xavier initialization [4]. The threshold of assigning a pseudo-label to unlabeled nodes is set to 0.8 for all the experiments. The parameter m is fixed to 2 for the large-margin softmax loss on all the experiments.

5.2 Compared Results

We compared the proposed RoGraph with many state-of-the-art methods [9], including label propagation (LP) [22], semi-supervised embedding (SemiEmb) [19], manifold regularization (ManiReg) [1], skip-gram based graph embeddings (DeepWalk) [14], iterative classification algorithm (ICA) [15] and Planetoid [20].

We further compare against with conventional GCN. For GCN, the hyperparameters are same with RoGraph. To validate the effectiveness of the two proposed technologies separately, we also compare with two variants of RoGraph, i.e., RoGraph-P (RoGraph with pseudo-label only) and RoGraph-M (RoGraph with L-Softmax only).

The experimental results are summarized in Table 2. For GCN, RoGraph-P, RoGraph-M, RoGraph, we reported the mean accuracy over five random splits. Results for all other methods are taken from [9].

From Table 2, we can see that, on all the datasets, RoGraph achieves a clear performance gain over the GCN method. It demonstrates the effectiveness of the proposed RoGraph. In addition, RoGraph-P and RoGraph-M also achieve better results than GCN but are not as good as RoGraph. It indicates that the two introduced technologies (label propagation and large-margin principle) are both useful to robust representation.

Table 2. Summary of results in terms of classification accuracy.

Method	Citeseer	Cora	Pubmed
MainReg	60.1	59.5	70.7
SemiEmb	59.6	59.0	71.1
LP	45.3	68.0	63.0
DeepWalk	43.2	67.2	65.3
ICA	69.1	75.1	73.9
Planetoid	64.7	75.7	77.2
GCN	69.6	80.8	77.8
RoGraph-P	72.4	82.4	77.8
RoGraph-M	71.4	83.2	78.1
RoGraph	**73.1**	**83.8**	**79.1**

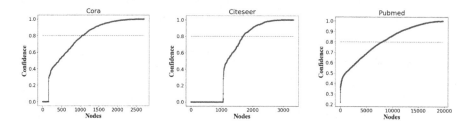

Fig. 4. The confidence of assigning a pseudo-label to unlabeled nodes.

5.3   The Confidence of Assigning Pseudo-Label

In this section, we show the confidence of assigning a pseudo-label to unlabeled nodes after label propagation. For each unlabeled node i, The confidence of assigning a pseudo-label is the maximum number in $\hat{Y}_i$ after row-normalization. The results for all three datasets are shown in Fig. 4. From Fig. 4, we can see that even we set the threshold as 0.8, we can still give a pseudo-label to about half of the unlabeled nodes. This demonstrates that the label-propagation process does help us exploit more label information encoded in the graph structure.

5.4   Loss vs. Epoch

Figure 5 illustrates the relationship between the loss and the epoch on Cora dataset (On the other two datasets, we achieve similar results). One can see that the proposed RoGraph not only achieves the lowest loss in both training set and validation set but also needs fewer epochs to converge than traditional GCN, though we adopt a harder loss function. This is consistent with the numerical results in Table 2 and also verifies the effectiveness of the proposal.

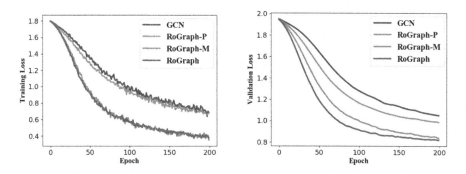

Fig. 5. Loss vs. epoch on Cora. The left (right) presents training (validation) loss.

6   Discussion

Graph representation learning has attracted significant attention in recent years. Meantime, the large-margin principle is also widely used to train a robust classifier and help avoid overfitting issues. However, to our best knowledge, the large-margin principle has rarely been applied to graph representation learning. In this paper, we successfully fill this blank by applying the large-margin principle to GCN and show the effectiveness through empirical results. It demonstrates that the large-margin principle can work well with graph-based methods.

We think that one key reason for why the large-margin principle can work well with graph-based methods is that, the underlying assumption for the large-margin principle (large-margin assumption) and graph-based methods (manifold assumption or smooth assumption) are kind of complementary. Specifically, the manifold assumption requires the learned representation of nodes in the same classes are similar to each other. It emphasizes that the data closeness within the same classes, whereas ignores the data separability between different classes. By contrast, the large-margin assumption requires the learned representation of nodes between different classes have a large margin. It emphasizes the data separability between different classes but ignores the data closeness. Therefore, by taking the two assumptions into account simultaneously, one can encourage both the inter-class separability and intra-class compactness between learned representations for graphs and leads to a better decision boundary. It is innovative for the algorithm design of graph representation learning.

7   Conclusion

We have introduced a novel and easy-to-implement approach RoGraph for robust semi-supervised representation learning on graph-structured data. RoGraph leverages label propagation to obtain high-confidence pseudo-label for unlabeled nodes, which can exploit label information encoded in the graph structure sufficiently. Besides, we adopt the large-margin softmax cross-entropy loss function instead of the traditional softmax function to produce a more rigorous decision boundary. Both of these technologies can help produce a robust representation. Experiments on a number of benchmark network datasets suggest that the proposed RoGraph achieves better results than many state-of-the-art methods. In future, we will extend the proposed strategy to edge representation rather than only node representation in this work.

Acknowledgments. This research was supported by the National Key R&D Program of China (2017YFB1001903) and the National Natural Science Foundation of China (61772262).

References

1. Belkin, M., Niyogi, P., Sindhwani, V.: Manifold regularization: a geometric framework for learning from labeled and unlabeled examples. J. Mach. Learn. Res. **7**, 2399–2434 (2006)
2. Bengio, Y., Courville, A., Vincent, P.: Representation learning: a review and new perspectives. IEEE Trans. Pattern Anal. Mach. Intell. **35**(8), 1798–1828 (2013)
3. Blei, D.M., Ng, A.Y., Jordan, M.I.: Latent dirichlet allocation. J. Mach. Learn. Res. **3**, 993–1022 (2003)
4. Glorot, X., Bengio, Y.: Understanding the difficulty of training deep feedforward neural networks. In: Proceedings of the 13th International Conference on Artificial Intelligence and Statistics, pp. 249–256 (2010)
5. Goodfellow, I., Bengio, Y., Courville, A., Bengio, Y.: Deep Learning. MIT Press, Cambridge (2016)
6. Grover, A., Leskovec, J.: Node2vec: scalable feature learning for networks. In: Proceedings of the 22nd ACM SIGKDD International Conference on Knowledge Discovery and Data Mining, pp. 855–864 (2016)
7. Hamilton, W.L., Ying, R., Leskovec, J.: Representation learning on graphs: methods and applications. IEEE Data Eng. Bull. 52–74 (2017)
8. Kingma, D.P., Ba, J.: Adam: a method for stochastic optimization. In: International Conference on Learning Representations (2015)
9. Kipf, T.N., Welling, M.: Semi-supervised classification with graph convolutional networks. In: International Conference on Learning Representations (2017)
10. Li, Y.F., Wang, S.B., Zhou, Z.H.: Graph quality judgement: a large margin expedition. In: Proceedings of the 25th International Joint Conference on Artificial Intelligence, pp. 1725–1731 (2016)
11. Liang, D.M., Li, Y.F.: Lightweight label propagation for large-scale network data. In: Proceedings of 27th International Joint Conference on Artificial Intelligence, pp. 3421–3427 (2018)
12. Liu, W., Wen, Y., Yu, Z., Yang, M.: Large-margin softmax loss for convolutional neural networks. In: Proceedings of the 33rd International Conference on Machine Learning, pp. 507–516 (2016)
13. Mikolov, T., Sutskever, I., Chen, K., Corrado, G.S., Dean, J.: Distributed representations of words and phrases and their compositionality. In: Advances in Neural Information Processing Systems, pp. 3111–3119 (2013)
14. Perozzi, B., Al-Rfou, R., Skiena, S.: DeepWalk: online learning of social representations. In: Proceedings of the 20th ACM SIGKDD International Conference on Knowledge Discovery and Data Mining, pp. 701–710 (2014)
15. Sen, P., Namata, G., Bilgic, M., Getoor, L., Galligher, B., Eliassi-Rad, T.: Collective classification in network data. AI Mag. **29**, 93 (2008)
16. Tang, J., Qu, M., Wang, M., Zhang, M., Yan, J., Mei, Q.: LINE: large-scale information network embedding. In: Proceedings of the 24th International Conference on World Wide Web, pp. 1067–1077 (2015)
17. Taskar, B., Guestrin, C., Koller, D.: Max-margin Markov networks. In: Advances in Neural Information Processing Systems, pp. 25–32 (2004)
18. Tu, C., Zhang, W., Liu, Z., Sun, M.: Max-margin deepwalk: discriminative learning of network representation. In: Proceedings of the 25th International Joint Conference on Artificial Intelligence, pp. 3889–3895 (2016)

19. Weston, J., Ratle, F., Mobahi, H., Collobert, R.: Deep learning via semi-supervised embedding. In: Montavon, G., Orr, G.B., Müller, K.R. (eds.) Neural Networks: Tricks of the Trade. LNCS, vol. 7700, pp. 639–655. Springer, Heidelberg (2012). https://doi.org/10.1007/978-3-642-35289-8_34

20. Yang, Z., Cohen, W.W., Salakhutdinov, R.: Revisiting semi-supervised learning with graph embeddings. In: Proceedings of the 33rd International Conference on Machine Learning, pp. 40–48 (2016)

21. Zhu, J., Ahmed, A., Xing, E.P.: MedLDA: maximum margin supervised topic models. J. Mach. Learn. Res. **13**, 2237–2278 (2012)

22. Zhu, X., Ghahramani, Z., Lafferty, J.D.: Semi-supervised learning using Gaussian fields and harmonic functions. In: Proceedings of the 20th International Conference on Machine Learning, pp. 912–919 (2003)

Characterizing the SOM Feature Detectors Under Various Input Conditions

Macario O. Cordel II[(⊠)] and Arnulfo P. Azcarraga

College of Computer Studies, De La Salle University,
2401 Taft Avenue, 1004 Manila, Philippines
{macario.cordel,arnulfo.azcarraga}@dlsu.edu.ph

Abstract. A classifier with self-organizing maps (SOM) as feature detectors resembles the biological visual system learning mechanism. Each SOM feature detector is defined over a limited domain of viewing condition, such that its nodes instantiate the presence of an object's part in the corresponding domain. The weights of the SOM nodes are trained via competition, similar to the development of the visual system. We argue that to approach human pattern recognition performance, we must look for a more accurate model of the visual system, not only in terms of the architecture, but also on how the node connections are developed, such as that of the SOM's feature detectors. This work characterizes SOM as feature detectors to test the similarity of its response vis-á-vis the response of the biological visual system, and to benchmark its performance vis-á-vis the performance of the traditional feature detector convolution filter. We use various input environments i.e. inputs with limited patterns, inputs with various input perturbation and inputs with complex objects, as test cases for evaluation.

Keywords: Feature detectors · Self-organizing maps ·
Multilayer perceptron · Pattern recognition

1 Introduction

The ability of living organisms to detect salient or target objects regardless of the background or lighting condition inspires most of the recent computational models for pattern recognition. For example, the results of the experiments to map the functional architecture of the monkey and cat's visual system [3,14,16] have been the bases for the layered architecture of successful machine vision systems [8,12,23]. Specifically, the Neocognitron [9,10] and the convolutional neural network (CNN) [17] networks rely on their layered architecture that are directly analog of the complex connection of neurons in the visual system.

With the success of CNN in pattern recognition [20], several applications have been proposed which leverage on the representational power of the trained feature detectors and address pattern recognition problems e.g. natural face

© Springer Nature Switzerland AG 2019
Q. Yang et al. (Eds.): PAKDD 2019, LNAI 11441, pp. 144–155, 2019.
https://doi.org/10.1007/978-3-030-16142-2_12

detection and recognition [19,22] and action recognition [2,24]. Recently, the Mask Regional-CNN [13] has shown to have promising performance in automatic object detection and segmentation. These current successes in such tasks raise the question: is pattern recognition a solved problem?

In this work, we argue that to approach the visual system pattern recognition performance, the computational model should consider the visual system behavior, including the manner in which the receptive fields are developed. The work in [5] proposes self-organizing maps (SOM)-based feature detectors for pattern recognition which exhibit competition-based development of its weight akin to the development of connection between neurons of the visual cortex [3,15]. They showed that indeed SOM feature detectors could be used in pattern recognition.

However, SOM feature detectors [5] performance remains to be inconclusive as its response to various cases e.g. constrained input environment, input perturbation and complex patterns has not been investigated, thus the proposed network is more of a blackbox. Evidences [3,15] show that the development of the visual system receptive fields, both simple and complex, have significant dependence on the kind of environment during its early development. That is, receptors which are exposed to a specific pattern, e.g. horizontal lines of different width, color, length, small distortion and the like, for a long time will develop receptors which are highly specialized to detect horizontal lines. The previous work [5] also has no verification that their proposed SOM feature detectors indeed capture the pattern information, i.e. the spatial relationship of pixels that form the patterns in the input image, as opposed to the convolution filter which was verified and quantified in [6].

This work examines the performance of SOM as the basic feature detectors for pattern recognition under various constrained and perturbed input environment. The classification performance of the feature detectors are determined and misclassification of patterns are carefully observed vis-á-vis the type of the training input patterns. In addition, the SOM feature detectors ability to extract pattern information is also verified by gradually removing the pixel spatial relationship of the input pattern. The classification performance of the two feature detectors are evaluated using simple (MNIST [17]) and complex patterns (vehicle dataset). MNIST is commonly used to evaluate the potential of several proposed pattern recognition algorithms, e.g. [4,17,21]. The following are the contributions of our work:

– *We verified that the proposed SOM feature detectors exhibit similar decline in pattern recognition performance of the visual system, when trained with limited input pattern during the training phase.* Towards biological visual system pattern recognition performance, we argue that it is important to have a similar visual system characteristics, not only in terms of the layered architecture, but also in terms of how the connections are developed. We discovered that increasing the number and proper positioning of the receptive fields result in robustness of the classifier from a set of feature detectors exposed to limited input pattern.

– *We show that for various input perturbation, the classifier using SOM feature detector was able to correctly classify the input.* We showed that the canonic forms help in the equivariance as the input conditions change and the entity is rotated over the appearance manifold or the SOM receptive field. Although convolution filter has better performance, the accuracy of SOM feature detectors at this early stage shows potential.

2   Related Works

Early studies [3, 15] revealed that the development of the visual system exhibits competition by limiting the input patterns that pass through the visual pathway during their development. For example in an experiment [15] which sutured the left eyelid of a kitten from birth, results show profound cell atrophy in layers receiving light input from the covered eye. A similar experiment was performed [3] which sutured the left eye of an infant monkey to observe the effect in the development of its striate cortex. Their experiment showed that the population of cells favors the open eye and the earlier and longer the exposure of the open eye, the greater shift in eye preference is observed. Further investigation reveals that eyes which are exposed to a certain pattern e.g. vertical stripes have more cells orientation like the input pattern.

Several feature detectors which exhibit competition during their development have been proposed as an alternative to convolution filters. SOM nodes are used as the feature map nodes in CNN architecture in [18]. They additionally introduce algorithm which determine the locations of high density data. Another work is proposed by Arevalo et al. [1] which uses topographic independent component analysis (TICA) for CNN learning. Their system is used in detecting basal cell carcinoma in medical images. A semi-supervised learning proposed by Dong et al. [7] uses Sparse Laplacian Filtering learning (SLFL) during the training of the convolution layer. This allows much less data during the network training. Their proposed work is applied to vehicle type classification which takes in high resolution video frames. These systems, however, presented the feature detectors in black box, i.e. did not show any verification as to how these extract the pattern information.

Current image pattern recognition systems [13, 22, 24] based on convolution filters show promising performance and demonstrate applicability to several pattern recognition tasks e.g. speech and text recognition. We argue that to continue to approach human-level pattern recognition performance, we should probe various computational models that better represent the biological visual system. We investigate further in this paper SOM as competition-based feature detectors for pattern recognition.

3   Test Setup

Three experiments are conducted in this work. The first experiment simulates the constrained input environment conducted to observe the response of the

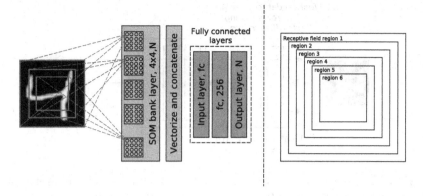

Fig. 1. Pattern recognition with SOM feature detectors (left). Six (not explicitly shown) SOMs are used as feature detector. The output of the feature detector is a similarity vector whose elements are equal to the cosine similarity of the node weights and the pixels in the corresponding limited domain. Shown at the right is the arrangement of the limited domains for each SOM feature detector.

visual system. It uses limited input pattern in training the feature detectors namely (a) vertical only, (b) horizontal only, and (c) circular only. The second experiment determines the robustness of the SOM feature detectors to input perturbation particularly (a) rotation, (b) variation of pattern size and stroke thickness and (c) affine transformation. The third experiment observes the pattern extraction capability of the feature detectors. The spatial relationships of the pixels which form the digit, e.g. curves and edges, are gradually removed by randomly repositioning the pixels. Note that the repositioning is random with respect to the image, but the new pixel arrangement is fixed for the training and test images [6].

3.1   Architecture

Shown in Fig. 1 (left) the architecture with SOM feature detectors under test. The map size of these SOM feature detectors is set to have 4×4 nodes. Six SOMs are used as the feature detectors which are trained separately from the classifier. The SOMs are arranged such that the limited domain or the receptive field for each SOM are focused at the center of the image as illustrated in Fig. 1 (right). This type of limited domain arrangement is used as visual system receptive fields encode the information at the center of the scene [11].

The performance of this architecture is then compared to the conventional feature detector with a single layer of convolution filters and mean pooling kernels. Twenty (20) 9×9 convolution kernels and 20 mean pooling kernel are used for the conventional architecture. A simple CNN architecture is used as we are after the comparison of the basic behavior of the two feature detectors, i.e. comparing deeper CNN will be unfair for the SOM feature detector. For both networks using convolution filters (CNN) and SOM feature detectors, the

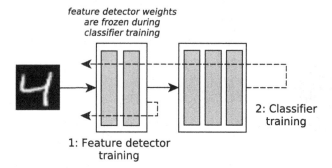

Fig. 2. Training sequence for the evaluation of the feature detector trained in limited input environment. The feature detectors are first trained using specific patterns. Afterwards, the feature detector weights are frozen and the classifier training proceeds. In the classifier training, input patterns are unconstrained.

number of hidden nodes in the fully-connected layer is 256. For digit classification task, the output node is 10 while for vehicle classification task, the output node is 3.

3.2   Training and Testing

In the first experiment, the feature detectors are trained separately from the classifier to allow the learning on limited pattern i.e. vertical, horizontal and circular patterns (first training). Afterwards, the learned feature detector weights are frozen, and the classifier are then trained to classify the handwritten digits (second training) as illustrated in Fig. 2. MNIST test set is used to evaluate the performance of this experiment.

In the second experiment, the feature detectors are trained together with the classifier using the MNIST training set for 60 epochs and batch size of 256. Afterwards, we use the test sets called the rotated-NIST, the size-NIST and the Affine-NIST[1] to evaluate the robustness of the SOM feature detectors and to compare it with the convolution filter. The rotated-NIST are simply the MNIST test set randomly rotated from $-15°$ to $15°$, the size-NIST are generated from the test set with varying dilation and resizing of the digit.

Finally, in the third experiment, the feature detectors are trained and the classifier using the MNIST for the digit classification and the vehicle training set for the vehicle classification. For each randomization of the pixel position (please refer to [6]), the two architectures are retrained using the new set repositioned pixels. In this way, we ensure that the pixel values as attributes of the dataset are preserved, and only the spatial correlation of the pixel forming the patterns are removed.

---

[1] Downloaded from: https://www.cs.toronto.edu/~tijmen/affNIST/.

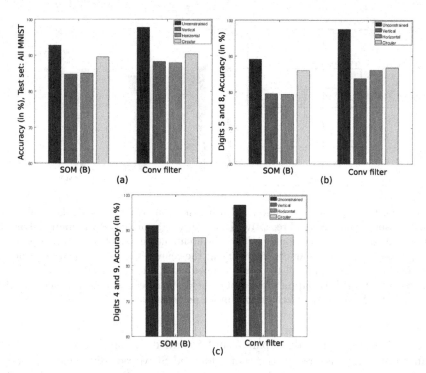

Fig. 3. The performance of the pattern recognition models with SOM feature detectors whose receptive fields are the SOM and the convolution filters. The average accuracies in classifying the MNIST digits using unconstrained input pattern are significantly higher when the feature detectors are trained using limited patterns. Also shown the commonly misclassified pairs digits 5 and 8 and digits 4 and 9.

4   Results and Analysis

4.1   Performance in Constrained Input Environment

Generally, as summarized in Fig. 3(a), the accuracies of SOM feature detectors and the convolution filter feature detector show expected decrease in classification ability when the feature detectors are trained with limited patterns. This was first observed in the visual system which manifests as cell atrophy when the visual receptors are exposed to limited environment during its early development stage [3,15]. The results also imply that both types of feature detectors exhibit similar response as that of the biological system visual receptors developed in a constrained environment.

Interestingly, however, for SOM feature detectors whose receptive fields are distributed all over the input image (shown in Fig. 4 (left)), the ability to identify the input pattern seems to be robust, see Fig. 4 (right), despite the limited training pattern of the feature detectors, presumably due to the richness of the sampled viewing domain or receptive fields. In the confusion matrix of these

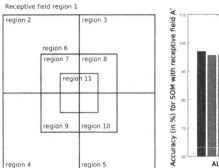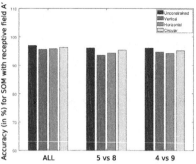

Fig. 4. Another experiment is conducted using distributed receptive fields as opposed to the previous center-focused receptive fields. Although each SOM has limited domain in the input, the distributed domain allows the architecture to look at the other parts of the input pattern rather than concentrating at the center. The respective average accuracies (right) for classifying all the digits trained using unconstrained and constrained patterns show that distributing the receptive field of the feature detector increases the robustness to the dependency to limited input patterns.

three evaluation setups, digit pair 5 and 8; and digit pair 4 and 9 are often confused and misclassified in the center-focused SOM receptive fields and convolution filter receptive fields, but not in distributed SOM receptive fields. The misclassification between the digit pairs are more common in center-focused SOM feature detectors and convolution filter (refer to Fig. 3(b) and (c)). Intuitively, these pairs (digits 5 and 8, digits 4 and 9) are very similar when the horizontal or vertical components are missing.

Both SOM and the convolution filter as feature detector exhibits similar behavior as biological detectors, such that when the detectors are exposed to limited pattern, the pattern recognition ability drops significantly. However, when SOM feature detectors have distributed receptive field, the classifier does not experience significant decrease in pattern recognition (see the three graphs of Fig. 3 A') – which implies that this type of receptive field arrangement overcomes the limitation of the biological visual system which only develops feature detectors depending on the input environment. For our experiments, the classifier whose feature detectors are arranged to focus on the center details, shows dependence on what the feature detectors have seen. For the case where the SOM feature detectors are distributed across the input pattern, the classifier has more viewing points or spatial sampling which add to the classifier input information.

Using the commonly confused input digit pair 5 and 8, we compared the feature maps of the digit pairs for the two receptive field arrangements, shown in Fig. 5 for the distributed and in Fig. 6 center-focused receptive fields. These feature maps are rendered from the element-by-element multiplication of the input pattern and the node's weight vector with the highest response or activation.

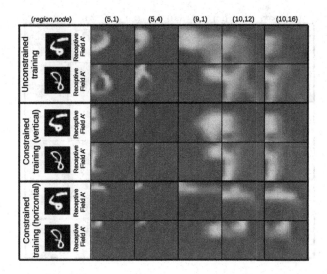

Fig. 5. The feature maps from distributed receptive fields (or arrangement A') trained using the 10 digits from MNIST (first two rows), trained using vertical patterns only (3rd to 4th rows) and trained using horizontal patterns only (5th to last row). The feature maps are formed by rendering the input image (digits 5 and 8) and the weight vector of the corresponding receptive field node with the highest response. The unconstrained feature maps of 8 and 5 has significant difference as compared with the corresponding constrained feature maps.

Note that rendered portion in Figs. 5 and 6 which are near yellow implies high value of similarity of the input pattern and the canonical weight value, while the feature map portions which are near blue means approaching zero similarity. In the test setup when the training is unconstrained, using both receptive field arrangements, the feature maps of 5 and 8 with significant activation differ visually when rendered, see the first two rows of Figs. 5 and 6.

For the constrained input pattern however, the difference between the rendered feature maps of 5 and 8 decreases significantly for the center-focused receptive field (see Fig. 6 trained with vertical and horizontal patterns) but not in distributed receptive field (see Fig. 5 trained with vertical and horizontal patterns) – thus, the classifier with center-focused receptive fields frequently fails to discriminate the input pattern of digits 5 and 8.

4.2 Performance in Perturbed Input Patterns

SOM as feature detectors was able to allow the classifier to detect the input pattern with small random rotation from $-15°$ to $15°$. For distributed receptive fields the accuracy is 97.97% and for the center-focused receptive fields the accuracy is 89.39% both of which are comparable to the performance when there is input no rotation. This implies that SOM feature detectors performance is robust to small rotation. For various stroke size and thickness, however, a significant

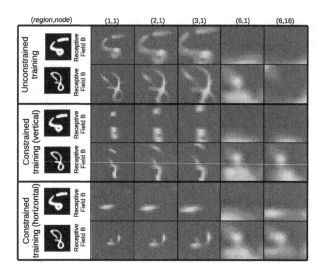

Fig. 6. The feature maps from center-focused feature detectors (or arrangement B) trained using the 10 digits from MNIST (first two rows), trained using vertical patterns only (3rd to 4th rows) and trained using horizontal patterns only (5th to last row). The unconstrained feature maps of 8 and 5 has significant difference as compared with the corresponding constrained feature maps.

Table 1. Average accuracy of 4×4 SOM feature detectors with distributed receptive field (A'), center-focused receptive field (B) and convolution filter feature detectors, under for various perturbed and complex patterns

Dataset	SOM (A')	SOM (B)	Conv filter
Perturbed NIST (small rotation)	97.97%	89.39%	96.08%
Perturbed NIST (stroke thickness)	72.59%	62.80%	96.98%
Affine-NIST	10.41%	9.25%	39.38%

decrease in classification accuracy is seen for both receptive field arrangements. For the distributed receptive fields, the accuracy drops to 72.59% and for the center-focused receptive fields, the accuracy becomes 62.80% only. This performance becomes even worse when the input patterns went through affine transformation. For both arrangements of receptive fields for SOM feature detectors, the accuracy was no better than chance (Table 1).

For small rotation the convolution filter was able to detect 96.08%, which is lower than the accuracy of the distributed SOM receptive field. For various size and thickness however, the convolution filters were able to extract the needed information for the classifier to achieve the accuracy of 96.98%. Finally, for the Affine-NIST, convolution filters show accuracy of 39.38% which is much better than the SOM feature detectors.

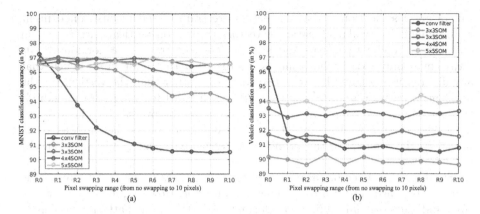

Fig. 7. As the spatial correlation of the (a) MNIST and the (b) vehicle datasets are gradually removed, the primitive pattern information e.g. edges and corners are also gradually removed. The plot shows gradual decrease in CNN performance using convolution filter as the pattern information from left to right are gradually removed for both datasets. The classifier using SOM as feature detectors shows consistent accuracy even in the absence of the input pattern.

4.3   Performance in Complex Vehicle Dataset

We also verify the performance of SOM feature detectors in complex dataset i.e. vehicle dataset. In addition to this, we gradually remove the pattern from the input image and observe the classification performance of SOM A' and SOM B. Previous work [6] shows that by gradually randomizing the position of the pixels in an image, while fixing these new randomized position of the pixels for all the images in the dataset, the primitive pattern information e.g. edges and corners, are removed while retaining the pixel value information as the only image attribute.

Figure 7 shows the classification accuracy of SOM feature detectors as for (a) MNIST and (b) vehicle datasets as the pixel positions are randomized to remove the resemblance of the object, from left to right. The convolution filter feature detector of the CNN shows this dependence to the spatial correlation of the pixel forming the edges as shown by the gradual decrease in the classification accuracy for both the MNIST and vehicle dataset. The consistent performance of the classifier using SOM as feature detector implies that SOM feature detector is robust to such removal of spatial correlation of pixels. However, as R increases from left to right, no canonical information, as to the kind of input patterns, could be obtained when the weights of the SOM nodes are rendered.

5   Discussion

We performed the evaluation of SOM as feature detectors for different input environment conditions. We showed that both the SOM and convolution filters suffers misclassification if these detectors are trained under constrained input environment. Particularly, feature detectors trained on vertical patterns could only extract vertical patterns and feature detectors trained on horizontal patterns could only extract horizontal patterns from the input image, such that any differentiating traits between two categories other than the vertical (or the horizontal) pattern, are not regarded. Although SOM feature detectors exhibit this behavior of the biological feature detectors, we discovered that the arrangement of the receptive fields of SOM feature detectors allows the classifier to be robust to the removal of the primitive patterns, e.g. edges and curves, in the input image.

SOM feature detectors also have better robustness to small rotation of the input pattern as compared to the convolution filter. However, for various stroke sizes and thickness and affine transformation of the input pattern, the convolution filter shows better performance. SOM feature detectors show promising results however when the resemblance of primitive patterns e.g. edges and curves, are slowly removed.

The remarkable performance of SOM feature detectors over the conventional convolution filters exhibits its potential. SOM feature detectors are still far from perfect. Examining the different receptive field arrangements and tweaking the connection of this feature detector to fit the dynamic routing algorithm [21] could help SOM feature detectors reach its full potential.

References

1. Arevalo, J., Cruz-Roa, A., Arias, V., Romero, E., Gonzalez, F.: An unsupervised feature learning framework for basal cell carcinoma image analysis. Artif. Intell. Med. **64**, 131–145 (2015)
2. Bilen, H., Fernando, B., Gavves, E., Vedaldi, A., Gould, S.: Dynamic image networks for action recognition. In: 2016 IEEE Conference on Computer Vision and Pattern Recognition (CVPR), pp. 3034–3042, June 2016
3. Carlson, M., Hubel, D.H., Wiesel, T.N.: Effects of monocular exposure to oriented lines on monkey striate cortex. Dev. Brain Res. **25**(1), 71–81 (1986)
4. Ciresan, D., Meier, U., Schmidhuber, J.: Multi-column deep neural networks for image classification. In: 2012 IEEE Conference on Computer Vision and Pattern Recognition (CVPR), pp. 3642–3649 (2012)
5. Cordel, M.O., Antioquia, A.M.C., Azcarraga, A.P.: Self-organizing maps as feature detectors for supervised neural network pattern recognition. In: Hirose, A., Ozawa, S., Doya, K., Ikeda, K., Lee, M., Liu, D. (eds.) ICONIP 2016. LNCS, vol. 9950, pp. 618–625. Springer, Cham (2016). https://doi.org/10.1007/978-3-319-46681-1_73
6. Cordel, M.O., Azcarraga, A.P.: Measuring the contribution of filter bank layer to performance of convolutional neural networks. Int. J. Knowl.-Based Intell. Eng. Syst. **21**(1), 15–27 (2017)
7. Dong, Z., Wu, Y., Pei, M., Jia, Y.: Vehicle type classification using semisupervised convolutional neural network. IEEE Trans. Intell. Transp. Syst. **16**, 2247–2256 (2015)
8. Fu, M., Xu, P., Li, X., Liu, Q., Ye, M., Zhu, C.: Fast crowd density estimation with convolutional neural networks. Eng. Appl. Artif. Intell. **43**, 81–88 (2015)

9. Fukushima, K.: Neocognitron: a self-organizing neural network model for a mechanism of pattern recognition unaffected by shift in position. Biol. Cybern. **36**, 193–202 (1980)
10. Fukushima, K.: Artificial vision by multi-layered neural networks: neocognitron and its advances. Neural Netw. **37**, 103–119 (2013)
11. Haines, D.E., Mihailoff, G.A.: The visual system. In: Fundamental Neuroscience for Basic and Clinical Applications, Chap. 20. Elsevier (2018)
12. Haoxiang, L., Zhe, L., Xiaohui, S., Jonathan, B., Gang, H.: A convolutional neural network cascade for face detection. In: 2015 IEEE Conference on Computer Vision and Pattern Recognition (CVPR), pp. 5325–5334 (2015)
13. He, K., Gkioxari, G., Dollár, P., Girshick, R.: Mask R-CNN. In: Proceedings of the International Conference on Computer Vision (ICCV) (2017)
14. Hubel, D.H., Wiesel, T.N.: Receptive fields, binocular interaction, and functional architecture in the cat's visual cortex. J. Physiol. **106**, 106–154 (1962)
15. Hubel, D.H., Wiesel, T.N.: Effects of visual deprivation on morphology and physiology of cells in the cats lateral geniculate body. J. Neurophysiol. **26**, 978–993 (1963)
16. Hubel, D.H., Wiesel, T.N.: Receptive fields and functional architecture in two nonstriate visual areas of the cat. J. Neurophysiol. **28**, 229–289 (1965)
17. Lecun, Y., Bottou, L., Bengio, Y., Haffner, P.: Gradient-based learning applied to document recognition. Proc. IEEE **86**, 2278–2324 (1998)
18. Mohebi, E., Bagirov, A.: A convolutional recursive modified self organizing map for handwritten digits recognition. Neural Netw. **60**, 104–118 (2014)
19. Parkhi, O.M., Vedaldi, A., Zisserman, A.: Deep face recognition. In: British Machine Vision Conference (2015)
20. Russakovsky, O., et al.: ImageNet large scale visual recognition challenge. Int. J. Comput. Vis. (IJCV) **115**(3), 211–252 (2015)
21. Sabour, S., Frosst, N., Hinton, G.: Dynamic routing for between capsules. In: Advances in Neural Information Processing Systems, pp. 3859–3869 (2017)
22. Schroff, F., Kalenichenko, D., Philbin, J.: FaceNet: a unified embedding for face recognition and clustering. In: CVPR, pp. 815–823. IEEE Computer Society (2015)
23. Sermanet, P., Eigen, D., Zhang, X., Mathieu, M., Fergus, R., LeCun, Y.: OverFeat: integrated recognition, localization and detection using convolution networks. In: International Conference on Learning Representations (2014)
24. Tu, Z., et al.: Multi-stream CNN: learning representations based on human-related regions for action recognition. Pattern Recogn. **79**, 32–43 (2018)

PCANE: Preserving Context Attributes
for Network Embedding

Danhao Zhu[1,2], Xin-yu Dai[1(✉)], Kaijia Yang[1], Jiajun Chen[1], and Yong He[2]

[1] Nanjing University, Nanjing 210031, Jiangsu, People's Republic of China
{zhudh,yangkj}@nlp.nju.edu.cn, {daixinyu,chenjj}@nju.edu.cn
[2] Jiangsu Police Institute, Nanjing 210093, Jiangsu, People's Republic of China
{zhudanhao,heyong}@jspi.cn

Abstract. Through mapping network nodes into low-dimensional vectors, network embedding methods have shown promising results for many downstream tasks, such as link prediction and node classification. Recently, attributed network embedding obtained progress on the network associated with node attributes. However, it is insufficient to ignore the attributes of the context nodes, which are also helpful for node proximity. In this paper, we propose a new attributed network embedding method named PCANE (Preserving Context Attributes for Network Embedding). PCANE preserves both network structure and the context attributes by optimizing new object functions, and further produces more informative node representations. PCANE++ is also proposed to represent the isolated nodes, and is better to represent high degree nodes. Experiments on 3 real-world attributed networks show that our methods outperform the other network embedding methods on link prediction and node classification tasks.

1 Introduction

Recently, some plain network (with only links and nodes) embedding methods were proposed and achieved substantial improvements, such as DeepWalk [12] and Node2Vec [1]. These methods first sampled a number of node sequences from the network based on random walk, which presented structural regularities of the network. Then they tried to preserve the context nodes to the source nodes in the sequences. Besides links, the rich information associated to the nodes can also help to produce more informative network embeddings, including attribute [6], label [10] and so on [16,17]. These auxiliary information can be considered as some types of attributes.

However, the existing attributed network embedding methods [6,10,17] can only utilize the attributes of the source node, so called source attributes, but ignored the attributes of the context nodes, named context attributes. Figure 1 shows some examples of a citation network. With the title attributes of A along, we only acquire that A is about feature learning of network. If the attributes of B, C, D and E are used, we can reasonable infer A may be related to random walk and social networks, and it's true. Note that a more recently research SEANO [5]

© Springer Nature Switzerland AG 2019
Q. Yang et al. (Eds.): PAKDD 2019, LNAI 11441, pp. 156–168, 2019.
https://doi.org/10.1007/978-3-030-16142-2_13

can model the attributes of the adjacent nodes, such as B and C. However, they are not able to utilize the attributes of other context nodes, such as D and E.

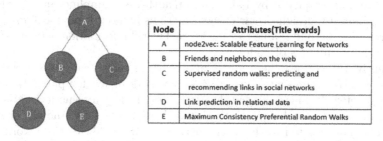

Node	Attributes(Title words)
A	node2vec: Scalable Feature Learning for Networks
B	Friends and neighbors on the web
C	Supervised random walks: predicting and recommending links in social networks
D	Link prediction in relational data
E	Maximum Consistency Preferential Random Walks

Fig. 1. Samples of a citation network. The citation relationship and the paper are regarded as the link and the node respectively. The title words are the attributes of the node.

In this paper, we propose two models called PCANE (Preserving Context Attributes for Network Embedding) and PCANE++ to learn node representations of attributed network. We first apply random walk based strategy to generate context nodes for the source node, which contain the high order structure information of the network to the source node. Second, we propose PCANE, which has two objective functions to preserve the context nodes and the context attributes respectively. Finally, some networks may contain isolated nodes that have only attributes but no links. These nodes will not be trained by PCANE. We therefore propose PCANE++, which directly encodes attributes to the source vectors. PCANE++ can thus cope with isolated nodes and enhance the effect of the source attributes.

In summary, the contributions of this paper are as follows.

- We propose PCANE and PCANE++ for attributed network embedding. The ability of preserving context attributes can help to produce better node embeddings.
- We conduct extensive experiments on 3 open datasets with two tasks of link prediction and node classification. Empirical results demonstrate the effectiveness and rationality of PCANE and PCANE++.

The rest of the paper is organized as follows. Section 2 discusses related works and our motivation. Section 3 introduces problem definition. In Sect. 4, we present our methods for attribute network embedding. The experiments and analysis are outlined in Sect. 5.

2   Related Works and Our Motivation

2.1   Related Work

One of the most fundamental problems in network analysis is network embedding, that focuses on embedding a network into a low-dimensional vector space. The plain network was investigated first. DeepWalk [12] used local node sequences obtained from truncated random walks to learn latent representations by treating walks as the equivalent of sentences. Node2Vec [1] refined the way to generate node sequence by balancing breadth-first sampling and depth-first sampling. Line [13] designed new object functions by preserving local and global structures. The method is able to scale for real world information networks which usually contain millions of nodes.

Recently, some researchers found attributes of nodes can help to produce more informative representations. SEANO [5] is designed for partially labeled attributed network. They encoded the attributes of the source node and its adjacent neighbor nodes, and then jointly decoded the source label and the neighbor nodes. The most related work to our method is ASNE [6]. In their work, each node was mapped to two vectors, an ID vector for encoding the structure information and an attribute vector for encoding its attributes. Then the two vectors were jointed optimized to maximize the likelihood of preserving neighborhood nodes. However, SEANO [5] cannot utilize high-order context attributes and ASNE [6] cannot use any context attributes at all.

2.2 Our Motivation

Most plain network embedding methods aim to preserve the structure of the original network, by maximizing the likelihood of the context nodes to their source node, as in the top of Fig. 2(a). Afterwards, some attributed network embedding methods (such as ASNE [6] followed the objective, but integrated the source attributes when projecting the source vector, as in the bottom of Fig. 2(a).

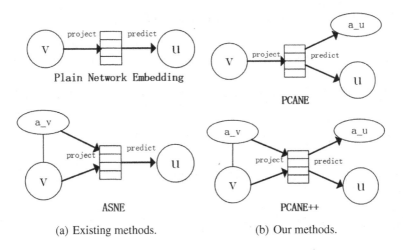

(a) Existing methods.   (b) Our methods.

Fig. 2. An illustration of the existing network embedding methods and our methods. v and u denotes a source node and one of its context nodes respectively. a_v and a_u denote the attributes of v and u respectively.

We can refer to the underlying idea of network embedding: if the high-dimensional information of the original network can be recovered from the learned embeddings, then the embeddings should contain all necessary network information for downstream tasks. For a plain network, preserving the context nodes indeed aims to recover the node topology structure. ASNE still follows the objective when embedding attributed network. Hence, no matter how strong the learning algorithm is, the global attributes will be forever lost since it is not able to recover any attribute information from the vectors. As in Fig. 2(b), PCANE and PCANE++ try to predict both the context nodes and their attributes. The additional objective to predict the context attributes can be considered as applying some extra constrains on v's embedding. As a result, our methods are able to learn higher quality vectors with richer information.

3   Problem Definition

We formally define the attributed network in Definition 1.

Definition 1 (Attributed Network). *An attributed network is defined as* $G = (V, E, T, A)$, *where* V *is the set of nodes, each corresponding to a data object;* E *is the set of links between the nodes, each corresponding to a relationship between two nodes;* T *is the attribute type set;* A *is the set of all discrete attribute values; Each attribute value* $a \in A$ *is corresponding to a type* $T(a) \in T$. *Each node* $v \in V$ *is associated with several attribute values* $A(v)$. *Each* $e \in E$ *is an ordered pair* $e = (u, v)$ *where* $u, v \in V$ *and is associated with a weight* $w_{uv} > 0$, *which denotes the strength of the edge.*

The notation differs from [6]'s definition by adding type T, which will be used to preserve the context attributes. PCANE and PCANE++ can be applied to any (un)directed, (un)weighted network. For undirected network, for $e = (u, v) \in E$, there must be an $e' = (v, u) \in E$. For unweighted network, every w_{uv} is constant 1.

We define the problem of attributed network embedding as follows.

Definition 2 (Attributed Network Embedding). *Given an attributed network* $G = (V, E, T, A)$, *the problem of attributed network embedding is to embed each* $v \in V$ *to a low-dimensional vector space* R^d, *where* $d << |V|$. *In the mapping, the network structure and network attributes are preserved.*

4   Proposed Method

In this section, we first introduce the PCANE model which is able to preserve the network structure and context attributes. Second, we describe the PCANE++ model, an modification to the PCANE model, which can cope with isolated nodes and enhance the importance of source attributes.

4.1   The PCANE Model

Structure Modeling. First, we apply random walks to the network V to obtain truncated node sequences. The node sequences contain the structure information of the network. In the node sequences, for a source node $v \in V$, its neighborhood nodes within certain steps are regarded as the context nodes of v, denoted as $N(v) \subset V$. We use simple uniform sampling strategy, the same as Deepwalk [12]. Our method can also use more complex biased sampling strategy, such as Node2vec [1]. But in practice, we find minor improvements.

Similar to [1,8,12], we propose to maximize the likelihood of the context nodes given a source node. The underline idea is that the source nodes share similar context nodes should be close in vector space. By assuming conditional independence of the source-context node pairs, we maximize the following objective, as in Eq. 1.

$$O_1 = \prod_{v \in V} \prod_{u \in N(v)} p_1(u|v) \tag{1}$$

We define the conditional probability of source-context nodes with a softmax function.

$$p_1(u|v) = \frac{exp(f(v) \cdot s(u))}{\sum_{k \in V} exp(f(v) \cdot s(k))} \tag{2}$$

f and s are the mapping functions from nodes to the source embeddings and context structure embeddings respectively. Equivalently, f and s are real-value matrices with size $|V| * d$, where each row is corresponding a node and d denotes the dimension of the vectors.

Context Attribute Modeling. We aim to preserve the attributes of the context nodes here. The attributes of the source vector is of course very important too. However, a source node could be the context node of itself during random walk sampling. Hence, for linked nodes, we have already preserve the source attributes implicitly. Since the context nodes is obtained via random walk, our methods can preserve the high-order context attributes.

Similar to structure modeling, we aim to maximize the likelihood of preserving the attributes of all the context nodes.

$$O_2 = \prod_{v \in V} \prod_{u \in N(v)} \prod_{a \in A(u)} p_2(a|v) \tag{3}$$

By optimizing Eq. 2, source nodes with similar context attributes will be closer in vector space. We define the conditional probability of $p(a|v)$ as the softmax function below, where g is the mapping function from attributes to the context attribute embeddings. Equivalently, g is a real-value matrix with size $|A| * d$, where each row is corresponding to a node.

$$p_2(a|v) = \frac{exp(f(v) \cdot g(a))}{\sum_{k \in V \ and \ T(k)=T(a)} exp(f(v) \cdot g(k))} \tag{4}$$

Maximizing Eq. 3 actually has two effects: to enhance the similarity between a source node v and its context attributes, as well as weaken that between v and other attributes. Note that the denominator is not the whole attribute set, but the attributes of the same type to a. In general, to enhance the similarity of one type of attributes should not affect that of the other types. For example, there are two types of attributes in a friendship network, the gender and the career. If a person has the attribute gender-male, then the probability of gender-female should be zero, but the probability of career-teacher should not be influenced. The other advantage of the design is to reduce the calculation cost in the denominator.

4.2   PCANE++: Encoding the Source Attributes Explicitly

Since a source node can be the context node of itself, PCANE has already modeled the source attributes implicitly. However, sometimes it is still necessary to encode the source attributes explicitly. First, some attributed networks may contain some isolated nodes with only attributes but no links. The random walk sampling strategy will produce no context nodes, and hence no context attributes for these nodes. Therefore, PCANE can not even model the source attributes for these nodes. Second, it is difficult to sample the source node as its own context node if the node degree is large or the network is dense, since the walker can easily walks faraway from the source node. Hence, source attributes may not be utilized efficiently. Based on the above considerations, we propose PCANE++, which integrates the source attributes explicitly to the source node. PCANE++ can better cope with isolated nodes and highlight the source attributes.

The basic idea is to project the source attributes of v to a separate vector $c(v) \in \mathbb{R}^{d_2}$, and then concatenate it with original source vector $f(v)$ to obtain the new source vector $f_{++}(v)$.

$$f_{++}(v) = \begin{bmatrix} f(v) \\ c(v) \end{bmatrix}$$

Then we replace $f(v)$ with $f_{++}(v)$ in Eqs. 2 and 4, and leave Eqs. 1 and 3 unchanged. To make the dot product in Eqs. 2 and 4 plausible, we define the dimension of f and c in PCANE++ as d_1 and d_2 where $d_1 + d_2 = d$.

To build $c(v)$, we first define a mapping h from attribute $a \in A$ to source attribute vector $h(a) \in \mathbb{R}^{d_2}$. Equivalently, h is a real-value matrix with dimension $|A| * d_2$. $c(v)$ is define as the summation of v's source attribute vectors.

$$c(v) = \sum_{a \in A(v)} h(a)$$

Considering a node t without any links, $f_{++}(t)$ will not be trained either. Hence, $f(t)$ will be the same as it was initialized for ever. However, $c(t)$ is built based on the t's attributes, which provides the attribute information of t. Moreover, the other training instances will project similar attributes to similar $h(.)$, which makes $c(t)$ more reasonable.

4.3   Optimization

Optimization for PCANE. For PCANE, the optimization of O_1 can be simplified to:

$$\arg\max_{f,s} \sum_{v \in V} \sum_{u \in N(v)} \log p_1(u|v) \tag{5}$$

The calculation of the denominator in $p_1(u|v)$ is computational expensive since it is required to traverse the entire node set. We approximate it using negative sampling [9]. For each sampled context node, we randomly select several other nodes as negative context nodes. The optimization of O_2 can be simplified to:

$$\arg\max_{f,g} \sum_{v \in V} \sum_{u \in N(v)} \sum_{a \in A(u)} \log p_2(a|v) \tag{6}$$

and we will also approximate it with negative sampling if the denominator of $p_2(a|v)$ requires too much calculation. We optimize the two objective functions using stochastic gradient ascent over the model parameters defining the features f, s, g. Specifically, we apply the Adaptive Moment Estimation (Adam) [3], which adapts the learning rate according to parameter frequency.

Optimizing the two objective functions is indeed a kind of multi-task learning. In practice, we will alternatively train the two objectives. Once a batch of source-context nodes are trained, we will feed the model the corresponding batch of source-context attributes. Since a node has multiple attributes in general, the latter batch size is larger and is not fixed.

Optimization for PCANE++. The optimization of PCANE++ is similar to PCANE. The only difference is that the learned parameters are f, h, s, g.

Final Embeddings. After optimization, previous wisdom shows using $f + s$ as the final embeddings will bring improvement [4,6,11]. However, we optimize an additional context attribute objective. f receives more training opportunities than s in PCANE, since both objective functions will update f. f is thereby expected to be more informative than g. We propose $f + \alpha s$ as the final embeddings, where α is a real value weight parameter between $[0, 1]$ and hence can turn down the impact of g. With the same consideration, we propose $f_{++} + \alpha s$ as the final embeddings of PCANE++.

5   Experiments

5.1   Experiment Setup

Dataset. We use one social network: the friendship network of students from University of North Carolina at Chapel Hill (UNC) [14], and two citation networks: DBLP[1] and CITESEER[2]. Code and preprocessed data to reproduce our results is in github page[3].

UNC data has 7 types of discrete attributes, 18163 nodes and 766800 links. For the citation, 4732 edges, and only the title is provided. We apply TF-IDF to extracted 5 most important words for the titles, and the words are used as discrete attributes. Each paper is labeled with research area, which we will used for node classification task later.

Baseline Methods. We compare our methods with several state of art network embedding methods. We set the final embedding size of all the methods to 128 and the parameters generally follow the settings in the original papers.

Node2vec [1]. Node2vec is a plain network embedding method. We set $p = 1$, $q = 0.25$, $window\_size = 10$, $walk\_length = 10$ and $number\_walks = 80$.

Line [13]. Line is also an embedding method for plain network. We set $order = 3$.

TADW [16]. TADW is an embedding method for network where the nodes are associated with text. We set $lambda = 0.2$.

ASNE [6]. ASNE is for attributed network embedding. We set the ID vector size to 100 and attribute vector size to 28. The rest of parameters are the same as Node2vec.

Some other methods related to attributed network embedding, e.g. AANE [2], TriDNR [10] and SLR [7], are excluded from comparison, as [6] has demonstrated that they were outperformed by ASNE. We excluded SEANO [5] since it is designed for partial labeled attributed networks. TADW is applied to the two citation datasets, while publication venue is not used on CITESEER, since TADW can utilize only text attributes.

Training Details. The random walk sampling parameters of PCANE and PCANE++ are similar to those used in Node2vec and Line. Specifically, we set $k = 10$, $l = 10$ and $r = 80$. The dimension of both PCANE and PCANE++ vectors is 128. The dimension of source structure vector and source attribute vector in PCANE++ are 100 and 28 respectively. We randomly initialize the parameters of the matrices with a Gaussian distribution whose mean is 0.0 and standard deviation is 0.01. We train the models with mini-batch Adam [3] whose batch size is 1024. The number of negative sampling is 64. We set $\alpha \in \{0, 0.05, 0.1, 0.15, 0.2\}$ when the best result is obtained on the validation set of link prediction task. We repeated our experiments for 10 random seed initializations and our results are statistically significant with a p-value of less than 0.01.

---

[1] https://www.aminer.cn/citation (V4 version).
[2] http://citeseerx.ist.psu.edu/.
[3] https://github.com/zhudanhao/PCANE.

5.2   Link Prediction

Task Description. The link prediction task aims to predict whether two nodes are linked in the test set, when they are not linked in the training set. Each dataset of links is divided to a training set and a test set with training ratio in $\{0.1, 0.3, 0.5, 0.7, 0.9\}$. We use normalized Cosine angle to measure the similarity between two vectors, and Area Under the ROC Curve (AUROC) [18] to evaluate the similarities. We train the models until the best results are obtained on the test set. Since the datasets have only positive edges, for the test set, we have to add the same number of random fake links as negative samples. It is worth noting that we design two test sets: a standard test set and a filtered test set. The design of the standard test set is the same as previous studies [1,6,15]. The standard test set may contain some isolated nodes with no links appearing in the training set. In the filtered test set, we filter out all links with isolated nodes.

Results. The results of link prediction task is shown in Fig. 3. Note that on UNC, the results on the standard and the test sets make minor differences. The reason is that even 10% training links has already covered almost the entire node set in UNC. Hence, the filtered and the standard test sets on UNC are nearly the same.

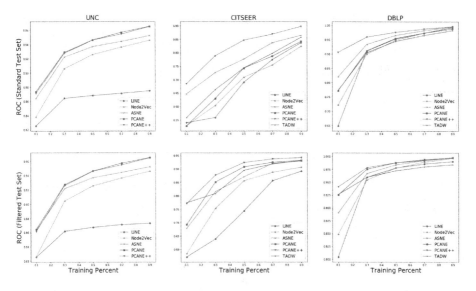

Fig. 3. The results of link prediction. The x axis denotes the fraction of training links, whereas the y axis in the top and bottom rows denote the ROC value on the standard test set and the filtered test set respectively.

Overall, the proposed PCANE++ consistently outperforms all baseline methods on both the standard test set and the filtered test set. For example, given

10% of training links on the standard test set of DBLP, PCANE++ achieves 9.95% improvement of ROC value. PCANE can not outperform TADW and ASNE on the standard test of CITESEER and DBLP, since the standard test set contains a lot of isolated nodes where PCANE cannot make use of. However, on the filtered test set without isolated nodes, both PCANE and PCANE++ can achieve better results than the baseline methods. The results show the modeling of context attributes enables our methods to learn better representations.

Since PCANE++ can cope with the isolated nodes, PCANE++ easily beats PCANE on the standard test sets of all the networks except UNC. However, the advantage is not that strong as on the filtered test sets. With training percent larger than 30% on DBLP, PCANE and PCANE++ achieve similar performance. On UNC, PCANE even slightly outperforms PCANE++. On Citeseer, the advantage is weaken as the training percent arising. The results indicate that for networks without isolated nodes, PCANE and PCANE++ performs similarly.

5.3 Node Classification

Task Description. Node classification aims to predict the label of node in the test set. Therefore, the task can assess if the learned vectors contains sufficient useful information for the downstream tasks. We conduct node classification on Citeseer and DBLP where the research area is used as labels. We exclude UNC for evaluation since no labels consistently exist on all nodes. We train each entire network for one epoch. Hence there are no isolated nodes in both networks. The trained node embeddings are split to training, development and test set with ratio 8:1:1. A simple softmax classifier is trained on the training set until the best result is obtained on the validation set, and we report the results on the test set.

Results. The results of the node classification task is in Table 1. From the results, we can find:

(1) PCANE++ significantly outperforms other baseline methods. On CITESEER, PCANE++ get an improvement of 2.2% on Macro-F1. On DBLP, PCANE++ achieves an improvement of 7.01% (with 62.23% relative error reduction) on Macro-F1. The result shows modeling context attributes can help to produce more informative network embeddings.
(2) PCANE achieves similar performance to PCANE++ on Citeseer. However, PCANE performs even slightly weaker than ASNE on DBLP. The key reason is that DBLP has much more high degree nodes and is denser, which makes PCANE more difficult to utilize the source attributes. We give detail analysis next.

Table 1. Results of node classification.

	CITESEER		DBLP	
	Micro-F1	Macro-F1	Micro-F1	Macro-F1
LINE	0.4796	0.4354	0.7932	0.7445
Node2vec	0.4887	0.4359	0.7712	0.7039
TADW	0.5746	0.5105	0.8299	0.783
ASNE	0.5053	0.453	0.8691	0.8401
PCANE	0.5822	0.5321	0.8578	0.8212
PCANE++	**0.5897**	**0.5335**	**0.9520**	**0.9396**

5.4   Discussion

To understand why our methods can produce better node representations, we present the classification performance w.r.t. the degree of nodes in Fig. 3.

The precisions of both Node2vec and ASNE are sharply increasing when the degree is getting larger. However, the precisions of PCANE and PCANE++ are relatively more stable w.r.t. different degrees. The result shows that utilizing only source attributes and network structure is not sufficient enough for embedding the sparser parts of networks. Preserving context attributes can greatly alleviate the data sparse problem, and enhance the learning of low degree nodes.

ASNE outperforms PCANE when degree is larger than 3. We believe the reason is that PCANE cannot sample source attributes as context attributes efficiently for high degree nodes. Since DBLP is dense, PCANE falls behind PCANE++ even on low degree nodes. Hence, it is necessary to use PCANE++ rather than PCANE when the network is dense or contains many high degree nodes.

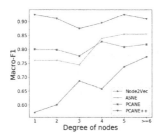

Fig. 4. Classification performance w.r.t. the degree of nodes on DBLP.

5.5   Parameter Sensitivity

Next, we investigate the parameter sensitivity of α and vector dimension. Figure 4(a) presents classification result on UNC w.r.t. different α, and the best result is obtained when $\alpha = 0.1$, which gives evidence to the necessity of turning down the impact of s. Figure 4(b) gives the result of link prediction w.r.t. different dimensions. The training ratio is 10% and we set the source attribute size of PCANE++ as 28/128% of the total dimension. The result indicates that it is not effective to use too large dimensions (Fig. 5).

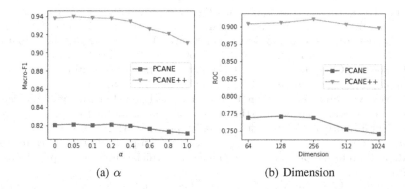

(a) α (b) Dimension

Fig. 5. Sensitivity w.r.t. α and dimension.

6 Conclusion

We introduced novel methods to preserve context attributes to improve attributed network embeddings. Our methods can outperform state of art attributed network embedding methods on link prediction and node classification tasks. A number of extensions and potential improvements are possible, such as sampling the context attributes to reduce training time, and improving the walking strategy to balance the effect of high degree nodes.

Acknowledgement. This work is sponsored, in part, by The Natural Science Foundation of the Jiangsu Higher Education Institutions of China under grant number 18KJB510010 and National Nature Science Foundation of China (NSFC) under grant number 61472183.

References

1. Grover, A., Leskovec, J.: node2vec: scalable feature learning for networks. In: Proceedings of the 22nd ACM SIGKDD International Conference on Knowledge Discovery and Data Mining, pp. 855–864. ACM (2016)
2. Huang, X., Li, J., Hu, X.: Accelerated attributed network embedding. In: Proceedings of the 2017 SIAM International Conference on Data Mining, pp. 633–641. SIAM (2017)
3. Kingma, D.P., Ba, J.: Adam: a method for stochastic optimization. Comput. Sci. (2014)

4. Levy, O., Goldberg, Y., Dagan, I.: Improving distributional similarity with lessons learned from word embeddings. Bulletin De La Socit Botanique De France **75**(3), 552–555 (2015)

5. Liang, J., Jacobs, P., Sun, J., Parthasarathy, S.: Semi-supervised embedding in attributed networks with outliers. In: Proceedings of the 2018 SIAM International Conference on Data Mining, pp. 153–161. SIAM (2018)

6. Liao, L., He, X., Zhang, H., Chua, T.S.: Attributed social network embedding. IEEE Trans. Knowl. Data Eng. **30**, 2257–2270 (2018). (Early access)

7. Liao, L., Ho, Q., Jiang, J., Lim, E.P.: SLR: a scalable latent role model for attribute completion and tie prediction in social networks. In: 2016 IEEE 32nd International Conference on Data Engineering, ICDE, pp. 1062–1073. IEEE (2016)

8. Mikolov, T., Chen, K., Corrado, G., Dean, J.: Efficient estimation of word representations in vector space. arXiv preprint arXiv:1301.3781 (2013)

9. Mikolov, T., Sutskever, I., Chen, K., Corrado, G., Dean, J.: Distributed representations of words and phrases and their compositionality. In: International Conference on Neural Information Processing Systems, pp. 3111–3119 (2013)

10. Pan, S., Wu, J., Zhu, X., Zhang, C., Wang, Y.: Tri-party deep network representation. In: International Joint Conference on Artificial Intelligence, pp. 1895–1901 (2016)

11. Pennington, J., Socher, R., Manning, C.: GloVe: global vectors for word representation. In: Conference on Empirical Methods in Natural Language Processing, pp. 1532–1543 (2014)

12. Perozzi, B., Al-Rfou, R., Skiena, S.: DeepWalk: online learning of social representations. In: Proceedings of the 20th ACM SIGKDD International Conference on Knowledge Discovery and Data Mining, pp. 701–710. ACM (2014)

13. Tang, J., Qu, M., Wang, M., Zhang, M., Yan, J., Mei, Q.: LINE: large-scale information network embedding. In: Proceedings of the 24th International Conference on World Wide Web, pp. 1067–1077. International World Wide Web Conferences Steering Committee (2015)

14. Traud, A.L., Mucha, P.J., Porter, M.A.: Social structure of facebook networks. Phys. A: Stat. Mech. Appl. **391**(16), 4165–4180 (2012). Social Science Electronic Publishing

15. Wang, D., Cui, P., Zhu, W.: Structural deep network embedding. In: ACM SIGKDD International Conference on Knowledge Discovery and Data Mining, pp. 1225–1234 (2016)

16. Yang, C., Liu, Z., Zhao, D., Sun, M., Chang, E.Y.: Network representation learning with rich text information. In: IJCAI, pp. 2111–2117 (2015)

17. Zhang, D., Yin, J., Zhu, X., Zhang, C.: User profile preserving social network embedding. In: Proceedings of IJCAI, pp. 3378–3384 (2017)

18. Zou, K., O'Malley, A.J., Mauri, L.: Receiver-operating characteristic analysis for evaluating diagnostic tests and predictive models. Circulation **115**(5), 654–657 (2007)

A Novel Framework for Node/Edge Attributed Graph Embedding

Guolei Sun$^{(\boxtimes)}$ and Xiangliang Zhang

King Abdullah University of Science and Technology, Thuwal, Saudi Arabia
{guolei.sun,xiangliang.zhang}@kaust.edu.sa

Abstract. Graph embedding has attracted increasing attention due to its critical application in social network analysis. Most existing algorithms for graph embedding utilize only the topology information, while recently several methods are proposed to consider *node content information*. However, the copious information on *edges* has not been explored. In this paper, we study the problem of representation learning in *node/edge attributed graph*, which differs from normal attributed graph in that *edges* can also be contented with attributes. We propose **GERI**, which learns graph embedding with rich information in node/edge attributed graph through constructing a heterogeneous graph. GERI includes three steps: construct a heterogeneous graph, take a novel and biased random walk to explore the constructed heterogeneous graph and finally use modified heterogeneous skip-gram to learn embedding. Furthermore, we upgrade GERI to semi-supervised GERI (named SGERI) by incorporating label information on nodes. The effectiveness of our methods is demonstrated by extensive comparison experiments with strong baselines on various datasets.

Keywords: Graph embedding · Node/edge attributed graphs · Network analysis

1 Introduction

Graph embedding, aiming to learn low-dimensional representations for nodes in graphs, has attracted a lot of attention recently due to its success in network learning tasks such as node classification [14], and link prediction [10]. Inspired by natural language models [9], Deepwalk is proposed to learn node embedding from network topology [11]. Then LINE [16] proposed to learn embedding by encoding first-order proximity and second-order proximity between nodes. Node2vec [2] improved Deepwalk [11] by introducing a more flexible random walk.

There is a new trend to integrate multiple types of input information including network topology and node content [3,20], neighbors homophily [22] or node labels [7,19,21]. In reality, networks are complex in terms that not only *nodes* but also *edges* contain rich information. For example, in a coauthor network, the nodes representing authors can be associated with a feature vector, which

Q. Yang et al. (Eds.): PAKDD 2019, LNAI 11441, pp. 169–182, 2019.
https://doi.org/10.1007/978-3-030-16142-2_14

contains information like affiliations or education background or research interest. Also, the edges indicating co-author relationships can be contented by the jointly published papers, which include key-words like classification, matrix completion etc. It is essential that graph embedding should learn from both *topology information* and *node/edge content information*.

There are several previous works considering attributed network embedding, where generally attributed network [5,12] is the network only with *node content information*. PPNE [6] uses node content information by enforcing representations to preserve the similarities between nodes. LANE [3] learns embedding by modelling node proximity in both attributed network space and label space. PLANETOID [21] uses deep neural networks to do semi-supervised representation learning, which utilizes text information as well as label information, and it considers multi-class classification problem. Generally, existing approaches have a common limitation: they cannot incorporate the *edge content information* and only consider *node attributes*.

In this paper, we extend the problem of attributed network embedding to a more general case, named *node/edge* attributed graph embedding, where not only *node*, but also *edges* can contain rich information. We propose a general framework for **graph embedding with rich information** (called GERI), which can learn scalable representations for nodes in networks with rich text information on *nodes/edges*. By incorporating label information during representation learning process, we extend GERI to semi-supervised GERI (named SGERI). GERI and its variant are composed of three steps. Firstly, a homogeneous graph with text information on nodes/edges is converted into a heterogeneous one. The main advantage of this conversion is that it naturally integrates graph topology with node/edge content information or label information (only for SGERI), giving us an opportunity to exploit all such information and enhance the performance of learned representations. Then, a novel discriminant and flexible random-walk method is proposed to preserve the high-order similarity between nodes targeted for embedding, by exploring the constructed network in a mixture of the breadth first search (BFS) and the depth first search (DFS) manner. Finally, modified heterogeneous skip-gram model is used to learn the embedding for the nodes in the original network.

The evaluation of obtained graph embedding is conducted with multi-label/multi-class classification task on three datasets with *nodes/edges* information. The results show that GERI consistently and significantly outperforms state-of-the-art algorithms for various dimensions on all datasets in the unsupervised setting. SGERI in semi-supervised setting has the better performance than the semi-supervised methods including LANE and PTE. What's more, GERI, and SGERI are also computationally efficient since its major sections can be easily parallelized.

2   Related Work

The study in *unsupervised* representation learning with *only the topology* information has a big family of developed approaches [13]. The network topology is usually represented by an adjacency matrix, $A_{(|V|*|V|)}$. To obtain node representation in $\mathbb{R}^d$, dimensionality reduction techniques like singular value decomposition or principal component analysis can be applied on graph Laplacian matrix and Modularity matrix [18]. However, the poor scalability and efficacy of these approaches makes them difficult to be applied to large-scale networks. Recently, another stream of work addresses the unsupervised representation learning of nodes in large-scale graph with an inspiration from neural language processing. Deepwalk [11] and node2vec [2] exploit word2vec [8,9] to learn embedding from word-context pairs sampled by random walks in the graph. And LINE [16] is proposed to explicitly preserve the first-order and second-order proximity between nodes.

The methods for *semi-supervised* representation learning with *only the topology* are also developed in order to incorporate label information. MMDW [19] jointly optimizes the max-margin classifier and the embedding learning model formulated as matrix factorization. Similarly, DDRW [7] jointly learns a classifier and vertex representation by combining the loss of SVM and Skip-gram model.

Then we introduce works which can exploit both *network topology* and *node features information*. TADW [20] considers node content information by decomposing an approximated word-context matrix, with the help of node information matrix as side information. HSCA [22] also follows matrix decomposition model and proposes to enforce homophily between nodes. An obvious weakness of both methods is that they require matrix operation like SVD decomposition, which prohibits them from dealing with large scale graphs. PPNE [6] is another method which belongs to this category. It proposes to preserve property similarity between nodes by adding inequality constraints or numeric constraints. Other works in this topic are semi-supervised. Yang et al. propose Planetoid for learning the representation for each graph node to jointly predict the class label and the neighborhood context in the graph [21], but the model is only designed for multi-class classification problem. LANE [3] learns embedding by modelling node proximity in both attributed network space and label space.

However, all the above-mentioned approaches are not able to incorporate information on edges, which can be integrated by our proposed model. In the first step of our model, we construct a heterogeneous graph to integrate information in both node and edges seamlessly.

3 Problem Formulation

Formally, let $G = (V, E, \mathbf{T}_V, \mathbf{T}_E)$ denotes a network with rich content information for nodes and edges. More specifically, $V = \{v_1, v_2, \ldots, v_{|V|}\}$ is a set of nodes, and $E = \{e = (v_i, v_j) : v_i, v_j \in V\}$ is a set of edges linking two nodes. $\mathbf{T}_V$ is node content attributes, e.g., the word occurrence matrix for nodes, where each entry $\mathbf{T}_V(i, k)$ indicates the occurrence of word w_k associating with node v_i, and $\mathbf{T}_V(i, k) = 0$ for the absence of w_k in v_i's content. $\mathbf{T}_E$ is the edge content attributes, e.g., word occurrence matrix for edges, where each entry $\mathbf{T}_E(i, j, k)$ indicates the occurrence of word w_k on edge connecting node v_i and v_j, and

$\mathbf{T}_E(i, j, k) = 0$ for the absence. The purpose of our work is to learn a low-dimensional representation vector $v \in \mathbb{R}^d$ for each node $v \in V$, by considering the network topology and rich text information on nodes ($\mathbf{T}_V$) and edges ($\mathbf{T}_E$). Note that $\mathbf{T}_V$ and $\mathbf{T}_E$ can be constructed by any attributes, not just text words that are used for simplifying model explanation.

Definition 1: Node/edge attributed graph: It differs from normal attributed graph in that not only nodes, but also edges can be associated with attributes.

Definition 2: Target nodes, Bridge nodes and Label nodes: *Target* nodes are the nodes in V in the original homogeneous network G, for which embedding will be learned. When converting G into a heterogeneous one, *bridge* nodes are created to incorporate the text information on nodes/edges, for assisting the embedding learning of target nodes. An example is shown in Fig. 1. The details of bridge nodes construction will be introduced in Sect. 4.1.

4   Method

4.1   GERI

Heterogeneous Network Construction. Given an attributed network $G = (V, E, \mathbf{T}_V, \mathbf{T}_E)$, we construct a bipartite heterogeneous network (V, U, E_{he}), where V includes target nodes, U contains bridge nodes, E_{he} are edges between V and U. Bridge nodes are the set of words, $U = \{w_1, w_2, \ldots, w_{|U|}\}$, existing in node and edge text information.

An edge in E_{he} connects a target node v_i and a bridge node w_k under two circumstances: (1) v_i and w_k are connected when $\mathbf{T}_V(i, k) \neq 0$. That is to say, a target node v_i is connected with a bridge node w_k if word w_k occurs in the content information of node v_i. The weight associating with the edge is the value of $\mathbf{T}_V(i, k)$. (2) v_i and v_j are both connected to w_k, when $\mathbf{T}_E(i, j, k) \neq 0$. In other words, target node v_i and v_j are both connected with a bridge node w_k if word w_k occurs in the content information of the edge connecting v_i and v_j.

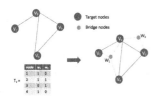

Fig. 1. Example of converting a homogeneous network (left) to a heterogeneous network (right) with bridge nodes.

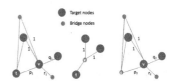

Fig. 2. Three cases of random walk in heterogeneous network, giving that random walk just reached v from t.

Then, the network our algorithm works on is $G_{he} = (V, U, E_{he}, E)$, which includes E_{he} and original E, and is associated with a mapping function $\varphi(v)$: $V \& U \to T = \{target, bridge\}$. A toy example is shown in Fig. 1. One prominent advantage of using the constructed heterogeneous network is that the text information is integrated seamlessly with the original network. There is no loss of information.

Modified Heterogeneous Skip-Gram. Given $G_{he} = (V, U, E_{he}, E)$, our goal is to learn a mapping: $v \to \mathbb{R}^d$ for target node $v \in V$. Besides, a bridge node can be also mapped to feature vectors in $\mathbb{R}^d$. We use $\mathbf{X}$ to represent the latent feature vector for $V \& U$ and $\mathbf{X} \in \mathbb{R}^{(|V|+|U|)*d}$. Inspired by metapath2vec [1], which formulated heterogeneous skip-gram and learn representation for nodes from meta-path. We maximizes the log-probability of observing network neighborhoods for all the nodes conditioned on their feature representation, and we formulate modified heterogeneous skip-gram in our constructed heterogeneous network as a maximum likelihood optimization problem with objective function defined as follows:

$$\underset{\mathbf{X}}{\arg\max} \sum_{v \in V} \sum_{t \in T} \sum_{n \in N_t(v)} \log(P(n|v; \mathbf{X})) + \lambda \sum_{v \in U} \sum_{t \in T} \sum_{n \in N_t(v)} \log(P(n|v; \mathbf{X})) \quad (1)$$

where $N_t(v)$ is the neighborhoods of node v, and has the type of t. As mentioned before, $t \in T = \{target, bridge\}$. The first and second part of the objective is the log-probability of observing network neighborhoods for target nodes and bridge nodes, respectively. λ is a balance parameter, controlling the weight of second part. It shows that λ does have a significant influence on the performance. We approximate $P(n|v; \mathbf{X})$ by negative sampling [9]. Then we use stochastic gradient ascent to get the $\mathbf{X}$. We formulate $\log(P(n|v; \mathbf{X}))$ as:

$$\log(P(n|v; \mathbf{X})) = \log(\sigma(\mathbf{X}_n \cdot \mathbf{X}_v)) + \sum_{m=1}^{M} \mathbb{E}_{u^m \sim P(u)}[\log(\sigma(-\mathbf{X}_{u^m} \cdot \mathbf{X}_v))] \quad (2)$$

where $\sigma(x) = \frac{1}{1+\exp(-x)}$, and $P(u)$ is the empirical unigram distribution defined on all nodes by viewing both target and bridge nodes homogeneously, where negative samples u^m will be drawn M times regardless of their types. Combining Eqs. (1) and (2), we can get the objective function for GERI.

An important component in the objectives of GERI is neighborhood $N_t(u)$, which has a significant influence on the embedding results. Inspired by node2vec [2], which proposed a concept of flexible neighborhood in homogeneous network, we propose a novel randomized procedure that can sample neighborhood of a source node in our constructed heterogeneous network.

Novel Sampling Strategy. Following but differing from node2vec, our proposed sampling method can explore the heterogeneous graph in a mixture of breadth first search (BFS) and depth first search (DFS), such that better neighbors of nodes can be obtained. Our sampling method is superior to the state-of-the-art sampling methods because the search method in node2vec [2] is designed for homogeneous network and the existing sampling strategy in PTE [15] can only preserve the low proximity between nodes, which is usually not desirable.

Consider a random walk that just reached node v from node t in Fig. 2. Then it needs to decide where to go in the next step, which depends on the transition probability β_{vx} between node v and next node x, and the types of previously visited node v and t.

We define the transition probability β_{vx} in three cases:

Case 1: node t and v are both *target* nodes, as shown in the left example of Fig. 2. The next node to visit from v can be a target node, or a bridge node. We introduce three parameters p_1, q_1, and r_1 to guide the walk, and discuss their meanings later. Given the weight e_{vx} between node v and x, the transition probabilities β_{vx} is defined as:

$$\beta_{vx} = \begin{cases} p_1 * e_{vx} & \text{if } d_{tx} = 0 \\ 1 * e_{vx} & \text{if } d_{tx} = 1 \\ q_1 * e_{vx} & \text{if } d_{tx} = 2,\ x \in V \text{ target nodes} \\ r_1 * e_{vx} & \text{if } d_{tx} = 2,\ x \in U \text{ bridge nodes} \end{cases}$$

where d_{tx} denotes the shortest path distance between t and x.

Case 2: node t is a *target* node and v is a *bridge* node (the middle example in Fig. 2). In this case, we don't allow the walk to go back and expect the walk to explore more target nodes because we focus more on the relationship between a target node and other target nodes. β_{vx} is defined as:

$$\beta_{vx} = \begin{cases} 0 & \text{if } d_{tx} = 0 \\ 1 * e_{vx} & \text{if } d_{tx} \neq 0 \end{cases}$$

Case 3: node t is a *bridge* node and node v is a *target* node (the right example in Fig. 2). We introduce three parameters p_2, q_2, and r_2 to guide the walk. The transition probabilities β_{vx} is as follows:

$$\beta_{vx} = \begin{cases} p_2 * e_{vx} & \text{if } d_{tx} = 0 \\ 1 * e_{vx} & \text{if } d_{tx} = 1 \\ q_2 * e_{vx} & \text{if } d_{tx} = 2,\ x \in V \text{ target nodes} \\ r_2 * e_{vx} & \text{if } d_{tx} = 2,\ x \in U \text{ bridge nodes} \end{cases}$$

In the following, we discuss the meaning of the parameters and their implications.

Back parameter p. p_1 and p_2 control the probability to revisit the node that has been visited in the second last step. Setting it to a small value means that the walk is less likely to go back. However, setting it to a large value (>1) means the walk is more likely to visit the local neighbors of the source node. Then, it is more like the BFS search.

Out-target parameter q. q_1 and q_2, on the one hand, control the likelihood of visiting target nodes in the random walk. If $q_1(q_2)$ is greater than $r_1(r_2)$, then

---

Algorithm 1. GERI algorithm

Require: $G = (V, E, \mathbf{T}_V, \mathbf{T}_E)$, Dimensions d, walks per vertex γ, window size τ, walk
    length l, λ, and p, q,r
Ensure: matrix of nodes representation $\Theta \in \mathbb{R}^{|V|*d}$
  1: Initialize Θ by standard normal distribution
  2: Construct $G_{he} = (V, U, E_{he}, E)$
  3: β=PreprocessBiasWeight(G_{he}, p, q, r)
  4: **for** $iter = 1$ to γ **do**
  5:     ϕ=shuffle(V)
  6:     **for** all nodes $v \in \phi$ **do**
  7:         walk=BiasedRandomWalk(G_{he}, β, v, l)
  8:         trainpairs=GenerateSkipGramTraining(walk, τ)
  9:         **for** $(v_1, v_2) \in$ trainpairs **do**
10:            **if** v_1 is a target node **then**
11:                SGD(k,d,(v_1, v_2),η)
12:            **else**
13:                SGD(k,d,(v_1, v_2),$\lambda\eta$)
14: **return** Θ

---

the random walk is more likely to visit target nodes, which means target nodes play a more important role in the random walk. On the other hand, q_1 and q_2 control the depth of exploring the graph. If q_1 and q_2 are large, then the random walk is more likely to go as deep as possible, which is like DFS search.

Out-bridge parameter r. Contrary to q, r controls the likelihood of visiting bridge nodes in the random walk. If $q_1(q_2)$ is less than $r_1(r_2)$, then the random walk is more likely to visit target nodes. Similar to q, r controls the probability to explore the graph deeply. If it's high (>1), the walk is more like DFS. Otherwise, the walk is more like BFS.

In practice, since each pair of $p_1(p_2)$, $q_1(q_2)$ and $r_1(r_2)$ has the same meaning, we set $p_1 = p_2 = p$, $q_1 = q_2 = q$ and $r_1 = r_2 = r$.

GERI Algorithm and Complexity. We show the pseudo-code of GERI in Algorithm 1. It shows that GERI includes three steps: construct heterogeneous graph, conduct biased random walk, and then use modified heterogeneous skip-gram to learn embedding. The overall complexity of GERI is $O(|V|*\gamma*l^2)$, linear w.r.t. $|V|$.

4.2 SGERI

GERI can be easily extended to consider node label information, resulting a semi-supervised GERI (named SGERI), which works on $G'_{he} = (V, U, L, E_{he}, E'_{he}, E)$, where $L = (l_1, l_2, \ldots, l_k)$ represents the labels of nodes (training data) in V, k denotes the number of labels for V and E'_{he} represents the edges between V and L.

Similar to GERI, the complexity of SGERI is also linear with respect to V and is also easily parallelizable and can be executed asynchronously.

5 Experiments

5.1 Dataset

We employ three benchmark networks with text information on *nodes/edges*. The first two networks, which are publicly accessible, contain *node information*. The last network which contains *edge information* was extracted from the source in Aminer [17].

Cora [20] contains 2708 publications from 7 classes and 5429 links. Each publication is described by a binary 1433-dimension feature vector.

DBLP [4] contains 27199 authors and 66832 links, representing co-authorship. Each node has some labels out of 4 labels, representing research areas of the author. Each author is described by a 3000-dimension feature vector.

Aminer: we constructed a co-author network from the source in Aminer [17], containing 20105 authors and 48944 links. Each link corresponds to a co-authored paper. After processing paper abstracts by removing stop words and stemming, we have each edge is associated with an 897-dimension feature vector. The labels of nodes are research fields of the author.

5.2 Comparison Algorithm

The proposed methods are compared with several sate-of-the-art embedding algorithms, which can be divided into four groups. Firstly, to investigate the contribution of node/edge information, we compare GERI++ with Deepwalk [11], Line [16], and node2vec [2]. Secondly, we also include node feature information and naive combination of node2vec feature with node feature information as baselines. Thirdly, to evaluate the power of constructed heterogeneous graph, we feed constructed heterogeneous graph directly to Deepwalk, Line and node2vec. Fourth, we compare GERI and SGERI with PTE [15], and LANE [3], which are regarded as state-of-the-art algorithms in attributed network embedding. The detailed descriptions are listed as follows.

Deepwalk & LINE & Node2vec [11]: apply on the original homogeneous graph and set length of random walk as 150, # of walk as 10 and # of negative sampling as 5.

Naive Combination: combine node2vec embedding and text information.

Deepwalk(hete) & LINE(hete) & Node2vec(hete): feed the constructed heterogeneous graph to Deepwalk, LINE and node2vec.

TADW [20]: the embedding is learned from matrix decomposition.

PTE(unsupervised) & PTE [15]: For PTE (unsupervised), we construct two bipartite heterogeneous networks(target-target, target-bridge) and restrain it as an unsupervised method; For PTE, we construct three bipartite heterogeneous networks (target-target, target-bridge, target-label) and thus it remains as a semi-supervised method.

LANE(unsupervised) & LANE [3]: LANE(unsupervised) uses network and node content information, while LANE not only uses network and node

Table 1. Comparison of Micro-F1 and Macro-F1 score on Cora datasets for different dimensions

Algorithm	Micro-F1				Macro-F1			
	d = 16	d = 32	d = 64	d = 128	d = 16	d = 32	d = 64	d = 128
Deepwalk	0.7569	0.7757	0.8013	0.8151	0.7421	0.7645	0.7917	0.8041
Line	0.7323	0.7179	0.7090	0.7127	0.7142	0.7080	0.7048	0.7045
Node2vec	0.7762	0.7936	0.8096	0.8206	0.7651	0.7829	0.8000	0.81
Text only	0.7242	0.7399	0.7344	0.6957	0.6989	0.718	0.7097	0.6651
Naive combination	0.7864	0.8070	0.8198	0.8148	0.7629	0.7898	0.8033	0.8012
Deepwalk(hete)	0.7858	0.8065	0.7951	0.7962	0.7648	0.7867	0.7757	0.7790
Line(hete)	0.7928	0.8131	0.7903	0.7866	0.7928	0.7927	0.7663	0.7703
Node2vec(hete)	0.8172	0.8131	0.8064	0.7920	0.7957	0.7948	0.7836	0.7689
TADW	0.6732	0.7736	0.825	0.8279	0.5676	0.7400	0.808	0.8093
PTE(unsupervised)	0.7256	0.6959	0.7293	0.7275	0.6931	0.6669	0.7058	0.7048
LANE(unsupervised)	0.6948	0.7843	0.8266	0.8371	0.6098	0.7549	0.8136	0.8275
GERI	**0.8639**	**0.8698**	**0.8699**	**0.8655**	**0.8501**	**0.8604**	**0.8563**	**0.8526**

content (if available), but also uses label information of training data. We did extensive grid search on parameters. For α_1, we search from 0.1 to 1, with step 0.1, and for α_2, we search over [0.01 0.1 1.0]. And for LANE, we also search over δ_1, and δ_2.

GERI & SGERI: we set # of walk, length of walk, # of walk and # of negative sampling, to be the same as Deepwalk and Node2vec, for fair comparisons. The balance coefficient λ is 1 (default) and we use grid search to tune only on p, q, and r.

All the representation vectors are finally normalized such that their L2-norm as 1. We use logistic classification to evaluate all the embeddings.

5.3 Performance of GERI

We report the performance of different methods under various embedding dimensions on Cora, DBLP and Aminer in Tables 1, 2 and 3, respectively. We use 50% data as training and another 50% as testing. In Table 3, LANE(unsupervised) uses only network structure because it can't use edge content. And we don't show Text-only and Naive Combination, because they are not applicable in Aminer, which contains *edge content*.

First, GERI consistently outperforms all baselines for various dimensions on three datasets. For Cora, its performance improvement over PTE(unsupervised) is at least 19% for all dimensions. And it outperforms unsupervised LANE by 24%, 11%, 5.2% and 3.4% for dimension 16, 32, 64, 128, respectively. For DBLP, it is better than PTE(unsupervised) and largely improve LANE(unsupervised) by at least 9.5% over all dimensions. For Aminer, it outperforms PTE(unsupervised) by 6.0%, 5.9%, 5.8%, and 4.3% on $d = 16, 32, 64$, and 128, respectively.

Table 2. Comparison of Micro-F1 and Macro-F1 score on DBLP datasets for different dimensions

Algorithm	Micro-F1				Macro-F1			
	d = 16	d = 32	d = 64	d = 128	d = 16	d = 32	d = 64	d = 128
Deepwalk	0.5600	0.5769	0.5839	0.6027	0.4552	0.4896	0.5114	0.5386
Line	0.5220	0.4939	0.4895	0.5080	0.4193	0.3920	0.3946	0.4291
Node2vec	0.5760	0.5860	0.5952	0.6112	0.4858	0.5040	0.525	0.5466
Text only	0.6113	0.6472	0.6698	0.6894	0.6044	0.6333	0.6521	0.6721
Naive combination	0.7440	0.7476	0.7524	0.7511	0.718	0.7233	0.7284	0.7300
Deepwalk(hete)	0.7555	0.7582	0.7684	0.7771	0.7299	0.7319	0.7451	0.7556
Line(hete)	0.7669	0.7703	*0.7792*	*0.7853*	0.7442	0.7479	*0.7578*	*0.7648*
Node2vec(hete)	0.7553	0.7623	0.7716	0.7787	0.7294	0.7387	0.7495	0.7569
TADW	0.5023	0.6031	0.6657	0.7179	0.4925	0.5904	0.6497	0.697
PTE(unsupervised)	0.7575	0.7585	0.7698	0.7848	0.7383	0.7393	0.7509	0.7664
LANE(unsupervised)	0.1894	0.2462	0.6745	0.7246	0.1377	0.1800	0.6287	0.6790
GERI	**0.7725**	**0.7791**	**0.7891**	**0.7939**	**0.7488**	**0.7586**	**0.7687**	**0.7742**

Table 3. Comparison of Micro-F1 and Macro-F1 score on Aminer datasets for different dimensions

Algorithm	Micro-F1				Macro-F1			
	d = 16	d = 32	d = 64	d = 128	d = 16	d = 32	d = 64	d = 128
Deepwalk	0.4564	0.4643	0.5015	0.5089	0.3354	0.3632	0.4109	0.4373
Line	0.2890	0.2902	0.3839	0.4356	0.1922	0.1995	0.2734	0.3368
Node2vec	0.4759	0.4968	0.5111	0.5335	0.3537	0.3961	0.4181	0.4582
Deepwalk(hete)	0.6600	0.6625	0.6696	0.6729	0.5951	0.6014	0.6113	0.6191
Line(hete)	0.6564	0.6646	0.6687	0.677	0.5849	0.6024	0.6121	0.6227
PTE(unsupervised)	0.6419	0.6485	0.6551	0.6728	0.5665	0.5805	0.5974	0.6209
LANE(unsupervised)	0.2571	0.2940	0.3617	0.4631	0.1412	0.1684	0.2476	0.3756
GERI	**0.6801**	**0.6867**	**0.6932**	**0.7027**	**0.6159**	**0.6241**	**0.6330**	**0.6514**

Second, Deepwalk(hete), Line(hete) and Node2vec(hete) all have very competitive performance and are better than Deepwalk, Line and Node2vec that are applied to the original homogeneous graph. It thus verifies that our constructed heterogeneous graph effectively integrates the network topology and rich text information. But since they are all inferior to GERI, we get that our proposed biased sampling method is better than the sampling methods in these approaches.

Last, we find that TADW and LANE(unsupervised) (both methods use matrix optimization to learn embeddings) perform very poorly with low dimensions such as $d = 16$ and 32, but perform well when dimension of embeddings increase to 64 or 128. However, PTE(unsupervised), Deepwalk, Line and Node2vec have consistent performance for all dimensions. For example, the performance of TADW and LANE(unsupervised) with low dimension of 16 or 32 is worse than other baselines in both Cora and DBLP data set. But these two methods perform well when d increases to 64 and 128.

5.4   Performance of SGERI

We compare SGERI with GERI, and other semi-supervised methods such as LANE and PTE on dataset DBLP and Aminer. For fair comparisons, we used the same set of training and testing data for all methods and did grid search over parameters.

We show the results on DBLP in Fig. 3. It shows that SGERI improved GERI by more than 4% in Micro-F1 and 5% in Macro-F1 score for all dimensions. From the comparisons between semi-supervised methods, we see that PTE outperforms LANE, and SGERI improved PTE by 14%, 12%, 4%, and 2% in Micro-F1 score for dimension of 16, 32, 64, and 128, respectively. For Macro-F1, SGERI improved PTE by 14%, 12%, 5%, and 2% for dimension of 16, 32, 64 and 128, respectively. Conclusively, we get that SGERI consistently outperforms PTE and LANE in all dimensions, and interestingly its superiority is more obvious in the setting of low dimension. The reason why SGERI is better than PTE is that it can better preserve proximity between nodes, which uses novel biased random walk and can take advantage of the high-order proximity while PTE only uses low-order proximity. For Aminer, the results are shown in Fig. 3. Similarly, the use of label information of training data really helps and largely improves the performance of our proposed methods. SGERI outperforms GERI by nearly 20% in both Micro-F1 and Macro-F1 score. Also, SGERI is better than PTE, with performance gain as least 7.1% for Micro-F1 and 9.8% for Macro-F1, for all dimensions. It further verifies that our sampling methods is better than the one in PTE.

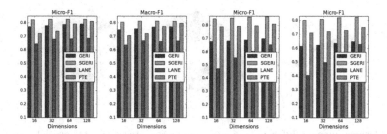

Fig. 3. Comparison between SGERI, GERI, LANE and PTE on DBLP dataset (Left two) and Aminer Dataset (Right two) over various dimensions

5.5   Parameter Analysis

We show effects of parameters in GERI. All experiments are done by setting d as 128.

Firstly, Fig. 4 shows that p, q, and r do have a significant influence on the performance. From the left plot in Fig. 4, we see that the setting of middle value for $1/p$ (from 1.0 to 2.0) and small value for $1/q$ (around 0.25) lead to better performance, which means a relatively high probability to explore target nodes when doing random walk better preserve the proximity between nodes. From the middle and right plots in Fig. 4, we find that a relatively small probability to explore the bridge nodes can give better performance. The underlying reason is that the information that bridge nodes contain is less important than target nodes. Considering the bridge nodes are the terms associated with nodes for this dataset, we can explain this from two aspects. First, the original homogeneous graph represents the co-authorship between authors, whose topology in an implicit way indicates the common research areas among authors. That's to say, the terms are supplementary for graph topology information even though they are the source of performance gain for our methods. Second, terms can be noisy. By limiting the probability of visiting bridge nodes, less noise will be brought to the embedding.

Next, we show how performance changes w.r.t. λ, walk length, # of walks and window size in Fig. 5. For λ, we see that good performance is obtained when lambda is a small value, i.e., 0.01 or 0.1. When λ further increases, F1 score drops dramatically. This is because λ controls the weight of loss function targeted on bridge nodes, and the information in bridge nodes is not as important as target nodes, following our discussion in above. For walk length, # of walks and window size, we see that the performance of node classification w.r.t. these three parameters follows similar pattern: performance increases sharply at the very beginning, increases slightly when we further increase parameter values, and fluctuates or converges or even decreases slightly in the later period.

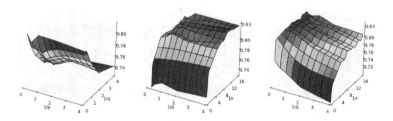

Fig. 4. Performance on different p, q, and r on Cora dataset: left (Micro-F1 score w.r.t. 1/p and 1/q); middle (Micro-F1 score w.r.t. 1/p and 1/r); right (Micro-F1 score w.r.t. 1/q and 1/r).

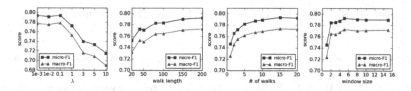

Fig. 5. Performance on different λ, walk length, the number of walks, and window size

6 Conclusion

We studied node/edge attributed graph embedding. GERI is proposed to firstly integrate original graph and copious information in *node/edges* into a heterogeneous graph, and then sample neighborhoods of nodes through the newly designed biased random walk. Finally, GERI learns embedding by modified heterogeneous skip-gram with negative samples. Furthermore, we develop SGERI which improves GERI by exploiting label information. For the future work, there are several possible directions. (1) consider dynamic nature of real graphs and the real-time changes of node/edge content information. (2) As is also the case with other attributed network embedding, we haven't considered the cases when node/edge content is not complete or contaminated.

Acknowledgement. This work is supported by King Abdullah University of Science and Technology (KAUST) Office of Sponsored Research (OSR) under Award No. 2639.

References

1. Dong, Y., Chawla, N.V., Swami, A.: metapath2vec: scalable representation learning for heterogeneous networks. In: KDD, pp. 135–144 (2017)
2. Grover, A., Leskovec, J.: node2vec: scalable feature learning for networks. In: KDD, pp. 855–864 (2016)
3. Huang, X., Li, J., Hu, X.: Label informed attributed network embedding. In: Proceedings of the Tenth International Conference on Web Search and Data Mining, pp. 731–739 (2017)
4. Ji, M., Sun, Y., Danilevsky, M., Han, J., Gao, J.: Graph regularized transductive classification on heterogeneous information networks. In: Balcázar, J.L., Bonchi, F., Gionis, A., Sebag, M. (eds.) ECML PKDD 2010. LNCS, vol. 6321, pp. 570–586. Springer, Heidelberg (2010). https://doi.org/10.1007/978-3-642-15880-3_42
5. Le, T.M., Lauw, H.W.: Probabilistic latent document network embedding. In: ICDM, pp. 270–279 (2014)
6. Li, C., et al.: PPNE: property preserving network embedding. In: Candan, S., Chen, L., Pedersen, T.B., Chang, L., Hua, W. (eds.) DASFAA 2017. LNCS, vol. 10177, pp. 163–179. Springer, Cham (2017). https://doi.org/10.1007/978-3-319-55753-3_11
7. Li, J., Zhu, J., Zhang, B.: Discriminative deep random walk for network classification. In: ACL, pp. 1004–1013 (2016)
8. Mikolov, T., Chen, K., Corrado, G., Dean, J.: Efficient estimation of word representations in vector space. arXiv preprint arXiv:1301.3781 (2013)

9. Mikolov, T., Sutskever, I., Chen, K., Corrado, G.S., Dean, J.: Distributed representations of words and phrases and their compositionality. In: NIPS, pp. 3111–3119 (2013)
10. Pachev, B., Webb, B.: Fast link prediction for large networks using spectral embedding. arXiv preprint arXiv:1703.09693 (2017)
11. Perozzi, B., Al-Rfou, R., Skiena, S.: DeepWalk: online learning of social representations. In: KDD, pp. 701–710 (2014)
12. Qi, G.J., Aggarwal, C., Tian, Q., Ji, H., Huang, T.: Exploring context and content links in social media: a latent space method. IEEE Trans. Pattern Anal. Mach. Intell. **34**, 850–862 (2012)
13. Roweis, S.T., Saul, L.K.: Nonlinear dimensionality reduction by locally linear embedding. Science **290**, 2323–2326 (2000)
14. Sen, P., Namata, G., Bilgic, M., Getoor, L., Galligher, B., Eliassi-Rad, T.: Collective classification in network data. AI Mag. **29**, 93 (2008)
15. Tang, J., Qu, M., Mei, Q.: PTE: predictive text embedding through large-scale heterogeneous text networks. In: KDD, pp. 1165–1174 (2015)
16. Tang, J., Qu, M., Wang, M., Zhang, M., Yan, J., Mei, Q.: LINE: large-scale information network embedding. In: WWW, pp. 1067–1077 (2015)
17. Tang, J., Zhang, J., Yao, L., Li, J., Zhang, L., Su, Z.: ArnetMiner: extraction and mining of academic social networks. In: KDD, pp. 990–998 (2008)
18. Tang, L., Liu, H.: Relational learning via latent social dimensions. In: KDD, pp. 817–826 (2009)
19. Tu, C., Zhang, W., Liu, Z., Sun, M.: Max-margin deepwalk: discriminative learning of network representation. In: IJCAI, pp. 3889–3895 (2016)
20. Yang, C., Liu, Z., Zhao, D., Sun, M., Chang, E.Y.: Network representation learning with rich text information. In: IJCAI, pp. 2111–2117 (2015)
21. Yang, Z., Cohen, W.W., Salakhutdinov, R.: Revisiting semi-supervised learning with graph embeddings. In: ICML, pp. 40–48 (2016)
22. Zhang, D., Yin, J., Zhu, X., Zhang, C.: Homophily, structure, and content augmented network representation learning. In: ICDM, pp. 609–618 (2016)

Mining Unstructured and Semi-structured Data

Context-Aware Dual-Attention Network
for Natural Language Inference

Kun Zhang[1], Guangyi Lv[1], Enhong Chen[1(✉)], Le Wu[2], Qi Liu[1],
and C. L. Philip Chen[3]

[1] Anhui Province Key Laboratory of Big Data Analysis and Application,
School of Computer Science and Technology,
University of Science and Technology of China, Hefei, China
{zhkun,gylv}@mail.ustc.edu.cn, {cheneh,qiliuql}@ustc.edu.cn
[2] Hefei University of Technology, Hefei, China
lewu@hfut.edu.cn
[3] University of Macau, Macau, China
philip.chen@ieee.org

Abstract. Natural Language Inference (NLI) is a fundamental task in natural language understanding. In spite of the importance of existing research on NLI, the problem of how to exploit the contexts of sentences for more precisely capturing the inference relations (i.e. by addressing the issues such as polysemy and ambiguity) is still much open. In this paper, we introduce the corresponding image into inference process. Along this line, we design a novel *Context-Aware Dual-Attention Network (CADAN)* for tackling NLI task. To be specific, we first utilize the corresponding images as the *Image Attention* to construct an enriched representation for sentences. Then, we use the enriched representation as the *Sentence Attention* to analyze the inference relations from detailed perspectives. Finally, a sentence matching method is designed to determine the inference relation in sentence pairs. Experimental results on large-scale NLI corpora and real-world NLI alike corpus demonstrate the superior effectiveness of our *CADAN* model.

1 Introduction

Natural Language Inference (NLI), also named as Recognizing Textual Entailment (RTE), requires an agent to determine the semantic relation between two sentences among *entailment* (if the semantic of hypothesis can be concluded from the premise), *contradiction* (if the semantic of hypothesis cannot be concluded from the premise) and *neutral* (neither entailment nor contradiction), as depicted in the following example from [19], where the semantic of hypothesis can be concluded from the premise:

p: *Several airlines polled saw costs grow more than expected, even after adjusting for inflation.*
h: *Some of the companies in the poll reported cost increases.*

© Springer Nature Switzerland AG 2019
Q. Yang et al. (Eds.): PAKDD 2019, LNAI 11441, pp. 185–198, 2019.
https://doi.org/10.1007/978-3-030-16142-2_15

p : People shopping at outside market

h : People are enjoying the sunny day at the market.

gold-label: Entailment

Fig. 1. Example from SNLI dataset.

Indeed, NLI not only is concerned with the key parts of natural language understanding, i.e. reasoning and inference [4], but also has broad applications, e.g. question answering [27] and automatic summarization [31]. Many research efforts have been conducted in this area. Generally, the main idea of these works can be summarized into two categories: sentence representation and words matching. Sentence representation models focus on extracting semantic representations for sentences by various network structures [3,9,21]. In contrast, words matching models express more concern about the interactions among aligned words between the premise and hypothesis, such as word-by-word matching model [34] and decomposable attention model [25].

To the best of our knowledge, most of existing research assumed that the hypothesis inference is independent of any context. The contexts (e.g. the corresponding images), however, are actually critical for natural language understanding [1]. Figure 1 gives an example. Both the premise and hypothesis sentences describe that people are shopping at the market. Without the image as context, we might conclude the inference relation is neutral since the weather in premise is unclear. However, when we know the context, it's easy to find out the relation is entailment, which indicates the importance of context. Non-literal contexts, like images, can be useful to clarify these issues such as polysemy, ambiguity, as well as fuzziness of words and sentences [39]. Therefore, it's urgent to take into consideration the image contexts for NLI.

In fact, researchers have converged that images convey important information about the associated sentences [14,18]. Much progress has been made on the image and sentence retrieval [13], image captioning [24], and visual question answer [28], e.g. m-RNN model [20] and NIC model [33]. However, these works focused more on the alignments between images and sentences rather than the interactions between sentences, which made it unsuitable for applying them to the conditional NLI task directly.

Inspired by these works, we introduce the corresponding image of the sentence pair as the context into inference process. The key challenge along this line is how to incorporate images into the inference processing effectively. Thus, in this paper, we propose a novel *Context-Aware Dual-Attention Network (CADAN)* to tackle NLI task. To be specific, we propose *Image Attention* layer to utilize the correlated image to enhance the sentence representations. The enhanced sentence representations are further sent to *Sentence Attention* layer to analyze the inference relations from detailed perspectives. With the help of this dual-attention,

CADAN can better evaluate sentence semantic and achieve better performance on NLI task. Finally, the extensive evaluations on the large-scale NLI corpus and real-world NLI alike corpus demonstrate the superior effectiveness of *CADAN*.

2   Related Work

In this section, we introduce the related works, which can be classified into two parts: methods about NLI and methods about image captioning.

Natural Language Inference Methods. With the help of large annotated datasets, such as Stanford Natural Language Inference (SNLI) [2] and Multi-Genre NLI [36], a variety of methods have been developed for NLI. These models can be classified into two frameworks: sentence representation framework and words matching framework.

The representation framework focused on the sentence representation and interaction. Bowman et al. [2] encoded the premise and hypothesis with different LSTMs. Munkhdalai et al. [22] proposed a memory augmented method, which understood the sentence through *read*, *compose* and *write* operation. In addition to network and sentence structures, inner information of sentences also attracted researchers' interests, such as TBCNN [21], bi-directional LSTM with inner-attention [16].

The second framework concentrated more on words matching. Rocktäschel et al. [29] proposed a word-by-word attention model to capture the attention information among words and sentences. Cheng et al. [5] proposed an LSTM with deep attention fusion model to process text incrementally from left to right. However, most of them assumed that the hypothesis inference was independent of any context, which is actually critical for natural language understanding and should be highly considered.

Image Captioning Methods. It has been observed that using the intermediate representation from Convolutional Neural Network (CNN) as an image descriptor significantly boosts subsequent tasks such as object detection, fine-grained recognition [6]. Moreover, researchers have found that using image descriptors from a pre-trained CNN benefited the image captioning [33]. For example, Karpathy et al. [10] proposed an alignment model to learn about the inter-model correspondences between images and texts. Then they utilize the alignments to learn to generate novel descriptions of images.

3   Problem Statement and Model Structure

In this section, we formulate the conditional NLI task as a supervised conditional classification problem and introduce the structure and technical details of the *Context-Aware Dual-Attention Network (CADAN)* for the task.

3.1   Problem Statement

The inputs of this problem are two sentences $s_a = \{s_1^a, s_2^a, \ldots, s_l^a\}$, $s_b = \{s_1^b, s_2^b, \ldots, s_l^b\}$, as well as one corresponding image c as the given context, where s_a and s_b denote the premise and hypothesis sentence. l represents the length of sentences. Note that s_i^a or s_i^b here denotes the one-hot representation of the ith word in the premise or hypothesis sentence. c is the feature representation of the image. The goal is to predict a label y that indicates the inference relation between the premise a and the hypothesis b.

Our task in this paper is to learn an accurate classification model, to predict y given a sentence pair with the associated image (s_a, s_b, c). To this end, we propose the *Context-Aware Dual-Attention Network (CADAN)* to tackle this issue.

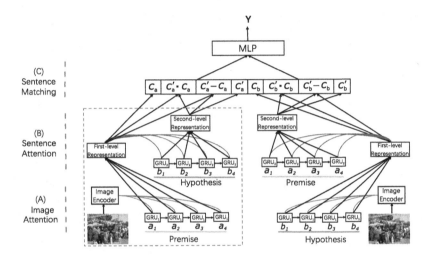

Fig. 2. Architecture of the *Context-Aware Dual-Attention Network (CADAN)*.

3.2   Context-Aware Dual-Attention Network

Our model can be divided into two parts; (1) The preprocessing part: generating the feature representations of sentences and images. (2) The inference part: utilizing the *Context-Aware Dual-Attention Network (CADAN)* to understand the sentences semantics and classify the inference relations between premise and hypothesis.

The Preprocessing Part. Since the inputs of the task are sentence pairs and corresponding images, we utilize different models to represent these different types of data.

For sentences, we utilize the concatenation of pre-trained word embedding (840B Glove) [26] and character feature for English words. The character feature is obtained by applying a convolutional neural network and a max pooling to the learned character embeddings. For Chinese words, we utilize AutoEncoder [12] to perform the representations of words in sentences. Thus, we get the word embedding E for further use.

For images, we choose the pre-trained VGG19 [32] to process the images. Then we extract the outputs of the last convolution layer of VGG19 and send them to a fully-connected layer to get feature representations of images.

The Inference Part. Figure 2 shows the overall framework of *CADAN*, which consists of three components: (1) Image Attention layer; (2) Sentence Attention layer; (3) Sentence Matching layer. In the following part, we take the premise processing as an example to describe technical details of these three components. The same method will be applied to the hypothesis processing.

(A) Image Attention Layer: The images contain the non-literal context of sentences. However, how to utilize the information effectively for sentence semantic is still challenging. Thus, we propose Image Attention layer to integrate them effectively.

In this layer, we first multiply the one-hot representations of the premise $s_a = \{s_1^a, s_2^a, \ldots, s_l^a\}$ and the hypothesis $s_b = \{s_1^b, s_2^b, \ldots, s_l^b\}$ by the word embedding E from the preprocessing part. Then we get the $\{a\}_{j=1}^l$ for premise and $\{b\}_{j=1}^l$ for hypothesis. Next, we leverage Gated Recurrent Units (GRU) [7] to encode these representations. The GRU hidden states below, i.e., $\{\bar{a}\}_{i=1}^l$ and $\{\bar{b}\}_{i=1}^l$ encode each word and sentence context around it:

$$\bar{a}_i = \mathrm{GRU}_1(\{a_{j=1}^i\}), \quad \bar{b}_i = \mathrm{GRU}_1(\{b_{j=1}^i\}), \quad i = 1, 2, \ldots, l. \tag{1}$$

After getting the hidden state of each word, we aim to identify the content of each sentence. Since sentences are both related to the image, the words that are more relevant to the image should get more attention. The attention mechanism can help the model focus on the most relevant part of the input [6,37]. Thus, we utilize VGG19 to get the feature representation c of the corresponding image and send it to the attention cell:

$$\bar{A} = [\bar{a}_1, \bar{a}_2, \ldots, \bar{a}_l], \quad M = \tanh(W\bar{A} + Uc \otimes e_l), \quad W, U \in \mathbb{R}^{k*k},$$
$$\alpha = \mathrm{softmax}(\omega^{\mathrm{T}} M), \quad c_a = \bar{A}\alpha^{\mathrm{T}}, \quad \omega \in \mathbb{R}^k, \tag{2}$$

here W, U, ω are trained parameters. k is the state size of GRU cell in Eq. (1), α is the attention weights vector of hidden states for words, c_a is the first-level representation for premise, and $e_l \in \mathbb{R}^l$ is a row vector of 1. The outer product $Uc \otimes e_l$ means repeating Uc as many times as the number of words in the premise (i.e. l times).

To be specific, the Image Attention representation m_i (i-th column vector in M) of the i-th word in the premise is obtained from a non-linear combination of the premise's hidden state $\bar{a}_i$ and the transformation of image representation c [29]. With the guidance of the image, the relevant words are selected to form the first-level sentence representation c_a. Therefore *CADAN* can understand what the sentence is discussing under the image context information and model the inference relation in term of contents.

(B) Sentence Attention Layer: However, knowing what exactly each sentence discusses is still not enough. What NLI is concerned with is the relations between two sentences. Thus, we also need to model the interaction between two sentences. Since sentence interaction can obtain mutual valued information of the premise and hypothesis, it will help to grasp the local relations in the premise and hypothesis. In order to further characterize the relationship between sentences, we propose Sentence Attention layer to analyze the interaction and local relations from detailed perspectives.

In this layer, we first send the $\{a\}_{j=1}^{l}$ for premise sentence and $\{b\}_{j=1}^{l}$ for hypothesis sentence to another GRU:

$$\bar{a}'_i = \text{GRU}_2(\{a_{j=1}^i\}), \quad \bar{b}'_i = \text{GRU}_2(\{b_{j=1}^i\}), \quad i = 1, 2, \ldots, l. \tag{3}$$

After getting hidden states $\{\bar{a}'\}_{i=1}^{l}$ and $\{\bar{b}'\}_{i=1}^{l}$, we utilize Sentence Attention to model the local relations between hypothesis and premise. Since the first-level sentence representation c_a contains the information that the image is concerned with, it can help to model the local interaction between the premise and hypothesis sentences on the same aspect. Therefore, we treat the first-level sentence representation as the input of Sentence Attention to figure out the local relations between two sentences in this layer.

In other words, with the help of Sentence Attention, the words in the hypothesis that are more important to the premise will get higher weights. We can use these concerned words to generate the second-level representation of premise, which contains enriched information from textual information and image information. We perform attention again and take the same mechanism like Image Attention as follows:

$$\bar{B}' = [\bar{b}'_1, \bar{b}'_2, \ldots, \bar{b}'_l], \quad M' = \tanh(W'\bar{B}' + U'c_a \otimes e_l), \quad W', U' \in \mathbb{R}^{k*k},$$
$$\alpha' = \text{softmax}(\omega'^{\mathrm{T}} M'), \quad c'_a = \bar{B}\alpha'^{\mathrm{T}}, \quad \omega' \in \mathbb{R}^k, \tag{4}$$

Different from Image Attention, here we treat the hidden states $\{\bar{b}'\}_{i=1}^{l}$ of hypothesis sentence and first-level premise representation c_a as the inputs. In this way, the content in $\{\bar{b}'\}_{i=1}^{l}$ that is relevant to c_a will be selected and represented as the second-level premise representation c'_a.

(C) Sentence Matching Layer: In order to determine the overall inference between two sentences, we leverage heuristic matching [4] between first-level sentence representations c_a, c_b and second-level sentence representations c'_a, c'_b after attention operation. Specifically, we use the element-wise product, their difference, and concatenation. Then we concatenate two calculated vectors v_a and v_b and send the result v to multi-layer perceptron (MLP) to calculate the probability of inference relation's existence between these sentence pairs. The MLP has two hidden layers with ReLu activation and a softmax output layer.

$$v_a = (c_a, \ c'_a \odot c_a, \ c'_a - c_a, \ c'_a), \quad v_b = (c_b, \ c'_b \odot c_b, \ c'_b - c_b, \ c'_b),$$
$$v = (v_a, v_b), \quad P(y|(s_a, s_b, c)) = \text{MLP}(v). \tag{5}$$

In this layer, concatenation can retain all the information [38]. The element-wise product is a certain measure of "similarity" of premise and hypothesis [21]. Their difference can capture the degree of distributional inclusion on each dimension [35].

3.3   Model Learning

In this section, we introduce the details about the model learning. Recalling the model description, the training processing can also be divided into two parts: (1) The preprocessing part: We separately train the AutoEncoder and fine-tune VGG19. (2) The inference part: The loss function we use in this part is softmax cross-entropy function.

To be specific, in both stages, mini-batch gradient descent is utilized to optimize the models, where the batch size is 64. The dimensions of feature representation of the image and the words are all 300. The lengths of premise and hypothesis are all set as 15. The state sizes of two GRU cells are set as 200, the dimensions of the parameters W, U, w are also set as 200. To initialize the model, we randomly set the weights W, U, w following the uniform distribution in the range between $-\sqrt{6/(nin + nout)}$ and $\sqrt{6/(nin + nout)}$ as suggested by [23]. We use SGD with momentum [30], where the learning rate and momentum are separately set as 0.05 and 0.6, and gradient clipping is performed to constrain the L2 norm of the global gradients do not exceed 1.0.

4   Experiments

In this section, we provide empirical validation on the large-scale NLI corpus and real-world NLI alike corpus, and utilize the parameter size and accuracy on different test sets to evaluate the models.

4.1   Dataset Description

SNLI. Stanford Natural Language Inference (SNLI) [2] has 570k human annotated sentence pairs with labels "entailment", "neutral", "contradiction". The premise data is drawn from the captions of the Flickr30k corpus. Thus, we can treat the corresponding images as the context. Since the hypothesis data is manually composed, annotation artifacts will lead the model correctly classify the hypothesis alone, Gururangan et al. [8] proposed a challenging hard subset, in which the premise-oblivious model cannot classify accurately, to better evaluate the models' ability to understand sentences. We also evaluate the models' performance on this test set.

DanMu. Different from SNLI that has been synthesized specifically for NLI task [11], DanMu data comes from the real world with labels "entailed" and "not-entailed". Both the premise and hypothesis data are user-generated time-sync comments on videos. Therefore, the corresponding video frames can be treated as the context information. Moreover, these sentences are highly diverse in various aspects (length, complexity, expression, etc.), posing linguistic challenges for the task. By the nature of its construction, DanMu focuses on what a good context-aware NLI system needs to find out inference relation between sentence pairs.

To be specific, DanMu contains 120,650 sentence pairs with associated video frames from more than 4,000 movie videos, including 42,527 positive and 78,123 negative pairs with the labels "entailed" and "not-entailed". Each item contains one premise sentence p, one hypothesis sentence h, and the corresponding video frame.

Following [15], we extract the premise and the corresponding image from a short period [17], the hypothesis sentence is a modified variant of one of the comments from either the same period or a random, unrelated one. The instances that have high word overlap are removed. Then, each remaining instance is modified by three annotators. The annotator was given the instance and asked *"whether he can conclude the hypothesis from the premise and the image"*. The majority of the answers from annotators was treated as the label of the instance. Figure 3(A) show some examples of this dataset.

Baselines. In order to better verify the performance of *CADAN*, we choose some sentence encoding-based NLI models and image captioning models as baselines.

- **LSTM encoders** [2]: encoding the premise and hypothesis with two different LSTMs.
- **W-by-W Attention** [29]: checking for inference relations of word-pairs and phrase-pairs between the premise and hypothesis.
- **BiLSTM with Inner-Attention** [16]: using bidirectional LSTM with inner attention mechanism to generating sentence representation for NLI.
- **CENN** [38]: utilizing different sentence vectors to determine the inference relation.
- **Gated-Att BiLSTM** [3]: employing intra-sentence gated-attention component to encodes a sentence to a fixed-length vector for NLI.
- **m-RNN** [20]: utilizing a deep RNN for sentences and a CNN for images to model the probability distribution of words.
- **NIC** [33]: utilizing a vision CNN and a language RNN for image captioning.

For these two models, we add the premise and hypothesis as inputs to RNN module separately and treat the final state of models as sentence representations. After getting sentences representations, we use Sentence Matching layer in *CADAN* to determine the inference relation in sentences pairs. Note that all the models use the same pre-trained word and image representations.

4.2   Overall Performance

We evaluate the performance of models and baselines from the following aspects: (A) The parameter size (*#Para.*); (B) The accuracy in (1) SNLI Full test set (*SNLI Full*); (2) SNLI Hard test set (*SNLI Hard*); (3) DanMu test set (*DanMu Test*).

Table 1. Performance (accuracy) of models for NLI.

Model	#Para.	SNLI full	SNLI hard	DanMu test
(1) LSTM encoders	3.0M	80.6	58.5	64.9
(2) CENN	<700K	82.1	60.4	65.2
(3) W-by-W Attention model	3.9M	83.5	61.7	66.9
(4) BiLSTM with Inner-Attention	2.8M	84.2	62.7	66.3
(5) Gated-Att BiLSTM	12m	85.5	65.5	67.3
(6) CENN with image	<700K	83.1	61.7	66.6
(7) NIC	-	84.3	63.6	67.9
(8) m-RNN	-	84.9	64.9	68.2
(9) *CADAN*	2.6M	**85.7**	**67.9**	**71.8**

The overall results are summarized in Table 1. We can observe that *CADAN* achieves comparable performance. To be specific, *CADAN* utilizes Image Attention layer to generate first-level sentence representation, thus it can understand sentences in terms of content accurately. Then Sentence Attention layer is employed to model the interaction and local relations from detailed perspectives. Therefore, our model achieved the best performance on SNLI full test. Since SNLI hard test remove those examples that premise-oblivious model can classify correctly, the performance on this test set can better evaluate the models' ability. We can observe that *CADAN* outperformed all the baselines by a large margin, e.g. Gated-Att BiLSTM (+2.4%), BiLSTM with Inner-Attention (+5.2%).

Compared with NLI Models. LSTM encoders [2] encode sentences with different LSTMs and lead many related works to use different neural networks as encoders. Thus, we choose it as one of the baselines. However, 58.5% in hard test and 64.9% in DanMu test prove that simply encoding a sentence with its own textual information is not enough. CENN and its variant have less than 700K parameters, but they achieve comparable results with Word-by-Word Attention model [29], which have 3.9M parameters. It proves that context is really helpful for sentence understanding and NLI indeed. BiLSTM with Inner-Attention [16] uses intra-attention on top of BiLSTM to generate sentence representation, and Gated-Att BiLSTM [3] leverages the gate information in LSTM to calculate the importance of states of words. Thus, they can understand sentences with a finer granularity. However, when sentence semantics become obscure, like the hard test, their performances is not so good, which proves the context is essential for sentence semantic understanding and NLI.

Compared with Variants of Image Captioning Models. Since *CADAN* introduces the image, we want to figure out whether image captioning works can have good performance on this task. We choose NIC [33] and m-RNN [20] as baselines. They can generate sentence representation and adapt to the NLI

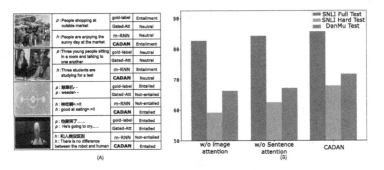

Fig. 3. Classification results of different models and ablation result of *CADAN*.

task through slight changes. From the results, we can conclude that they achieve comparable performance on the full test with the help of images. When sentences become complex, i.e. instances in hard test, their performances are still steady, which indicates the importance of images. However, their original purpose is generation rather than classification. They are good at aligning the images and sentence, but poor at modeling the interaction between sentences from detailed perspectives.

4.3   Ablation Performance

To investigate the effectiveness of the major components of *CADAN*, Fig. 3(B) provides additional analysis. From the best model, we remove the Image Attention layer, in which images are removed, to verify the performance of the model. We also remove the Sentence Attention layer to verify whether only Image Attention layer was enough. Without Image Attention layer, the performance drops to 59.2% (−8.7%) for hard test and 66.3% (−5.5%) for DanMu test, showing that incorporating image as the context is essential. Without Sentence Attention layer, the performance drops to 62.5% (−5.4%) for hard test and 67.2% (−4.6%) for DanMu test, proving that it's important to consider local relations between sentences from detailed perspectives. Based on these observations, we can summarize that contexts and sentence interaction are both very important for sentences semantic understanding.

4.4   Qualitative Evaluation

Evaluation of the Results. Here we choose the Gated-Att BiLSTM and m-RNN as they perform the best of the baselines in the NLI-related baselines and image captioning-related baselines for qualitative evaluation. The results are shown in Fig. 3(A). The first two examples come from SNLI hard test and the rest come from DanMu test. Taking the last instance as an example, this instance describes that a robot looks very sad like human beings. Without the image, the description of the premise will be ambiguous. We don't realize that

'He' is referring to an object rather than a person until we know the non-literal context. With the image, we could understand that the meaning of premise is that this robot may have the same emotions as humans since ordinary robots cannot be able to cry. Then it's easy to infer that this robot has no difference with humans. *CADAN* makes a right choice, while the other two misclassify it.

The rest examples also show the importance of images as contexts in Fig. 3(A). All of them indicate that context is essential for NLI.

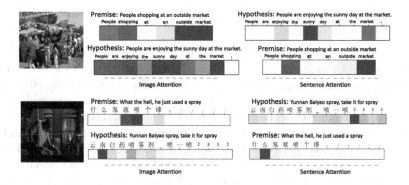

Fig. 4. Visualization of attention on two examples.

Evaluation of the Attention. Here we visualize the attention in our model. There are two kinds of attention: (1) The image's attention to each sentence: (2) The sentence's attention on each other. Figure 4 shows to what extent the Image Attention and Sentence Attention focus on the hidden states of two sentences respectively.

The example above may be confusing without the image. As described before, both premise and hypothesis describe that people are shopping at the market. However, the weather in the premise is unclear. We may conclude that the weather is sunny since the premise describes that the market is outside, which is hard for machines and we are not sure about the conclusion. However, with the image's help, it's easy for us to find out the relation is *entailment*. Moreover, *CADAN* focuses on the word "outside" in the premise and "sunny day" in hypothesis sentence. On this basis, *CADAN* also pay attention to "people shopping" in premise and "enjoying, market" in hypothesis. Therefore, our model not only makes the right classification, but also gives a clear explanation about the inference relation between the sentence pairs.

The example below, which comes from the movie "I, Robot"[1], also indicates that our model not only makes the right classification, but also gives a clear explanation.

With the information of the image, *CADAN* finds the alignment between "spray" in premise and "Yunnan Baiyao spray"[2] in hypothesis. Moreover,

---

[1] https://en.wikipedia.org/wiki/I,_Robot_(film).
[2] Yunnan Baiyao is a kind of healing spray.

CADAN finds that "2333" in hypothesis and "what the hell" in the premise both express the same feeling about the image. All these indicate "entailed" relation between the sentence pair.

In conclusion, when semantic meanings of sentences are clear, *CADAN* can make the right choice and give a detailed explanation about the inference relation. When sentence semantics are obscure, *CADAN* can utilize image as context to understand its meaning precisely and make the correct classification.

5   Conclusion and Future Work

In this paper, we argued that *context* is crucial for sentence understanding. We proposed a novel *Context-Aware Dual-Attention Network (CADAN)* to incorporate both textual and image information into the inference processing effectively. To be specific, we utilized Image Attention to incorporate image to understand the semantic meaning of sentences in terms of contents. Then Sentence Attention was employed to model the interaction and local relations of sentences from detailed perspectives. With the help of two-level representations and dual-attention mechanisms, our model could better understand sentence semantic and make correct decision. Experimental results demonstrated the superiority of our proposed model. In the future, we will explore more effective ways to process the context and finer grained methods to understand sentences semantics more precisely.

Acknowledgements. This research was partially supported by grants from the National Key Research and Development Program of China (No. 2016YFB1000904) and the National Natural Science Foundation of China (Grants No. 61727809, U1605251, 61572540, and 61751202).

References

1. Altmann, G., Steedman, M.: Interaction with context during human sentence processing. Cognition **30**(3), 191–238 (1988)
2. Bowman, S.R., Angeli, G., Potts, C., Manning, C.D.: A large annotated corpus for learning natural language inference. In: EMNLP (2015)
3. Chen, Q., Zhu, X., Ling, Z.H., Wei, S., Jiang, H., Inkpen, D.: Recurrent neural network-based sentence encoder with gated attention for natural language inference. arXiv preprint arXiv:1708.01353 (2017)
4. Chen, Q., Zhu, X., Ling, Z., Wei, S., Jiang, H., Inkpen, D.: Enhanced LSTM for natural language inference. In: ACL. ACL, Vancouver, July 2017
5. Cheng, J., Dong, L., Lapata, M.: Long short-term memory-networks for machine reading. In: EMNLP (2016)
6. Cho, K., Courville, A.C., Bengio, Y.: Describing multimedia content using attention-based encoder-decoder networks. IEEE Trans. Multimed. **17**, 1875–1886 (2015)
7. Chung, J., Gulcehre, C., Cho, K., Bengio, Y.: Empirical evaluation of gated recurrent neural networks on sequence modeling. CoRR abs/1412.3555 (2014)

8. Gururangan, S., Swayamdipta, S., Levy, O., Schwartz, R., Bowman, S.R., Smith, N.A.: Annotation artifacts in natural language inference data. arXiv preprint arXiv:1803.02324 (2018)

9. Huang, Z., et al.: Question difficulty prediction for READING problems in standard tests. In: AAAI (2017)

10. Karpathy, A., Fei-Fei, L.: Deep visual-semantic alignments for generating image descriptions. In: CVPR, pp. 3128–3137 (2015)

11. Khot, T., Sabharwal, A., Clark, P.: SciTail: a textual entailment dataset from science question answering. In: AAAI (2018)

12. Kingma, D.P., Welling, M.: Auto-encoding variational bayes. CoRR abs/1312.6114 (2013)

13. Klein, B., Lev, G., Sadeh, G., Wolf, L.: Associating neural word embeddings with deep image representations using Fisher Vectors. In: CVPR, pp. 4437–4446 (2015)

14. Kun, Z., Guangyi, L., Le, W., Enhong, C., Qi, L., Han, W.: Image-enhanced multi-level sentence representation net for natural language inference. In: ICDM (2018)

15. Lai, A., Bisk, Y., Hockenmaier, J.: Natural language inference from multiple premises. In: IJCNLP (2017)

16. Liu, Y., Sun, C., Lin, L., Wang, X.: Learning natural language inference using bidirectional LSTM model and inner-attention. CoRR abs/1605.09090 (2016)

17. Lv, G., Xu, T., Chen, E., Liu, Q., Zheng, Y.: Reading the videos: temporal labeling for crowdsourced time-sync videos based on semantic embedding. In: AAAI (2016)

18. Ma, L., Lu, Z., Shang, L., Li, H.: Multimodal convolutional neural networks for matching image and sentence. In: ICCV, pp. 2623–2631 (2015)

19. MacCartney, B.: Natural Language Inference. Stanford University, Stanford (2009)

20. Mao, J., Xu, W., Yang, Y., Wang, J., Yuille, A.L.: Deep captioning with multimodal recurrent neural networks (m-RNN). CoRR abs/1412.6632 (2014)

21. Mou, L., et al.: Natural language inference by tree-based convolution and heuristic matching. In: ACL (2016)

22. Munkhdalai, T., Yu, H.: Neural tree indexers for text understanding. CoRR abs/1607.04492 (2016)

23. Orr, G.B., Müller, K.R.: Neural Networks: Tricks of the Trade. Springer, Heidelberg (2003)

24. Pan, Y., Mei, T., Yao, T., Li, H., Rui, Y.: Jointly modeling embedding and translation to bridge video and language. In: CVPR, pp. 4594–4602 (2016)

25. Parikh, A.P., Täckström, O., Das, D., Uszkoreit, J.: A decomposable attention model for natural language inference. In: EMNLP (2016)

26. Pennington, J., Socher, R., Manning, C.: GloVe: global vectors for word representation. In: EMNLP, pp. 1532–1543 (2014)

27. Clark, P., et al.: Combining retrieval, statistics, and inference to answer elementary science questions. In: AAAI (2016)

28. Ren, M., Kiros, R., Zemel, R.S.: Exploring models and data for image question answering. In: NIPS (2015)

29. Rocktäschel, T., Grefenstette, E., Hermann, K.M., Kociský, T., Blunsom, P.: Reasoning about entailment with neural attention. CoRR abs/1509.06664 (2015)

30. Ruder, S.: An overview of gradient descent optimization algorithms. CoRR abs/1609.04747 (2016)

31. Rush, A.M., Chopra, S., Weston, J.: A neural attention model for abstractive sentence summarization. In: EMNLP (2015)

32. Simonyan, K., Zisserman, A.: Very deep convolutional networks for large-scale image recognition. CoRR abs/1409.1556 (2014)

33. Vinyals, O., Toshev, A., Bengio, S., Erhan, D.: Show and tell: a neural image caption generator. In: CVPR, pp. 3156–3164 (2015)
34. Wang, S., Jiang, J.: Learning natural language inference with LSTM. In: HLT-NAACL (2016)
35. Weeds, J., Clarke, D., Reffin, J., Weir, D.J., Keller, B.: Learning to distinguish hypernyms and co-hyponyms. In: COLING, pp. 2249–2259 (2014)
36. Williams, A., Nangia, N., Bowman, S.R.: A broad-coverage challenge corpus for sentence understanding through inference. CoRR abs/1704.05426 (2017)
37. Yin, Y., et al.: Transcribing content from structural images with spotlight mechanism. In: KDD (2018)
38. Zhang, K., Chen, E., Liu, Q., Liu, C., Lv, G.: A context-enriched neural network method for recognizing lexical entailment. In: AAAI (2017)
39. Zheng, X., Feng, J., Chen, Y., Peng, H., Zhang, W.: Learning context-specific word/character embeddings. In: AAAI (2017)

Best from Top k Versus Top 1: Improving Distant Supervision Relation Extraction with Deep Reinforcement Learning

Yaocheng Gui[1], Qian Liu[2], Tingming Lu[1], and Zhiqiang Gao[1(✉)]

[1] School of Computer Science and Engineering, Southeast University, Nanjing, China
{yaochgui,zqgao}@seu.edu.cn, lutingming@163.com
[2] School of Computer Science and Technology,
Nanjing University of Posts and Telecommunications, Nanjing, China
qianliu@njupt.edu.cn

Abstract. Distant supervision relation extraction is a promising approach to find new relation instances from large text corpora. Most previous works employ the *top 1* strategy, i.e., predicting the relation of a sentence with the highest confidence score, which is not always the optimal solution. To improve distant supervision relation extraction, this work applies the *best from top k* strategy to explore the possibility of relations with lower confidence scores. We approach the *best from top k* strategy using a deep reinforcement learning framework, where the model learns to select the optimal relation among the top k candidates for better predictions. Specifically, we employ a deep Q-network, trained to optimize a reward function that reflects the extraction performance under distant supervision. The experiments on three public datasets - of news articles, Wikipedia and biomedical papers - demonstrate that the proposed strategy improves the performance of traditional state-of-the-art relation extractors significantly. We achieve an improvement of 5.13% in average F_1-score over four competitive baselines.

Keywords: Distant supervision · Relation extraction ·
Deep reinforcement learning · Deep Q-networks

1 Introduction

Relation extraction aims to predict the relation for entities in a sentence [20]. It is an important task in information extraction and natural language understanding. However, for the early development of relation extraction applications, a major issue is creating human labeled training sets which is both time-consuming and expensive.

Therefore, a new task in terms of distant supervision relation extraction [2, 4, 8, 13, 15, 18] becomes popular, since it uses entity pairs and their relations from knowledge bases to heuristically create training sets. The definition of distant supervision relation extraction is as follows:

© Springer Nature Switzerland AG 2019
Q. Yang et al. (Eds.): PAKDD 2019, LNAI 11441, pp. 199–211, 2019.
https://doi.org/10.1007/978-3-030-16142-2_16

Definition 1. *Let $\mathcal{X}$ be the sentence space and $\mathcal{Y}$ the set of relations,* **distant supervision relation extraction** *aims to learn a function $f : 2^{\mathcal{X}} \to 2^{\mathcal{Y}}$ from a given data set $\{(X_1, Y_1), (X_2, Y_2), \dots, (X_N, Y_N)\}$, where $X_i \subseteq \mathcal{X}$ is a set of sentences $\{\mathbf{x}_1, \mathbf{x}_2, \dots, \mathbf{x}_{|X_i|}\}$, $Y_i \subseteq \mathcal{Y}$ is a set of relations $\{y_1, y_2, \dots, y_{|Y_i|}\}$.*

Here X_i denotes the set of sentences that relates to the ith entity pair and Y_i its relations, $|X_i|$ denotes the number of sentences in X_i and $|Y_i|$ the number of relations in Y_i.

Strategy *Top 1*. Most previous works resolve distant supervision relation extraction by a sentence-level extractor along with an entity-pair-level predicator to make the final decision [2,4,15,18]. The sentence-level extractor outputs a set of real-valued scores for each sentence $\mathbf{x}$, the score $h(\mathbf{x}, y)$ indicates the confidence of sentence $\mathbf{x}$ describes relation y. For each sentence, at least one relation should be selected and fed to the entity-pair-level predictor, which will make the final prediction based on all the selected relations for all the sentences. Existing distant supervision relation extraction models usually employ the *top 1* strategy, i.e., selecting $\arg\max_{y \in \mathcal{Y}} h(\mathbf{x}, y)$ as the predicted relation for $\mathbf{x}$. However, the relation with the highest confidence score, i.e., $\arg\max_{y \in \mathcal{Y}} h(\mathbf{x}, y)$ is not always the optimal option, existing models have not explored the possibility of other relations with lower confidence scores.

For example, Fig. 1 shows a sample sentence that describes the relation instance (Ernst Haefliger, *place_of_birth*, Davos). As shown in the bottom of the figure, a sentence-level extractor outputs the confidence score for each relation. Obviously, the relation with the highest confidence score (i.e., *place_lived*) is not the best choice for the sentence.

Strategy *Best from Top k*. This paper proposes a strategy to address the issue in existing models. Instead of employing the *top 1* strategy, we investigate the possibility of improving distant supervision relation extraction by using the *best from top k* strategy, i.e., we choose the best prediction from the top k candidates $\{y | \forall y \in \mathcal{Y}, rank(h(\mathbf{x}, y)) \le k\}$, where $rank(h(\mathbf{x}, y))$ returns the rank of y derived from $h(\mathbf{x}, y)$, and then feed it to the entity-pair-level predictor. For example, there is a chance to make an optimal selection for the sentence in Fig. 1 (i.e., *place_of_birth*) using the *best from top k* strategy ($k = 3$).

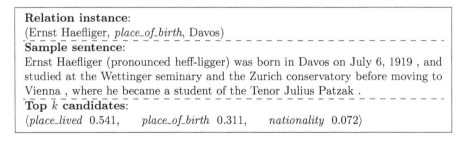

Fig. 1. The top k ($k = 3$) outputs of the sentence-level extractor.

Specifically, we address the *best from top k* strategy using a deep reinforcement learning (RL) framework that learns to predict a set of most possible relations for each entity pair based on the top k candidate relations from the sentence-level extractor. To effectively select among the top k candidate relations, the state representation encodes information about the confidence scores and the context in which the entity pair appears. We train the RL model using a deep Q-network (DQN) [9], whose goal is to learn to select good actions in order to optimize the reward function, which reflects the extraction performance under distant supervision.

While we use the sentence-level extractors of four state-of-the-art models in the experiments, i.e., MultiR [2], MIMLRE [15], CNN+ATT and PCNN+ATT [4], this method can be inherently applied to other models. The experiments on three public datasets from different domains, the New York Times news articles, the Wikipedia articles, and the PubMed paper abstracts, demonstrate that the proposed method outperforms four comparative baselines significantly. The average F_1-score has an improvement of 5.13% compared with baseline models.

The contributions in this work include:

- This work proposes the *best from top k* strategy, which is implemented with a novel deep reinforcement learning framework, to improve existing distant supervision relation extraction models.
- The proposed strategy can be applied to any distant supervision relation extractors that output confidence scores for predicted relations.

2   Related Work

Pioneer work in distant supervision relation extraction used a set of frequent relations in Freebase to train relation extractors over Wikipedia without labeled data [8]. Since then, a lot of works focused on relation extraction using distant supervision. However, using distant supervision to annotate training data would introduce a lot of false positive labels [13].

To alleviate the wrong label issue, a series of graphical models have been proposed based on hand-craft features. A joint model was proposed to learn with multiple relations [2]. Later, a multi-instance multi-label learning (MIML) framework was proposed to further improve the performance [15]. Additional information has been employed to reduce wrong labels of training data upon these models. For example, the fine-grained entity types [3], the document structure [6], the side information about rare entities [14], and the human labeled data [7].

Neural network models have shown superior performance over approaches using hand-crafted features in distant supervision relation extraction [4,18]. Convolutional neural networks (CNN) and piecewise convolutional neural networks (PCNN) are among the first deep neural network models that have been applied to this task [18]. An instance-level selective attention mechanism was introduced for multi-instance multi-label learning [4], and has significantly improved the prediction accuracy for several of these base deep neural network models.

Recently, deep reinforcement learning have been applied to distant supervision relation extraction [1,12,19]. The relation extractor is regarded as a reinforcement learning agent and the goal is to achieve higher long-term reward [19]. To further improve the performance, an instance selector was proposed to cast the sentence selection task as a reinforcement learning problem to choose high-quality training sentence for a relation classifier [1], and a false-positive indicator was proposed to automatically recognize false positive labels and then redistribute them into negative examples [12].

This work relates to the previous works that based on graphical models and neural network models because their sentence-level extractors can be reused in our model. This work also relates to the previous works that based on deep RL methods as we also learn a RL agent. The main differences between our work and existing deep RL methods are that our RL agent tries to improve the testing process of relation extraction while theirs are designed for better training process, and our RL agent is based on deep Q-networks while theirs are mainly based on policy gradient. Considering the training cost of the model, we do not update the parameters of sentence-level extractors during training the RL agent. Thus the learning process tends to be faster than the previous works.

3   Framework

The task of improving relation extraction models under distant supervision can be modeled as a markov decision process (MDP), which learns to utilize the outputs of a sentence-level extractor to improve extractions. We represent the MDP as a tuple $\langle S, A, T, R \rangle$, where $S = \{s\}$ is the space of all possible states, $A = \{a\}$ is the set of all actions, $R(s, a)$ is the reward function, and $T(s'|s, a)$ is the transition function. The overall framework of the task is shown in Fig. 2. Given a set of sentences, the sentence-level extractor produces the predicated relations and their confidence scores. The RL agent selects one action for each state to produce the best relation, which is merged into the selected relation set.

States. The state s in our MDP consists of the sentence-level extractor's confidence scores of the predicted relations and the context in which the entity pair appears. As shown in Fig. 2 (the bottom boxes), we represent the state as a continuous real-valued vector incorporating the following pieces of information:

- Confidence scores of current selected relations between the entities.
- Confidence scores of the newly predicted relations between the entities in the new sentence.
- One-hot encoding of matches between current and newly predicted relations.
- TF-IDF counts[1] of context words, which occur in the neighborhood of the entities in a sentence.

---

[1] TF-IDF counts are computed based on the training sentences.

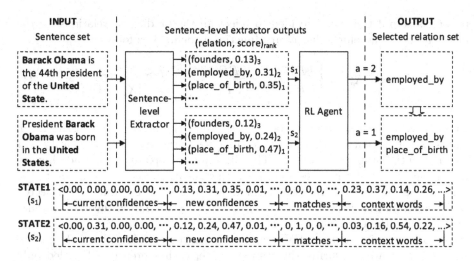

Fig. 2. Overall framework. The left boxes are input sentences. The middle boxes are predicted relations and their confidence scores of the sentence-level extractor. The right boxes are relations that selected by the RL agent step by step from the current episode. The bottom boxes are two sample states corresponding to the input sentences.

Actions. We define an action $a \in \{0, 1, 2, \ldots, k\}^2$ to indicates whether the predicted relations by the sentence-level extractor should be rejected or accepted. Here the number k corresponds to the k in the *best from top k* strategy. The decision can be one of the following types: (1) reject all the relations, i.e., $a = 0$, or (2) accept the lth ($1 \leq l \leq k$) relation according to the ranked predicted confidence scores, i.e., $a = l$. The agent continues to inspect more sentences until the episode ends. The current relation and confidence scores are simply updated with the accepted relation and the corresponding confidences.

Rewards. The reward function is an indicator of the quality of chosen relations. For a certain set of training sentences $X_i = \{\mathbf{x}_1, \mathbf{x}_2, \ldots, \mathbf{x}_{|X_i|}\}$ of an episode, the agent selects an action for each sentence to determine whether the sentence-level extractor's outputs should be accepted or not. We assume that the agent has a terminal reward when it finishes all the selection. Therefore we receive a delayed reward at the terminal state $s_{|X_i|+1}$ based on the performance of current selected relation set Y^{cur} for the ith entity pair on X_i (see the outputs in Fig. 2).

At other states, after an action is taken (i.e., a relation is chosen), the reward is computed immediately based on the agent's performance on the newly predicted relation set Y_j^{new} ($j \leq |X_i|$) for the new sentence. The performances of Y^{cur} and Y_j^{new} are computed using the number of true positive (i.e., TP) and false positive (i.e., FP) relations compared with the distantly annotated data. The intuition is that, the reward is positive if true positive relations are more than false positive ones, and the reward is negative vice versa. Note that, the

---

[2] We choose k by ranging it from 1 to 5 in our experiments, the model achieves the best performance in most cases when $k = 3$.

reward is zero if the agent decided to reject all the predicated relations for the new sentence. Therefore, the reward function is defined as follows:

$$r(s_j|X_i) = \begin{cases} TP(Y_j^{new}) - FP(Y_j^{new}) & j < |X_i| + 1 \\ TP(Y^{cur}) - FP(Y^{cur}) & j = |X_i| + 1 \end{cases} \quad (1)$$

Transitions. Each episode starts off with an initial state that consists of an empty set of current relation and its confidence score respect to the entity pair (see the initial state s_1 in Fig. 2). The subsequent steps in the episode involve traversing the set of sentences and integrating the extracted new relation to the current relation set. The transition function $T(s'|s, a)$ incorporates the selected decision a from the agent in state s along with the relation from the next sentence and produces the next state s', e.g., $s = s_1, s' = s_2$ in Fig. 2.

Algorithm 1 details the MDP framework for the training phase of the *best from top k* strategy. During the testing phase, each sentence is handled only once in a single episode. The training process of our agent contains M epochs. For the ith entity pair, we first initialize an empty set to the current relation set, denoted as Y^{cur} and set the initial reward r to 0 (line 3), then traverse all training sentences in X_i to update the current relation set Y^{cur} and the immediate reward r according to the action taken by the agent based on the state s_j (lines 4–13). The terminal state $s_{|X_i|+1}$ and the delayed reward r for the ith entity pair based on X_i is then sent to the agent (line 14). After the training, the agent learns a policy to further improve the relation extraction results of sentence-level extractors.

Algorithm 1. MDP framework for the *best from top k* strategy

---

1: **for** $epoch = 1, M$ **do**
2:     **for** $i = 1, N$ **do**
3:         $Y^{cur} \leftarrow \{\}, r \leftarrow 0$
4:         **for** $j = 1, |X_i|$ **do**
5:             Compute confidence score vector $\mathcal{F}(\mathbf{x}_j)$
6:             Compute context vector $\mathcal{C}(\mathbf{x}_j)$
7:             Form state s_j using Y^{cur}, $\mathcal{F}(\mathbf{x}_j)$ and $\mathcal{C}(\mathbf{x}_j)$
8:             Send (s_j, r) to agent
9:             Get action a from agent
10:             $Y_j^{new} \leftarrow \text{Select}(\mathcal{F}(\mathbf{x}_j), a)$
11:             $Y^{cur} \leftarrow \text{Reconcile}(Y^{cur}, Y_j^{new})$
12:             update r using Equation 1
13:         **end for**
14:         Send $(s_{|X_i|+1}, r)$ to agent
15:     **end for**
16: **end for**

---

4 DQN Parameter Learning

For the purpose of learning a good policy for an agent, we utilize the deep reinforcement learning framework described in the previous section. Following previous work [10], the MDP can be viewed in terms of a sequence of transitions (s, a, r, s'). The agent seeks to learn a policy to determine which action a to perform in state s. A commonly used technique for learning an optimal policy is Q-learning [17], in which the agent iteratively updates $Q(s, a)$ using the rewards obtained from experiences. The updates are derived from the recursive Bellman equation [16] for the optimal Q:

$$Q^*(s, a) = E\left[r + \gamma \max_{a'} Q^*(s', a') \Big| s, a\right] \tag{2}$$

where r is the reward and γ is a factor discounting the value of future rewards and the expectation is taken over all transitions involving state s and action a.

We use DQN [9] as a function approximator $Q(s, a) \approx Q(s, a; \theta)$, since our problem involves a continuous state space. The DQN has been shown to learn better value functions than linear approximators [9] and can capture non-linear interactions between different pieces of information in continuous state [10]. We use a DQN that consists of two linear layers (20 hidden units each) followed by rectified linear units (ReLU), along with a separate output layer.

The parameters θ of the DQN are learnt using stochastic gradient descent with RMSprop[3]. The parameter update aims to close the gap between the $Q(s, a; \theta)$ predicted by the DQN and the expected Q-value from the experiences. Following previous work [9], we make use of a (separate) target Q-network to calculate the expected Q-value, in order to have stable updates. The target Q-network parameters $\hat{\theta}$ is periodically updated with the current parameters θ. We also make use of an experience replay memory $\mathcal{D}$ to store transitions. To perform updates, we sample a batch of transitions (s, a, r, s') randomly from $\mathcal{D}$ and minimize the loss function:

$$\mathcal{L}(\theta) = E_{s,a,r,s'}\left[\left(r + \gamma \max_{a'} Q(s', a'; \hat{\theta}) - Q(s, a; \theta)\right)^2\right] \tag{3}$$

The learning updates are made every training step using the following gradients:

$$\nabla_\theta \mathcal{L}(\theta) = E_{s,a,r,s'}\left[2\left(r + \gamma \max_{a'} Q(s', a'; \hat{\theta}) - Q(s, a; \theta)\right) \nabla_\theta Q(s, a; \theta)\right] \tag{4}$$

5 Experimental Setup and Results

In our experiments, we first evaluate the performance of the proposed model compared with four state-of-the-art baseline models. Then to further illustrate the effectiveness of the *best from top k* strategy, we also evaluate the performances of the models that apply different strategies respectively.

---

[3] See http://www.cs.toronto.edu/~tijmen/csc321/slides/lecture_slides_lec6.pdf.

5.1 Dataset

Our experiments use three public datasets from different domains. (1) **NYT** [13] is constructed from New York Times news articles. It contains 522,611 sentences in the training set, and 172,448 sentences in the testing set. Among these data, there are 53 unique relations from Freebase including a special relation *NA* that signifies no relation between two entities in a sentence. (2) **Wiki-KBP** [5] is derived from Wikipedia articles. It contains 23,111 sentences in the training set, and 15,847 sentences in the testing set. There are 7 unique relations from the KBP 2013 slot filling database including a *NA* relation. (3) **BioInfer** [11] is sampled from PubMed paper abstracts. It contains 1,139 sentences in the training set, and 876 sentences in the testing set. There are 92 unique relations including a *NA* relation among these data.

5.2 Baseline Extractors

We compare the proposed model with the following distant supervision relation extraction models in our experiments. Note that the sentence-level extractors of these baseline models are also used in our model.

MultiR [2] is a typical work based on probabilistic graphical model for multi-instance learning. It uses the perceptron algorithm for learning and a greedy search algorithm for inference. We implemented this model using the publicly available code[4].

MIMLRE [15] is a graphical model for multiple instances and multiple relations. It is trained by using hard discriminative Expectation-Maximization. We use the publicly available code provided by the authors[5].

CNN+ATT and **PCNN+ATT** [4] are two state-of-the-art neural networks for relation extraction, which adopt a sentence-level attention over the sentences and thus can reduce the weights of noisy sentences. We implemented the two models using the publicly available code[6].

5.3 RL Models

We train a RL model using the proposed *best from top k* strategy based on each sentence-level extractor respectively. For example, **MultiR+RL** uses the same sentence-level extractor as in **MultiR**, then learn a RL model to generate the final predictions in the entity-pair-level.

We used the same network architecture, hyperparameter values and learning procedure throughout to demonstrate that our approach robustly learns successful policies over a variety of datasets based only on distant supervision knowledge. The RL models are trained for 10,000 steps every epoch using the sentence-level extractors, and evaluate the entire test set every epoch. The final

---

[4] http://www.cs.washington.edu/ai/raphaelh/mr/.

[5] http://nlp.stanford.edu/software/mimlre.shtml.

[6] https://github.com/thunlp/NRE/.

evaluation metrics reported are averaged over 20 epochs after 100 epochs of training. We used a replay memory $\mathcal{D}$ of size $500k$, and a discount (γ) of 0.8. We set the learning rate to $2.5E^{-5}$. The ϵ-greedy exploration is annealed from 1 to 0.1 over $500k$ transitions. The target-Q network is updated every $5k$ steps.

5.4 Evaluation Metrics

Similar to the previous works [13], we adopt the held-out evaluation to evaluate our models, which can provide an approximate measure of the classification ability without costly human evaluation. The held-out evaluation compares the predicted relations of the entity pair with the gold relations, which is automatically labeled by knowledge bases. It's an effective evaluation method for large dataset. Precision (P), recall (R), and F_1-score (F_1) are used as our evaluation metrics.

We compute the evaluation metrics based on the distinct occurrence of each relation instance, i.e., any occurrence of the extracted relation instance is considered as one extraction. All compared models are evaluated use the same method.

5.5 Experimental Results

The precisions, recalls and F_1-scores of eight compared models evaluated on three datasets are shown in Table 1. We observe from the table that all RL models yield obvious and steady improvements compared with baseline models on all datasets except the PCNN+ATT+RL model on the BioInfer dataset. It not only demonstrates the rationality of our *best from top k* strategy, but also verifies our hypothesis that the state-of-the-art distant supervision relation extractors can be further improved by the *best from top k* strategy.

Specifically, the F_1 score of CNN+ATT+RL has 18.2% improvement compared with CNN+ATT on the NYT dataset, and the average F_1 score of all RL models on all datasets has 5.13% improvement compared with that of all baseline models. These comparable results illustrate that our approach is capable in improving relation extraction based on distant supervision in different domains, and making RL models develop towards a good direction.

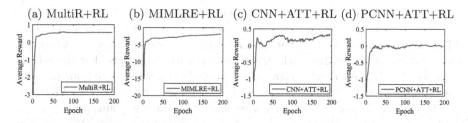

Fig. 3. Training curves tracking the RL model's average reward achieved per episode for models (a) MultiR+RL, (b) MIMLRL+RL, (c) CNN+ATT+RL and (d) PCNN+ATT+RL on the dataset NYT.

Table 1. Precision, Recall and F_1-score of the compared models on three datasets. The RL models use the same sentence-level extractor as in the baseline models, and apply the proposed *best from top k* strategy. The average F_1-score improvement of RL models over the baseline models is 5.13%.

System	NYT			Wiki-KBP			BioInfer		
	P	R	F_1	P	R	F_1	P	R	F_1
MultiR	0.756	0.371	0.497	0.444	0.427	0.435	0.102	0.087	0.094
MultiR+RL	0.731	0.412	**0.527**	0.421	0.620	**0.501**	0.117	0.114	**0.116**
MIMLRE	0.529	0.506	0.517	0.489	0.461	0.475	0.059	0.049	0.054
MIMLRE+RL	0.722	0.427	**0.537**	0.515	0.677	**0.585**	0.073	0.148	**0.097**
CNN+ATT	0.965	0.426	0.591	0.654	0.680	0.667	0.119	0.100	0.109
CNN+ATT+RL	0.773	0.773	**0.773**	0.652	0.700	**0.675**	0.113	0.119	**0.116**
PCNN+ATT	0.938	0.504	0.656	0.604	0.593	0.599	0.193	0.179	**0.186**
PCNN+ATT+RL	0.764	0.779	**0.772**	0.592	0.637	**0.614**	0.190	0.176	0.183

Figure 3 shows the training curves tracking the average reward achieved per episode for each RL model on the dataset NYT. We can see from the figures that our RL models are able to improve the performance of all the traditional distant supervision relation extraction models in a stable manner. The same conclusion can be derived from the results on other datasets.

5.6   Analysis and Case Study

We analyze the influence of different k values for RL models with *best from top k* strategy. Figure 4 shows precision, recall and F_1-score of the compared approaches on the NYT dataset. Sub-figure (a) shows the comparison results of MultiR, MultiR+RL ($k = 1$) and MultiR+RL ($k = 3$). The MultiR method uses the *top 1* strategy. The difference between MultiR and MultiR+RL ($k = 1$) is that they use different methods in entity-pair-level to make the final prediction. We can see from sub-figure (a) that MultiR+RL ($k = 3$) achieves the best F_1-score. The same observations can be derived from other sub-figures. It illustrates that models applying the *best from top k* strategy ($k = 1$) is as good as those applying the *top 1* strategy, and models applying the *best from top k* strategy ($k > 1$) can achieve significant improvements compared with the baselines.

Table 2 shows three examples of sentence-level relation extraction for PCNN+ ATT and PCNN+ATT+RL. For the first sentence, both models select the correct relation for the entities, which are labeled with subscripts. For the second sentence, PCNN+ATT selects the most possible but wrong relation, i.e., *NA* based on the predicted confidence scores in the brackets, while PCNN+ATT+RL selects the correct relation, i.e., *nationality*. This case explains why our RL models can achieve higher recalls than baseline models in most cases. For the third sentence, PCNN+ATT selects a wrong relation, i.e., *NA*, while PCNN+ATT+RL rejects all the predicted relations by PCNN+ATT since the

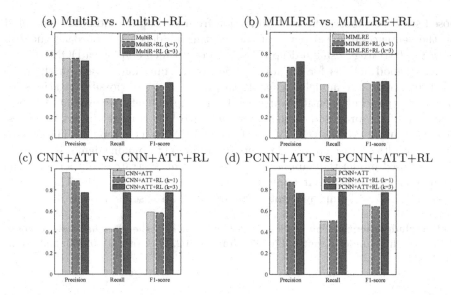

Fig. 4. Precision, Recall and F_1-score of the compared models that use the *top 1* strategy and the *best from top k* strategy ($k = 1$ and $k = 3$) on the NYT dataset.

Table 2. Relation extraction examples by different models. The correct relations between entities in the three sentences are *company*, *nationality* and *contains*.

Test sentence	PCNN+ATT	PCNN+ATT+RL
mel karmazin$_1$, the chief executive of **sirius satellite radio**$_2$, made a lot of ... radio on monday	**company(0.790)** NA(0.170) place_of_birth(0.011)	Company
a young cape verdean singer who was born in **portugal**$_2$, **lura**$_1$ specializes in bubbly, ... by cesaria evora	**NA(0.387)** nationality(0.159) place_lived(0.075)	Nationality
despite **madrid**$_2$'s efforts to catch up, barcelona arguably remains the design capital of **spain**$_1$, and vinçon ...	**NA(0.842)** nationality(0.042) place_of_birth(0.024)	/

correct relation, i.e., *contains* is not in the top 3 candidates. This case indicates that our RL models are able to prevent potential errors. It is clearly show that our model can do better relation extraction than traditional state-of-the-art distant supervision relation extraction models.

6   Conclusions

This paper proposed the *best from top k* strategy to improve existing distant supervision relation extraction models, which use the *top 1* strategy. The proposed strategy chooses the best prediction from the top k candidates generated by the sentence-level extractor of the existing models. We approach the *best from top k* strategy using a deep RL framework, which employs a DQN to learn to select good actions for optimizing the reward function. Based on the deep RL framework, our model is capable to predict a set of possible relations for each entity pair in the entity-pair-level. In the experiments, we evaluate the performance of the proposed model compared with four state-of-the-art baselines, i.e., the MultiR, MIMLRE, CNN+ATT and PCNN+ATT models. The experimental results on three public datasets from different domains demonstrate that the proposed model that applies the *best from top k* strategy outperforms the comparative baselines that apply the *top 1* strategy significantly. The average F_1-score has 5.13% improvement compared with all baseline models.

Acknowledgements. This work is partially funded by the National Science Foundation of China under Grant 61170165, Grant 61702279, Grant 61602260, and Grant 61502095.

References

1. Feng, J., Huang, M., Zhao, L., Yang, Y., Zhu, X.: Reinforcement learning for relation classification from noisy data. In: Proceedings of AAAI 2018 (2018)
2. Hoffmann, R., Zhang, C., Ling, X., Zettlemoyer, L., Weld, D.S.: Knowledge-based weak supervision for information extraction of overlapping relations. In: Proceedings of ACL 2011, pp. 541–550 (2011)
3. Koch, M., Gilmer, J., Soderland, S., Weld, D.S.: Type-aware distantly supervised relation extraction with linked arguments. In: Proceedings of EMNLP 2014, pp. 1891–1901 (2014)
4. Lin, Y., Shen, S., Liu, Z., Luan, H., Sun, M.: Neural relation extraction with selective attention over instances. In: Proceedings of ACL 2016, pp. 2124–2133 (2016)
5. Ling, X., Weld, D.S.: Fine-grained entity recognition. In: Proceedings AAAI 2012, vol. 12, pp. 94–100 (2012)
6. Lockard, C., Dong, X.L., Einolghozati, A., Shiralkar, P.: CERES: distantly supervised relation extraction from the semi-structured web. In: Proceedings of VLDB 2018, pp. 1084–1096 (2018)
7. Lourentzou, I., Alba, A., Coden, A., Gentile, A.L., Gruhl, D., Welch, S.: Mining relations from unstructured content. In: Phung, D., Tseng, V.S., Webb, G.I., Ho, B., Ganji, M., Rashidi, L. (eds.) PAKDD 2018. LNCS, vol. 10938, pp. 363–375. Springer, Cham (2018). https://doi.org/10.1007/978-3-319-93037-4_29
8. Mintz, M., Bills, S., Snow, R., Jurafsky, D.: Distant supervision for relation extraction without labeled data. In: Proceedings of ACL 2009, pp. 1003–1011 (2009)
9. Mnih, V., et al.: Human-level control through deep reinforcement learning. Nature **518**(7540), 529–533 (2015)
10. Narasimhan, K., Yala, A., Barzilay, R.: Improving information extraction by acquiring external evidence with reinforcement learning. In: Proceedings of EMNLP 2016, pp. 2355–2365 (2016)
11. Pyysalo, S., et al.: BioInfer: a corpus for information extraction in the biomedical domain. BMC Bioinform. **8**(1), 50 (2007)

12. Qin, P., Xu, W., Wang, W.Y.: Robust distant supervision relation extraction via deep reinforcement learning. In: Proceedings of ACL 2018, pp. 2137–2147 (2018)
13. Riedel, S., Yao, L., McCallum, A.: Modeling relations and their mentions without labeled text. In: Balcázar, J.L., Bonchi, F., Gionis, A., Sebag, M. (eds.) ECML PKDD 2010. LNCS, vol. 6323, pp. 148–163. Springer, Heidelberg (2010). https://doi.org/10.1007/978-3-642-15939-8_10
14. Ritter, A., Zettlemoyer, L., Etzioni, O., et al.: Modeling missing data in distant supervision for information extraction. Trans. Assoc. Comput. Linguist. 1, 367–378 (2013)
15. Surdeanu, M., Tibshirani, J., Nallapati, R., Manning, C.D.: Multi-instance multi-label learning for relation extraction. In: Proceedings of EMNLP 2012, pp. 455–465 (2012)
16. Sutton, R.S., Barto, A.G.: Introduction to Reinforcement Learning. MIT Press, Cambridge (1998)
17. Watkins, C.J., Dayan, P.: Q-learning. Mach. Learn. 8(3–4), 279–292 (1992)
18. Zeng, D., Liu, K., Chen, Y., Zhao, J.: Distant supervision for relation extraction via piecewise convolutional neural networks. In: Proceedings of EMNLP 2015, pp. 1753–1762 (2015)
19. Zeng, X., He, S., Liu, K., Zhao, J.: Large scaled relation extraction with reinforcement learning. In: Proceedings of AAAI 2018 (2018)
20. Zhou, G., Su, J., Jie, Z., Zhang, M.: Exploring various knowledge in relation extraction. In: Proceedings of ACL 2005, pp. 427–434 (2005)

Towards One Reusable Model for Various Software Defect Mining Tasks

Heng-Yi Li, Ming Li$^{(\boxtimes)}$, and Zhi-Hua Zhou

National Key Laboratory for Novel Software Technology, Nanjing University,
Nanjing 210023, China
{lihy,lim,zhouzh}@lamda.nju.edu.cn

Abstract. Software defect mining is playing an important role in software quality assurance. Many deep neural network based models have been proposed for software defect mining tasks, and have pushed forward the state-of-the-art mining performance. These deep models usually require a huge amount of task-specific source code for training to capture the code functionality to mine the defects. But such requirement is often hard to be satisfied in practice. On the other hand, lots of free source code and corresponding textual explanations are publicly available in the open source software repositories, which is potentially useful in modeling code functionality. However, no previous studies ever leverage these resources to help defect mining tasks. In this paper, we propose a novel framework to learn one reusable deep model for code functional representation using the huge amount of publicly available task-free source code as well as their textual explanations. And then reuse it for various software defect mining tasks. Experimental results on three major defect mining tasks with real world datasets indicate that by reusing this model in specific tasks, the mining performance outperforms its counterpart that learns deep models from scratch, especially when the training data is insufficient.

Keywords: Software defect mining · Machine learning · Model reuse

1 Introduction

Software Quality Assurance (SQA) is vital in software engineering and one of the biggest influencing factors is *software defects* (also referred as *bugs*). There have been many ways to find bugs, such as conducting software testing. Recently, software defect mining, which leverages data mining techniques to help identifying the software defects, has shown its advantages in reducing the software testing resources, and drawn significant attention.

Various software defect mining tasks can be employed to identify software defects. The major tasks are: *software clone detection*, *defect prediction* and *bug localization*. In software engineering, copy-pasting existing code snippets can usually cause bug propagation. If one code snippet contains a bug, all other

© Springer Nature Switzerland AG 2019
Q. Yang et al. (Eds.): PAKDD 2019, LNAI 11441, pp. 212–224, 2019.
https://doi.org/10.1007/978-3-030-16142-2_17

snippets similar to it may also exist the same bug [13]. Therefore, software clone detection aims to mine such bugs by identifying the cloned code snippets. Apart from that, defect prediction is to directly check if a certain software module contains bugs before a software system releasing, while bug localization refers to locate buggy source code based on bug reports written in natural language submitted by the users after the system releasing.

```
/*Code1: factorial*/          /*Code2: factorial*/          /*Code3: cumulative sum*/
public static void fac(int n) public static void fac(int n) public static void csum(int n)
{                             {                             {
    if(n==0)                      int i,re=1;                   int i,re=0;
        return 1;                 for(i=1;i<=n;i++)             for(i=1;i<=m;i++)
    else                              re=re*i;                      re=re+i;
        return n*fac(n-1);        return re;                    return re;
}                             }                             }
```

Fig. 1. An example of three Java code snippets with comments. Code1 and Code2 are similar with the same functionality shown by "factorial" though implemented in different ways (i.e., for-loop and recursion). Code2 and Code3 are dissimilar in functionality shown by "factorial" and "cumulative sum" though nearly the same in appearance.

Many methods have been proposed for these mining tasks. The most common way is to design hand-crafted features for specific mining tasks, such as sequence features, AST (Abstract Syntax Tree) features and PDG (Program Dependence Graph) features in clone detection [1,7,11], software metrics in defect prediction [3,9], bag-of-words features in bug localization [4,21]. Recently, deep neural networks have been applied to tackle software defect mining tasks. Wei and Li [18] address the clone detection problem with deep learning model equipped with AST-based LSTM (Long Short Term Memory) and learning to hash. Huo et al. [6] propose a novel deep model structure based on CNN (Convolutional Neural Network) to learn unified features from both bug reports and source code. They also improve it by taking LSTM to capture the sequential nature of source code [5]. All these deep models have significantly pushed forward the state-of-the-art performance in various software defect mining tasks.

To achieve such promising performance, deep models usually require a huge amount of training data. However, acquiring sufficient number of training data and their labels is usually difficult for software defect mining tasks. For example, after a software system releasing, it takes long time for underlying bugs to be exposed to users for firing bug reports, and hence the number of bug reports that can be used to train the model is small; additionally, much human effort is required to locate the buggy source code from the code bases. Similar problems hold for software clone detection and defect prediction. Therefore, these proposed deep models may not perform as well as they should be in practice.

On the other hand, there has been huge amounts of source code as well as their corresponding textual explanations in the open source software repositories (e.g., SourceForge[1]) and technical forums that discuss and share source code

---

[1] https://sourceforge.net/.

(e.g., Stack Overflow[2]). These data is publicly available, but is not collected and preprocessed for any particular software mining tasks. One question arises: can we leverage the huge amount of task-free data to help software defect mining tasks with insufficient training data?

Intuitively, if the source code functionality is correctly modeled, it would be apparent to determine whether the code behaves as it is expected to (i.e., whether it contains defects). Thus, the key is to effectively model the functionality of source code which can be reused in many software defect mining tasks to further assist to mine the defect. However, it is sometimes difficult even for software maintenance engineers to determine the source code functionality solely based on the code itself [15], since the same functionality can be implemented in various ways (e.g., summation implemented with for-loop and recursion) and source code similar in appearance may carry different meanings, especially when it is freely written. In this case, additional textual information (e.g., code comments, design documents) may be further referred to. An example of three code snippets with comments is given in Fig. 1 to show how the textual information helps.

In this paper, we propose a novel approach to learn one ReUsable deep Model RUM for the functional representation of source code, which is trained with the huge amount of publicly available source code resources. It is obvious that the code functionality can be well captured with the help of textual information. Unluckily, detailed textual information even comments for source code in specific task is always missing. Therefore, our approach first leverages both source code and their corresponding textual explanations which can be available in public source resources to derive a text-enriched code functionality space. Based on this space, a reusable code functional representation model RUM, which only leverages source code, is constructed by aligning the learned representation towards its counterpart in the text-enriched code functionality space. Such a reusable model can be plugged into different software defect mining tasks with moderate adaptation over the task-specific data to generate the text-enriched functional representations even if no additional textual information available for the specific task. The experimental results on three major software defect mining tasks (i.e., software clone detection, defect prediction and bug localization) with real world datasets indicate that by using this model to generate functional representations for task-specific source code, the mining performance outperforms that learns deep models from scratch, especially when the task-specific training data is insufficient.

2   The Proposed Approach

The goal of the proposed approach is to learn a good code functional representation model using the huge amount of publicly available task-free source code resources and then reuse it for many specific software defect mining tasks.

---

[2] https://stackoverflow.com/.

Let $\mathcal{O} = \{o_1, o_2, ..., o_N\}$ denotes the code-text set, where $o_i = (c_i, t_i)$, c_i and t_i denote the i-th raw code snippet and corresponding textual comment respectively, N is set size. Let $\mathcal{C} = \{c_1, c_2, ..., c_N\}$ denotes the code set from $\mathcal{O}$.

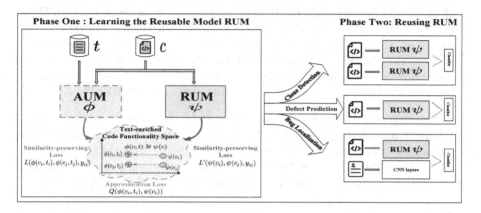

Fig. 2. The framework of learning and reusing RUM, containing two phases. In phase one, we learn a feature mapping ψ in RUM with the help of feature mapping ϕ in AUM. In phase two, we reuse ψ shown in red color in three software defect mining tasks. (Color figure online)

The framework of the proposal approach is shown in Fig. 2. It contains two phases. The first phase shown in the left part is to learn a ReUsable code functional representation Model RUM which accepts source code input. With only code information, semantic functionality is hard to model since the same functionality can be implemented with different lexical or syntactic ways and similar code functionalities are always with similar textual comments (e.g., factorial and cumulative sum shown in Fig. 1). Therefore, we first build an AUxiliary Model AUM to leverage both code and comments to learn the text-enriched code functionality representation space, resulting in the feature mapping $\phi(c, t)$. To further utilize the space, we design a approximation mechanism for RUM with feature mapping $\psi(c, t)$ to align the learned representation to its counterpart, i.e., $\psi(c_i) \cong \phi(c_i, t_i)$, such that it can implicitly encode textual information.

In the second phase shown in the right part, we plug the reusable feature mapping $\psi(c)$ in RUM into different task-specific deep models to replace the code feature extraction substructure, and adapt it towards the task with a small amount of task-specific data. Here, we employ simple fine-tuning technique to RUM in the purpose of verifying the feasibility of our approach. However, any advanced model adaptation techniques can be employed and better performance can be expected. To provide concrete examples on how to reuse RUM in specific defect mining tasks, we select the aforementioned three major software defect mining tasks, namely clone detection, defect prediction and bug localization:

- **RUM for Clone Detection.** Given a source code set $(c_1, ..., c_N)$, the goal is to predict if (c_i, c_j) belong to a clone pair. For this task, double substructures of $\phi(\cdot)$ are reused for pairwise input. The fully connected layers are followed as the classifier to make a prediction. We denote this model as RUM^{cd}.

- **RUM for Defect Prediction.** Given a source code set $(c_1, ..., c_N)$, the goal is to classify if c_i is defective. For this mining task, we reuse the substructure of $\psi(\cdot)$ and add fully connected layers as classifier. It is denoted as RUM^{dp}.
- **RUM for Bug Localization.** Given a source code set $(c_1, ..., c_M)$ and bug report set $(r_1, ..., r_N)$, the goal is to identify the association y_{ij} between c_i and r_j. For this mining task, we employ the deep model proposed in [6]. Especially, we replace the substructure responsible for source code with the reusable structure $\psi(\cdot)$ in RUM. We denote this model as RUM^{bl}.

It is noteworthy that any software mining tasks, even not for mining defects, may benefit from RUM if they need to model the code functionality. The key of our approach lies in how to derive the reusable text-enriched code functional representation model, i.e., how to learn the feature mapping $\phi(\cdot, \cdot)$ in AUM and the feature mapping $\psi(\cdot)$ in RUM, which will be discussed in the following.

2.1   Auxiliary Model

Auxiliary model AUM is designed to learn a feature mapping $\phi(\cdot, \cdot)$ from $o_i = (c_i, t_i)$ to a text-enriched functionality space where the source code with similar functionality should be mapped close to each other and dissimilar ones should be apart. According to [10], such learning task can be formalized as a binary classification problem that attempts to learn a prediction function $f \colon \mathcal{O} \times \mathcal{O} \mapsto \mathcal{Y}$. $y_{ij} \in \mathcal{Y} = \{0, 1\}$ indicates whether a pair of input $o_i, o_j \in \mathcal{O}$ is similar or not. Specifically, we employ L_1-distance to weight the affinity of input pairs, and the probability of a pair (o_i, o_j) to be similar can be computed as $f = \sigma(\boldsymbol{\alpha}^\mathsf{T} |\phi(o_i) - \phi(o_j)|)$, where σ is the sigmoid activation function and $\boldsymbol{\alpha}$ is the parameter to be learned. We solve the learning problem by optimizing the following regularized similarity-preserving loss function:

$$\min_f \quad \mathcal{L} + \lambda \Omega(f), \tag{1}$$

where $\mathcal{L}$ is the cross-entropy loss, $\Omega(f)$ is the L_2 regularization term and λ is the trade-off parameter. This objective can be effectively optimized using SGD.

Note that the number of source code with similar functionality is usually far less than that with dissimilar functionality. Such imbalanced distribution may severely affect the quality of learned code functionality space. To reduce the influence, we impose a larger cost for miss-classifying the similar code pairs (denoted by cost_{fn}) and a smaller cost for miss-classifying the dissimilar code pairs (denoted by cost_{fp}). Therefore, $\mathcal{L}$ can be defined as:

$$\mathcal{L} = \sum_{i,j} (\text{cost}_{fp}(1 - y_{ij}) \log(1 - f(o_i, o_j) + \text{cost}_{fn} y_{ij} \log f(o_i, o_j)). \tag{2}$$

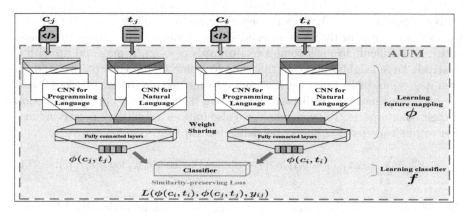

Fig. 3. The siamese structure of AUM than is weighted sharing for (c_i, t_i) and (c_j, t_j).

We instantiate the auxiliary network with siamese convolutional neural network. Siamese network [2], consisting of two identical neural networks with their weights tied, is usually employed to differentiate the paired input data points [10,12]. Thus, we leverage the siamese structure to help modeling the similarity of source code pair. The network structure of AUM is shown in Fig. 3. Source code is always written in programming language in which multiple continues statements is constructed in a block to convey the information, e.g., for-loops and while-loops. While text is written in natural language in a flat way that several words together can express the complete meanings. Thus, we design two different feature extraction modules for each in AUM. In the code feature extraction layers, since convolutional neural networks have shown great performance in [6], we use the same convolution layers with specific convolution operations for code structure to extract the semantic features for source code. In the text feature extraction layers, we use the standard approach in [8] to extract the text features. Next, we use fully connected layers to further fuse the code features and text features and get final representations. The above feature extraction layers are weight-shared for a pair of input (c_i, t_i) and (c_j, t_j) to get unbiased representations. In the end, fully connected layers followed by a sigmoid layer are constructed to build a classifier based on fused representations for optimization objective (i.e., similarity-preserving loss).

2.2 Reusable Model

Reusable model RUM aims to learn a feature mapping $\psi(\cdot)$ from c_i to the same text-enriched functionality space by aligning it to its text-enriched counterpart $o_i = (c_i, t_i)$. To achieve it, we force $\psi(c_i) \cong \phi(c_i, t_i)$ by imposing approximation loss over the distances between $\psi(c_i)$ and $\phi(c_i, t_i)$, as defined in Eq. (3):

$$\mathcal{Q} = \sum_i \|\psi(c_i) - \phi(c_i, t_i)\|_2^2. \tag{3}$$

By minimizing Eq. (3), we can squeeze $\psi(c_i)$ into the neighborhood of $\phi(c_i, t_i)$ from all directions in the text-enriched space. However, when some hard cases occur, it is difficult to push some $\psi(c_i)$ towards $\phi(c_i, t_i)$, it may end up with a relatively large neighborhood. In this case, for some similar code pairs c_i and c_j, even if $\psi(c_i)$ and $\psi(c_j)$ may be squeezed into the neighborhoods of $\phi(c_i, t_i)$ and $\phi(c_j, t_j)$, they may still be distant by mapping them from the opposite direction of their counterparts. To overcome it, we also impose a similarity-preserving loss over source code pairs c_i and c_j, as defined in Eq. (4):

$$\mathcal{L}' = \sum_{i,j} (\text{cost}_{fp}(1 - y_{ij}) \log(1 - g(c_i, c_j)) + \text{cost}_{fn} y_{ij} \log g(c_i, c_j)). \tag{4}$$

Therefore, we solve the problem of learning the reusable model RUM by optimizing the following regularized objective loss function,

$$\min_{g} \quad \mathcal{Q} + \beta \mathcal{L}' + \lambda \Omega(g), \tag{5}$$

where $\Omega(g)$ is the L_2 regularization term, β and λ' are the trade-off parameters.

We instantiate this learning task also by siamese convolutional neural network. The network structure of RUM is the same as that of AUM except for the text feature extraction layers since RUM only takes source code as the input. Similar to AUM, the feature extraction layers are weight-shared and a classifier is built based on the approximation representations for optimization objective.

3   Experiment

In the experiment, we first show how good the functional representation learned by RUM is and then we show the benefit of reusing RUM for various software defect mining tasks.

3.1   How Good Is RUM

In this section, we conduct experiments on the real-world dataset *Stack Overflow* downloaded from Stack Exchange[3] to evaluate the performance of identifying functional similar code pairs based on the learned representations by RUM.

The dataset contains 8237 questions (text) and 8237 answers (code), in which each question is along with a answer. In order to get the similarity label, we label dual problems that are with similar question as similar pairs and generate dissimilar pairs from non-dual problems, which totally get 16839 pairs.

Since RUM is benefit from the text-enriched functionality space, we compare RUM with RSiaCNN which learns functional representations only from code. In both, the network parameters are chosen as follow: the convolution filter size in code feature extraction layers is 3, 4 with 100 feature maps each and in text

---

[3] https://archive.org/details/stackexchange.

feature extraction layers is 2, 3 with 100 feature maps each. We set experiment dropout probability $p = 0.5$ and activation function $ReLU(x) = \max(x, 0)$.

The performance ratio of RSiaCNN and RUM over AUM in terms of AUC is 91.4% and 95.4%, respectively. Thus, the performance can be improved by nearly 4% if trained with the help of text-enriched space. The similar conclusion can be observed in terms of F1 which improved by 6%. Next, we will verify the effectiveness of RUM on three major defect mining tasks.

3.2 Reusing RUM for Clone Detection

BigCloneBench [16] is a widely used benchmark dataset with known true and false clones. Following [18], we extract 6282 code snippets as dataset. Since *BigCloneBench* is highly imbalanced, we measure the performance in terms of AUC, F1 and Recall. Besides, Top k Rank ($k = 10$) is recorded to measure the retrieval performance. We first compare RUM^{cd} to the state-of-the-art deep models DeepClone [19] and CDLH [18]. Further more, we compare with our variants SiaCNN, which is with the same structure as RSiaCNN but trained from scratch, and $RSiaCNN^{cd}$ to evaluate the effective of enriched text information.

Table 1. Top 10 Rank and AUC of all methods on training data with different sizes.

Methods	Top 10 Rank				AUC			
	50	100	250	500	50	100	250	500
DeepClone	.894 ○	.894 ○	.894 ○	.894 ○	.483 ○	.483 ○	.483 ○	.483 ○
CDLH	.795 ○	.774 ○	.794 ○	.858 ○	.500 ○	.500 ○	.500 ○	.500 ○
SiaCNN	.659 ○	.698 ○	.753 ○	.852	.507 ○	.533 ○	.650 ○	.757 ○
$RSiaCNN^{cd}$	.808 ○	.866 ○	.876 ○	.905	.609 ○	.665	.781	.829
RUM^{cd}	**.912**	**.917**	**.933**	**.935**	**.621**	**.666**	**.794**	**.838**

We randomly sample 5000 code for training and 500 code for testing, resulting in 25000000 (5000 × 5000) training pairs and 250000 (500 × 5000) test pairs. To evaluate the performance on small datasets, we use only small sizes of the train pairs, assuming N ($N = 50, 100, 250, 500, 750, 1000$ respectively), to train and test on all test pairs which is large enough to prove our performance.

All experiments are randomly repeated 30 times and we report the average results. The performance with respect to Top k Rank and AUC of all methods on different training samples are tabulated in Table 1 where the best performance on each dataset is boldfaced. The performance with respect to F1 score is depicted in Fig. 4. We conduct Pairwise t-test at 95% confidence level. The compared methods that are significant inferior than our approach will be marked with "○" and significant better will be marked with "●".

From the results in Table 1, we can observe that when training size is very small, e.g. 100, RUM^{cd} can achieve the best performance (0.917) in terms of

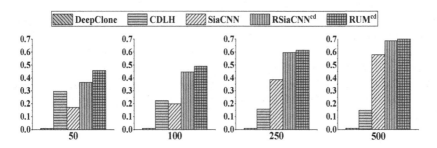

Fig. 4. F1 score of all methods on training data with different size.

Top 10 Rank which improves CDLH (0.774) by 14.3% and SiaCNN (0.698) by 21.9% since they are easy to overfit. When compared to the unsupervised method DeepClone which trained with all code, we still get 1.8% improvement with only 50 training samples. Similar traces can be found in Fig. 4. The superiority of RUM is obvious. We further evaluate effectiveness of text information. Indicted in Table 1, the performance of RUM^{cd} is better than $RSiaCNN^{cd}$ and improves by 3%–9% in terms of Top 10 Rank. It shows that encoded text information is beneficial when the code information is not enough. Taking a concrete example of the cloned pair Code1 and Code2 in Fig. 1, we can get the similar enriched representations by using RUM with the help of comments "factorial" though they are very dissimilar in lexical structure.

3.3   Reusing RUM for Defect Prediction

For defect prediction task, we conduct experiments on the widely used benchmark datasets [22]. It contains source code files and detailed software metric information, such as complexity metrics (e.g., number of methods calls, total lines of code), structure of AST and so on. It also gives the number of defects that are reported in the first six months before and after releasing, named as pre-release defects and post-release defects respectively. In this experiment, we use three different projects as our datasets and use the number of post-release defects as prediction label. The statistics of the datasets can be found in Table 2.

As indicated in Table 2, the number of defective source code is very imbalanced. Therefore, we use AUC to evaluate the performance. We compare RUM^{dp} with a baseline Logistic Regression LR and two state-of-the-art methods

Table 2. Statistics of three datasets in defect prediction.

Datasets	# attributes	# instances	# defective
Debug	198	194	13%
UI	198	1166	4%
SWT	198	841	17%

DBN [20] and AST-DBN [17]. Besides, we compare with two variants RSiaCNN^{dp} and P-CNN [6] which is with the same code feature extraction structure as RUM but trained from scratch. For each dataset, we randomly sample 30% data to train and the remaining data to test. All experiments are randomly repeated 30 times and the average results are reported in Fig. 5.

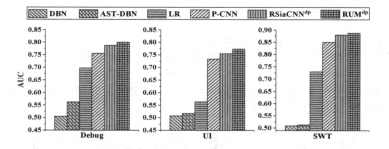

Fig. 5. AUC of compared methods on all datasets in defect prediction.

It can be observed from Fig. 5 that RUM^{dp} achieves the best average performance (0.821) among all compared approaches. Compared with LR trained with software metric features (0.664), RUM^{dp} can improve the average performance by 15.7%. When compared with deep models DBN (0.507) and AST-DBN (0.530), RUM^{dp} can also improve by 31.4% and 29.0%. It is notable that the performance of DBN and AST-DBN is even worse than LR, we explain that DBN extracts code features in an unsupervised way. To evaluate the effectiveness of reusable code functional representations, we use P-CNN for comparison. It is clearly that RUM^{dp} can improve the performance of P-CNN (0.780) by 4.1%. Also, we compare RUM^{dp} with RSiaCNN^{dp} to evaluate the effectiveness of encoded text information. Though RUM^{dp} is fine-tuned without any text input, it can still improve RSiaCNN^{dp} (0.807) by 1.4% on average. Therefore, the encoded text information is useful for finding more defective modules. Here we give a more intuitive explanation for the usefulness of text. If one aims to get factorial function but wrongly writes Code3 in Fig. 1, the encoded text information "cumulative summation" in the reusable representation can help the detection.

3.4 Reusing RUM for Bug Localization

In bug localization, we extract different well-known open source software projects and the ground truth of relevance of bug reports and source files using bug tracking system (Bugzilla) and version control system (Git), following [6]. We use matched code-report pair as positive instance. To generate the negative instance, we label the reports with irrelevant code files as negative. Table 3 shows the detailed information of used datasets.

We use Top k Rank ($k = 10$) to measure our performance, which has been widely applied for evaluation in information retrieval based bug localization

Table 3. Statistics of three datasets in bug localization.

Datasets	#source files	#bug reports	#total matches
Debug	249	132	301
UI	1152	314	698
JDT	1980	1005	1610

problems [14,21]. We compare our method with the state-of-the-art methods NP-CNN [6], LSTM-CNN [5] and RSiaCNN^{bl}. For each dataset, we randomly sample 30% data to train our model, and test on the remaining data. All experiments are randomly repeated 30 times and we report the average results. The performance is depicted in Fig. 6.

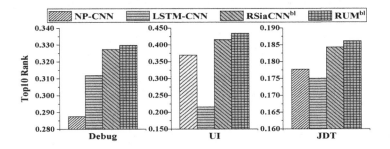

Fig. 6. Top 10 Rank of compared methods on all datasets in bug localization.

Figure 6 indicates that RUM^{bl} achieves the best average Top k Rank at 0.317, which improves the average performance of NP-CNN (0.279) by 3.8% and LSTM-CNN (0.234) by 8.3%. It should be notable that LSTM-CNN performs better than NP-CNN when the number of training samples is small, we explain that LSTM without any dropout layers is more easy to overfit. Further to evaluate the effectiveness of text information, we use RSiaCNN^{bl} for comparison, which is pretrained with only source code. It can be observed that RUM^{bl} improves RSiaCNN^{bl} (0.309) by 0.8% on average in terms of Top k Rank, indicating that encoded text information is also helpful to locate buggy source files. For a more clear explanation, assuming that we are given the bug report "I always get the same value for factorial of n" and one aims to locate some factorial function containing bugs, then the encoded text "factorial" in code representations is consistent to report description "factorial" and is useful in localization.

4   Conclusion

In this paper, we propose a novel framework to learn one deep model for the code functional representation using the huge amount of publicly available task-free source code and their textual comments, and reuse it for many software defect mining tasks. Experimental results on three major software defect mining tasks indicate that by reusing this model in specific task, the mining performance outperforms its counterpart that learns deep models from scratch, especially when the task-specific training data is insufficient.

Acknowledgment. This research was supported by National Key Research and Development Program (2017YFB1001903) and NSFC (61751306).

References

1. Alemi, M., Haghighi, H., Shahrivari, S.: CCFinder: using Spark to find clustering coefficient in big graphs. J. Supercomput. **73**(11), 4683–4710 (2017)
2. Bromley, J., Guyon, I., LeCun, Y., Säckinger, E., Shah, R.: Signature verification using a Siamese time delay neural network. In: Advances in Neural Information Processing Systems, pp. 737–744 (1993)
3. D'Ambros, M., Lanza, M., Robbes, R.: Evaluating defect prediction approaches: a benchmark and an extensive comparison. Empir. Softw. Eng. **17**(4–5), 531–577 (2012)
4. Gay, G., Haiduc, S., Marcus, A., Menzies, T.: On the use of relevance feedback in IR-based concept location. In: Proceedings of the 25th IEEE International Conference on Software Maintenance, pp. 351–360 (2009)
5. Huo, X., Li, M.: Enhancing the unified features to locate buggy files by exploiting the sequential nature of source code. In: Proceedings of the 26th International Joint Conference on Artificial Intelligence, pp. 1909–1915 (2017)
6. Huo, X., Li, M., Zhou, Z.H.: Learning unified features from natural and programming languages for locating buggy source code. In: Proceedings of the 25th International Joint Conference on Artificial Intelligence, pp. 1606–1612 (2016)
7. Jiang, L., Misherghi, G., Su, Z., Glondu, S.: DECKARD: scalable and accurate tree-based detection of code clones. In: Proceedings of the 29th International Conference on Software Engineering, pp. 96–105 (2007)
8. Johnson, R., Zhang, T.: Effective use of word order for text categorization with convolutional neural networks. In: Proceedings of the 2015 Conference of the North American Chapter of the Association for Computational Linguistics: Human Language Technologies, pp. 103–112 (2015)
9. Kim, S., Zimmermann, T., Whitehead Jr., E.J., Zeller, A.: Predicting faults from cached history. In: Proceedings of the 29th International Conference on Software Engineering, pp. 489–498 (2007)
10. Koch, G., Zemel, R., Salakhutdinov, R.: Siamese neural networks for one-shot image recognition. In: Proceedings of the 32nd International Conference on Machine Learning Deep Learning Workshop, vol. 2 (2015)
11. Komondoor, R., Horwitz, S.: Using slicing to identify duplication in source code. In: Cousot, P. (ed.) SAS 2001. LNCS, vol. 2126, pp. 40–56. Springer, Heidelberg (2001). https://doi.org/10.1007/3-540-47764-0_3
12. Mueller, J., Thyagarajan, A.: Siamese recurrent architectures for learning sentence similarity. In: Proceedings of the 30th AAAI Conference on Artificial Intelligence, pp. 2786–2792 (2016)
13. Roy, C.K., Cordy, J.R.: A survey on software clone detection research. Queen's Sch. Comput. TR **541**(115), 64–68 (2007)

14. Saha, R.K., Lease, M., Khurshid, S., Perry, D.E.: Improving bug localization using structured information retrieval. In: Proceedings of the 28th IEEE/ACM International Conference on Automated Software Engineering, pp. 345–355 (2013)
15. de Souza, S.C.B., Anquetil, N., de Oliveira, K.M.: A study of the documentation essential to software maintenance. In: Proceedings of the 23rd Annual International Conference on Design of Communication: Documenting & Designing for Pervasive Information, pp. 68–75 (2005)
16. Svajlenko, J., Islam, J.F., Keivanloo, I., Roy, C.K., Mia, M.M.: Towards a big data curated benchmark of inter-project code clones. In: Proceedings of the 30th IEEE International Conference on Software Maintenance and Evolution, pp. 476–480 (2014)
17. Wang, S., Liu, T., Tan, L.: Automatically learning semantic features for defect prediction. In: Proceedings of the 38th International Conference on Software Engineering, pp. 297–308 (2016)
18. Wei, H.H., Li, M.: Supervised deep features for software functional clone detection by exploiting lexical and syntactical information in source code. In: Proceedings of the 26th International Joint Conference on Artificial Intelligence, pp. 3034–3040 (2017)
19. White, M., Tufano, M., Vendome, C., Poshyvanyk, D.: Deep learning code fragments for code clone detection. In: Proceedings of the 31st IEEE/ACM International Conference on Automated Software Engineering, pp. 87–98 (2016)
20. Yang, X., Lo, D., Xia, X., Zhang, Y., Sun, J.: Deep learning for just-in-time defect prediction. In: Proceedings of the 2015 IEEE International Conference on Software Quality, Reliability and Security, pp. 17–26 (2015)
21. Zhou, J., Zhang, H., Lo, D.: Where should the bugs be fixed? More accurate information retrieval-based bug localization based on bug reports. In: Proceedings of the 34th International Conference on Software Engineering, pp. 14–24 (2012)
22. Zimmermann, T., Premraj, R., Zeller, A.: Predicting defects for Eclipse. In: Proceedings of the 3rd International Workshop on Predictor Models in Software Engineering, p. 9 (2007)

User Preference-Aware Review Generation

Wei Wang, Hai-Tao Zheng$^{(\boxtimes)}$, and Hao Liu

Tsinghua-Southampton Web Science Laboratory, Graduate School at Shenzhen,
Tsinghua University, Shenzhen, China
{w-w16,liuhao17}@mails.tsinghua.edu.cn, zheng.haitao@sz.tsinghua.edu.cn

Abstract. There are more and more online sites that allow users to express their sentiments by writing reviews. Recently, researchers have paid attention to review generation. They generate review text under specific contexts, such as rating, user ID or product ID. The encoder-attention-decoder based methods achieve impressive performance in this task. However, these methods do not consider user preference when generating reviews. Only considering numeric contexts such as user ID or product ID, these methods tend to generate generic and boring reviews, which results in a lack of diversity when generating reviews for different users or products. We propose a user preference-aware review generation model to take account of user preference. User preference reflects the characteristics of the user and has a great impact when the user writes reviews. Specifically, we extract keywords from users' reviews using a score function as user preference. The decoder generates words depending on not only the context vector but also user preference when decoding. Through considering users' preferred words explicitly, we generate diverse reviews. Experiments on a real review dataset from Amazon show that our model outperforms state-of-the-art baselines according to two evaluation metrics.

Keywords: Review generation · Natural language generation · Mining review data

1 Introduction

Natural Language Generation (NLG) belongs to the subtopic of artificial intelligence and computational linguistics. The aim of NLG is generating understandable texts in human languages [15]. The progress achieved in NLG will contribute to building strong intelligent systems that can comprehend and compose human languages. Recurrent Neural Networks (RNN) have shown promising performance in text generation [1,6,18]. This advantage makes an increasing number of researchers explore a variety of NLG tasks, such as image caption [3,16,17]. Recently, researchers have paid attention to review generation [2,4,9,19,24]. For that more and more online sites allow users to express their sentiments about products by writing reviews, such as Amazon, Taobao, and Yelp. This task is

© Springer Nature Switzerland AG 2019
Q. Yang et al. (Eds.): PAKDD 2019, LNAI 11441, pp. 225–236, 2019.
https://doi.org/10.1007/978-3-030-16142-2_18

very useful for explainable recommendation, which is aimed at generating explanations rather than only predicting a numerical rating for an item.

In this work we also focus on review generation, which tries to generate human reviews under specific contexts. Ratings of reviews are first used as context to generate reviews. Specifically, a character-level RNN is applied to generate reviews with the input of a review character concatenated with a rating [9]. Conditioning on multiple ratings in different aspects to generate reviews is also similar [2]. Both methods only use rating as context and merely contain an RNN based decoder. In order to condition on user ID, product ID and rating together, an encoder-decoder based framework was proposed, which first encodes contexts into a vector and then decodes it to generate reviews [19]. Recently attention mechanism was used to enhance the encoder-decoder method, which pays attention to different contexts in each time step [4]. These methods all apply a decoder based on RNN to generate reviews and the encoder-attention-decoder method [4] achieves the best performance under multiple kinds of contexts.

However, these methods do not consider user preference when generating reviews. They only consider numeric contexts such as rating, user ID or product ID. With limited information provided by numeric contexts, these methods tend to generate generic and boring reviews like "i loved this book" or "it was a good read", which results in a lack of diversity when generating reviews for different users or products. In addition, due to the high frequency of these patterns in data, the RNN based decoder learns these patterns easily. This aggravates the problem of generic reviews.

By taking account of user preference, we propose a User Preference-Aware Review Generation model (UPRG) to improve the diversity of generated reviews. There are many words in languages while people always have their own favorite words. These words can be seen as user preference. User preference reflects the characteristics of the user. Users tend to use these words when writing something, such as online reviews. Through considering user preference, we provide rich information as a supplement to numeric contexts. Furthermore, we can control the generated reviews and produce more personalized words. The generated reviews are better interpreted. Additionally, introducing user preference into decoding process is beneficial for the decoder to learn more patterns.

UPRG is consisted of two components: context encoder and user preference augmented decoder. The context encoder coverts one-hot representations of contexts into low dimensional vectors used for decoding. The decoder produces final review depending on the context vector computed by a context attention mechanism. To address the problem of generic reviews, we incorporate user preference into the decoder. Specifically, we extract important words from users' reviews in training set using a score function. These words are treated as user preference. When decoding we decide to generate a word or copy a word from user preference using a gate function. The copying probability of each word in user preference is computed by a user attention mechanism. We conduct various experiments on real review data to evaluate our model. Experimental results on a real review dataset show that UPRG achieves higher performance than baseline methods in two evaluation metrics. The main contributions of this paper are summarized as follows:

- We utilize user preference to improve the diversity of generated reviews which efficiently alleviates the problem of generic reviews.
- We design a user preference augmented decoder which dynamically decides when to select and which word to select in user preference.
- Experiments on a real review dataset from Amazon demonstrate that our model outperforms state-of-the-art baselines according to two evaluation metrics.

The remainder of this paper is organized as follows. We review related works in Sect. 2. Section 3 introduces the detail of our proposed model. Section 4 describes dataset, experimental setups and gives the result analysis. Finally, we give a summary of this paper and discuss the future direction of improvement in Sect. 5.

2   Related Work

Natural Language Generation (NLG) belongs to the subtopic of artificial intelligence and computational linguistics. The aim of NLG is generating understandable texts in human languages [15]. The RNN based approaches of text generation have drawn more and more attention in recent years. Compared to the traditional rule-based approaches, the RNN based approaches provide an end-to-end solution without much human participation. Mikolov *et al.* [11,12] improved the performance of language modeling through the long dependency of RNN. Their methods outperformed the n-gram language modeling significantly. These work prove the effectiveness of RNN in text generation. Since then RNN has become a standard component in many tasks which contain a module of text generation, such as poetry generation [5,22,23], image caption [3,16,17], and neural machine translation [21,25].

Review generation aims to generate realistic reviews under specific contexts, which is a kind of text generation. There are some work on review generation. Lipton *et al.* [9] proposed a character-level recurrent neural networks(RNN) to generate review given auxiliary information, such as a sentiment. They concatenate auxiliary information with character as input. Costa *et al.* [2] used a similar method to generate review according to user's rating in different aspects. Both methods use rating as context and only contain a RNN based decoder. In order to condition on user ID, product ID and rating together, Tang *et al.* [19] proposed an encoder-decoder based framework, which first encodes contexts into a vector and then decodes it to generate reviews. Dong *et al.* [4] enhanced Tang *et al.*'s method through adding encoder-side attention mechanism, which pays attention to different contexts in each time step. Our method is similar to [4] but we introduce user preference to improve the diversity. Comparing to these methods, we provide rich information related to users. Through explicitly adding bias to users' favorite words in decoding phase, our method generates more novel words rather than common words for each user.

3   Methodology

To begin with, we state the problem of review generation as follows: given the contexts $x = x_1, x_2, ..., x_m$ as input, the model needs to generate corresponding review $y = y_1, y_2, ..., y_n$ by maximizing the conditional probability $p(y|x)$. x_i is one-hot representation of context such as rating, user ID, or product ID. $|x_i|$ is the number of rating categories, or the number of users, m is the number of contexts. In this work we use three contexts: user ID, product ID and rating. $y_i \in \mathbb{R}^{|V|}$ is a review token, n denotes the number of review tokens.

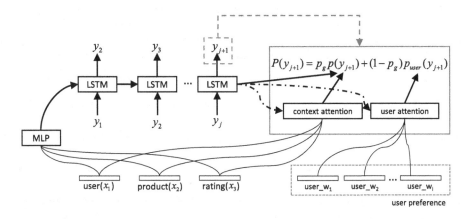

Fig. 1. User preference-aware review generation

As shown in Fig. 1, UPRG contains two parts: context encoder and user preference enhanced decoder. The encoder utilizes multilayer perceptrons (MLP) to covert one-hot representations of user, product and rating into a low dimensional vector. The vector is used for initializing the decoder. The decoder produces final review depending on two probability distributions, generating probability and copying probability. The former is conditioning on the context vector computed by a context attention. The latter is computed by a user attention on user preference. Next we first introduce some background knowledge, and then present two components respectively.

3.1   Background: Recurrent Neural Networks

Recurrent neural networks process sequence information well, which represents history information into a hidden state and then computes a probability distribution of next word according to the hidden state. However, training the basic RNN above suffers from the problem of gradient vanishing or exploding. The long-short term memory (LSTM) unit [7] addresses the problem effectively. The core idea of LSTM is introducing the memory state and multiple gating functions to control the information to be written to the memory sate, read from the memory state, and removed from the memory state. The computing process of LSTM is as follows:

$$
\begin{aligned}
i_t &= \sigma\left(W_{xi}x_t + W_{hi}h_{t-1}\right) \\
f_t &= \sigma\left(W_{xf}x_t + W_{hf}h_{t-1}\right) \\
o_t &= \sigma\left(W_{xo}x_t + W_{ho}h_{t-1}\right) \\
g_t &= \tanh\left(W_{xc}x_t + W_{hc}h_{t-1}\right) \\
c_t &= f_t \odot c_{t-1} + i_t \odot g_t \\
h_t &= o_t \odot \tanh\left(c_t\right),
\end{aligned}
\tag{1}
$$

where σ is the sigmoid function, i_t, f_t, o_t are input, forget, and output gates respectively, and g_t, c_t are proposed cell value and true cell value, h_t is the new hidden state.

3.2   User Preference-Aware Review Generation

Context Encoder: The encoder uses multilayer perceptrons with one hidden layer to encode contexts into a context vector. Each context is represented as a one-hot vector, e.g. user context. We use m different embedding matrices to convert $x = x_1, x_2, ..., x_m$ into the embedding representations. The embedding representation of x_i is computed as follows:

$$
e_i = E_i x_i,
\tag{2}
$$

where $E_i \in \mathbb{R}^{l \times |x_i|}$ is the embedding matrix for context i, l is the dimension of embedding, and $|x_i|$ is the length of one-hot vector, e.g. user numbers. Then we concatenate them and feed them into a hidden layer to get a context vector as follows:

$$
c = \tanh(W_c[e_1, e_2, .., e_m] + b_c),
\tag{3}
$$

where $c \in \mathbb{R}^{qk}$ is a qk-dimension context vector, $W_c \in \mathbb{R}^{qk \times ml}$ is a weight matrix, b_c is the bias, and q, k are the number of layers and hidden units in the decoder respectively. Next, vector c is used to initialize the decoder.

User Preference Enhanced Decoder: The decoder utilizes RNN to generated review tokens one by one according to weighted context vector computed by attention mechanism. We first get word representation e_j by the word embedding matrix $E_w \in \mathbb{R}^{k \times |V|}$ and then feed it into LSTM to get hidden vector of word j as follows:

$$
h_j = LSTM(e_j, h_{j-1}),
\tag{4}
$$

where $h_j \in \mathbb{R}^k$ is a hidden vector and $LSTM$ denotes the LSTM unit. A converting layer is applied on each context in order to transform these contexts into the same space. It is useful to make sense of the addition between contexts. The converting process of e_i is as follows:

$$
\tilde{e}_i = \tanh(W_i e_i + b_i),
\tag{5}
$$

where $W_i \in \mathbb{R}^{l \times l}$ and $b_i \in \mathbb{R}^l$ are the parameters. After that, *a context attention mechanism* is used to compute weight of each context as follows:

$$d_{ij} = v^T \tanh \left(W_e \tilde{e}_i + W_h h_j \right), \tag{6}$$

$$a_{ij} = \frac{\exp \left(d_{ij} \right)}{\sum_{k=1}^{m} \exp \left(d_{kj} \right)}, \tag{7}$$

where $v \in \mathbb{R}^k$ is a parameter vector, $W_e \in \mathbb{R}^{k \times l}$ and $W_h \in \mathbb{R}^{k \times k}$ are parameter matrices. The new context vector $\tilde{c}$ is obtained by:

$$\tilde{c}_j = \sum a_{ij} \tilde{e}_i, \tag{8}$$

where $\tilde{c}_j \in \mathbb{R}^l$ is the weighted context vector. Then the new context vector fuses with original hidden vector through a nonlinear operation as follows:

$$\tilde{h}_j = \tanh \left(W_f [h_j, \tilde{c}_j] + b_f \right), \tag{9}$$

where $W_f \in \mathbb{R}^{k \times (k+l)}$ is a parameter matrix and b_f is the bias. Now the model computes the probability of the next word via:

$$p(y_{j+1} | y \le j) = p(y_{j+1} | h_j) \propto \exp \left(W_o \tilde{h}_j + b_o \right), \tag{10}$$

where $W_o \in \mathbb{R}^{|V| \times k}$ is a parameter matrix and b_o is the bias.

In order to incorporate user preference, there are three questions to be answered: (1) how to construct user preference; (2) which word to select in user preference; (3) how to incorporate it into original decoder.

First, we need to construct user preference. In this task we use users' favorite words as user preference. We collect all words of each user from training set as candidates. Then TFIDF weight is used as the score function to grade words. We can use other methods to estimate the importance of words. We leave this as future work. Next two heuristic rules are used to filter out nonsense words. 200 words with the highest word frequency are removed for that these words are too common. In addition, words whose part of speech (POS) belongs to punctuation, article, pronoun or preposition are also removed. In the remaining words we choose up to 100 words for each user as user preference.

Second, *a user attention mechanism* is applied to compute copying probability of each word in user preference. For time j the copying probability of word w_i in user preference V_u is computed as follows:

$$u_{ij} = v_u^T \tanh \left(W_{ue} e_{wi} + W_{uh} h_j \right), \tag{11}$$

$$p_{user}(w_i) = \frac{\exp \left(u_{ij} \right)}{\sum_k \exp \left(u_{kj} \right)}, \tag{12}$$

where $v_u \in \mathbb{R}^k$ is a parameter vector, $W_{ue} \in \mathbb{R}^{k \times l}$ and $W_{uh} \in \mathbb{R}^{k \times k}$ are parameter matrices, e_{wi} is the embedding representation of w_i.

Third, in order to combine generating probability with copying probability, a gate function is computed as follows:

$$p_g = \sigma(w_c^T \tilde{c}_j + w_h^T h_j + w_y^T y_j + b_g), \tag{13}$$

where w_c, w_h, w_y are parameter vectors, b_g is the bias. p_g indicates the probability of selecting from two modes. Given p_g, the final probability of y_{j+1} is computed as follows:

$$p(y_{j+1}) = p_g p(y_{j+1}|y \leq j) + (1 - p_g)p_{user}(y_{j+1}), \tag{14}$$

where $p(y_{j+1}|y \leq j)$ computed by Eq. 10 is the generating probability, p_{user} (y_{j+1}) is the copying probability. Based on the probability of next word computed by Eq. 14 and the actual word we have, we can calculate the cross-entropy of the generated review sequence for model training.

4 Experiments

In this section, we describe our experiments and results on a real review dataset from Amazon. We first introduce the dataset for this task. Then we compare UPRG with two state-of-the-art baselines. We use two automated metrics to evaluate UPRG and detail results analysis.

4.1 Experimental Setup

Data Set. We use the processed review data[1] from [4]. It is built upon Amazon product data of book domain [10]. Every review is paired with three attributes, user ID, product ID and rating. We introduce the process of building the dataset briefly. We first filter books and users which appear less than 6 and 15 times, respectively. And then we filter reviews whose lengths are greater than 60 words. After that we get 937,033 reviews paired with attributes, which contain 80,256 books, 19,675 users, and 5 rating levels. The average length of reviews is about 35 words, and the average number of sentences is 3. We select the most popular words as the vocabulary and other words are replaced with UNK. The size of word vocabulary is 30K. Then, the whole dataset is randomly split into TRAIN, DEV and TEST (70%/10%/20%).

Setup. We use a one-layer MLP encoder and a two-layer LSTM decoder. The dimensions of all contexts are set to 64 and the dimension of word embeddings is set to 512. The dimension of hidden vectors is set to 512 in the decoder. The batch size is set to 16. All the parameters are randomly initialized by sampling from a uniform distribution $[-0.08, 0.08]$. We train the model using RMSProp [20] with learning rate 0.002 and an smoothing constant 0.95. We also clamp gradient values into the range $[-5.0, 5.0]$ to avoid the exploding gradient problem [14]. The

---

[1] The data is available at https://goo.gl/TFjEH4.

dropout layer is inserted between different LSTM layers and the dropout rate is set to 0.2 for regularization. We use early stopping on DEV set to determine the number of epochs and apply the greedy search algorithm to generate reviews at test time.

4.2   Results

We use the BLEU [13] and Distinct [8] score for automatic evaluation, which are applied in many text generation tasks. The BLEU score measures the precision of n-gram matching by comparing the generated results with references, and penalizes length using a brevity penalty term. We compute BLEU-1 (unigram) and BLEU-4 (up to 4 grams) in experiments. What's more, to evaluate the quality of diversity we compute Distinct-1/2, which is the percentage of distinct n-grams in all predicted results. Compared to the BLEU, the Distinct is related to the recall and higher scores indicate higher diversity. In Table 1 Reference denotes the score of reference reviews in TEST set, which indicates the expected higher bound. We describe the comparison methods as follows:

Random: For each test example this method randomly samples from all the reviews in the training set as generated result. This baseline method suggests the expected lower bound for this task.

Enc2Dec: This method first utilizes MLP to encode contexts into a context vector. Then a LSTM based decoder initialized by the context vector is used to generate reviews.

Enc2AttDec: This method improves Enc2Dec method with a context attention mechanism, which is similar to [4].

UPRG: This is our proposed method which is built on Enc2AttDec. The difference is that, UPRG enhances the decoder by user preference and adds a context converting layer. In order to verify the role of two components separately, we compare two versions of UPRG. Specifically, UPRGv1 uses a user preference enhanced decoder. UPRGv2 uses a context converting layer and an enhanced decoder concurrently.

The first three methods use user ID, product ID and rating as input. UPRG uses these three contexts and user preference as input.

As shown in Table 1, the result of Random is the worst in terms of BLEU score, whereas other methods taking account of contextual information are better. This shows that contextual information is important for generating reviews, especially for BLEU-4. Compared to Enc2Dec and Enc2AttDec, UPRGv1 improves the score of Distinct significantly. This demonstrates user preference contributes to generating diverse reviews. With the user preference enhanced decoder, UPRGv1 has access to more novel words directly and has a big learning capacity. Therefore, UPRGv1 learns to produce novel words easily. While Enc2Dec and Enc2AttDec only learn to produce some common words well. It is difficult for them to learn to produce more novel words. However, we notice that UPRGv1 is almost no gain in BLEU. Because the novel words produced by

Table 1. Evaluation results on TEST set

Method	BLEU-1(%)	BLEU-4(%)	Distinct-1(%)	Distinct-2(%)
Random	21.36	0.87	/	/
Enc2Dec	23.92	3.51	0.018	0.083
Enc2AttDec	24.26	3.49	0.027	0.141
UPRG(v1)	24.34	3.49	**0.114**	**0.435**
UPRG(v2)	**25.26**	**3.71**	0.078	0.269
Reference	/	/	1.112	10.847

UPRGv1 do not necessarily appear in the reference review. UPRGv2 achieves the best performance comparing to other methods in BLEU, which demonstrates the effectiveness of context converting layer. UPRGv2 transforms representation of three contexts into the same space so it gets better performance in BLEU. But the context converting layer results in a decline in Distinct. We claim that the context converting layer improves the weight of contexts between contexts and user preference when generating reviews.

4.3 Analysis of User Preference

In order to evaluate the effect of user preference, we analyze the probability of user mode when generating reviews. The probability of user mode is represented by $1 - p_g$ (Eq. 14). We find some typical phenomena in TEST set. The results are shown in Fig. 2. As shown in Fig. 2(a), "marie" and "force" have higher probability than other words. This demonstrates that UPRG can efficiently copy some noun words, such as authors' name. In addition, UPRG also chooses novel verb words as shown in Fig. 2(b), in which "considers" has the highest probability among all words. These examples show that the user preference enhanced decoder learns when to copy and which to copy successfully and user preference introduces various words and contributes to generating diverse reviews.

4.4 Case Study

We analyze generated results on TEST set produced by different methods to find the reason why BLEU and Distinct of different approaches behave differently. There are some representative examples of generated reviews in Table 2.

As we can see, reviews from two different methods are very similar in the mass. The results of two methods express the sentiment properly. For example, they generate "i loved this book" and "this book was very well written" for 5-rating, "i was disappointed in this book" and "this book was so hard to read" for 1-rating. However, UPRG generates more novel words, such as "i love the way *marina adair* writes" in Example 1 and "i was so *happy* to see how the story ended" in Example 2, while the result of Enc2AttDec does not. Because user preference enhanced decoder has access to more information that is closely

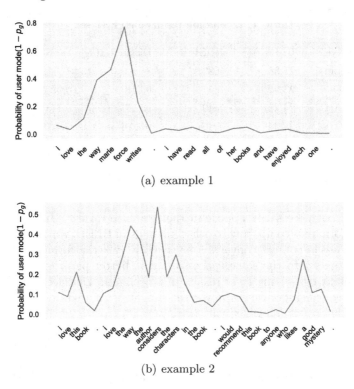

(a) example 1

(b) example 2

Fig. 2. Probability of user mode when generating each word of review

Table 2. Examples of generated reviews

Method	UID	PID	Rating	Generated Review
Enc2AttDec	A	X	5	i loved this book . it was a great story and i loved the characters . i would recommend this book to anyone who likes a good romance .
	B	Y	5	i loved this book . it was a great story and i could n't put it down . i would recommend this book to anyone who likes a good romance .
	C	Z	1	i was disappointed in this book . i was n't able to finish it . i was n't able to finish it . i 'm not sure i would read it again .
UPRG	A	X	5	i love the way marina adair writes . i have read all of her books and have n't been disappointed yet .
	B	Y	5	this book was very well written . i could n't put it down . i was so happy to see how the story ended .
	C	Z	1	this book was so hard to read . i could n't finish it . i could n't finish it . i do n't think i 'll read the other books .

related to user context and selects from them to produce reviews. Therefore, UPRG generates diverse and personalized reviews compared to Enc2AttDec.

5   Conclusion and Future Work

In this paper, we propose a user preference-aware review generation model (UPRG), which is consisted of context encoder and review decoder. The encoder first encodes the contexts into vectors, then the decoder generates review words one by one according to the context vector computed by a context attention mechanism. What's more, we introduce user preference when decoding. This gives priority to users' favorite words than common words and improves the diversity of generated reviews efficiently. Experimental results on a real review dataset show that UPRG is superior to baseline methods.

There are some directions of improvement. We will use more contextual condition to control the generated reviews, which is more suitable for real situations. We will explore transfer learning to generate reviews for new products that have no reviews.

Acknowledgements. This research is supported by National Natural Science Foundation of China (Grant No. 61773229), Basic Scientific Research Program of Shenzhen City (Grant No. JCYJ20160331184440545), and Overseas Cooperation Research Fund of Graduate School at Shenzhen, Tsinghua University (Grant No. HW2018002).

References

1. Bowman, S.R., Vilnis, L., Vinyals, O., Dai, A.M., Jozefowicz, R., Bengio, S.: Generating sentences from a continuous space. CoNLL 2016, p. 10 (2016)
2. Costa, F., Ouyang, S., Dolog, P., Lawlor, A.: Automatic generation of natural language explanations. arXiv preprint arXiv:1707.01561 (2017)
3. Dai, B., Fidler, S., Urtasun, R., Lin, D.: Towards diverse and natural image descriptions via a conditional GAN. In: Proceedings of the IEEE Conference on Computer Vision and Pattern Recognition, pp. 2970–2979 (2017)
4. Dong, L., Huang, S., Wei, F., Lapata, M., Zhou, M., Xu, K.: Learning to generate product reviews from attributes. In: Proceedings of the 15th Conference of the European Chapter of the Association for Computational Linguistics: Volume 1, Long Papers, vol. 1, pp. 623–632 (2017)
5. Ghazvininejad, M., Shi, X., Choi, Y., Knight, K.: Generating topical poetry. In: Proceedings of the 2016 Conference on Empirical Methods in Natural Language Processing, pp. 1183–1191 (2016)
6. Graves, A.: Generating sequences with recurrent neural networks. arXiv preprint arXiv:1308.0850 (2013)
7. Hochreiter, S., Schmidhuber, J.: Long short-term memory. Neural Comput. **9**(8), 1735–1780 (1997)
8. Li, J., Galley, M., Brockett, C., Spithourakis, G., Gao, J., Dolan, B.: A persona-based neural conversation model. In: Proceedings of the 54th Annual Meeting of the Association for Computational Linguistics (Volume 1: Long Papers), vol. 1, pp. 994–1003 (2016)

9. Lipton, Z.C., Vikram, S., McAuley, J.: Capturing meaning in product reviews with character-level generative text models. arXiv preprint arXiv:1511.03683 (2015)

10. McAuley, J., Pandey, R., Leskovec, J.: Inferring networks of substitutable and complementary products. In: Proceedings of the 21th ACM SIGKDD International Conference on Knowledge Discovery and Data Mining, pp. 785–794. ACM (2015)

11. Mikolov, T., Karafiát, M., Burget, L., Černockỳ, J., Khudanpur, S.: Recurrent neural network based language model. In: Eleventh Annual Conference of the International Speech Communication Association (2010)

12. Mikolov, T., Zweig, G.: Context dependent recurrent neural network language model. SLT **12**, 234–239 (2012)

13. Papineni, K., Roukos, S., Ward, T., Zhu, W.J.: BLEU: a method for automatic evaluation of machine translation. In: Proceedings of the 40th Annual Meeting on Association for Computational Linguistics, pp. 311–318. Association for Computational Linguistics (2002)

14. Pascanu, R., Mikolov, T., Bengio, Y.: On the difficulty of training recurrent neural networks. In: International Conference on Machine Learning, pp. 1310–1318 (2013)

15. Reiter, E., Dale, R.: Building Natural Language Generation Systems. Cambridge University Press (2000)

16. Ren, Z., Wang, X., Zhang, N., Lv, X., Li, L.J.: Deep reinforcement learning-based image captioning with embedding reward. In: Proceedings of the IEEE Conference on Computer Vision and Pattern Recognition, pp. 290–298 (2017)

17. Rennie, S.J., Marcheret, E., Mroueh, Y., Ross, J., Goel, V.: Self-critical sequence training for image captioning. In: Proceedings of the IEEE Conference on Computer Vision and Pattern Recognition, pp. 7008–7024 (2017)

18. Sutskever, I., Martens, J., Hinton, G.E.: Generating text with recurrent neural networks. In: Proceedings of the 28th International Conference on Machine Learning (ICML-11), pp. 1017–1024 (2011)

19. Tang, J., Yang, Y., Carton, S., Zhang, M., Mei, Q.: Context-aware natural language generation with recurrent neural networks. arXiv preprint arXiv:1611.09900 (2016)

20. Tieleman, T., Hinton, G.: Lecture 6.5-RMSProp: divide the gradient by a running average of its recent magnitude. COURSERA Neural Netw. Mach. Learn. **4**(2), 26–31 (2012)

21. Wu, S., Zhang, D., Yang, N., Li, M., Zhou, M.: Sequence-to-dependency neural machine translation. In: Proceedings of the 55th Annual Meeting of the Association for Computational Linguistics (Volume 1: Long Papers), vol. 1, pp. 698–707 (2017)

22. Zhang, J., et al.: Flexible and creative Chinese poetry generation using neural memory. In: Proceedings of the 55th Annual Meeting of the Association for Computational Linguistics (Volume 1: Long Papers), vol. 1, pp. 1364–1373 (2017)

23. Zhang, X., Lapata, M.: Chinese poetry generation with recurrent neural networks. In: EMNLP, pp. 670–680 (2014)

24. Zheng, H.T., Wang, W., Chen, W., Sangaiah, A.K.: Automatic generation of news comments based on gated attention neural networks. IEEE Access **6**, 702–710 (2018)

25. Zhou, H., Tu, Z., Huang, S., Liu, X., Li, H., Chen, J.: Chunk-based Bi-scale decoder for neural machine translation. In: Proceedings of the 55th Annual Meeting of the Association for Computational Linguistics (Volume 2: Short Papers), vol. 2, pp. 580–586 (2017)

Mining Cluster Patterns in XML Corpora via Latent Topic Models of Content and Structure

Gianni Costa and Riccardo Ortale[✉]

ICAR-CNR, Via P. Bucci 8/9C, Rende, CS, Italy
{costa,ortale}@icar.cnr.it

Abstract. We present two innovative machine-learning approaches to topic model clustering for the XML domain. The first approach consists in exploiting consolidated clustering techniques, in order to partition the input XML documents by their meaning. This is captured through a new Bayesian probabilistic topic model, whose novelty is the incorporation of Dirichlet-multinomial distributions for both content and structure. In the second approach, a novel Bayesian probabilistic generative model of XML corpora seamlessly integrates the foresaid topic model with clustering. Both are conceived as interacting latent factors, that govern the wording of the input XML documents. Experiments over real-world benchmark XML corpora reveal the overcoming effectiveness of the devised approaches in comparison to several state-of-the-art competitors.

Keywords: Bayesian probabilistic XML analysis · XML clustering ·
Latent topic modeling

1 Introduction

XML document clustering poses two major challenges. Firstly, the explicit manipulation of XML documents to catch content and structural resemblance embraces several research issues, namely the alignment of their (sub)structures, the identification of similarities between such (sub)structures and between the textual data nested therein, along with the discovery of possible mutual semantic relationships among textual data and (sub)structure labels. Secondly, resemblance between the structures and textual contents of XML documents should be caught at a semantic (i.e., topical) level.

In this paper, we focus on XML document clustering based on latent topic modeling for the purpose of avoiding the aforementioned issues. Our intuition is to partition a corpus of XML documents by topical similarity rather than by content and structure similarity. In particular, two are the proposed approaches.

The first approach consists in applying well-known clustering techniques to partition the semantic representations of the XML documents of an input corpus according to the MUESLI model. MUESLI (*xMl clUstErS from Latent topIcs*) is

© Springer Nature Switzerland AG 2019
Q. Yang et al. (Eds.): PAKDD 2019, LNAI 11441, pp. 237–248, 2019.
https://doi.org/10.1007/978-3-030-16142-2_19

an innovative XML topic model, that is conceived as an adaptation to the XML domain of the former LDA [6] topic model for unstructured text data. Under MUESLI, the semantics of the observed XML documents is modeled as a probability distribution over a number of latent (or, beforehand unknown) topics. In turn, each such a topic consists of two multinomial probability distributions placed over the word tokens and the root-to-leaf paths, respectively. Both probability distributions are randomly sampled from respective Dirichlet priors. The latent topics are inferred from the observed XML documents by conventional Bayesian reasoning. For this purpose, approximate posterior inference and parameter estimation are derived. Additionally, a Gibbs sampling algorithm implementing both is designed. MUESLI differs from previous topic models of documents with text and tags (e.g., [10, 20, 21, 23, 24]) primarily in the generation of document structure. In particular, [20, 21, 23, 24] are not explicitly meant for XML corpora. Instead, [10] proposes the only one previous topic model for the XML domain. However, the latter differs from MUESLI, in that topics are not also characterized by a specific probability distribution over the root-to-leaf paths.

The second approach combines XML document clustering and topic modeling into one unified process. For this purpose, a new generative model of XML corpora, named PAELLA (*toPicAl clustEr anaLysis of xmL corporA*), is presented. Essentially, PAELLA describes a generative process, in which XML document clustering and topic modeling act as interacting latent factors, that rule the formation of the observed XML documents. Technically, this is accomplished through the incorporation of MUESLI into an innovative Bayesian probabilistic model, that also associates a latent cluster-membership random variable with each XML document. To the best of our knowledge, the integration of document clustering and topic modeling is unprecedented in the XML domain and PAELLA is the first effort along this previously unexplored line of research.

A comparative evaluation on real-world XML corpora reveals the superior effectiveness of the devised approaches.

This paper proceeds as follows. Section 2 presents notation and preliminaries. Sections 3 and 4 cover the approaches based on MUESLI and PAELLA, respectively. Section 5 provides a comparative evaluation of our approaches on real-world benchmark XML corpora. Section 6 concludes and highlights future research.

2   Preliminaries

In this section, we introduce the adopted notation and some basic concepts.

2.1   Traditional Tree-Based XML Document Representation

The structure and content of an XML document with no references [1] can be modeled through a suitable XML tree representation, that refines the traditional notion of *rooted labeled tree* to also catch content and its nesting into structure.

An **XML tree** is a rooted, labeled tree, represented as a tuple $\mathbf{t} = (\mathbf{V_t}, r_t, \mathbf{E_t}, \lambda_t)$, whose elements have the following meaning. $\mathbf{V_t} \subseteq \mathbb{N}$ is a set of nodes and $r_t \in \mathbf{V_t}$ is the root of $\mathbf{t}$, i.e. the only node with no entering edges. $\mathbf{E_t} \subseteq \mathbf{V_t} \times \mathbf{V_t}$ is a set of edges, catching the parent-child relationships between nodes of $\mathbf{t}$. $\lambda_t : \mathbf{V_t} \mapsto \Sigma$ is a node labeling function, with Σ being an alphabet of node tags (i.e., labels).

Notice that the elements of XML documents are not distinguished from their attributes in an XML tree: both are mapped to nodes in the corresponding XML-tree representation.

Let $\mathbf{t}$ be a generic XML tree. Nodes in $\mathbf{V_t}$ can be divided into two disjoint subsets: the set $\mathbf{L_t}$ of *leaves* and the set $\mathbf{V_t} - \mathbf{L_t}$ of *inner nodes*. An inner node has at least one child. A leaf has no children and can only enclose textual items.

A root-to-leaf path $p_l^{r_t}$ in $\mathbf{t}$ is a sequence of nodes encountered along the path from the root r_t to a leaf node l in $\mathbf{L_t}$, i.e., $p_l^{r_t} = <r_t, \ldots, l>$. Notation $\lambda_t(p_l^{r_t})$ denotes the sequence of labels that are associated in the XML tree $\mathbf{t}$ with the nodes of path $p_l^{r_t}$, i.e., $\lambda_t(p_l^{r_t}) = <\lambda_t(r_t), \ldots, \lambda_t(l)>$. The set of all root-to-leaf paths in $\mathbf{t}$ is denoted as $paths(\mathbf{t}) = \{p_l^{r_t} | l \in \mathbf{L_t}\}$.

Let l be a leaf in $\mathbf{L_t}$. The set $text\text{-}items(l) = \{w_1, \ldots, w_h\}$ is a model of the text items provided by l. Elements w_i (with $i = 1 \ldots h$) are as many as the distinct text items in the context of l. The whole text content of the XML tree $\mathbf{t}$ is denoted as $text\text{-}items(\mathbf{t}) = \cup_{l \in \mathbf{L_t}} text\text{-}items(l)$.

Notation $\lambda_t(p_l^{r_t}).w_h$ indicates an enriched path and will be used to explicitly represent the nested occurrence of the text item w_h in the structural context of the labeled root-to-leaf path $p_l^{r_t}$. Notice that prefixing a content item with the sequence of labels of the respective root-to-leaf path is an instance of *tagging* [7, 28]. The collection of all enriched paths in $\mathbf{t}$ is instead indicated as $paths^{(e)}(\mathbf{t}) = \cup_{l \in \mathbf{L_t}, w \in text\text{-}items(l)} \{\lambda_t(p_l^{r_t}).w\}$.

Hereafter, the notions of XML documents and XML tree are used interchangeably. Moreover, the generic (labeled) root-to-leaf path and (labeled) enriched path are indicated as p and $p.w$, respectively, to avoid cluttering notation.

2.2   XML Features for Topic Modeling

The design of topic models for the XML domain benefits from the adoption of a flat representation for the XML documents, since the underlying generative process is relieved of nesting text items into arbitrarily complex tree structures.

The generic XML document $\mathbf{t}$ can be flattened into a collection $\mathbf{x}^{(\mathbf{t})}$ of XML features chosen from its tree-based model. In this paper, we represent $\mathbf{t}$ as a bag of enriched paths, since such XML features preserve the nesting of text items into root-to-leaf paths. Accordingly, we define $\mathbf{x}^{(\mathbf{t})} \triangleq \{p.w| \in paths^{(e)}(\mathbf{t})\}$.

3   MUESLI: A Topic Model for Clustering XML Corpora

MUESLI (*xMl clUstErS from Latent topIcs*) is a new hierarchical topic model of XML corpora, that is conceived as an adaptation of the basic LDA model [6] to

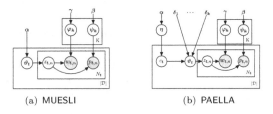

(a) MUESLI   (b) PAELLA

Fig. 1. Graphical representation of MUESLI (a) and PAELLA (b)

the XML domain. More precisely, let $\mathbf{D} = \{\mathbf{x}^{(\mathbf{t})} | \mathbf{t} \in \mathcal{D}\}$ be the *bag-of-enriched-paths* representation of an input XML corpus $\mathcal{D}$, in which the individual XML documents are characterized as discussed in Sect. 2.2. MUESLI is a Bayesian probabilistic model of the imaginary process, that generates $\mathbf{D}$.

Such a generative process is assumed to be influenced by K latent topics. Each XML document $\mathbf{x}^{(\mathbf{t})}$ in $\mathbf{D}$ (or, also, $\mathbf{t}$ in $\mathcal{D}$) exhibits the different topics to distinct degrees. This is captured by associating $\mathbf{x}^{(\mathbf{t})}$ with an unknown probability distribution $\vartheta_{\mathbf{t}}$ over the individual topics $k = 1, \ldots, K$, such that $\vartheta_{\mathbf{t},k}$ is the probability of topic k within $\mathbf{x}^{(\mathbf{t})}$. In turn, each topic consists of

- an unknown probability distribution φ_k over the text items in the vocabulary $\mathcal{I} \triangleq \cup_{\mathbf{t} \in \mathcal{D}} \textit{text-items}(\mathbf{t})$, such that $\varphi_{k,w}$ indicates the probability in topic k of the generic text item w from $\mathcal{I}$;
- an unknown probability distribution ψ_k over the root-to-leaf paths in the vocabulary $\mathcal{R} \triangleq \cup_{\mathbf{t} \in \mathcal{D}} \textit{paths}(\mathbf{t})$, such that $\psi_{k,p.w}$ captures the probability in topic k of the generic root-to-leaf path $p.w$ from $\mathcal{R}$.

Figure 1(a) formalizes the conditional (in)dependencies among the random variables of MUESLI through a graphical representation in plate notation. All random variables of MUESLI are represented as nodes. The shaded nodes mark observed random variables, whose values are the observed results of the generation process (i.e., the XML documents in their *bag-of-enriched path* representation). Instead, the unshaded nodes indicate hidden random variables, whose values correspond to latent (or unobserved) aspects (i.e., sampled distributions and topic assignments). Plates (or rectangles) indicate reiterations.

Based on the conditional (in)dependencies of Fig. 1(a), the generative probabilistic process assumed by MUESLI implements the realization of all random variables as algorithmically detailed in Fig. 2. Notice that α, β and γ are hyperparameters of the MUESLI model and their role is clarified in Sect. 3.1.

3.1   Observed-Data Likelihood and Prior Distributions

Let $\mathbf{x}^{(\mathbf{t})} = \{p^{(\mathbf{t},1)}.w^{(\mathbf{t},1)}, \ldots, p^{(\mathbf{t},N_{\mathbf{t}})}.w^{(\mathbf{t},N_{\mathbf{t}})}\}$ be the flattened representation of the XML tree $\mathbf{t}$ from the XML corpus $\mathcal{D}$, in which $N_{\mathbf{t}}$ stands for the number of enriched paths in $\mathbf{t}$. Moreover, let $\mathbf{z}^{(\mathbf{t})} \triangleq \{z_{\mathbf{t},1}, \ldots z_{\mathbf{t},N_{\mathbf{t}}}\}$ be the collection of topic assignments in $\mathbf{t}$, i.e., the generic element $z_{\mathbf{t},i}$ is the latent topic of the corresponding enriched path $p^{(\mathbf{t},i)}.w^{(\mathbf{t},i)}$ in $\mathbf{x}^{(\mathbf{t})}$ (with $i = 1, \ldots, N_{\mathbf{t}}$). In addition, assume that $\boldsymbol{P}$ and $\boldsymbol{W}$ denote, respectively, all observed root-to-leaf paths and text items, i.e., $\boldsymbol{P} \triangleq \cup_{t \in \mathcal{D}} \{p^{(t,1)}, \ldots, p^{(t,N_t)}\}$ and $\boldsymbol{W} \triangleq \cup_{t \in \mathcal{D}} \{w^{(t,1)}, \ldots, w^{(t,N_t)}\}$.

The data likelihood can be formalized as the following conditional probability distributions over $\boldsymbol{P}$ and $\boldsymbol{W}$

$$\Pr(\boldsymbol{P}|\boldsymbol{Z}, \boldsymbol{\Psi}) = \prod_{k=1}^{K} \prod_{p \in \mathcal{R}} \psi_{k,p}^{n_k^{(p)}} \qquad \Pr(\boldsymbol{W}|\boldsymbol{Z}, \boldsymbol{\Phi}) = \prod_{k=1}^{K} \prod_{w \in \mathcal{I}} \varphi_{k,w}^{n_k^{(w)}}$$

where

- $n_k^{(p)}$ stands for the occurrences of the root-to-leaf path p under the topic k;
- $n_k^{(w)}$ stands for the occurrences of the text item w under the topic k;
- $\boldsymbol{\Psi}$ is a compact notation denoting all topic-specific root-to-leaf path distributions, i.e., $\boldsymbol{\Psi} \triangleq \{\psi_1, \ldots, \psi_K\}$ (with K being the number of latent topics);
- $\boldsymbol{\Phi}$ is a compact notation denoting all topic-specific word distributions, i.e., $\boldsymbol{\Phi} \triangleq \{\varphi_1, \ldots, \varphi_K\}$ (with K being the number of latent topics);
- $\boldsymbol{Z}$ compactly denotes all topic assignments in $\mathcal{D}$, i.e., $\boldsymbol{Z} \triangleq \{\mathbf{z}^{(t)}|\mathbf{x}^{(t)} \in \mathbf{D}\}$.

Furthermore, the conditional probability distribution over $\boldsymbol{Z}$ is

$$\Pr(\boldsymbol{Z}|\boldsymbol{\Theta}) = \prod_{t \in \mathcal{D}} \prod_{k=1}^{K} \vartheta_{t,k}^{n_t^{(k)}}$$

where

- $n_t^{(k)}$ stands for the occurrences of the topic k in the XML document $\mathbf{t}$;
- $\boldsymbol{\Theta}$ is a compact notation, that stands for the whole set of the topic distributions associated with the individual XML documents, i.e., $\boldsymbol{\Theta} \triangleq \{\vartheta_t|t \in \mathcal{D}\}$.

In compliance with standard Bayesian modeling, under MUESLI, uncertainty on ψ, $\boldsymbol{\Phi}$ and $\boldsymbol{\Theta}$ is captured by means of the below conjugate Dirichlet priors

$$\Pr(\boldsymbol{\Psi}|\boldsymbol{\beta}) = \prod_{k=1}^{K} \frac{1}{\Delta(\boldsymbol{\beta})} \prod_{p \in \mathcal{R}} \psi_{k,p}^{\beta_p - 1} \quad \Pr(\boldsymbol{\Phi}|\gamma) = \prod_{k=1}^{K} \frac{1}{\Delta(\gamma)} \prod_{w \in \mathcal{I}} \varphi_{k,w}^{\gamma_w - 1} \quad \Pr(\boldsymbol{\Theta}|\alpha) = \prod_{t \in \mathcal{D}} \frac{1}{\Delta(\alpha)} \prod_{k=1}^{K} \vartheta_{t,k}^{\alpha_k - 1}$$

The above $\boldsymbol{\beta} = \{\beta_p | p \in \mathcal{R}\}$, $\boldsymbol{\alpha} = \{\alpha_k | k = 1, \ldots, K\}$ and $\boldsymbol{\gamma} = \{\gamma_w | w \in \mathcal{R}\}$ are three hyperparameters. Their generic elements β_p, α_k and γ_w represent suitable pseudo-counts, enabling the incorporation of domain-specific prior knowledge [17] into the exploratory analysis of the latent topics in $\mathcal{D}$.

- For each topic k
  - sample the probability distribution φ_k over the text items of vocabulary $\mathcal{I}$, i.e., $\varphi_k \sim Dirichlet(\gamma)$;
  - Sample the probability distribution ψ_k over the root-to-leaf paths of vocabulary $\mathcal{R}$, i.e., $\psi_k \sim Dirichlet(\beta)$.
- For each $\mathbf{t}$ in $\mathcal{D}$
  - sample the probability distribution ϑ_t over the latent topics, i.e., $\vartheta_t \sim Dirichlet(\alpha)$;
  - for each $n = 1, \ldots, N_t$
    - * choose a latent topic $z_{t,n} \sim Discrete(\vartheta_t)$;
    - * choose a root-to-leaf path $p_{t,n} \sim Discrete(\psi_{z_{t,n}})$;
    - * choose a text item $w_{t,n} \sim Discrete(\varphi_{z_{t,n}})$;

Fig. 2. The probabilistic generative process under MUESLI

3.2 Approximate Posterior Inference and Parameter Estimation

MUESLI is a generative model of XML corpora given their latent aspects. Essentially, it postulates assumptions explaining how such latent aspects govern the generation of the individual XML documents. Nonetheless, in order to cluster the XML documents by their latent topics, one has to infer the latent aspects (including the foresaid topic distributions) from the XML documents. Posterior inference is used for this purpose.

As it generally happens with probabilistic models of practical interest, under MUESLI, exact posterior inference is intractable, due to the complexity of the posterior distribution. Thus, we resort to collapsed Gibbs sampling, a *Markov-Chain Monte-Carlo* method for approximate inference [3,5], that enables simple inference algorithms, even if the number of hidden variables is very large [5,17]. The pseudo code of Gibbs sampling under MUESLI is sketched in Algorithm 1. The full conditional below is used for sampling (at step 10) any topic assignment $z_{t,n}$ given all other topic assignments $Z_{\neg(t,n)}$ and the observed data W and P

$$\Pr(z_{t,n} = k | Z_{\neg(t,n)}, W, P, \alpha, \beta, \gamma)$$

$$= \frac{n_k^{(w)} - 1 + \gamma_w}{(\sum_{w' \in \mathcal{I}} n_k^{(w')} + \gamma'_w) - 1} \cdot \frac{n_k^{(p)} - 1 + \beta_p}{(\sum_{p' \in \mathcal{R}} n_k^{(p')} + \beta'_p) - 1} \cdot \frac{n_t^{(k)} - 1 + \alpha_k}{\sum_{k'=1}^{K}(n_t^{(k')} + \alpha_{k'}) - 1}$$

$$(1)$$

Concerning parameter estimation, due to conjugacy, $\Pr(\vartheta_t | z_t, \alpha)$, $\Pr(\varphi_k | Z, W, \gamma)$ and $\Pr(\psi_k | Z, P, \beta)$ are Dirichlet distributions. Thus, by using the expectation of the Dirichlet distribution [17], one can calculate the below parameter estimates

$$\vartheta_{t,k} = \frac{n_t^{(k)} + \alpha_k}{\sum_{k'=1}^{K} n_t^{(k')} + \alpha_{k'}}, \quad t \in \mathcal{D} \wedge k = 1, \ldots, K \tag{2}$$

$$\varphi_{k,w} = \frac{n_k^{(w)} + \gamma_w}{\sum_{w' \in \mathcal{I}} n_k^{(w')} + \gamma_{w'}}, \quad k = 1, \ldots, K \wedge w \in \mathcal{I} \tag{3}$$

$$\psi_{k,p} = \frac{n_k^{(p)} + \beta_p}{\sum_{p' \in \mathcal{R}} n_k^{(p')} + \beta_{p'}}, \quad k = 1, \ldots, K \wedge \quad p \in \mathcal{R} \tag{4}$$

3.3 Partitioning Algorithms

The MUESLI topic model produces a lower-dimensional mixed-membership representation Θ of the XML corpus $\mathcal{D}$, by projecting the individual XML documents into a K-dimensional space of latent topics. The parameters Θ establish the degree of participation of the individual XML documents in the distinct latent topics. We next discuss two techniques for partitioning $\mathcal{D}$ based on Θ.

Naive Partitioning. This technique places each XML document t inside the cluster $C^* = argmax_{k=1,\ldots,K}\vartheta_{t,k}$, with C^* corresponding to the most representative topic of t according to MUESLI.

Algorithm 1. Collapsed Gibbs sampling with parameter estimation

GIBBS SAMPLING($\mathbf{D}, \alpha, \beta, \gamma, K$)

Input: The XML corpus $\mathcal{D}$ in its flat representation;
the hyperparameters α, β and γ;
the number K of latent topics;

Output: The topic assignments Z;
the multinomial parameters Θ, Φ and ψ;

1: zero all counts $n_t^{(k)}$, $n_k^{(w)}$, $n_k^{(p)}$;
2: randomly assign topics to the text items in the context of the enriched paths of the XML documents and set the related counts accordingly;
3: $iteration \leftarrow 1$;
4: $s \leftarrow 1$;
5: **while** $iteration \leq Maximum\_iteration\_number$ **do**
6:   **for each** t in $\mathcal{D}$ **do**
7:     **for each** $n = 1, \ldots, |paths^{(e)}(\mathbf{t})|$ **do**
8:       $k \leftarrow z_{\mathbf{t}, n}$;
9:       decrement counts $n_t^{(k)}$, $n_k^{(w^{(\mathbf{t}, n)})}$ and $n_k^{(p^{(\mathbf{t}, n)})}$ by 1;
10:       sample k' from Eq. (1);
11:       increment counts $n_t^{(k')}$, $n_{k'}^{(w^{(\mathbf{t}, n)})}$ and $n_{k'}^{(p^{(\mathbf{t}, n)})}$ by 1;
12:     **end for**
13:   **end for**
14:   **if** $(iteration > burn\text{-}in)$ **and** $(iteration \bmod lag == 0)$ **then**
15:     **for each** t in $\mathcal{D}$ **and each** $k = 1, \ldots, K$ **do**
16:       estimate the individual parameters $\vartheta_{\mathbf{t}, k}^{(s)}$ by Eq. (2);
17:     **end for**
18:     **for each** $k = 1, \ldots, K$ **and each** w **in** $\mathcal{I}$ **do**
19:       estimate the individual parameters $\varphi_{k, w}^{(s)}$ by Eq. (3);
20:     **end for**
21:     **for each** $k = 1, \ldots, K$ **and each** p **in** $\mathcal{R}$ **do**
22:       estimate the individual parameters $\psi_{k, p}^{(s)}$ by Eq. (4);
23:     **end for**
24:     $s \leftarrow s + 1$;
25:   **end if**
26:   $iteration \leftarrow iteration + 1$;
27: **end while**
28: **for each** t in $\mathcal{D}$ **and each** $k = 1, \ldots, K$ **do**
29:   $\vartheta_{\mathbf{t}, k} \leftarrow \frac{1}{s} \sum_{d=1}^{s} \vartheta_{\mathbf{t}, k}^{(d)}$;
30: **end for**
31: **for each** $k = 1, \ldots, K$ **and each** w **in** $\mathcal{I}$ **do**
32:   $\varphi_{k, w} \leftarrow \frac{1}{s} \sum_{d=1}^{s} \varphi_{k, w}^{(d)}$;
33: **end for**
34: **for each** $k = 1, \ldots, K$ **and each** p **in** $\mathcal{R}$ **do**
35:   $\psi_{k, p} \leftarrow \frac{1}{s} \sum_{d=1}^{s} \psi_{k, p}^{(d)}$;
36: **end for**

K-Medoids Partitioning. A more sophisticated technique for separating $\mathcal{D}$ based on MUESLI consists in partitioning the topic distributions Θ. This allows for grouping the XML documents through their cross-topic similarity along with using a number K of latent topics larger than the number $\overline{K}$ of clusters to find in $\mathcal{D}$. Both are expected to enable a more accurate separation of $\mathcal{D}$. k-medoids [16] is a well-known clustering algorithm, that can be chosen to partition Θ, because of its effectiveness and robustness to noise as well as outliers. k-medoids involves the computation of the intra-cluster divergences. To this end, we use the square root of the Jensen-Shannon distance, that was shown to be a metric [14].

4 PAELLA: Joint XML Clustering and Topic Modeling

PAELLA (*toPicAl clustEr anaLysis of xmL corporA*) is an innovative generative model of XML corpora, in which document clustering and topic modeling act as simultaneous and interdependent latent factors in the formation of the individual XML documents. Essentially, PAELLA envisages a scenario, in which each XML document $\mathbf{x}^{(t)}$ is associated with a corresponding latent cluster membership c_t as in [26]. c_t is randomly sampled from an unknown cluster distribution $\boldsymbol{\eta}$. Furthermore, the underlying semantics $\boldsymbol{\vartheta_t}$ of the XML document $\mathbf{x}^{(t)}$ is a unknown distribution over K latent topics. These are individually characterized as in the MUESLI topic model of Sect. 3. Figure 1(b) shows the graphical representation of PAELLA. Its generative process is detailed in Fig. 3.

Under PAELLA, collapsed Gibbs sampling is exploited to perform the approximate posterior inference of c_t and $\mathbf{z}^{(t)}$ for each XML document $\mathbf{x}^{(t)}$. Besides, parameter estimation is utilized to calculate the cluster distribution $\boldsymbol{\eta}$, the topic distribution $\boldsymbol{\vartheta_t}$ for each XML document $\mathbf{x}^{(t)}$ as well as the distributions $\boldsymbol{\varphi_k}$ and $\boldsymbol{\psi_k}$ for each topic $k = 1, \ldots, K$. The mathematical and algorithmic details of collapsed Gibbs sampling and parameter estimation under PAELLA are omitted for space limitations, being similar to the respective developments in Sect. 3.2.

5   Evaluation

In this section, we empirically assess the effectiveness of our approaches to XML clustering in comparison to various state-of-the-art competitors. In the following, the naive and K-Medoids clustering techniques adopted in conjunction with MUESLI are named, respectively, Naive and K-Medoids.

5.1   XML Corpora, Competitors and Evaluation Measures

All tests are carried out on *Wikipedia* and *Sigmod*. These are two real-world benchmark XML corpora, that are often used in the literature for the evaluation of techniques devoted to XML classification and clustering.

Wikipedia was adopted as the test-bed for the task of XML clustering by both content and structure, in the context of the XML Mining Track at INEX 2007 [13]. The overall corpus consists of $47,397$ articles from the online digital encyclopedia, that are organized into 19 classes (or thematic categories). Each such a class corresponds to a different Wikipedia Portal.

The 140 XML documents of the *Sigmod* corpus represent a portion of the SIGMOD Record issues. The documents comply with two different structural

---

- Draw the probability distribution over clusters, i.e., $\boldsymbol{\eta} \sim Dirichlet(\boldsymbol{\alpha})$;
- For each topic k
  - sample the probability distribution $\boldsymbol{\varphi_k}$ over the text items of vocabulary $\mathcal{I}$, i.e., $\boldsymbol{\varphi_k} \sim Dirichlet(\boldsymbol{\gamma})$;
  - Sample the probability distribution $\boldsymbol{\psi_k}$ over the root-to-leaf paths of vocabulary $\mathcal{R}$, i.e., $\boldsymbol{\psi_k} \sim Dirichlet(\boldsymbol{\beta})$.
- For each $\mathbf{t}$ in $\mathcal{D}$.
  - Draw cluster membership $c_{\mathbf{t}} \sim Discrete(\boldsymbol{\eta})$;
  - sample the probability distribution $\boldsymbol{\vartheta_t}$ over the latent topics, i.e., $\boldsymbol{\vartheta_t} \sim Dirichlet(\boldsymbol{\delta}_{c_{\mathbf{t}}})$;
  - for each $n = 1, \ldots, N_{\mathbf{t}}$
    * choose a latent topic $z_{\mathbf{t},n} \sim Discrete(\boldsymbol{\vartheta_t})$;
    * choose a root-to-leaf path $p_{\mathbf{t},n} \sim Discrete(\boldsymbol{\psi}_{z_{\mathbf{t},n}})$;
    * choose a text item $w_{\mathbf{t},n} \sim Discrete(\boldsymbol{\varphi}_{z_{\mathbf{t},n}})$;

Fig. 3. The probabilistic generative process under PAELLA

class DTDs and were, initially, used to evaluate the effectiveness of XML structural clustering techniques (e.g., in) [2]. However, the minimal number of structural classes makes this task not truly challenging. Thus, in our experimentation, we consider a rearrangement of *Sigmod* into 5 general classes proposed in [19]. These classes were formed, by means of expert knowledge, to reflect as many groups of structural and content features of the underlying XML documents.

Interestingly, the choice of *Wikipedia* and *Sigmod* allows for assessing the effectiveness of our approaches on XML corpora with diverging features. In particular, while the XML documents in *Wikipedia* can be viewed as schema-less XML trees with a deep structure and a high branching factor, *Sigmod* includes a much smaller number of XML trees with two distinct schema definitions [19]. Table 1 summarizes a selection of primary statistics of the chosen XML corpora.

Table 1. Characteristics of the chosen XML corpora

XML Corpus	Size	Classes	Max. out degree	Max. tree depth	Distinct paths	Terms	Distinct terms
Wikipedia	47,397	21	1,776	48	18,839	21,840,997	1,004,207
Sigmod	140	5	29	8	33	25,666	6,286

Naive, K-Medoids and PAELLA are compared on *Wikipedia* and *Sigmod* against several state-of-the-art competitors, i.e., HPXTD [10], MCXTD [10], XC-NMF [9], XPEC [8], XCFS [19], HCX [18], CRP [27], 4RP [27], SOM [15] and LSK [25].

The clustering effectiveness of all competitors is measured in terms of macro-averaged and micro-averaged purity, according to the standard evaluation guidelines of the of the Mining Track at the INEX 2007 competition [13].

5.2 Partitioning Effectiveness

All competitors are tested in the discovery of a number of clusters in the chosen XML corpora, that amounts to the actual number of natural classes.

Cluster discovery through MUESLI and PAELLA also involves setting a reasonable number of underlying topics. In the context of the Naive clustering strategy, MUESLI was trained to unveil both in *Sigmod* and in *Wikipedia* as many latent topics as the number of natural classes within the respective XML corpora. Instead, a preliminary sensitivity analysis was conducted to determine the number of topics under K-Medoids and PAELLA. This was accomplished by ranging the number of topics in the interval [5, 30] over *Sigmod* and [10, 60] over *Wikipedia*. Figure 4 shows the sensitivity of clustering effectiveness under K-Medoids and PAELLA to the number of topics. We fixed the number of topics under K-Medoids and PAELLA, so that to maximize their clustering effectiveness.

The clustering effectiveness of all competitors is compared in Fig. 5. PAELLA and K-Medoids deliver an overcoming clustering effectiveness, being aware of the whole semantics of the individual XML documents. Naive achieves a lower effectiveness compared to both PAELLA and K-Medoids, since cluster assignment is determined for each XML document only on the basis of its most pertinent topic. This does not allow for grouping the XML documents on an actual cross-topic similarity basis. Moreover, with Naive, inference under MUESLI is subject to the constraint on the number of topics, that is required to equal the number of clusters. Such limitations affect neither PAELLA nor K-Medoids, that naturally exploit the specificity of MUESLI (i.e., modeling the semantics of an XML corpus with no prior restrictions on the actual number of underlying topics), in order to group the XML documents by their respective topic mixtures.

The superiority of PAELLA with respect to K-Medoids is due to the fact that the former conceives and seamlessly integrates MUESLI as a natural complement, with which to enhance XML document clustering.

Noticeably, the better clustering performance delivered by Naive and K-Medoids in comparison with HPXTD and MCXTD, respectively, substantiates the rationality of enriching topics under MUESLI through the incorporation of probability distributions over root-to-leaf paths. Clearly, such a modeling choice also contributes to the performance gain attained by PAELLA.

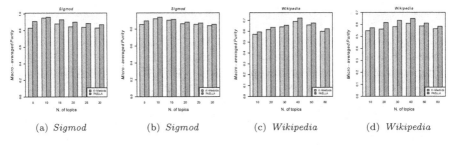

(a) *Sigmod*   (b) *Sigmod*   (c) *Wikipedia*   (d) *Wikipedia*

Fig. 4. Sensitivity of K-Medoids and PAELLA to the number of topics.

Next, we demonstrate the behavior of the proposed approaches on XML corpora, by inspecting the results outputted by PAELLA over *Sigmod*.

Figure 6 shows the topic mixtures associated with the 5 uncovered clusters. These are obtained by averaging the topic distributions of the individual docu-

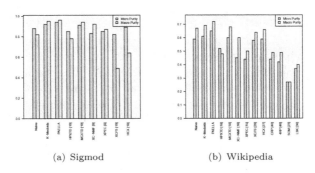

(a) Sigmod   (b) Wikipedia

Fig. 5. Macro-averaged and micro-averaged purity on Sigmod (a) and Wikipedia (b)

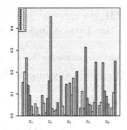

Fig. 6. Topic distribution across clusters.

Table 2. Two *Sigmod* topics.

Topic 1 (`Mobile`)	Topic 2 (`DBMS`)
Wireless	Memory
Mobile	Architecture
Communication	Database
Access	Node
Data	Transaction

ments therein. Additionally, Table 2 details two inferred word topics. Each topic is summarized by its top-5 most relevant words, whose clarity, specificity and coherence enable the intuitive interpretations in brackets.

6   Conclusions and Future Research

We proposed two innovative machine-learning approaches for clustering XML corpora by latent topic homogeneity. The empirical evidence from experiments on real-world benchmark XML corpora showed the effectiveness of the devised approaches against several state-of-the-art competitors.

It is interesting to refine MUESLI and PAELLA, in order to also account for the syntactic and semantic relationships among words [4,22], which is expected to improve XML clustering effectiveness. Finally, the incorporation of an n-gram topic model for text items is likely beneficial to more accurately catch the meaning of the textual content of the XML documents. In turn, this may be useful to further increase clustering effectiveness [11,12].

References

1. Abiteboul, S., Buneman, P., Suciu, D.: Data on the Web: From Relations to Semistructured Data and XML. Morgan Kaufmann, Burlington (2000)
2. Aggarwal, C.C., Ta, N., Wang, J., Feng, J., Zaki, M.: XProj: a framework for projected structural clustering of XML documents. In: Proceedings of ACM KDD, pp. 46–55 (2007)
3. Andrieu, C., De Freitas, N., Doucet, A., Jordan, M.I.: An introduction to MCMC for machine learning. Mach. Learn. **50**(1–2), 5–43 (2003)
4. Bengio, Y., Ducharme, R., Vincent, P., Janvin, C.: A neural probabilistic language model. J. Mach. Learn. Res. **3**, 1137–1155 (2003)
5. Bishop, C.M.: Pattern Recognition and Machine Learning. Springer, New York (2006)
6. Blei, D.M., Ng, A.Y., Jordan, M.I.: Latent dirichlet allocation. J. Mach. Learn. Res. **3**, 993–1022 (2003)
7. Bratko, A., Filipič, B.: Exploiting structural information for semi-structured document categorization. Inf. Process. Manag. **42**(3), 679–694 (2006)

8. Costa, G., Ortale, R.: Developments in partitioning XML documents by content and structure based on combining multiple clusterings. In: Proceedings of IEEE ICTAI, pp. 477–482 (2013)
9. Costa, G., Ortale, R.: A latent semantic approach to XML clustering by content and structure based on non-negative matrix factorization. In: Proceedings of IEEE ICMLA, pp. 179–184 (2013)
10. Costa, G., Ortale, R.: Mining clusters in XML corpora based on Bayesian generative topic modeling. In: Proceedings of IEEE ICMLA, pp. 515–520 (2015)
11. Costa, G., Ortale, R.: XML clustering by structure-constrained phrases: a fully-automatic approach using contextualized N-grams. Int. J. Artif. Intell. Tools **26**(1), 1–24 (2017)
12. Costa, G., Ortale, R.: Machine learning techniques for XML (co-)clustering by structure-constrained phrases. Inf. Retr. J. **21**(1), 24–55 (2018)
13. Denoyer, L., Gallinari, P.: Report on the XML mining track at INEX 2007. ACM SIGIR Forum **42**(1), 22–28 (2008)
14. Endres, D.M., Schindelin, J.E.: A new metric for probability distributions. IEEE Trans. Inf. Theory **49**(7), 1858–1860 (2003)
15. Hagenbuchner, M., Tsoi, A.C., Sperduti, A., Kc, M.: Efficient clustering of structured documents using graph self-organizing maps. In: Fuhr, N., Kamps, J., Lalmas, M., Trotman, A. (eds.) INEX 2007. LNCS, vol. 4862, pp. 207–221. Springer, Heidelberg (2008). https://doi.org/10.1007/978-3-540-85902-4_19
16. Han, J., Kamber, M., Pei, J.: Data Mining: Concepts and Techniques, 3rd edn. Morgan Kaufmann, Burlington (2011)
17. Heinrich, G.: Parameter estimation for text analysis. Technical report, University of Leipzig (2008). http://www.arbylon.net/publications/text-est.pdf
18. Kutty, S., Nayak, R., Li, Y.: HCX: an efficient hybrid clustering approach for XML documents. In: Proceedings of ACM DocEng, pp. 94–97 (2009)
19. Kutty, S., Nayak, R., Li, Y.: XCFS: an XML documents clustering approach using both the structure and the content. In: Proceedings of ACM CIKM, pp. 1729–1732 (2009)
20. Li, S., Huang, G., Tan, R., Pan, R.: Tag-weighted dirichlet allocation. In: Proceedings of IEEE ICDM, pp. 438–447 (2013)
21. Li, S., Li, J., Pan, R.: Tag-weighted topic model for mining semi-structured documents. In: Proceedings of IJCAI, pp. 2855–2861 (2013)
22. Mikolov, T., Sutskever, I., Chen, K., Corrado, G., Dean, J.: Distributed representations of words and phrases and their compositionality. In: Proceedings of NIPS, pp. 3111–3119 (2013)
23. Mimno, D.M., McCallum, A.: Topic models conditioned on arbitrary features with dirichlet-multinomial regression. In: Proceedings of UAI, pp. 411–418 (2008)
24. Ramage, D., Manning, C.D., Dumais, S.: Partially labeled topic models for interpretable text mining. In: Proceedings of ACM KDD, pp. 457–465 (2011)
25. Tran, T., Nayak, R., Bruza, P.: Document clustering using incremental and pairwise approaches. In: Fuhr, N., Kamps, J., Lalmas, M., Trotman, A. (eds.) INEX 2007. LNCS, vol. 4862, pp. 222–233. Springer, Heidelberg (2008). https://doi.org/10.1007/978-3-540-85902-4_20
26. Xie, P., Xing, E.P.: Integrating document clustering and topic modeling. In: Proceedings of UAI (2013)
27. Yao, J., Zerida, N.: Rare patterns to improve path-based clustering of Wikipedia articles. In: Pre-proceedings of INEX, pp. 224–231 (2007)
28. Yi, J., Sundaresan, N.: A classifier for semi-structured documents. In: Proceedings of ACM KDD, pp. 340–344 (2000)

A Large-Scale Repository of Deterministic Regular Expression Patterns and Its Applications

Haiming Chen[1(✉)], Yeting Li[1,2], Chunmei Dong[1,2], Xinyu Chu[1,2], Xiaoying Mou[1,2], and Weidong Min[3]

[1] State Key Laboratory of Computer Science, ISCAS, Beijing 100190, China
{chm,liyt,dongcm,chuxy,mouxy}@ios.ac.cn
[2] University of Chinese Academy of Sciences, Beijing, China
[3] School of Software, Nanchang University, Nanchang, China
minweidong@ncu.edu.cn

Abstract. Deterministic regular expressions (DREs) have been used in a myriad of areas in data management. However, to the best of our knowledge, presently there has been no large-scale repository of DREs in the literature. In this paper, based on a large corpus of data that we harvested from the Web, we build a large-scale repository of DREs by first collecting a repository after analyzing determinism of the real data; and then further processing the data by using normalized DREs to construct a compact repository of DREs, called DRE pattern set. At last we use our DRE patterns as benchmark datasets in several algorithms that have lacked experiments on real DRE data before. Experimental results demonstrate the usefulness of the repository.

Keywords: Deterministic regular expressions · Repository · Evaluation

1 Introduction

Regular expressions (REs) occur naturally in the definition of database (for example in the schema definition of structured and certain semi-structured data sets) and as fragments of most tree and graph query languages. This paper focuses on deterministic regular expressions (DREs), which have been used and studied in a myriad of areas in data management.

Roughly speaking, determinism means that, when matching a word from left to right against an expression, a symbol can be matched to only one position in the expression without looking ahead. For example, $a(a)^*$ is deterministic but $(a)^*a$ is not, although they define the same language. One immediate benefit

Work supported by the National Natural Science Foundation of China under Grant Nos. 61872339, 61472405, 61762061 and the Natural Science Foundation of Jiangxi Province, China under Grant 20161ACB20004.

© Springer Nature Switzerland AG 2019
Q. Yang et al. (Eds.): PAKDD 2019, LNAI 11441, pp. 249–261, 2019.
https://doi.org/10.1007/978-3-030-16142-2_20

of using DREs is efficient parsing. Indeed, it gives a natural manner to define determinism in REs. As a result, several decision problems behave better for DREs than for general ones. For example, language inclusion for REs is PSPACE-complete but is tractable when the expressions are deterministic. It is known that DREs are strictly less expressive than REs and not every non-deterministic RE can be defined by a DRE [11].

There have been many applications of DREs in practice, here we mention a few examples. First, DREs have been used in different kinds of applications such as the SPARQL query language for RDF [26], efficiently evaluating regular path queries [20], AXML [4], etc. Second, DREs are commonly appeared in RegExLib [2] which is the main regular expression repository available on the Web. It contains multiple kinds of expressions for matching URIs, markup code, C style strings, pieces of Java code, SQL queries, spam, etc. Third, DREs are widely used in the popular schema languages of XML, such as DTD and XSD, which are recommended by the World Wide Web Consortium (W3C) [30], and Relax NG, which is a standard of ISO (International Organization for Standardization) [27].

However, the lack of benchmark datasets is a problem for many research areas, including the research of DREs. In detail, a large-scale repository of DREs would be quite useful in the research of DREs for many purposes, such as the testing, experiments of programs having input of DREs, etc. Nevertheless, to the best of our knowledge, currently there has been no such repository of DREs in the literature. Actually, the real data set has been a weak point in the literature. The related work is summarized in Table 1 (we use RNG to represent Relax NG in this and following tables), from which we can clearly observe that the data obtained contains only from several dozens of to several thousands of schemas, which are quite insufficient. For example, many researches on different subclasses of DREs were based just on several hundreds of XSDs or DTDs (i.e., very small set). This is because harvesting a large-scale DTD, XSD and Relax NG schema files from the Web is not an easy task (we give discussions in Sect. 3). Consequently, many researches have lacked experiments on large-scale real DRE data, although these experiments should be very important. The following are some examples, disjunctive multiplicity expressions (DME) [21], the membership and inclusion checkers for conflict-free REs [16,17] and the DRE inclusion checkers [12].

Furthermore, one may think that we can use generators of REs to generate large-scale DREs by first generating REs then selecting the deterministic ones using a determinism checker for DREs. However, as our previous experiments have shown, since the ratio of DREs in REs is very low (e.g, the ratio of DREs is less than 1.2% for REs of length 50 with alphabet size 26), and as the given length of DREs increases, the probability that the generated RE is deterministic decreases very quickly, repeatedly generating arbitrary REs until we obtain a DRE is not a feasible option to generate DREs. For instance, when the given length is much larger than the alphabet size, it cannot generate a DRE even in one day. This further shows the significance of having a large-scale repository of DREs.

Considering the above problem, in this paper, we get a large-scale repository of DREs, called DRE pattern set. We further show the usefulness of our repository by using it in some applications that have lacked experiments on real DRE data before.

In detail, a large-scale data set is the prerequisite for analyzing DREs. For this purpose, we have obtained data files from the Web, including RegExLib, Relax NG, XSD and DTD. The data set that we collected is sufficiently large compared with previous work, and is also representative to investigate the practical usage of DREs

Table 1. DTD, XSD AND RELAX NG SCHEMA FILES OBTAINED OVER THE YEARS

Year	DTDs	XSDs	RNGs	Total	Work
2002	60	N/A	N/A	60	Choi et al. [14]
2004	109	93	N/A	202	Bex et al. [7]
2005	N/A	819	N/A	819	Bex et al. [6]
2005	N/A	199	N/A	199	Laender et al. [23]
2006	75	N/A	N/A	75	Barbosa et al. [5]
2007	N/A	697	N/A	697	Bex et al. [9]
2008	N/A	223	N/A	223	Laender et al. [23]
2011	3087	4141	337	7565	Grijzenhout et al. [19]
2015	N/A	8000+	N/A	8000+	Björklund et al. [10]
2016	2427	4859	N/A	7286	Li et al. [25]

with counting and interleaving (see Sect. 3 for details). Starting from the data set we constructed a large-scale repository of DREs.

Then we use our repository of DREs in some algorithms involving DREs. The first experiments are to evaluate the correctness and efficiency of some inference algorithms. Since DREs are widely used in schema definitions, inference algorithms are quite common. The second experiments are to compare the performance and efficiency of two algorithms for checking inclusion of DREs [12], which can be used in many applications such as query processing, schema update and so on. Experiments illustrate that it is possible to effectively apply DRE algorithms on this data set. We call algorithms that deal with DREs as DRE algorithms for simplicity. It should be noted that actually every subclass of DREs, besides the ones used in the experiments, and their algorithms, can be evaluated using our repository similarly in this way. The results demonstrate the usefulness of the repository.

Contributions. To the best of our knowledge, there has been no large-scale repository of DREs in the literature. In this paper we build such a repository of DREs as follows. (1) We harvest a large corpus of data from the Web, including RegExLib, Relax NG, XSD and DTD, which forms the basis of our approach (Sect. 3). (2) We collect a large-scale repository of DREs from this data set after analyzing determinism of all data in this set (Sect. 4). (3) We use *normalized DREs* to get a compact repository of DREs, called DRE pattern set (Sect. 4). The lack of benchmark datasets is a problem for the research of DREs. We show that the DRE pattern set is an important step to fix this problem by using the DRE patterns in several algorithms that have lacked experiments on real DRE data before. Experimental results give some insight into the performance of the algorithms, and thus demonstrate that the repository can be a helpful tool for research in this research area (Sect. 5).

2   Preliminaries

Let Σ be an alphabet of symbols. The set of all finite words over Σ is denoted by Σ^*. The empty word is denoted by ε. A (standard) regular expression over Σ is defined as: $\emptyset, \varepsilon$ or $a \in \Sigma$ is a regular expression, the union $E_1|E_2$, the concatenation E_1E_2, the Kleene-star E_1^*, the option $E_1^?$, or the plus E_1^+ is a regular expression for regular expressions E_1 and E_2. Note $E^?$ and E^+ are actually redundant, science $E^? = \varepsilon|E$ and $E^+ = EE^*$. We include them in the definition because they are used in practice. Also, the concatenation E_1E_2 is often written as E_1, E_2 in practice.

Let $\mathbb{N}$ denotes the set $\{0, 1, 2, \ldots\}$. A regular expression with *counting and interleaving* is extended from RE by further using the *numerical iteration operator* $E^{[m,n]}$ and the *interleaving operator* $E_1 \& E_2$. The bounds m and n satisfy: $m \in \mathbb{N}$, $n \in \mathbb{N}\backslash\{0\} \cup \{\infty\}$, and $m \leq n$. For $s, s_1, s_2 \in \Sigma^*$ and $a, b \in \Sigma$, $s\&\varepsilon = \varepsilon\&s = \{s\}$ and $as_1\&bs_2 = \{a(s_1\&bs_2)\} \cup \{b(as_1\&s_2)\}$.

For a regular expression we can mark symbols with subscripts so that in the marked expression each marked symbol occurs only once. For example $(a_1 + b_1)^*a_2$ is a marking of the expression $(a + b)^*a$. We use $\overline{E}$ to denote the marking of E, and $\overline{\Sigma}$ to denote the alphabet of subscripted symbols. The same notation will also be used for dropping of subscripts from the marked symbols: $\overline{\overline{E}} = E$. Now we can give the definition of deterministic regular expressions (DREs for short):

Definition 1 [11]. *An expression E is* deterministic *if and only if, for all words $uxv, uyw \in L(\overline{E})$ where $u, v, w \in \overline{\Sigma}^*$ and $x, y \in \overline{\Sigma}$, if $x \neq y$ then $\overline{x} \neq \overline{y}$. A regular language is* deterministic *if it is denoted by some deterministic regular expression.*

For example, $E = aa^*$ is a DRE while $E = a^*a$ is not. Because for $E = a^*a$ a marking of E is $\overline{E} = a_1^*a_2$, then we have $a_1, a_2 \in L(\overline{E})$ and $a_1 \neq a_2$ but $\overline{a_1} = a = \overline{a_2}$, where $x = a_1, y = a_2, u, v, w = \varepsilon$, thus $E = a^*a$ is not deterministic.

Furthermore, it is known that DREs are strictly less expressive than REs, which means that not every RE can be rewritten to an equivalent DRE [11]. It is nontrivial for ordinary users to decide whether an expression is DRE or not, and if not, whether it has an equivalent DRE. For instance, $(a + b)^*a + \varepsilon$ is not a DRE. Nevertheless, the language denoted by $(a + b)^*a + \varepsilon$ is a deterministic language, since it is also denoted by $(b^*a)^*$, which is a DRE. Clearly it is not easy for an ordinary user to find the DRE $(b^*a)^*$ to replace $(a + b)^*a + \varepsilon$.

3   The Data Set

One basis of our work is the large-scale data set, which we introduce in this section.

3.1   The Data Sources

The data set should be representative to investigate DREs. For this purpose, as we have mentioned, we have obtained data files from the Web, including RegExLib, Relax NG, XSD and DTD. DTD and XSD are recommended by W3C [30], Relax NG is a standard of ISO [27], and RegExLib is the main regular expression repository available on the Web [2]. W3C specification requires that the content models of DTDs and XSDs must be DREs, while both Relax NG and RegExLib do not have determinism restrictions. Both XSD and Relax NG support interleaving, in which the interleaving supported by XSD is very limited and the interleaving supported by Relax NG is unlimited. And both XSD and RegExLib support counting. So it is representative to take them as examples to investigate the practical usage of DREs with counting and interleaving.

3.2   Harvesting Schema Files from the Web

In order to investigate DREs, the data set should also be sufficiently large. One feature of our work is that the data set is sufficiently large compared with previous work. Harvesting a large corpus of DTD, XSD and Relax NG schema files from the Web is not an easy task, because although there are many schema files on the Web, they exist in different forms and locations so they cannot be directly obtained in batches. Previous research gained schema files usually from some local sources, e.g. Bex et al. studied 109 DTDs and 93 XSDs downloaded from the XML Cover pages repository [7]. However, we have made good use of search engines and project platforms (such as GitHub and Maven) to obtain data without source restriction, thus the data obtained are much larger and representative than those of previous research. In detail, we proposed four data collection strategies: comprehensively utilizing Google search engine, path-ascending crawling, downloading and analyzing the Web sites and finding potential data, to attain more schema files from the Web. Figure 1 shows the data collection process and the strategies used.

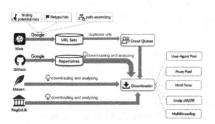

Fig. 1. The process of data collection and strategies used

Finally, we obtained 276371 data files including 124326 DTDs, 134816 XSDs, 13946 RELAX NGs and 3283 RegExLib expressions. And previous work is summarized in Table 1. For example, the total number of data files is 34 times of Björklund et al.'s [10] (see Table 1). Such an extensive and large-scale data set is significant in analyzing DREs, because studying the practicability of DRE requires data to reflect practical applications as far as possible, and large-scale random schemas are in line with this requirement. Our repository can be found at https://github.com/yetingli/IDEAS18.

3.3 A Practical Study of DREs

For the data collected, we preprocess with the steps such as schema normalization, duplicate file removal, well-formedness and validity checking, and so on. And then we parse the content models into REs.

Then we conducted an extensive study to investigate the practical usage of DREs using the data. For example, we studied the percentage of DREs of the data set, and the usages of various subclasses of DREs; we analyzed the complexity of the data set, including star height, nesting depth, density, and so on; we investigated counting and interleaving used in the data set. Details about the practical study can be found in [24].

4 The Repository

4.1 Getting the DRE Set from the Data Set

To build the DRE set, we need techniques and tools that can decide determinism of REs with interleaving. Furthermore, they are also necessary for using the DRE set. Fortunately, we have solved this problem and have tools for deciding determinism of REs with counting and unlimited interleaving [29]. This forms the basis of the present work.

Determinism. Using our data set and our determinism checking tools, we studied the determinism of REs generated by schema files and the REs from RegExLib respectively. The results are shown in Table 2, where DREs (%) denotes the percentage of DREs in each type of files in our data set, which demonstrate that large proportions of expressions in Relax NG and expressions from RegExLib are deterministic.

The DRE Set. Then we collect deterministic REs from the data set to build the initial repository of DREs, in which after removing duplicate DREs the total number of DREs is 222163. This repository then is used in the following.

Table 2. DETERMINISM OF REs

Source	DTD	XSD	RNG	RegExLib
DREs (%)	0.9883	0.9993	0.9825	0.5656

4.2 From the DRE Set to the DRE Pattern Set

Now we start to consider how to make the repository of DREs more efficient. To this end, we further process the data which are introduced in this section.

Theoretical Preparations. To further processing the data, we first need the following results.

From the theoretical studies of DREs [11,13,22,29], we have that *any subexpression of a DER is a DRE.*

According to this statement, for a DRE E in the repository, a part of E may still be a DRE if this part is a subexpression of E. The subexpressions of an expression E can be easily found, e.g., using the syntax tree of E. In fact, any subtree of the syntax tree of E is a subexpression of E. And a DRE given by the user can start from a position that is the starting position of a subexpression of a DRE in the repository, not necessarily the starting position of a DRE in the repository. That is, *a DRE given by the user can be a subexpression of a DRE in the repository of DREs.*

From formal language theory we know that two grammars which only differs in using different alphabets are isomorphic. Applying this result to DREs, we have that *for any DRE r over the alphabet Σ, when replacing Σ by another alphabet Σ', the result is also a DRE.*

Constructing the DRE Pattern Set from the DRE Set. Based on the above results, we further use normalized DREs to construct a more compact repository, called DRE pattern set. We normalize the DREs as follows.

Definition 2. *A DRE is normalized if the symbols in the DRE, in the order from left to right, are uniformly substituted by symbols $a_1, a_2, a_3, \ldots$. Note that a repeatedly occurring symbol in the DRE will be substituted by a same symbol.*

Example 1. Suppose we have three original DREs: (1) red, green, blue; (2) SORE, CHARE, eCHARE; and (3) red, green, green. The normalized DREs are as follows: (1) a_1, a_2, a_3; (2) a_1, a_2, a_3; and (3) a_1, a_2, a_2.

Thus two different DREs in the repository of DREs that only differ in symbols become the same in their normalized forms. In Example 1, the first two DREs become the same normalized DREs, and in the third DRE the repeatedly occurring symbol 'green' is substituted by the same symbol a_2.

Normalized DREs make it possible for us to concentrate on the structures of the expressions, regardless of the alphabets. As a result, expressions with same or similar structures can be merged, thus we get a more compact set. Moreover, the normalized DREs cover more DREs than the original repository of DREs in the following sense: actually any DRE that has the same structure as (a subtree of) a normalized DRE but is not contained in the original repository of DREs will be covered. These observations are formalized as follows.

Fact 1. For normalized DREs E_1 and E_2, if $E_1 = E_2$, or E_2 is isomorphic to a subtree of E_1 up to a renaming of symbols, then E_2 is redundant and can be removed. And any DRE that has the same structure as (a subtree of) E_1 is covered by E_1. Notice here = means identical.

Actually, it is easy to see that any DRE covered by E_2 will also be covered by E_1, so we can safely remove E_2. So we get a more compact set that covers more DREs than the original DRE repository, which is called *DRE pattern set*.

For example, the first two normalized DREs in *Example 1* are identical, i.e., a_1, a_2, a_3, so one of them is removed. Suppose we also have a fourth normalized DRE in *Example 1*, which is a_1, a_2. Since it is isomorphic to a subtree of a_1, a_2, a_3, then it is removed. To be mentioned, it is also isomorphic to a subtree of a_1, a_2, a_2. Furthermore, other DREs, for example the ones that can be normalized as a_1, a_2, a_3 but are not contained in the original DREs, such as $yellow, seagreen, pink$, will also be covered by the normalized DRE a_1, a_2, a_3.

Since in the repository of DREs that we have obtained there is a large portion of DREs satisfying the above conditions, this will effectively get a compact set (see Sect. 4.4 for details).

4.3 Dynamically Increasing the Power of the Repository

If a DRE that the user wanted is not in the repository, we also allow the user to add new DREs that meet her needs into the repository. That is, we allow the user to input an arbitrary expression. Then, we check if the expression is deterministic by using our determinism check tools [29]. There will be two cases. If the expression is deterministic, then we normalize this new DRE, add it into the DRE pattern set and rebuild the pattern set. So the repository knows the new pattern from now on. Otherwise, our tools will hint the possible reasons for nondeterminism, which can be used by the user for redesigning of the expression.

The diagnostic information for nondeterminism that our tools can report include locating the nondeterministic subexpression, giving competing positions, and so on. The interface of one of our tools is shown in Fig. 2.

In this manner, the repository is able to increase its power dynamically.

4.4 Getting the Repository of DREs

We processed the data set according to the above as follows. First, we get the set of DREs from the original data set after determinism checking and duplicate DREs removal. The total number of DREs is 222163. Then, we get the DRE pattern set from the set of DREs, in which the number of expressions in the pattern set is only about 0.145% of the set of DREs. The results for the DRE set and the DRE pattern set are shown in Table 3[1].

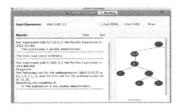

Table 3. Number of DREs

Type	DTD	XSD	RNG	RegExLib	Total
DRE set	25142	155475	42968	1402	222163
Pattern set	3767	14771	2791	724	20339

Fig. 2. The interface of our tool

---

[1] The number of total is not equal to the sum of DTD, XSD, RNG and RegExLib, because there exist duplicate DREs among the different types of files.

5   Experiments

We experimentally evaluate our repository on several algorithms that have lacked experiments on real DRE data before, showing that it is possible to effectively apply DRE algorithms on this data set. All experiments were conducted on a machine with a Intel(R) Xeon(R) CPU (3.19 GHz) and 48G memory.

5.1   Algorithm Selection

The first experiments are to evaluate the correctness and efficiency of some inference algorithms, including the inference algorithms for chain regular expression (CHARE) [8], the subset of regular expression with interleaving (SIRE) [28], and DME [21]. All of them are subclasses of DREs, where DME supports unordered concatenation, a weaker form of interleaving, SIRE supports interleaving, and CHARE supports standard REs. Since DREs are widely used in schema definitions, inference algorithms are quite common for DREs and their subclasses. And the subclasses we selected in the experiments have different features so they are quite representative. The second experiments are to compare the performance and efficiency of two algorithms for checking inclusion of DREs [12], which is another kind of application of DREs that can be widely used.

5.2   Experiment1

We give experiments to evaluate the correctness and efficiency of the inference algorithms for three subclasses of DREs: CHARE, SIRE, and DME. Roughly speaking, an inference algorithm for a subclass of DREs $\mathcal{D}$ is to, given a sample set S, infer a DRE e satisfying $S \subseteq L(e)$, where S and e belong to $\mathcal{D}$. By selecting different DREs that satisfy different definitions of the subclasses from our repository, we performed test of the correctness and efficiency of their inference algorithms using the data.

In detail, we randomly selected three groups of DREs from our repository with each group containing 200 DREs, and the groups respectively belong to CHARE, SIRE and DME. The alphabet sizes of the selected DREs range from 5 to 100. For each selected DRE, we randomly generated three sample sets with size of 500, 1000 and 3000 respectively. For each sample set, we run its corresponding inference algorithms and record the average runtime.

First, using the inferred DREs, we verified whether the sample sets can be generated by the corresponding inferred DREs, that is, whether we have $S \subseteq L(e)$ for each sample set S and the corresponding inferred DRE e. Results show all the sample sets can be generated by the corresponding inferred DREs. This gives a test of the correctness of the inference algorithms using the data as the test set.

Next, we compare the efficiency of the inference algorithms. We use the algorithm $Soa2Chare$ [18] for the inference of CHARE. Results are shown in Fig. 3, where the horizontal axis denotes alphabet size and the vertical axis denotes average runtime of $Soa2Chare$ for samples of different sizes in milliseconds, $|S|$

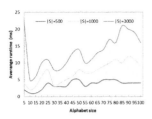

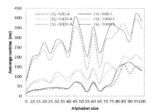

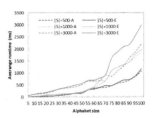

Fig. 3. Average runtime for learning CHARE of varying alphabet size

Fig. 4. Average runtime for learning SIRE of varying alphabet size

Fig. 5. Average runtime for learning DME of varying alphabet size

denotes the size of sample. We can see the sample size $|S|$ has a significant influence on the inference time, that is, average runtime increases with the increase of sample size.

We run the approximation algorithm and the exact algorithm in [28] respectively for inference of SIRE. The main difference between the two algorithms is the solution to the maximum independent set problem, which is NP-hard, thus approximation algorithms are necessary to find approximate solutions to this problem. In detail, the exact algorithm calls the igraph package [1], and the approximation algorithm uses the NetworkX package [3] to get the maximum independent set. The experimental results are shown in Fig. 4, where the horizontal and vertical axes are same as in Fig. 3, in addition we use marks $-E$ and $-A$ respectively to denote the exact and approximation algorithms. On the whole, the approximation algorithm is faster than the exact algorithm. Still, average runtime increases with the increase of sample size.

Figure 5 shows the result of DME inference algorithm [15], where the horizontal and vertical axes and the marks are same as in Fig. 4. The difference between the exact and the approximation algorithms lies in the solution of maximum clique problem, which is also NP-hard. Actually, the independent set problem and the clique problem is complementary. It should be noted that the DME inference algorithm described in [15] is not accompanied with source code. Therefore we implemented their algorithm according to the pseudo-code given in [15], in which the exact algorithm calls the igraph package and the approximation algorithm uses the NetworkX package. These algorithms are used in the present experiments. Compared with the sample size, the alphabet has a greater impact on the average runtime, that is, the average runtime increases with the increase of alphabet size. Moreover, the average runtime of approximation algorithm is faster than the average runtime of exact algorithm.

From the above experiments we can also have the following conclusions. First, the inference algorithm for CHARE is more efficient than that for SIRE and DME. This is mainly because CHARE is defined on standard REs while both SIRE and DME support interleaving. Second, both approximation algorithms for SIRE and DME are efficient than their exact versions, which are the same as expected. Third, exceeding the authors' expectations, the inference algorithm for DME is much slower than that for SIRE, but indeed DME only supports unordered concatenation, a weaker form of interleaving, while SIRE supports interleaving.

5.3 Experiment2

We give experiments to evaluate two algorithms for checking inclusion of DREs [12] using the repository. An algorithm for checking inclusion of DREs is to decide whether $e_1 \subseteq e_2$ for DREs e_1 and e_2. In [12] there are two algorithms for checking inclusion of DREs, namely the DFA-based and the derivative-based algorithms. Please notice that there is a necessary condition for $e_1 \subseteq e_2$ holds, that is the alphabet of e_2 must include the alphabet of e_1.

To evaluate the two algorithms, we randomly selected 5680 DREs from our repository. Then we get 11463823 pairs of DREs from these DREs, satisfying the alphabet of e_2 includes the alphabet of e_1 for each pair (e_1, e_2). Figure 6 shows the distribution of the pairs of DREs, where the horizontal axis denotes the sum of the lengths of DRE pairs (e_1, e_2), and the vertical axis denotes the number of DRE pairs in each length interval.

We randomly selected 10,000 DRE pairs in each length interval, totally get 120,000 DRE pairs. We run each inclusion algorithm on the pairs to determine whether $e_1 \subseteq e_2$, and account the runtime for each pairs. From these DRE pairs, we get 8396 pairs of DREs satisfying the inclusion relation, which form the positive sample, and 111604 pairs of DREs not satisfying the inclusion relation, which form the negative sample.

We compare the two algorithms using both positive and negative samples. Figure 7 shows the results on positive sample, where the horizontal axis denotes the lengths of DRE pairs, and the vertical axis denotes the average runtime of each of the algorithms, i.e., running each DRE pairs 100 times for totally 10000 DRE pairs in each length interval then divided by 10000. We show the runtime in milliseconds. Results show that the derivative-based algorithm is more efficient on smaller expressions (i.e., when the length of the DRE pairs is smaller than 100), while the DFA-based algorithm performs better on larger expressions (i.e., when the length of the DRE pairs is larger than 100). This is because that the derivative-based algorithm performs, in a sense, a breadth-first exploration of the two compared expressions, and smaller expressions tend to have a smaller amount of such exploration.

Figure 8 shows the results on negative sample, from which we can see that the derivative-based algorithm performs better (due to that looking for a reason to fail is easy to spot) then the DFA-based algorithm (due to the construction of Glushkov DFAs).

Discussion of the Experiments. The repository makes it possible to evaluate DRE algorithms on large-scale real DRE data, and the experiments give some insight into the performance of the algorithms, thus demonstrate the usefulness of our repository.

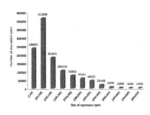 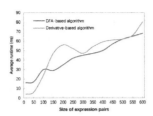 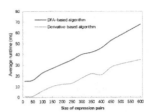

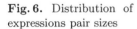

Fig. 6. Distribution of expressions pair sizes

Fig. 7. Positive sample experiment

Fig. 8. Negative sample experiment

6   Conclusion

Based on a large corpus of real data that we harvested from the Web, we collected a large-scale repository of DREs after analyzing determinism of the real data. Then we further process the collected repository by using normalized DREs to compact the DRE set into the pattern set. At last we use the DRE patterns in several algorithms that have lacked experiments on real DRE data before, including some inference algorithms, and DRE inclusion checkers. Experimental results demonstrate the usefulness of the repository. We can further enhance the repository by using pattern generalization and machine learning techniques, which remain as future work.

References

1. igraph - the network analysis package. http://igraph.org/
2. RegExLib. www.regexlib.com
3. Software for complex networks. http://networkx.github.io/
4. Abiteboul, S., Milo, T., Benjelloun, O.: Regular rewriting of active XML and unambiguity. In: PODS 2005, pp. 295–303. ACM (2005)
5. Barbosa, D., Mignet, L., Veltri, P.: Studying the XML Web: gathering statistics from an XML sample. World Wide Web **9**(2), 187–212 (2006)
6. Bex, G.J., Martens, W., Neven, F., Schwentick, T.: Expressiveness of XSDs: from practice to theory, there and back again. In: WWW 2005, pp. 712–721. ACM (2005)
7. Bex, G.J., Neven, F., Van den Bussche, J.: DTDs versus XML schema: a practical study. In: WebDB 2004, pp. 79–84. ACM (2004)
8. Bex, G.J., Neven, F., Schwentick, T., Tuyls, K.: Inference of concise DTDs from XML data. In: VLDB 2006, pp. 115–126. VLDB Endowment (2006)
9. Bex, G.J., Neven, F., Vansummeren, S.: Inferring XML schema definitions from XML data. In: VLDB 2007, pp. 998–1009 (2007)
10. Björklund, H., Martens, W., Timm, T.: Efficient incremental evaluation of succinct regular expressions. In: CIKM 2015, pp. 1541–1550. ACM (2015)
11. Brüggemann-Klein, A., Wood, D.: One-unambiguous regular languages. Inf. Comput. **142**(2), 182–206 (1998)

12. Chen, H., Chen, L.: Inclusion test algorithms for one-unambiguous regular expressions. In: Fitzgerald, J.S., Haxthausen, A.E., Yenigun, H. (eds.) ICTAC 2008. LNCS, vol. 5160, pp. 96–110. Springer, Heidelberg (2008). https://doi.org/10.1007/978-3-540-85762-4_7
13. Chen, H., Lu, P.: Checking determinism of regular expressions with counting. Inf. Comput. **241**, 302–320 (2015)
14. Choi, B.: What are real DTDs like. Technical reports (CIS), p. 17 (2002)
15. Ciucanu, R., Staworko, S.: Learning schemas for unordered XML. arXiv:1307.6348 [cs.DB] (2013)
16. Colazzo, D., Ghelli, G., Pardini, L., Sartiani, C.: Efficient asymmetric inclusion of regular expressions with interleaving and counting for XML type-checking. Theor. Comput. Sci. **492**(2013), 88–116 (2013)
17. Colazzo, D., Ghelli, G., Sartiani, C.: Linear time membership in a class of regular expressions with counting, interleaving, and unordered concatenation. ACM Trans. Database Syst. (TODS) **42**(4), 24 (2017)
18. Freydenberger, D.D., Kötzing, T.: Fast learning of restricted regular expressions and DTDs. Theory Comput. Syst. **57**, 1114–1158 (2015)
19. Grijzenhout, S., Marx, M.: The quality of the XML web. In: CIKM 2011, pp. 1719–1724 (2011)
20. Huang, X., Bao, Z., Davidson, S.B., Milo, T., Yuan, X.: Answering regular path queries on workflow provenance, pp. 375–386. IEEE (2015)
21. Boneva, I., Ciucanu, R., Staworko, S.: Simple schemas for unordered XML. In: WebDB 2013, pp. 13–18 (2013)
22. Kilpeläinen, P.: Checking determinism of XML Schema content models in optimal time. Inf. Syst. **36**(3), 596–617 (2011)
23. Laender, A.H., Moro, M.M., Nascimento, C., Martins, P.: An X-ray on web-available XML schemas. ACM SIGMOD Rec. **38**(1), 37–42 (2009)
24. Li, Y., Chu, X., Mou, X., Dong, C., Chen, H.: Practical study of deterministic regular expressions from large-scale XML and schema files. In: IDEAS 2018, pp. 45–53. ACM (2018)
25. Li, Y., Zhang, X., Peng, F., Chen, H.: Practical study of subclasses of regular expressions in DTD and XML schema. In: Li, F., Shim, K., Zheng, K., Liu, G. (eds.) APWeb 2016. LNCS, vol. 9932, pp. 368–382. Springer, Cham (2016). https://doi.org/10.1007/978-3-319-45817-5_29
26. Losemann, K., Martens, W.: The complexity of regular expressions and property paths in SPARQL. ACM Trans. Database Syst. **38**(4), 24:1–24:39 (2013)
27. Makoto, M.: RELAX NG home page (2014). http://relaxng.org/. Accessed 25 Feb 2014
28. Peng, F., Chen, H.: Discovering restricted regular expressions with interleaving. In: Cheng, R., Cui, B., Zhang, Z., Cai, R., Xu, J. (eds.) APWeb 2015. LNCS, vol. 9313, pp. 104–115. Springer, Cham (2015). https://doi.org/10.1007/978-3-319-25255-1_9
29. Peng, F., Chen, H., Mou, X.: Deterministic regular expressions with interleaving. In: Leucker, M., Rueda, C., Valencia, F.D. (eds.) ICTAC 2015. LNCS, vol. 9399, pp. 203–220. Springer, Cham (2015). https://doi.org/10.1007/978-3-319-25150-9_13
30. Thompson, H.S., Beech, D., Maloney, M., Mendelsohn, N.: XML Schema part 1: structures second edition. W3C Recommendation (2004)

Determining the Impact of Missing Values on Blocking in Record Linkage

Imrul Chowdhury Anindya[1]([⊠]), Murat Kantarcioglu[1], and Bradley Malin[2]

[1] The University of Texas at Dallas, Richardson, TX 75080, USA
{icanindya,muratk}@utdallas.edu
[2] Vanderbilt University, Nashville, TN 37235, USA
b.malin@vanderbilt.edu

Abstract. Record linkage is the process of integrating information from the same underlying entity across disparate data sets. This process, which is increasingly utilized to build accurate representations of individuals and organizations for a variety of applications, ranging from credit worthiness assessments to continuity of medical care, can be computationally intensive because it requires comparing large quantities of records over a range of attributes. To reduce the amount of computation in record linkage in big data settings, *blocking* methods, which are designed to limit the number of record pair comparisons that needs to be performed, are critical for scaling up the record linkage process. These methods group together potential matches into blocks, often using a subset of attributes before a final comparator function predicts which record pairs within the blocks correspond to matches. Yet data corruption and missing values adversely influence the performance of blocking methods (e.g., it may cause some matching records not to be placed in the same block). While there has been some investigation into the impact of missing values on general record linkage techniques (e.g., the comparator function), no study has addressed the impact of the missing values on blocking methods. To address this issue, in this work, we systematically perform a detailed empirical analysis of the individual and joint impact of missing values and data corruption on different blocking methods using realistic data sets. Our results show that blocking approaches that do not depend on one type of blocking attributes are more robust against missing values. In addition, our results indicate that blocking parameters must be chosen carefully for different blocking techniques.

Keywords: Record linkage · Deduplication · Missing values · Blocking methods · Data corruption

1 Introduction

Record linkage is the process of identifying the same entity across different and possibly dispersed data sources. It is a task of paramount importance in many domains where linking and combining data related to the same entity is of vital

© Springer Nature Switzerland AG 2019
Q. Yang et al. (Eds.): PAKDD 2019, LNAI 11441, pp. 262–274, 2019.
https://doi.org/10.1007/978-3-030-16142-2_21

necessity. Occasionally, data belonging to different entities such as clients, consumers, social network users, patients, tax payers and travelers remain dispersed among different sources [6]. Linking relevant data sets from these sources produces concise, but comprehensive, high quality data and ensures better analytics and business intelligence. Deduplication can be considered a special type of record linkage where a data set is linked to its own in order to retrieve the duplicate pairs belonging to the same entity. It is a crucial step in the data cleaning process which offers more accurate statistics by removing the redundancies.

Regardless of the type, record linkage procedures require quadratic time complexity. This is because each record from a data set is compared to all the records in the other data set. Consequently, blocking methods are often applied to reduce the number of record pair comparisons by grouping together similar records based on some blocking key value (BKV). Afterwards, a comparator function can be applied to compare record pairs only within the groups (i.e., the blocks). Unfortunately, a common phenomenon in real world data sets is the existence of missing values. Understanding the impact of missing values on blocking methods is important because these methods partition the records into blocks based on BKV, which in turn is generated from the attribute values. Generally, blocking methods cannot handle empty key values and discard the record from comparison. To address the missing attribute values' problem in the blocking step, there are several options.

First, a straw-man approach is to place the records with an empty key into another block, termed as the "empty key block" and then to compare all the records within the block. This naïve approach, while seemingly effective in terms of finding the matches that would otherwise be missed, may be computationally infeasible if the percentage of records having empty key is high enough to require a huge number of comparisons for the empty key block. A second idea is to define multiple blocking keys and then to iteratively perform comparisons within the blocks generated by each of these keys and finally merging the matches. Yet another idea is to construct the blocking key from several different attributes such that the probability of getting an empty key is very low even if the record contains one or more missing attribute values. An additional step applicable to the previous two ideas is to define the parameters of the blocking method in a loose way such that matches having very dissimilar blocking key values due to having one or more missing attributes can still be placed into the same block. There is no systematic analysis to understand the tradeoffs of these potential approaches and how they work under different existing blocking schemes.

1.1 Our Contributions

In this research, we explore what impact the missing attribute values have on the blocking methods. Our contributions are as follows:

- To the best of our knowledge, this is the first work that tries to evaluate the impact of missing values on blocking techniques for record linkage.

- Extensive empirical analysis are conducted to understand how different settings and techniques perform when we have both missing and noisy attributes. Our results indicate that multiple blocking steps that combine different sets of attributes can enhance blocking performance with little overhead.
- This work also provides guidance for choosing the length of the blocking key and the type of the parameters for different blocking techniques.

The remainder of this paper is organized as follows. In Sect. 2, we briefly describe prior research related to our own. In Sect. 3, we provide a general overview of blocking methods and their performance metrics. Then in Sect. 4, we summarize the technical details of several widely used blocking methods that we incorporate in our study. The data sets, experimental design, and empirical results are presented in Sect. 5. Section 6 reports on our conclusion and highlights future directions.

2   Related Work

To provide context, we summarize the works most relevant to our research.

Christen et al. published a detailed theoretical and experimental study of six blocking methods with a total of twelve variations of those [6]. This study found that the most important factor behind effective and efficient blocking for record linkage and deduplication is the selection of blocking keys. Also, it was shown that the performance of the blocking methods may change significantly based on the chosen blocking parameters while finding the optimal parameters is a difficult problem as it depends on the quality and characteristics of the data to be linked or deduplicated. While this study provides important insights into the performance of the blocking methods on general data sets, the performance degradation caused by missing values is not explicitly addressed.

Ong et al. introduced three methods, namely (i) Weight Redistribution, (ii) Distance Imputation, and (iii) Linkage Expansion to solve the missing value problem in record linkage [11]. Weight Redistribution omits missing value attributes for linkage and redistributes their weights to other available attributes in proportion to the missing attributes' weights. Distance Imputation infers the distance accounted for the missing value attribute instead of inferring the missing value itself. Linkage Expansion adds more attributes to the set of linkage attributes to compensate for the missing values. Please note that, *their methods deal with the final comparator* in the record linkage process and are not applicable to the blocking methods.

Prasad et al. proposed a data driven approach of selecting blocking key attribute(s) automatically by considering the amount of missing values along with two other characteristics of the attributes [12]. Unfortunately, their method is applicable only to the simplest type of blocking method known as the Standard Blocking. It further suffers from complexity issues when multiple attributes are collectively chosen for blocking key construction in high-dimensional data sets.

3   Background

Effective and scalable Record Linkage is not a single step procedure, but a combination of three. Data comes in different formats, shapes and qualities. So, the first step in linking data sets is to remove these idiosyncrasies that make the task difficult. The chance of a successful linkage depends heavily on how much endeavor is exerted on cleaning and standardizing the data [7]. The second step is known as blocking. The general idea of blocking is to distribute the records into overlapping or non-overlapping sets of blocks based on the blocking key value. The blocking key is constructed by choosing a subset of the attributes, defining some transformation functions for each of these attributes (optionally) and defining a combination strategy for the transformed attributes. In the third step, a classifier or comparator function compares the records within the generated blocks from the previous step using deterministic or probabilistic (e.g., Fellegi Santur EM algorithm [14]) techniques and outputs the matches, possible-matches, and non-matches.

The performance of a blocking method depends on three different metrics known as pair completeness, reduction ratio, and blocking time. Suppose, A and B are two data sets with size n_A and n_B respectively. Then for record linkage, the total number of record pair comparisons needed to link the two data sets is $n_c = n_A \times n_B$. In case of deduplication of data set A, since a record is compared to every other record in the same data set, the total number of comparisons becomes $n_c = n_A \times (n_A - 1)/2$.

Pair completeness (PC), also referred to as detection rate or recall, provides the ratio of total number of distinct true matches found in any of the generated blocks, d_m to the actual number of true matches, n_m. So, $PC = d_M/n_M$.

Reduction ratio (RR) indicates the reduction in the number of record pair comparisons required after applying blocking method. If the particular blocking method distributes the records into k blocks, each requiring of n_i comparisons for $i \in \{1, ..., k\}$, then $RR = 1 - \frac{1}{n_c} \cdot \sum_{i=1}^{k} n_i$.

Blocking time is another important metric to measure the performance of the blocking methods. Note that, the complexity of a blocking method depends not only on the length and properties of the defined blocking key, but also on the parameters related to the method. As a result, complexity expressions are often incomprehensible when there are lots of parameters involved.

4   Major Blocking Methods

In this section, we summarize the most common blocking techniques implemented in popular record linkage software packages such as Febrl [5].

Standard Blocking (STD): The Standard Blocking method [4] places records having the same blocking key value into the same blocks. Only the records within the same block are compared by the comparator function.

Sorted-Neighborhood (SRT): The Sorted-Neighborhood method [8] starts by combining all the records into a sequential list of n records. The records in the list are then sorted based on the blocking key value. Then a window of size ω is slid along the list and the last record in the window is compared to the previous $\omega - 1$ records.

Q-gram Indexing (QGM): The Q-gram Indexing method, implemented in Febrl [5], converts the blocking key of a record into a list of q-grams. Then all possible sublists down to a minimum length are generated, where that length is set by multiplying a user-defined threshold ($\tau \in [0,1]$) to the size of the q-gram list. The resulting sublists are then sorted and placed into an inverted index data structure in which all sublists keep track of the particular record's identifier. Here, a record is compared to all other records that have at least a single q-gram sublist in common. This method works efficiently only if the length of the blocking key is small since large ones generate too many sublists (and hence blocks) in the above process for lower values of τ.

Canopy Clustering (CNP): Canopy clustering method [10] depends on two user-defined distance thresholds, namely a tight threshold, T_{tight} and a loose threshold, T_{loose} ($T_{tight} < T_{loose}$). It begins by selecting a particular record randomly from the list of all records as the center record of a cluster (known as *canopy*) against which other records in the list are compared. The records that are within the T_{loose} distance are placed into the same canopy with the center record and the records that are within T_{tight} distance are removed from the list. This process continues until the list is empty. Here, the distance is often measured by the TF-IDF or Jaccard distance based on the q-grams available in the blocking keys.

String-Map (STM): The string-Map method maps the blocking key values into a multi-dimensional Euclidean space that preserves the distances over strings using a modified FastMap algorithm [9]. While the original method uses an R-tree to retrieve similar pairs of strings from the high-dimensional space, Febrl uses an inverted index based approach similar to the iGrid index [5].

Suffix Array (SFX): The Suffix Array method [3] extracts the suffixes of the blocking key values down to a minimum length. Then all the suffixes are alphabetically sorted to form the suffix array. Each of these suffixes generates a block by grouping together the records having the suffix in their blocking key values. If the size of a block is more than a user-defined threshold, the corresponding suffix is considered too general and removed from the suffix array. A variation of this method can be obtained by extracting all substrings, instead of only the suffixes, which our experimental analysis shows achieves better results.

5   Experimental Evaluation

To conduct an experimental evaluation, we perform deduplication tasks on data sets with known ground truth about matches to measure the impact of missing attribute values on blocking methods. This section begins with a description of the data sets and experimental setup. We then report on the results of the experiments.

5.1   Data Set Construction

To construct realistic data sets, we rely upon the publicly-available state-wide voter registration records of Florida (FL) and North Carolina (NC) [1,2], both having more than 10 million records. Based on these data sources, we generate data sets for our experiments as follows.

For each data source, we randomly sample 35,000 records defined over 16 attributes (e.g., first name, year of birth, sex etc.) that can be useful for overall record linkage procedure. We then randomly choose 15,000 records from the sample to serve as duplicates. We merge these two sets of data to generate a sample of size 50,000 having 30% (i.e., 15,000) duplicates. This is done for each data source.

For each sample, we choose 7 attributes as candidates for blocking keys. We modify $x\%$ ($x \in \{10, 20\}$) records of the sample to generate missing values for up to three of these attributes such that the percentages of *missing value records* having 1, 2, or 3 missing attributes are 50%, 30%, and 20% respectively. In addition, we generate the corrupted versions of the data sets using the GeCo [13] data corrupter, such that 5% of the records are corrupted on the candidate blocking attributes. An attribute is modified at most twice and a maximum of three modifications is applied to a record. The rate of *corrupted records* having 1, 2, or 3 modifications are 50%, 30%, and 20% respectively. Different corruption functions defined in GeCo, such as edit_corruptor2, ocr_corruptor, keyboard_corruptor, phonetic_corruptor and a custom-defined function for sex (gender_corrputor), are used on the attributes according to the probability distribution in Table 1.

Table 1. Corruption functions and their probabilities.

Attributes	Corruption function	Probability
First name, middle name, last name	edit_corruptor2	0.1
	ocr_corruptor	0.1
	keyboard_corruptor	0.1
	phonetic_corruptor	0.7
Month of birth, day of birth, year of birth, age, zip code	edit_corruptor2	0.5
	keyboard_corruptor	0.5
Address	edit_corruptor2	0.2
	ocr_corruptor	0.2
	keyboard_corruptor	0.3
	phonetic_corruptor	0.3
Sex	gender_corruptor	1.0

Table 2. Blocking keys defined for this study.

Data set	Key	Value*	Max length
FL	F1	`ln[:4] + yy[2:4] + s`	7
	F2	`fn[:4] + mn[:1] + mm + dd`	9
	F3	`F1 + F2`	16
NC	N1	`dmph(ln,4) + mn[:1] + s + zip[2:5]`	9
	N2	`dmph(fn,4) + age + dmph(ad,4)`	11
	N3	`N1 + N2`	20

* Here `fn`, `ln`, `mn`, `s`, `mm`, `dd`, `yy`, `zip` and `ad` represent the values of first name, middle name, last name, sex, month of birth, day of birth, year of birth, zip code and address respectively. The function `dmph(x,y)` returns the Double Metaphone encoding of `x` upto the length of `y`.

Table 3. Parameters involved in different methods

Method	Parameter list*
SRT	Window size
QGM	(q-gram length, threshold)
CNP	(Canopy method, thresholds/nearest, tight threshold/remove nearest, loose threshold/cluster nearest, q-gram length)
STM	(Grid resolution, dimension, sub-dimension, thresholds/nearest, tight threshold/remove nearest, loose threshold/cluster nearest)
SFX	(Suffix method, minimum suffix length, maximum block size)

*A detailed description of these parameters can be found in the `indexing.py` module of Febrl [5].

5.2 Experimental Setup

To experiment with the various strategies mentioned earlier, we use the following experimental settings. We define three blocking keys for each data set as shown in Table 2 (using Python notation). Note that the first two keys (F1, F2 or N1, N2) are based on a smaller number of attributes, while the third key (F3 or N3) is the concatenation of the two and thus larger. We also use the first two keys iteratively (F1 & F2 or N1 & N2) and merge the matches to perform multi-key blocking. Next, we define two sets of parameter values for each of the blocking methods, which we refer to as tight and loose. Tight values are expected to encompass less records in the same block, while loose values are expected to encompass more. The reason is that duplicates may have very dissimilar blocking key values if one or more component attributes are missing and thus may not be placed in the same block due to the application of the tight parameter values. Tables 3 and 4 provide the parameter list and the corresponding tight and loose values for different methods, respectively. It should be noted that the Standard Blocking method does not require any parameter and, thus its results remain unchanged for the same blocking key.

Table 4. Parameter values for different methods

Method	Tight parameters	Loose parameters
SRT	2, 3, 5, 7, 10	100, 125, 150, 175, 200
QGM	(2, 0.95), (2, 0.9), (3, 0.95), (3, 0.9)	(2, 0.85), (2, 0.8), (3, 0.85), (3, 0.8)
CNP	('jaccard', 'threshold', 0.9, 0.8, 2), ('jaccard', 'threshold', 0.8, 0.7, 2), ('jaccard', 'nearest', 5, 10, 2), ('jaccard', 'nearest', 10, 20, 2), ('tfidf', 'threshold', 0.9, 0.8, 2), ('tfidf', 'threshold', 0.8, 0.7, 2), ('tfidf', 'nearest', 5, 10, 2), ('tfidf', 'nearest', 10, 20, 2)	('jaccard', 'threshold', 0.6, 0.4, 2), ('jaccard', 'threshold', 0.7, 0.5, 2), ('jaccard', 'nearest', 50, 100, 2), ('jaccard', 'nearest', 100, 200, 2), ('tfidf', 'threshold', 0.6, 0.4, 2), ('tfidf', 'threshold', 0.7, 0.5, 2), ('tfidf', 'nearest', 50, 100, 2), ('tfidf', 'nearest', 100, 200, 2)
STM	(100, 20, 18, 'nearest', 20, 40), (100, 20, 18, 'nearest', 50, 100), (100, 20, 18, 'nearest', 10, 20)	(100, 20, 18, 'nearest', 20, 80), (100, 20, 18, 'nearest', 50, 200), (100, 20, 18, 'nearest', 10, 40)
SFX	('allsubstr', 3, 5), ('allsubstr', 3, 10), ('allsubstr', 3, 15), ('allsubstr', 5, 5), ('allsubstr', 5, 10), ('allsubstr', 5, 15), ('suffixonly', 3, 5), ('suffixonly', 3, 10), ('suffixonly', 3, 15), ('suffixonly', 5, 5), ('suffixonly', 5, 10), ('suffixonly', 5, 15)	('allsubstr', 3, 100), ('allsubstr', 4, 100), ('allsubstr', 5, 100), ('allsubstr', 3, 200), ('allsubstr', 4, 200), ('allsubstr', 5, 200), ('suffixonly', 3, 100), ('suffixonly', 4, 100), ('suffixonly', 5, 100), ('suffixonly', 3, 200), ('suffixonly', 4, 200), ('suffixonly', 5, 200)

We utilize Febrl [5] to perform the experiments on an Intel Core i7 3.40 GHz machine with 16GB of RAM. For different data set qualities (% missing, % corrupted), we run the deduplication tasks for all of the methods described in Sect. 4 using the defined blocking keys and the multi-keys.

5.3 Experimental Results

In this section, we present the results of the experiments. We use the notation $\langle M = x\%,\ C = y\%\rangle$ to represent that x% records have missing values and y% records have corruption in the data set. Also note that, instead of reporting the average value of the performance metric for different parameters in the figures, we provide the range of the values.

Tight vs. Loose Parameters. Figures 1 and 2 depict the three performance metrics for the FL and NC data sets, respectively, under the setting $\langle M = 10\%,$ $C = 0\%\rangle$. It can be seen that the detection rate increases by 1–3% when using the loose parameters for SRT, QGM, CNP and SFX, but remain unchanged for STM. When we increase the missing value rate with the setting $\langle M = 20\%,$ $C = 0\%\rangle$, it can be seen that the detection rates are much higher (3–7%) for the loose parameters in comparison to the previous setting. This is depicted for the FL data set in Fig. 3. The same applies to NC data set (not shown due to space limitations). Similarly, when we add corruption while keeping the missing value rate intact with the setting $\langle M = 10\%,\ C = 5\%\rangle$, we achieve much higher

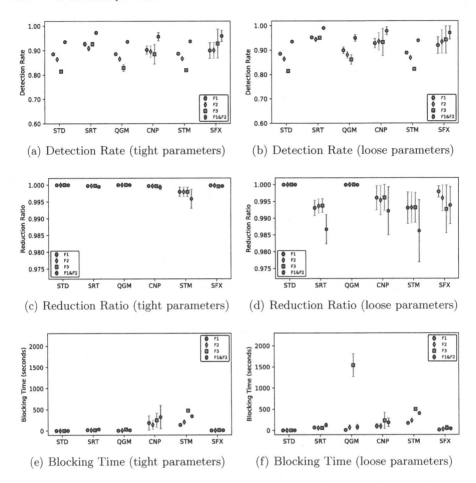

(a) Detection Rate (tight parameters)   (b) Detection Rate (loose parameters)

(c) Reduction Ratio (tight parameters)   (d) Reduction Ratio (loose parameters)

(e) Blocking Time (tight parameters)   (f) Blocking Time (loose parameters)

Fig. 1. Results for FL data set under $\langle M = 10\%,\ C = 0\% \rangle$.

detection rates for loose parameters compared to the first setting as depicted in Fig. 4 for the FL data set (and the same applies to the NC data set). This finding reveals that loose parameters are highly successful in detecting the duplicates in the presence of missing values and corruption.

In terms of reduction ratio, we find that loose parameters cause small overhead. In Fig. 1, while the lowest reduction ratio for tight parameters is 0.992, it is 0.977 for the loose parameters. This means that for loose parameters, it requires 2.3% of the total number of comparisons that would be required if no blocking step was incorporated into the linkage procedure. We consider this to be an acceptable number of comparisons in practice.

Finally, in terms for blocking time we see that for most methods (i.e., SRT, QGM, CNP, STM, and SFX) the blocking time increases only negligibly while using loose parameters. But for some special cases of QGM, the blocking time

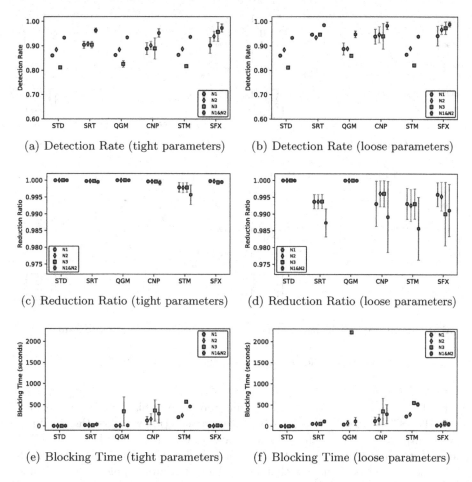

Fig. 2. Results for NC data set under $\langle M = 10\%, C = 0\% \rangle$.

may be mischievously higher, particularly when large blocking key is used as evident from Figs. 1(f) and 2(f) (the green outliers). Actually for QGM, a large key coupled with lower value of τ (see Sect. 4) may result in generating too many q-gram sublists per record (and hence blocks) as already mentioned earlier. Consequently, it may require an impractical amount of time to finish the blocking process.

In view of the above discussion, we argue that using loose parameters, in general, improves the detection rate while keeping the blocking time intact and decreasing the reduction ratio tolerably. Also note that, data quality (% missing, % corrupted) has only negligible impact on reduction ratio and blocking time.

Small Key vs. Large Key vs. Multi-key. In Table 2, it can be seen that the keys for each data set are sorted according to increasing length. Now, it can

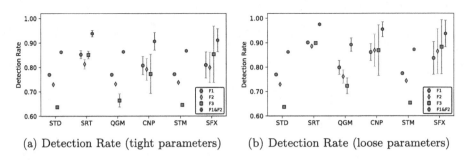

(a) Detection Rate (tight parameters)   (b) Detection Rate (loose parameters)

Fig. 3. Results for FL data set under $\langle M = 20\%, C = 0\% \rangle$.

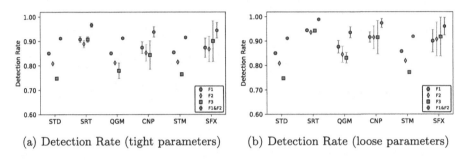

(a) Detection Rate (tight parameters)   (b) Detection Rate (loose parameters)

Fig. 4. Results for FL data set under $\langle M = 10\%, C = 5\% \rangle$.

further be seen in Figs. 1 and 2 that having too large of blocking key is suitable for neither STD and GQM. This is because the detection rates for the third key (F3 or N3) are always much lower in comparison to the other keys. For CNP and SFX, the highest value of detection rate increases while increasing the length of the blocking keys. It means that larger blocking keys coupled with suitable parameters are expected to increase the detection rates for these methods. For SRT, there is no obvious impact of the size of the blocking key on the detection rate.

However, for all methods, except for SFX, it can be seen that multi-key blocking (see Sect. 5.2) always achieves significantly higher detection rates. Surprisingly for SFX, using large keys (F3 or N3) improves the detection rate slightly (0.5–1.0%) over multi-key. It is clear that, for multi-key, the reduction ratio decreases and the blocking time increases linearly according to number of component keys. Based on the above discussion, we argue for limiting the length of the blocking keys. Instead, iteratively applying multiple smaller keys leads to a better rate of duplicate detection while maintaining an acceptable level of overhead in the reduction ratio and blocking time.

Performance of Different Methods. In the Figs. 1, 2, 3, and 4, it can be seen that SRT, CNP, and SFX achieve superior detection rates over alternative methods. SFX, in particular, achieves higher upper bounds of detection rate for all blocking key configurations. While SRT and CNP achieve their best for multi-key configuration, it can be seen that SFX achieves its best rate with one large single key (coupled with "allsubstr" option). Upon closer inspection, we find that SFX with one large single key and tuned parameters achieves the best detection rate among all the methods, as well as key and parameter configurations. Moreover, SFX has a significantly lower blocking time than that of SRT and CNP while maintaining an acceptable reduction ratio.

In summary, these findings lend support to the belief that the Suffix Array (SFX) is the best method for duplicate detection under missing value constraints. In addition, using multiple blocking keys provides significant performance improvement over using a single key for all of the methods except SFX.

6   Conclusion

In this work, we perform a systematic study to understand the impact of missing values on different blocking methods. From our extensive set of experiments, we figure out the particular blocking method, blocking key, and parameter configurations that can subdue this impact quite effectively and efficiently. We believe that, the results presented here would provide guidelines to anyone performing record linkage tasks under missing value constraint. As part of future task, we would also like to see how multi-method blocking performs against the missing value problem and how to fine-tune the parameters in a systematic way.

Acknowledgements. The research reported herein was supported in part by NIH awards 1R01HG006844, RM1HG009034, NSF awards CICI- 1547324, IIS-1633331, CNS-1837627, OAC-1828467 and ARO award W911NF-17-1-0356.

References

1. Florida Voter Registration Records. http://flvoters.com/downloads.html. Accessed 10 July 2018
2. North Carolina Voter Registration Records. https://dl.ncsbe.gov/index.html?prefix=data/Snapshots. Accessed 10 July 2018
3. Aizawa, A.N., Oyama, K.: A fast linkage detection scheme for multi-source information integration. In: Proceedings of International Workshop on Challenges in Web Information Retrieval and Integration, pp. 30–39 (2005)
4. Baxter, R., Christen, P., Churches, T.: A comparison of fast blocking methods for record linkage. In: Proceedings of ACM SIGKDD Conference on Knowledge Discovery and Data Mining, pp. 25–27 (2003)
5. Christen, P.: Febrl-a open source data cleaning, deduplication and record linkage system with a graphical user interface. In: Proceedings of ACM SIGKDD Conference on Knowledge Discovery and Data Mining, pp. 1065–1068 (2008)
6. Christen, P.: A survey of indexing techniques for scalable record linkage and deduplication. IEEE Trans. Knowl. Data Eng. **24**(9), 1537–1555 (2012)
7. Dusetzina, S.B., Tyree, S., Meyer, A.M., Meyer, A., Green, L., Carpenter, W.R.: Linking data for health services research: a framework and instructional guide. Agency for Healthcare Research and Quality (US), Rockville (MD) (2014)

8. Hernández, M.A., Stolfo, S.J.: Real-world data is dirty: data cleansing and the merge/purge problem. Data Mini. Knowl. Discov. **2**(1), 9–37 (1998)

9. Jin, L., Li, C., Mehrotra, S.: Efficient record linkage in large data sets. In: Proceedings of International Conference on Database Systems for Advanced Applications, pp. 137–146. IEEE (2003)

10. McCallum, A., Nigam, K., Ungar, L.H.: Efficient clustering of high-dimensional data sets with application to reference matching. In: Proceedings of ACM SIGKDD Conference on Knowledge Discovery and Data Mining, pp. 169–178 (2000)

11. Ong, T.C., Mannino, M.V., Schilling, L.M., Kahn, M.G.: Improving record linkage performance in the presence of missing linkage data. J. Biomed. Inf. **52**, 43–54 (2014)

12. Prasad, K.H., Chaturvedi, S., Faruquie, T.A., Subramaniam, L.V., Mohania, M.K.: Automated selection of blocking columns for record linkage. In: Proceedings of International Conference on Service Operations and Logistics, and Informatics, pp. 78–83. IEEE (2012)

13. Tran, K.N., Vatsalan, D., Christen, P.: GeCo: an online personal data generator and corruptor. In: Proceedings of ACM International Conference on Information and Knowledge Management, pp. 2473–2476 (2013)

14. Winkler, W.E.: Using the EM algorithm for weight computation in the Fellegi-Sunter model of record linkage. In: Proceedings of the Section on Survey Research Methods, American Statistical Association, vol. 667, p. 671 (1988)

Behavioral Data Mining

Bridging the Gap Between Research and Production with CODE

Yiping Jin[1](✉), Dittaya Wanvarie[1], and Phu T. V. Le[2]

[1] Department of Mathematics and Computer Science, Chulalongkorn University,
Bangkok 10300, Thailand
Dittaya.W@chula.ac.th
[2] Knorex Pte. Ltd., 8 Cross St, Singapore 048424, Singapore
{jinyiping,le_phu}@knorex.com

Abstract. Despite the ever-increasing enthusiasm from the industry, artificial intelligence or machine learning is a much-hyped area where the results tend to be exaggerated or misunderstood. Many novel models proposed in research papers never end up being deployed to production. The goal of this paper is to highlight four important aspects which are often neglected in real-world machine learning projects, namely **C**ommunication, **O**bjectives, **D**eliverables, **E**valuations (CODE). By carefully considering these aspects, we can avoid common pitfalls and carry out a smoother technology transfer to real-world applications. We draw from a priori experiences and mistakes while building a real-world online advertising platform powered by machine learning technology, aiming to provide general guidelines for translating ML research results to successful industry projects.

Keywords: Machine learning · Project management ·
Online advertising · Real-time bidding

1 Introduction

Modern machine learning approaches achieved impressive results on many challenging tasks, such as image recognition [21], machine translation [2] and speech recognition [3]. However, most machine learning research is concerned with building high-quality classification methods in isolation. While they are essential in advancing the research frontier, many proposed methods are never productionised and applied to real-world problems.

Having a high accuracy on the test set does not guarantee ML models can be applied in production. We need to consider multiple factors, such as latency, scalability, cost and the accuracy on real-world data. Even if all of these conditions are satisfied, the research team still need to convince the CTO and the engineering team to give their support before the model can be integrated into production systems. All of these complications will cause ML models built by

© Springer Nature Switzerland AG 2019
Q. Yang et al. (Eds.): PAKDD 2019, LNAI 11441, pp. 277–288, 2019.
https://doi.org/10.1007/978-3-030-16142-2_22

the research team to remain on the shelf and do not have an impact on the real-world problem. A natural question then arises: what is the gap between research outputs and real-world applications and how can we bridge this gap?

The goal of this paper is to propose four important aspects as a framework for technology transfer that will bridge the above gap, namely, **C**ommunication, **O**bjectives, **D**eliverables, and **E**valuations (CODE). By giving careful considerations to these aspects, we can align the ML projects better with the business goals and accelerate the productionisation process. We demonstrate the CODE framework with the help of past industrial experiences building a large-scale machine learning-powered online advertising platform. While most cases studies used in this paper are related to online advertising, the framework we propose is not limited to a particular domain but is applicable to a wide range of machine learning projects in the production.

2   Related Works

Most attention to machine learning was paid to the modelling step with numerous new model architectures being proposed every year. However, applying machine learning models to real-life problems is far beyond that. The data mining community has been trying to standardise the data mining workflow and establish a common methodology since the 1990s. The resultant cross-industry standard process for data mining (CRISP-DM) [23] breaks the process of data mining into six major phrases, namely business understanding, data understanding, data preparation, modeling, evaluation and deployment. Modern large-scale ML frameworks share a similar workflow [12,18]. While a common methodology is critical to the success of machine learning projects, most research papers only focus on the modeling part, leaving little discussion on the rest of the steps.

Sculley et al.'s seminal work [22] highlighted technical debts in machine learning systems that can incur massive maintenance cost and make future changes forbiddingly tricky. They commented that entanglement is in some sense "innate" to machine learning because it aims to mix the information sources to make a more accurate prediction. Raeder et al. [20] is also concerned with building large-scale machine learning systems. They proposed an end-to-end ML system that has been designed with maintenance and quality control in mind. They formulated three design principles for building massive and robust prediction systems, namely *yield fail-safe predictions*, *scale and easily extend* and *minimise human intervention*. They demonstrated the application of these principles with a large-scale online advertising platform.

Thomas [26] listed several scenarios where machine learning projects fail. Based on the article, none of the failure cases is due to the incompetency of machine learning practitioners. In contrary, almost all cases are related to miscommunication, either with the management team or with the engineering team. Some example scenarios are (1) the machine learning team produces models faster than engineering team can put them in production. (2) The model built by the machine learning team does not align with the business priority or logic.

Most recently, Ng [13] and Hermann and Del Balso [8] published blog articles sharing their experience executing AI projects in large Internet companies. Ng [13] drew from his experience leading the AI transformation in Google and Baidu and provided five recommendations for large enterprises who wish to become an AI company. Hermann and Del Balso [8] reflected on the ML evolution at Uber and the process of developing Michelangelo, a platform which helped the company to scale ML services in production.

Our work is closest to [13] and [8] in that they also drew insights and recommendations from their experience working on ML projects. Both articles were published after we submitted our manuscript and they were developed independently from our work. We were not surprised to see that they share some common ideas with our work. All in all, we are not trying to propose a totally new methodology, but to draw from our success and failure working on ML projects and formalise a simple and easy-to-follow framework for the ML research community.

3   Background of Online Advertising

Real-time bidding (RTB) is an emerging business model of online advertising markets. In RTB, the Ad exchange will consolidate opportunities for showing ads, namely *impressions*, and send them to eligible demand-side platforms (DSPs). DSPs act as an agents for multiple advertisers to run ad campaigns across different platforms and ad formats. Each DSP will evaluate the value of the impression and submit its bid. The highest bidder among all DSPs will win the impression and display their ad. The whole process takes place within 100 ms and hence is called "real-time bidding" [27].

Usually, advertisers want to optimise the number of clicks on their ads and the number of *conversions*, which is a specific user action they define a priori, such as user booking a hotel, purchasing a product or signing up for a newsletter. Standard metrics of online advertising are cost per 1,000 impressions (CPM), click-through rate (CTR), cost per click (CPC), conversion rate (CVR) and cost per acquisition (conversion) (CPA).

4   The CODE Framework

In this section, we present the CODE framework, which summarises four important yet often neglected aspects of machine learning projects. CODE stands for Communication, Objectives, Deliverables, Experiments. Besides illustrating each aspect, we will also provide related case studies based on our experiences while building a large-scale online advertising platform.

4.1   Communication

In ML research, people give strong emphasis on novel models and approaches. Almost no attention has been given to the human and organisational aspects of ML projects. In the real-world scenario, teams across different departments need to work together in synergy to deploy a large-scale ML model to production. Each team has their own priorities. If we ignore the people aspect, we will almost certainly run into obstacles trying to push the progress.

The large team size and the management cost are the major causes of communication overhead in software engineering [5]. As a result, the total productivity increases sub-linearly with additional manpower added. In machine learning projects, there is another challenge: the technical knowledge. Machine learning is difficult for people outside the field to understand, even on an intuitive level [6]. In the industry, machine learning teams often need to work closely with the rest of the organisation, such as the management team, the engineering team or even the business team. For a ML model to be productionised, the rest of the organisation need to "buy" the idea and offer their support (e.g. the management team needs to allocate sufficient budget and time; the engineering team needs to build the supporting data pipeline). However, it's hard for them to be willing to support without sufficient understanding of the newly proposed model.

In a recent project, we worked on optimising clicks on online ads. The team proposed two approaches. The first one is a K-nearest neighbour approach [16], which finds the users who are most similar to the users who clicked on the ads and then displays the ads to them. The second approach is a recent state-of-the-art click-through rate prediction model based on the interaction of feature embeddings [19]. When we communicated the proposed methods to the management team, the first approach was immediately embraced because it was straightforward to understand. In fact, it is the core idea behind an advertising strategy called look-alike model [1]. However, the management team was sceptical about the second approach despite the impressive reported accuracy.

Having learned from this experience, we urge our researchers to be able to "sell" their models. Doing novel research and building highly-accurate models are important, yet being able to communicate the idea succinctly with the decision makers who have little ML background can be more critical to productionise the ML models. To this end, we initiated an internal technical blog and host knowledge sharing sessions where the researchers share the ideas behind their models in simple language. Other teams can also ask questions to better understand the models and their potential impact on the business. We believe that in machine learning projects, although external teams do not have to understand the technical details of machine learning models, it is critical to communicate to them the intuition behind the models and their implications and impact. A clear communication and a common ground will drastically boost people's trust in the ML models and the willingness to adopt them.

4.2 Objectives

A large proportion of machine learning problems can be modelled as optimisation problems [25]. Thus, the careful selection of the objective function (also known as cost function) plays an essential role in the success of machine learning projects. Machine learning researchers tend to be eager to jump into the modelling part without spending time trying to understand the business metric. The specific business metric may differ subtly from the commonly used objective functions in machine learning. Failing to notice this difference will lead to misalignment of the produced model and makes it unable to achieve the business objective.

In the user click prediction example, the business requirement is to optimise the click-through rates (CTR) of ad campaigns. However, the final observed CTR is affected by many other factors, including the campaign budget, the market competitiveness and the bidding strategy, which takes the estimated CTR as an input and returns the bid price. After conducting a literature survey, the team decided to follow the simple linear bidding approach proposed by Perlich et al. [15]. The system consists of two components. The first component is a CTR estimation model, which predicts the likelihood that the user will click on the ad. The second component is a simple bid price logic which bids proportional to the predicted CTR. The intuition is that we bid at a higher price for the ad impressions which are more likely to be clicked.

Since only the CTR estimation part is an ML model, the team focused on improving the model performance as measured by the area under the ROC curve (AUC), which is the standard evaluation metric for CTR estimation in the literature. AUC is a number between 0 and 1 with a larger value indicating a more accurate prediction. An AUC of 1 means the model can perfectly predict clicks and non-clicks. An AUC of 0.5 means the prediction is equivalent to random. After deploying the new model, we observed that although AUC improved consistently between 26th Aug and 3rd Sep, the CTR did not show clear improvement as in Fig. 1. We finally resolved the problem by building a bidding simulation environment and testing each component in the pipeline (CTR estimator, bidding strategy, budget allocation) individually. This helped us to improve the average CTR by 10%, which is a huge improvement in the context of online advertising and saves us millions of dollars of advertising cost per year. The improvement would not be likely if we focused on CTR estimation model only, because previous work showed that even by applying a very complex state-of-the-art CTR estimation model, the improvement in CTR is no more than 2.2%–6.3% [11].

This example demonstrates that although AUC is the standard evaluation metric for CTR estimation in the literature, optimising it alone does not guarantee a better business metric because the final business metric is affected by multiple other factors and components, not the ML model alone.

4.3 Deliverables

The performance indicator of an industrial R&D team is the models that it delivers and the overall impact on the business and products, not the amount of novel research which remains a proof-of-concept. In the IT industry, the business pressure is always high. Companies may lose their business if their competitors provide a feature which they do not provide. Therefore, the management often proposes a new problem to the machine learning team and expect them to deliver a solution within a short time frame, usually within six months. However, it

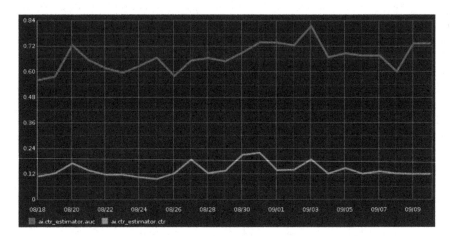

Fig. 1. Area-under-curve (AUC) and click-through rates (CTR) of an ad campaign.

takes considerable time to conduct an in-depth literature survey, to build the data pipeline, to evaluate competitive methods and to fine-tune the model. The given time is usually not sufficient to deliver a model with a good accuracy.

Instead of delaying the delivery date, the ML team should wrap up a functioning v0.1 and deliver it first. From the business point of view, whether the ML model has a good accuracy is less critical than whether the model is in production or not. After the first deliverable, the business pressure will reduce. We will have time to refactor the model and deliver on a better v1.0. Another advantage of deploying an initial version of the ML model fast is that we can gather feedbacks from internal and external users, which will shape the directions for the subsequent effort.

As part of the global expansion effort, our team were requested to extend the text classifiers for contextual advertising [10] to ten more languages within a quarter. For each language, the number of categories is around 400, and we have millions of training documents. Even the engineering effort is tremendous to train all the models within the given time frame. Our team initially surveyed semi-supervised learning [24] or cross-lingual deep learning methods [14]. After a shallow exploration, we concluded that we would not be able to use these approaches in the first version. They require either a bi-lingual dictionary or multi-lingual word embeddings aligned to the same semantic space. Such resources may be available for high-resource languages such as German, Chinese and French, but not for low-resource languages such as Bahasa Malay and Thai. To meet the project deadline, we decided to make use of Google Translate API[1] to translate the training documents from English to the target language. The translated corpus is then used to train the text classifiers. With this, we were able to deliver the text classifiers for ten languages with an average accuracy of 2–3% lower than English, which is acceptable for the application.

---

[1] https://cloud.google.com/translate/.

Besides the tight timeline, the management team often formulates requirements for the ML models based on the business requirements, instead of based on the state of the technology development. For example, the business may require that a chatbot to give a meaningful response 95% of the time while the state-of-the-art may only achieve 80% on a similar benchmark. While keeping pushing the research frontier is an obvious solution, it may not be feasible within the timeframe or the target may not be achievable at all. A more immediate and promising solution is to propose a new scope of the problem. While we cannot solve the general problem with 95% accuracy, it is possible that we can solve a subset of relatively easy problems accurately. In the same line, Goodfellow et al. [7] presented a case study on Street View address number transcription system. The goal was to automatically transcribe 95% of the address numbers at 98% accuracy. The rest 5% hard cases will be transcribed by human annotators.

4.4   Evaluations

While the "Experiments and Results" section is in almost every ML research paper, the result can sometimes be difficult to interpret. It is especially true for intrinsic evaluation metrics (such as mean-squared error) and artificially designed metrics (such as ROUGE score for machine translation). Even the straight-forward accuracy measure can sometimes have discrepancy between the reported figure and the accuracy the user perceives.

Although *automatic evaluations* are essential for quick experimentation, we believe that *human evaluation* cannot be neglected, especially for ML models in production systems. Automatic evaluation has certain limitations, such as it does not reflect which type of mistakes the model tends to make and it does not guarantee that the test data and the real-world data are similar enough. Exhaustive human evaluation takes a long time, but a small-scale "smoke testing" can already help us to identify most obvious problems of the ML model.

Evaluate with Simple Examples. In a project detecting the language of web pages. We made use of the Optimaize library[2], which claimed to be the best open-source language detection library. The author of the library reported a 99% accuracy for 53 languages. We also evaluated the library using sampled Wikipedia pages and obtained similar results. When we delivered the project, the engineering manager randomly tried the API with a few sentences. One example input was "today is wednesday", where the library wrongly predicted as *Somali* instead of *English*. He then concluded that the accuracy of the library is bad and it cannot even classify a simple case correctly. After days' of investigation, the team found out that the problem is because the model uses a Naive Bayes classifier with character n-gram features. When the input text is short, it may not contain sufficient unique n-grams to distinguish the languages. A seemingly trivial example turned out to be the weakness of the ML model.

---

[2] https://github.com/optimaize/language-detector.

The team then worked on improving the accuracy for short text input by extending the unigram, bigram, trigram features to 4, 5, 6-grams. By Adding longer n-gram features, we hope to capture n-gram features unique to each language (We do not use word dictionaries for each language because we need to know the language before we can tokenise the text into words). We selected two most ambiguous language pairs to conduct the evaluation, namely English-French and Bahasa Indonesia-Bahasa Malay[3]. The results in Table 1 shows that the simple treatment drastically improved the language detection accuracy.

Table 1. Short text language detection accuracy for ambiguous language pairs before and after the improvement $(P/R/F_1)$.

Language	Original	After improvement
English	0.76/0.52/0.62	0.91/0.98/0.94
French	0.67/0.58/0.62	0.97/0.95/0.96
Indonesian	0.62/0.36/0.46	0.94/0.80/0.87
Malay	0.75/0.58/0.66	0.83/0.94/0.88
Macro avg.	0.70/0.51/0.59	**0.92/0.92/0.91**

From this experience, we learned that a high accuracy does not necessarily mean a good user perception. We need to analyse the actual user input and ensure that the model can deliver the promised accuracy on the real-world data.

Compare with Simple Baselines. One of our data scientists recently proposed to use a multi-layer Long Short-Term Memory (LSTM) [9] model to predict the real-time bidding traffic coming to our system. LSTM model is a specific type of recurrent neural networks architecture which can model long-term dependencies effectively. The model has become the de facto approach for sequence prediction tasks such as speech recognition, and part-of-speech tagging. It is therefore natural to expect that it will yield good performance in time series prediction.

The model we used contains two layers of LSTM units, with dropout layers to prevent over-fitting. At each time step, it takes a scalar number as input, which is the traffic volume of the current minute. The last layer is a linear layer with one output unit predicts traffic volume for the following minute.

The model was trained using a single epoch consisting roughly 5,000 training sequences. The point-by-point prediction achieves a mean-squared error (MSE) of 0.0333 on the test set. The prediction of the model is also shown in Fig. 2(b). At this point, we may conclude that the model is effective. However, the model

---

[3] Adding more languages will actually inflate the average accuracy because most other languages can be easily identified by looking at the character alone and have an accuracy close to 1 (e.g. Chinese, Korean).

may output "accurate" predictions simply because it predicts a value similar to the value of the previous time step.

To validate our assumption, we implemented a naive baseline of persistent prediction, a model which simply predicts the same value as the previous time step. The simple baseline turned out to achieve an MSE of 0.0316, which is lower than the proposed LSTM model. We can also observe clearly from Fig. 2 that the predictions of the baseline are closer to the actual values. This example demonstrates that the effectiveness of a ML model cannot be validated without a meaningful benchmark and a comparison with (possibly rule-based) baselines.

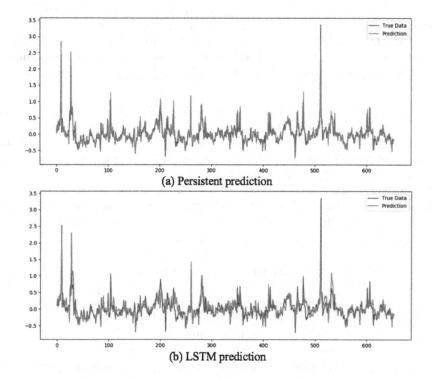

(a) Persistent prediction

(b) LSTM prediction

Fig. 2. Comparison of persistent prediction (loss = 0.0316) and LSTM prediction (loss = 0.0333).

Ensure Fair Evaluation. For ML models in production, another challenge is that sometimes it is impossible to conduct a head-to-head comparison. In online advertising, the most popular way to compare two competing strategies is to perform A/B testing[4]. Advertisers will run two models with the same setting except for the different bidding strategies. After the evaluation period, they

---

[4] https://vwo.com/ab-testing/.

collect the performance metrics such as clicks and conversions to compare which strategy is better. However, a pitfall of this approach is that the two strategies will never display ads to the same user at the same moment. Therefore, the samples used to evaluate the two strategies are different. This problem was observed in [4] as well. They proposed to split the real-time bidding traffic based on geography and allocate two subpopulations for each competing strategy.

When we first deployed our ML models, our system could not serve multiple versions of models simultaneously to conduct A/B testing yet. Therefore, we had *model A* running during *period* 1 and *model B* running during *period* 2 (there is no overlap between the two periods). Nevertheless, to conclude whether *model A* and *model B* yield statistically different performance, we calculate the confidence interval for two independent samples [17]. We first compute the sample sizes (n_1 and n_2, the number of days we run each strategy), means ($\overline{x}_1$ and $\overline{x}_2$, the average click-through rates) and standard deviations (s_1 and s_2) of each sample. The pooled estimate of the common standard deviation S_p is computed as:

$$S_p = \sqrt{\frac{(n_1 - 1)s_1^2 + (n_2 - 1)s_2^2}{n_1 + n_2 - 2}} \tag{1}$$

Depending on the sample size, we use either z-table or t-table to compute the final confidence interval, which is used to evaluate whether the performance difference between *model A* and *model B* is significant.

Regardless of the evaluation metrics, we should always try to identify and eliminate potential bias and make the evaluation as fair as possible. In the case where it is not possible, we should also take note of the bias and understand its impact on the possible conclusions we can derive.

Table 2. Summary of the CODE Framework.

Communication	Communicate the intuition and implication of ML models with external teams to facilitate decision-making
Objectives	Optimise for business metrics instead of focusing on objective functions of ML models only
Deliverables	Deliver the first version of the ML model fast without worrying too much about the accuracy. Carefully scope the problem to make it feasible and useful to the business
Experiments	Make sure the model predicts correctly for simple examples and beats naive baselines. Try to ensure the evaluation is as fair as possible

5   Conclusions and Future Work

In this work, we highlighted the gap between academic research and industry applications of machine learning technologies. We proposed the CODE framework, which summarises four essential aspects for machine learning projects to succeed, namely **C**ommunication, **O**bjectives, **D**eliverables, and **E**xperiments. We summarise the key takeaway from this paper in Table 2. We wish that the recommendations in this paper will be helpful for machine learning researchers who want to productionise their models.

In future work, we want to re-examine established frameworks in software engineering such as Agile or Scrum and adapt them for machine learning projects. We also want to establish evaluation methods to quantitatively evaluate the effectiveness of the proposed framework. This work is just a tip of the iceberg, and we believe much more effort needs to be invested to establish a general framework for machine learning projects and to help the community to translate the success in academic research into real-world applications.

Acknowledgement. The first author is supported the scholarship from "The 100th Anniversary Chulalongkorn University Fund for Doctoral Scholarship" and also "The 90th Anniversary Chulalongkorn University Fund (Ratchadaphiseksomphot Endowment Fund)". We would like to thank Assoc. Prof. Peraphon Sophatsathit and the anonymous reviewers for their careful reading and their insightful suggestions.

References

1. Bagherjeiran, A., Tang, R., Zhang, Z., Hatch, A., Ratnaparkhi, A., Parekh, R.: Adaptive targeting for finding look-alike users. US Patent 9,087,332, 21 July 2015
2. Bahdanau, D., Cho, K., Bengio, Y.: Neural machine translation by jointly learning to align and translate. arXiv preprint arXiv:1409.0473 (2014)
3. Barker, J., Watanabe, S., Vincent, E., Trmal, J.: The fifth 'CHiME' speech separation and recognition challenge: dataset, task and baselines. arXiv preprint arXiv:1803.10609 (2018)
4. Boyko, A., Harchaoui, Z., Nedelec, T., Perchet, V.: A protocol to reduce bias and variance in head-to-head tests. Criteo Internal Report (2015)
5. Brooks, F.P.: The mythical man-month. Datamation **20**(12), 44–52 (1974)
6. Enam, S.Z.: Why is machine learning 'hard'? (2016). http://ai.stanford.edu/~zayd/why-is-machine-learning-hard.html. Accessed 10 Sept 2018
7. Goodfellow, I., Bengio, Y., Courville, A., Bengio, Y.: Deep Learning, vol. 1. MIT press, Cambridge (2016)
8. Hermann, J., Del Balso, M.: Scaling machine learning at uber with michelangelo (2018). https://eng.uber.com/scaling-michelangelo/
9. Hochreiter, S., Schmidhuber, J.: Long short-term memory. Neural Comput. **9**(8), 1735–1780 (1997)
10. Jin, Y., Wanvarie, D., Le, P.: Combining lightly-supervised text classification models for accurate contextual advertising. In: Proceedings of the Eighth International Joint Conference on Natural Language Processing (Volume 1: Long Papers), vol. 1, pp. 545–554 (2017)
11. Juan, Y., Lefortier, D., Chapelle, O.: Field-aware factorization machines in a real-world online advertising system. In: Proceedings of the 26th International Conference on World Wide Web Companion, pp. 680–688. International World Wide Web Conferences Steering Committee (2017)
12. Modi, A.N., et al.: TFX: a tensorflow-based production-scale machine learning platform. In: KDD 2017 (2017)

13. Ng, A.: AI transformation playbook: how to lead your company into the AI era (2018). https://landing.ai/ai-transformation-playbook/
14. Pappas, N., Popescu-Belis, A.: Multilingual hierarchical attention networks for document classification. arXiv preprint arXiv:1707.00896 (2017)
15. Perlich, C., Dalessandro, B., Hook, R., Stitelman, O., Raeder, T., Provost, F.: Bid optimizing and inventory scoring in targeted online advertising. In: Proceedings of the 18th ACM SIGKDD International Conference on Knowledge Discovery and Data Mining, pp. 804–812. ACM (2012)
16. Peterson, L.E.: K-nearest neighbor. Scholarpedia 4(2), 1883 (2009)
17. Pfister, R., Janczyk, M.: Confidence intervals for two sample means: calculation, interpretation, and a few simple rules. Adv. Cogn. Psychol. 9(2), 74 (2013)
18. Polyzotis, N., Roy, S., Whang, S.E., Zinkevich, M.: Data management challenges in production machine learning. In: Proceedings of the 2017 ACM International Conference on Management of Data, pp. 1723–1726. ACM (2017)
19. Qu, Y., et al.: Product-based neural networks for user response prediction. In: 2016 IEEE 16th International Conference on Data Mining (ICDM), pp. 1149–1154. IEEE (2016)
20. Raeder, T., Stitelman, O., Dalessandro, B., Perlich, C., Provost, F.: Design principles of massive, robust prediction systems. In: Proceedings of the 18th ACM SIGKDD International Conference on Knowledge Discovery and Data Mining, pp. 1357–1365. ACM (2012)
21. Russakovsky, O., et al.: ImageNet large scale visual recognition challenge. Int. J. Comput. Vis. (IJCV) 115(3), 211–252 (2015). https://doi.org/10.1007/s11263-015-0816-y
22. Sculley, D., Phillips, T., Ebner, D., Chaudhary, V., Young, M.: Machine learning: the high-interest credit card of technical debt (2014)
23. Shearer, C.: The CRISP-DM model: the new blueprint for data mining. J. Data Warehous. 5(4), 13–22 (2000)
24. Shi, L., Mihalcea, R., Tian, M.: Cross language text classification by model translation and semi-supervised learning. In: Proceedings of the 2010 Conference on Empirical Methods in Natural Language Processing, pp. 1057–1067. Association for Computational Linguistics (2010)
25. Sra, S., Nowozin, S., Wright, S.J.: Optimization for Machine Learning. MIT Press, Cambridge (2012)
26. Thomas, R.: What do machine learning practitioners actually do? (2018). http://www.fast.ai/2018/07/12/auto-ml-1/. Accessed 10 Sept 2018
27. Yuan, Y., Wang, F., Li, J., Qin, R.: A survey on real time bidding advertising. In: 2014 IEEE International Conference on Service Operations and Logistics, and Informatics (SOLI), pp. 418–423. IEEE (2014)

Distance2Pre: Personalized Spatial Preference for Next Point-of-Interest Prediction

Qiang Cui[1], Yuyuan Tang[2], Shu Wu[1(✉)], and Liang Wang[1]

[1] Center for Research on Intelligent Perception and Computing,
National Laboratory of Pattern Recognition, Institute of Automation,
Chinese Academy of Sciences, Beijing, China
cuiqiang2013@ia.ac.cn, {shu.wu,wangliang}@nlpr.ia.ac.cn
[2] University of Science and Technology Beijing, Beijing, China
tangyyuanr@gmail.com

Abstract. Point-of-interest (POI) prediction is a key task in location-based social networks. It captures the user preference to predict POIs. Recent studies demonstrate that spatial influence is significant for prediction. The distance can be converted to a weight reflecting the relevance of two POIs or can be utilized to find nearby locations. However, previous studies almost ignore the correlation between user and distance. When people choose the next POI, they will consider the distance at the same time. Besides, spatial influence varies greatly for different users. In this work, we propose a Distance-to-Preference (Distance2Pre) network for the next POI prediction. We first acquire the user's sequential preference by modeling check-in sequences. Then, we propose to acquire the spatial preference by modeling distances between successive POIs. This is a personalized process and can capture the relationship in user-distance interactions. Moreover, we propose two preference encoders which are a linear fusion and a non-linear fusion. Such encoders explore different ways to fuse the above two preferences. Experiments on two real-world datasets show the superiority of our proposed network.

Keywords: POI · Sequential preference · Spatial preference · Non-linear

1 Introduction

Point-of-interest (POI) prediction is one of the most important tasks in location-based social networks (LBSNs). With rich check-ins and contextual information, physical movements of users can be predicted, which is beneficial to explore POIs for users, launch advertisements, and so on. In this work, we focus on successive POI prediction by modeling check-in sequences and incorporating spatial influence in a personalized way.

Q. Cui and Y. Tang—These authors contributed equally to this paper.

Q. Yang et al. (Eds.): PAKDD 2019, LNAI 11441, pp. 289–301, 2019.
https://doi.org/10.1007/978-3-030-16142-2_23

Spatial influence has been considered in lots of works and mostly modeled by computing the distance between two POIs. The distance can be computed as a weight to reflect the relevance of two POIs [5,11]. Usually, the smaller the distance, the stronger the relevance. Besides, people can apply the distance to find nearby locations. Neighbors around a visited POI can be considered as negative samples for BPR optimization criterion [1], used to construct a hierarchical preference [20], and so on. People can also divide multiple locations close to each other into the same region [4]. Furthermore, recent works try to acquire spatial influence between POIs in other formats. Wang et al. [17] apply three factors to model spatial influence: geo-influence, geo-susceptibility, and distance. Geo-influence acquires a POI's ability to spread its spatial influence to other POIs. Geo-susceptibility captures how a POI is spatially influenced by others.

Although the aforementioned studies achieve successful results, they still have a critical limitation. These modelings of spatial influence are conducted within POIs and do not consider the relationship with users. They capture the sequential preference by modeling user-poi check-in sequences, but people have preferences for distances. For example, if a user wants some spicy food in a restaurant, how far would he want to go? It is likely that there will be several restaurants which all satisfy the user interest at different distances. Under such a situation, it is beneficial to predict user's preference for the distance that a user would take at next time. Previous works almost ignore the user's personalized choice of distance, while we propose to model the spatial preference.

In this paper, we propose a Distance-to-Preference (Distance2Pre) network to predict the next POI. First, we apply the recurrent neural network to model check-in sequences and construct the sequential preference. Then, based on distances of successive POIs, spatial preference can be computed to indicate the probabilities of different distances for the next time. This preference can explore the relationship between user and distance. Then, we devise different preference encoders, which can explore the influence of different combinations of the two preferences on the performance of POI prediction. Specifically, we propose a linear fusion and a non-linear fusion. Next, a pair-wise ranking framework is used to optimize the two preferences. The contributions are as follows:

- We first introduce and compute the personalized spatial preference, which can effectively capture the relationship between the user and spatial distance.
- We propose a linear way and a non-linear way as preference encoders to combine sequential preference with spatial preference.
- Experiments on two real-world datasets reveal that our network is effective and outperforms the state-of-the-art methods.

2 Related Work

In this section, we briefly review the POI prediction, including modeling successive POIs and incorporating spatial influence.

We can arrange a user's successive POIs into a check-in sequence and it is important to model the sequential pattern. Many studies apply the Markov chain to predict POIs. Cheng et al. [1] recommend POIs based on first-order Markov chain. Recently, the neural network is also investigated to model sequences. The work [14] applies the word2vec to model context of locations (Fig. 1).

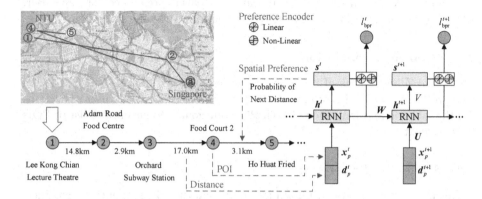

Fig. 1. A user's POIs in Singapore and the framework of our Distance2Pre network. At t-th time, the input is a POI vector $\boldsymbol{x}_p^t$ and a distance vector $\boldsymbol{d}_p^t$. The hidden state $\boldsymbol{h}^t$ is used to compute the spatial preference $\boldsymbol{s}^t$ for next time. User's sequential preference and spatial preference are fused by our preference encoders.

Liu et al. [12,13] employs recurrent neural network (RNN) to model POIs by using different contexts. RNN can model the recent check-ins. Because of the gradient vanishing and exploding problem, gated activation function like gated recurrent unit (GRU) [2] and long short-term memory (LSTM) [8] are developed to better capture the long-term dependency.

The spatial influence has been proven to be a significant factor in POI prediction. Firstly, some works convert the distance to a weight. Feng et al. [5] incorporate the spatial influence by using the weight of distance. The smaller the distance between the last POI and a POI, the more likely this POI to be recommended. Li et al. [11] build a Rank-GeoFM model to capture the user preference as well as spatial influence score, but the distance is still used as a weight between a POI and its neighbors. Secondly, people apply distance to find neighbors for a visited POI. The study [4] builds a binary tree by distances. Nearby POIs are clustered into the same region in this POI2Vec model because they are highly relevant. Zhao et al. apply the POIs that are nearby and far away to construct a hierarchical pairwise preference relation [20]. Thirdly, some studies other spatial information in addition to distance. Wang et al. [17] model a POI's ability to spread its visited users to other POIs (geo-influence) and receive users from other POIs (geo-susceptibility). However, spatial influence mostly works between POIs now and no work has studied the user's spatial preference.

3   The Distance2Pre Network

In this section, we begin with the problem formulation of the next POI prediction, then introduce the proposed Distance-to-Preference (Distance2Pre) network. In detail, we model the check-in sequence to obtain sequential preference and model the distance sequence to capture spatial preference for each user. Then, we fuse sequential and spatial preferences linearly and non-linearly.

3.1   Problem Formulation

Let $\mathcal{U}$ and $\mathcal{I}$ be the sets of users and POIs respectively. Use $\mathcal{I}_u = (\mathcal{I}_u^1, ..., \mathcal{I}_u^{|\mathcal{I}^u|})$ to represent the check-ins of user u in the time order. Given each user's check-ins $\mathcal{I}_u$, the latitude and longitude of each POI, our goal is to generate a list of POIs for u at next time.

3.2   Sequential Preference

In this part, we model user-POI sequences and capture the sequential preference. Previous work has indicated that the sequential pattern is important for POI prediction [5].

Instead of using traditional Markov chain, we apply RNN to model each user's check-ins $\mathcal{I}_u$.

$$h^t = f\left(Ux_p^t, Wh^{t-1}, b\right), \qquad h^t \in \mathbb{R}^d, \tag{1}$$

where h^t is the hidden state, U, W are transition matrices and b is the bias. A vector $x_p^t \in \mathbb{R}^d$ is used to represent the POI at the t-th time, where the subscript p indicates this POI is in $\mathcal{I}_u$. Function $f(\cdot)$ is non-linear, and we choose the gated recurrent unit (GRU) in order to better capture the long-term dependency [2].

$$\begin{aligned}
z^t &= \sigma\left(U_1 x_p^t + W_1 h^{t-1} + b_1\right) \\
r^t &= \sigma\left(U_2 x_p^t + W_2 h^{t-1} + b_2\right) \\
\tilde{h}^t &= \tanh\left(U_3 x_p^t + W_3\left(r^t \odot h^{t-1}\right) + b_3\right) \\
h^t &= \left(1 - z^t\right) \odot h^{t-1} + z^t \odot \tilde{h}^t
\end{aligned} \tag{2}$$

where $U_{1\sim3}, W_{1\sim3} \in \mathbb{R}^{d\times d}$ and $b_{1\sim3} \in \mathbb{R}^d$. GRU has an *update* gate z^t and a *reset* gate r^t to control the flow of information. $\tilde{h}^t$ is the candidate state.

In our network, we consider h and x as latent vectors for a user and a POI respectively. Inspired by matrix factorization, a user's preference for a POI by considering sequential preference is denoted as

$$\hat{x}_{up}^t = \left(h^t\right)^{\mathrm{T}} x_p^{t+1} \tag{3}$$

where h^t is used as the current user latent vector when modeling $\mathcal{I}_u$. The preference $\hat{x}_{up}^t$ is an inner product between h^t and the next POI vector x_p^{t+1}.

3.3   Spatial Preference

We acquire the spatial preference from user-distance sequences. Previous works show that spatial influence is helpful, but it is usually modeled only among POIs [1,4,5,11,17,20]. We go a step further and build the relationship between users and distances. In previous studies, we find that people define a return time prediction problem and propose to apply survival analysis [3,9,10]. These works model time gaps from a user's visiting sequence to predict when a user will return to the service. However, they usually predict a certain time value as a single regression task which does not help to recommend items. Inspired by these works but different from them, we model a user's spatial preference for a wide range of distances and promote the task of recommending POIs.

To model the spatial preference, we map each distance value to an interval. Firstly, all distances between two successive POIs in each $\mathcal{I}_u$ are computed. We define two values δd and M_D to represent the minimum interval and maximum interval. Then, we have a vector $[0, \delta d, 2\delta d, ..., M_D]$ to indicate all intervals. Each distance is converted to an interval. If a distance is bigger than M_D, it is also represented by M_D. Then, the modeling of distance is converted to the modeling of interval. Just like each POI has a vector x, we define a latent vector $d \in \mathbb{R}^d$ for each interval d, and this operation forms a latent matrix $D \in \mathbb{R}^{(M_D+1)\times d}$ for all intervals $[0, \delta d, 2\delta d, ..., M_D]$. The non-bold d is a value, while the bold d is a vector. Given $\mathcal{I}_u$ for each user, we will have a sequence of intervals $[d_p^1, d_p^2, ...]$ and a sequence of vectors $[d_p^1, d_p^2, ...]$. Next, we update the computation of h^t.

$$h^t = f\left(U\left[x_p^t; d_p^t\right], Wh^{t-1}, b\right), \qquad h^t \in \mathbb{R}^d, \tag{4}$$

where $U \in \mathbb{R}^{d\times 2d}$, d_p^t is concatenated with x_p^t.

At each time, we calculate spatial preference of all intervals for next time.

$$\begin{aligned} s^t &= \text{SoftReLU}\left(V_s h^t + b_s\right) \\ &= \left[s^t(0), s^t(\delta d), s^t(2\delta d), ..., s^t(M_D)\right] \end{aligned} \tag{5}$$

where each value in s^t is the spatial preference for a certain interval. Accordingly, the spatial preference for next ground truth interval is $s^t\left(d_p^{t+1}\right)$, where the value d_p^{t+1} is the distance interval between x_p^t and x_p^{t+1}.

3.4   Preference Encoders

As we have two preferences, we need to encode them together and we propose a linear way and a non-linear way. Our network considers not only which POIs a user would like at next time, but also how far he wants to go.

By introducing a weight w_d, the sequential preference and spatial preference can be combined together linearly.

$$\hat{x}_{up}^t = \left(h^t\right)^{\mathrm{T}} x_p^{t+1} + w_d s^t\left(d_p^{t+1}\right) \tag{6}$$

However, linear fusion is natural. It is worth exploring non-linearity to investigate correlations between two preferences. Inspired from the attention mechanism, our innovative strategy is

$$\hat{x}_{up}^t = \boldsymbol{v}_a^T \tanh\left(\boldsymbol{r}_a\left(\boldsymbol{h}^t\right)^T \boldsymbol{x}_p^{t+1} + \boldsymbol{e}_a \boldsymbol{s}^t\left(d_p^{t+1}\right)\right) \tag{7}$$

where $\boldsymbol{v}_a, \boldsymbol{r}_a, \boldsymbol{e}_a \in \mathbb{R}^{d \times d}$ are weight vectors. The attention mechanism enables a model to concentrate on critical parts and it has been widely in many tasks and fields, such as image classification [15], next item recommendation [18], and so on. However, previous works using attention usually assign an appropriate weight for each factor to tell its importance. Therefore, the attention previously models one-to-many problems. In our work, we change the attention to capture one-to-one relationship and replace two commonly used weight matrices in attention with two weight vectors $\boldsymbol{r}_a, \boldsymbol{e}_a$. By innovatively using the attention, we create a non-linear combination Eq. (7) to encode two preferences.

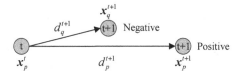

Fig. 2. Illustration of positive and negative POIs and distances from t-th time to $(t+1)$-th time. The $\boldsymbol{x}_p^{t+1} \in \mathcal{I}_u$ is positive. The negative POI $\boldsymbol{x}_p^{t+1} \notin \mathcal{I}_u$ is randomly chosen. The negative distance (interval) d_q^{t+1} is computed between $\boldsymbol{x}_p^t$ and $\boldsymbol{x}_q^{t+1}$.

3.5 Training Framework

In this subsection, we apply the widely-used pair-wise Bayesian Personalized Ranking (BPR) [1,5,16] to train the model.

$$l_{\text{bpr}}^t = -\ln \sigma\left(\hat{x}_{up}^t - \hat{x}_{uq}^t\right) \tag{8}$$

where $\hat{x}_{up}^t$ and $\hat{x}_{uq}^t$ are positive and negative preferences. At each time, a negative POI $\boldsymbol{x}_q \notin \mathcal{I}_u$ is randomly chosen from $\mathcal{I}$. Illustrated in Fig. 2, the negative distance d_q^{t+1} is calculated between $\boldsymbol{x}_p^t$ and $\boldsymbol{x}_q^{t+1}$. Finally, the loss function is

$$\Theta^* = \underset{\Theta}{\arg\min} \sum_u \sum_{t=1}^{t=|\mathcal{I}_u|} l_{\text{bpr}}^t + \frac{\lambda_\Theta}{2}\|\Theta\|^2 \tag{9}$$

where Θ denotes a set of parameters $\Theta = \{\boldsymbol{X}, \boldsymbol{D}, \boldsymbol{U}, \boldsymbol{W}, \boldsymbol{b}, \boldsymbol{V}_s, \boldsymbol{b}_s, w_d, \boldsymbol{v}_a, \boldsymbol{r}_a, \boldsymbol{e}_a\}$, where $\boldsymbol{X}, \boldsymbol{D}$ are sets of all POI vectors and all distance vectors respectively. Stochastic gradient descent (SGD) is used to learn the parameters.

4 Experiments

4.1 Experimental Settings

Datasets. We apply two widely-used datasets called Foursquare and Gowalla which are preprocessed in [19]. Specifically, all the information used in our work includes each user's chronological check-in sequence and the corresponding distance sequence, except for the time of check-ins. Following previous works [6,7,16], we employ the leave-one-out evaluation. For each user's check-in sequence, we treat the last POI as the test data and apply the rest POIs for training.

Comparison Methods. Our Distance2Pre network is compared with the following methods. (1) **BPR** [16]: This method refers to the BPR-MF for implicit feedback. It optimizes the difference of the user's preferences for positive and negative items. (2) **GRU** [2]: RNN is effective for successive POI prediction. We apply GRU in this work. (3) **FPMC-LR** [1]: This work is based on first-order Markov chain and uses neighbors as negative samples. (4) **PRME-G** [5]: It is a metric embedding method, and the spatial distance is considered as the weight. (5) **CA-RNN** [12]: A novel model incorporates input and transition contexts. Accordingly, we apply GRU to implement CA-RNN and compute the transition context by using distance intervals. (6) **POI2Vec** [4]: A binary tree is used to cluster the nearby POIs into the same region. Moreover, a POI is assigned to multiple regions in this model to strengthen the spatial influences of POIs. As our proposed network has a linear fusion and a non-linear fusion, it has two variants: **Distance2Pre (Linear)** and **Distance2Pre (Non-Linear)**.

Evaluation Metrics. The top-k metrics are popular for POI prediction [1,4, 5,11,20]. In this work, we apply metrics called Recall and F_1-score. Values of metrics in our work are all expressed as percentages. During the test, each user's training sequence $(x_p^1, ..., x_p^n)$ is recomputed by using GRU to obtain h^n and s^n. Then, h^n is applied to acquire user's sequential preferences for all items X. Meanwhile, all distances between x_i^n and each item in X are calculated because we do not know any information about the test set. These distances are fixed in each epoch and converted to spatial preferences for X by using s^n. Then, we obtain each user's final preferences for all items and recommend top-k items with the highest preference.

Additionally, parameters Θ are initialized to the same range, e.g., uniform distribution $[-0.5, 0.5]$. The learning rate, regularization λ_Θ and the dimension are set as 0.01, 0.001 and 20 for all methods. Weight w_d is initialized by a positive value 1.0 and is also updated by SGD. Details of w_d are illustrated in Fig. 4. The code is written by using Theano and is available on GitHub[1].

---

[1] https://github.com/cuiqiang1990/Distance2Pre.

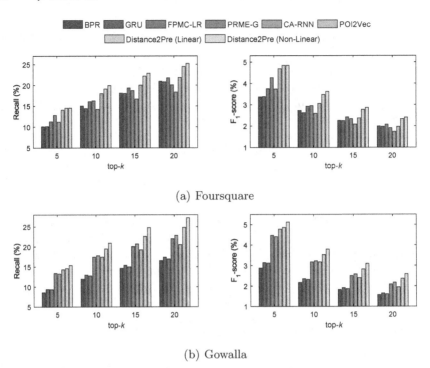

(a) Foursquare

(b) Gowalla

Fig. 3. Experimental results on two datasets.

4.2   Performance Comparison

Performances of all methods are illustrated in Fig. 3. First, we explore baselines BPR, GRU, and FPMC-LR. They are comparable but perform differently on two datasets. FPMC-LR is always better than BPR and proves the effectiveness of the spatial influence. GRU performs worst among them on Foursquare. Perhaps because there is more than one behavior at a certain time and we are unable to know the true order of these multiple check-ins. This disordered property in Foursquare hinders the sequential modeling of GRU. Fortunately, GRU is the best on Gowalla which has the right time order. This meaningful result indicates that correct sequential modeling is important for POI prediction.

In the following, we compare PRME-G, CA-RNN, POI2Vec with our Distance2Pre network. First, performances of these four methods are also adversely affected by the disordered property. Their performance on Foursquare is close to or even below the performance of BPR, GRU and FPMC-LR, especially CA-RNN. On the contrary, their performance on Gowalla is obviously better. The CA-RNN treats distance intervals as transition contexts. Accurately, CA-RNN acquires a transition matrix for every possible interval. Such a precise modeling will result in great improvement as well as great decline, which depends on whether the order is correct or not. POI2Vec has a comparable performance of *top*-5 with our Distance2Pre on both datasets, while it is obviously weaker than our network on Recall@20 and F_1-score@20. Actually, POI2Vec clusters nearby POIs into the same region, which causes a strong local correlation of POIs. Therefore, POI2vec is good at recommending a small quantity of POIs.

Overall, our Distance2Pre is optimal on two datasets. Our spatial preference is powerful for predicting next POI and robust to disordered property in Foursquare. We have a visualization of spatial preference in Sect. 4.5 and find that people have personalized moving pattern. Such a pattern is a kind of user interest which is regular and does not change dramatically. Therefore, our Distance2Pre can still get good performance on Foursquare.

4.3 Settings of Max Distance M_D and Distance Interval δd

In this part, we explore the effect of max distance $M_D(km)$ and distance interval $\delta d(km)$ on the spatial preference s in Eq. (5). The M_D and δd reflect the range and granularity of s. Results are in Table 1.

The proper M_D and δd are chosen based on all the distances between successive POIs in user sequences. The distribution of distances is different on two datasets. For example, $M_D = 20$ km covers 97.6% and 79.9% distances on Foursquare and Gowalla respectively. Gowalla has a greater proportion of large distance. Finally, we set $M_D = [2.5, 5, 10], \delta d = [0.10, 0.15, 0.20]$ for Foursquare and $M_D = [10, 20, 40], \delta d = [0.10, 0.20, 0.30]$ for Gowalla. Obviously, $M_D = 5, \delta d = 0.10$ and $M_D = 5, \delta d = 0.15$ are best for our linear fusion and non-linear fusion on Foursquare. $M_D = 20, \delta d = 0.20$ and $M_D = 20, \delta d = 0.10$ are the best on Gowalla. We can see that if a dataset covers more larger distances, setting larger $M_D, \delta d$ may be more suitable. The comparison between linear fusion and non-linear fusion is discussed in the next subsection.

Table 1. Performance evaluated by Recall@10 with varying max distance M_D (km) and distance interval δd (km).

Method			Distance2Pre (linear)						Distance2Pre (non-linear)					
Evaluation			Recall@10			F1-score@10			Recall@10			F1-score@10		
		δd	0.10	0.15	0.20	0.10	0.15	0.20	0.10	0.15	0.20	0.10	0.15	0.20
Foursquare	M_D	2.5	18.23	18.53	18.44	3.31	3.37	3.35	19.86	19.74	19.26	3.61	3.59	3.50
		5	**19.13**	18.92	18.48	**3.48**	3.44	3.36	19.13	**19.96**	19.39	3.47	**3.63**	3.53
		10	18.70	18.57	18.48	3.40	3.38	3.36	19.57	19.65	19.70	3.56	3.57	3.58
		δd	0.10	0.20	0.30	0.10	0.20	0.30	0.10	0.20	0.30	0.10	0.20	0.30
Gowalla	M_D	10	19.23	19.23	19.01	3.50	3.50	3.46	16.79	20.71	16.06	3.05	3.77	2.92
		20	19.33	**19.44**	19.04	3.51	**3.53**	3.46	**20.89**	19.57	19.77	**3.80**	3.56	3.59
		40	19.15	19.33	19.14	3.48	3.51	3.48	19.96	20.28	20.28	3.63	3.69	3.69

Table 2. Performance of our proposed Distance2Pre network on two datasets.

Method	Distance2Pre (linear)								Distance2Pre (non-linear)							
Evaluation	Recall@				F1-score@				Recall@				F1-score@			
	5	10	15	20	5	10	15	20	5	10	15	20	5	10	15	20
Foursquare	14.55	19.13	22.21	24.50	4.85	3.48	2.78	2.33	14.55	19.96	22.90	25.24	4.85	3.63	2.86	2.40
Gowalla	14.59	19.44	22.57	24.82	4.86	3.53	2.82	2.36	15.37	20.89	24.79	27.21	5.12	3.80	3.10	2.59

4.4   Linear Fusion Vs. Non-Linear Fusion

In this subsection, we investigate the linear fusion and non-linear fusion by analysis values in Tables 1 and 2 and Fig. 4. Assuredly, non-linear fusion is more powerful to handle two different preferences.

The performance is higher under non-linear fusion. In Table 1, there is a big difference in values between linear fusion and non-linear fusion on both datasets. Most values under non-linear fusion are obviously larger than those under linear fusion. Moreover, many values of Recall@10 are one percentage point higher than corresponding values in the left half of the table. By using best parameters, the performance of our two Distance2Pre networks is shown in Table 2. It is interesting that the difference between the two kinds of fusions on Gowalla is greater than that on Foursquare. Perhaps non-linearity is more powerful to deal with more complex situations as Gowalla has much more data than Foursquare.

Illustrated in Fig. 4, we analysis changes of weight w_d. This parameter is also updated by SGD and we preserve the value of w_d after each epoch. (1) On both datasets, w_d changes from a large value to a smaller one. Because search space of sequential preference is much bigger than that of spatial preference, it is not easy to obtain a good representation of sequential preference and spatial preference plays a major role in the beginning. During the later period of training, the effect of sequential preference gradually appears and w_d eventually stabilizes near a certain value, 1.1 and 2.1 on two datasets respectively. (2) Gowalla's curve is steeper than Foursquare's because the search space of sequential preference on Gowalla is obviously larger. (3) Both curves are concussion drops, not smooth ones. The relationship between the two preferences is actually complicated. We do not know which preference will play a bigger role when choosing the next POI. Therefore, when modeling each pair of two preferences, non-linear fusion tends to be a better fit rather than linear fusion.

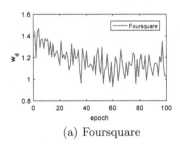
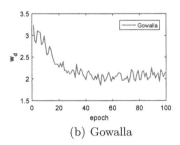

(a) Foursquare   (b) Gowalla

Fig. 4. Changes of weight w_d in Eq. (6) from epoch 1 to epoch 100.

4.5   Visualization of Spatial Preference

We make visualization to study different spatial preferences on Foursquare. At the end of each user's training sequence, we compute each user's spatial preference s^n for the test set. We choose Distance2Pre (Non-Linear) as the sample. Because we have $M_D = 5\,\mathrm{km}$, $\delta d = 0.15\,\mathrm{km}$ in this network, each s^n is a 34-dimensional vector and the horizontal axis length is 34. First, we convert vector s^n by softmax to cause the sum of it to be 1. Then, based on all spatial preferences, we obtain 10 clusters by the k-means method. Multiple vectors of spatial preferences within one cluster are reduced to one vector by averaging. We illustrate three representative clusters cluster-[3, 4, 9] in Fig. 5(a). In order to distinguish three curves, horizontal axis uses log coordinate in Fig. 5(a). Besides, we select a user from each cluster to show its own spatial preference in Fig. 5(b) and draw his historical POIs in Fig. 5(c).

Different groups of people may have different moving patterns. Cluster-3 has large probabilities for small intervals and the probability reduces rapidly with the increase of interval. This pattern is likely to be around a point. User-688 may be a retired people and his POIs are almost around the center of Singapore. In cluster-4, there are two crests. Probabilities on large intervals are also almost zero. Such POIs of a user are mainly distributed around two points. POIs of user-1591 focus on the Nanyang Technological University and the center of Singapore. She might be a student. Cluster-9 is obviously different from cluster-3/4 because it has probabilities for many large intervals. These users may often need to go to different places for business. User-1537 prefers the center of Singapore but he also goes everywhere. By clustering, we find that patterns of movement are personalized. By looking at users one by one, the learned spatial preference in our network can effectively reflect the distribution of user's historical POIs.

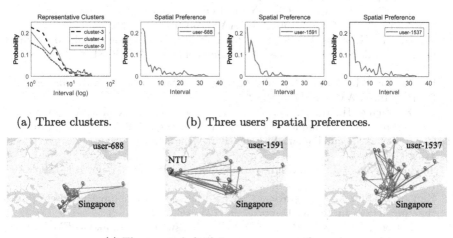

(a) Three clusters.   (b) Three users' spatial preferences.

(c) Three users' check-in sequences on the map.

Fig. 5. Visualization of spatial preference on the test set. (a) shows three representative clusters. (b) are three users' spatial preferences sequentially chosen from each cluster. (c) are three users' historical POIs. Each straight line links two successive POIs and each node is a POI located by its longitude and latitude.

5   Conclusion

In this work, we have proposed a Distance2Pre network for the next POI prediction. It can mine spatial preference to model the correlation of the user-distance. Besides, we propose two preference encoders which are a linear fusion and a non-linear fusion. Both encoders can capture the relationship between two preferences and the non-linear fusion is better. Experiments demonstrate the effectiveness of our network. In the future, we will incorporate more information, like the time of check-ins and time interval.

Acknowledgment. This work is jointly supported by National Natural Science Foundation of China (61772528), National Key Research and Development Program (2016YFB1001000), National Natural Science Foundation of China (U1435221).

References

1. Cheng, C., Yang, H., Lyu, M.R., King, I.: Where you like to go next: successive point-of-interest recommendation. In: IJCAI, vol. 13, pp. 2605–2611 (2013)
2. Cho, K., et al.: Learning phrase representations using RNN encoder-decoder for statistical machine translation. In: EMNLP, pp. 1724–1734 (2014)
3. Du, N., Dai, H., Trivedi, R., Upadhyay, U., Gomez-Rodriguez, M., Song, L.: Recurrent marked temporal point processes: embedding event history to vector. In: SIGKDD, pp. 1555–1564. ACM (2016)
4. Feng, S., Cong, G., An, B., Chee, Y.M.: POI2Vec: geographical latent representation for predicting future visitors. In: AAAI, pp. 102–108 (2017)
5. Feng, S., Li, X., Zeng, Y., Cong, G., Chee, Y.M., Yuan, Q.: Personalized ranking metric embedding for next new POI recommendation. In: IJCAI, pp. 2069–2075 (2015)
6. He, X., Liao, L., Zhang, H., Nie, L., Hu, X., Chua, T.S.: Neural collaborative filtering. In: WWW, pp. 173–182 (2017)
7. He, X., Zhang, H., Kan, M.Y., Chua, T.S.: Fast matrix factorization for online recommendation with implicit feedback. In: SIGIR, pp. 549–558. ACM (2016)
8. Hochreiter, S., Schmidhuber, J.: Long short-term memory. Neural Comput. **9**(8), 1735–1780 (1997)
9. Jing, H., Smola, A.J.: Neural survival recommender. In: WSDM, pp. 515–524. ACM (2017)
10. Kapoor, K., Sun, M., Srivastava, J., Ye, T.: A hazard based approach to user return time prediction. In: SIGKDD, pp. 1719–1728. ACM (2014)
11. Li, X., Cong, G., Li, X.L., Pham, T.A.N., Krishnaswamy, S.: Rank-GeoFM: a ranking based geographical factorization method for point of interest recommendation. In: SIGIR, pp. 433–442. ACM (2015)
12. Liu, Q., Wu, S., Wang, D., Li, Z., Wang, L.: Context-aware sequential recommendation. In: ICDM, pp. 1053–1058 (2016)
13. Liu, Q., Wu, S., Wang, L., Tan, T.: Predicting the next location: a recurrent model with spatial and temporal contexts. In: AAAI, pp. 194–200 (2016)
14. Liu, X., Liu, Y., Li, X.: Exploring the context of locations for personalized location recommendations. In: IJCAI, pp. 1188–1194 (2016)
15. Mnih, V., Heess, N., Graves, A., et al.: Recurrent models of visual attention. In: NIPS, pp. 2204–2212 (2014)

16. Rendle, S., Freudenthaler, C., Gantner, Z., Schmidt-Thieme, L.: BPR: Bayesian personalized ranking from implicit feedback. In: UAI, pp. 452–461 (2009)
17. Wang, H., Shen, H., Ouyang, W., Cheng, X.: Exploiting poi-specific geographical influence for point-of-interest recommendation. In: IJCAI, pp. 3877–3883 (2018)
18. Wang, S., Hu, L., Cao, L., Huang, X., Lian, D., Liu, W.: Attention-based transactional context embedding for next-item recommendation. In: AAAI, pp. 2532–2539 (2018)
19. Yuan, Q., Cong, G., Ma, Z., Sun, A., Thalmann, N.M.: Time-aware point-of-interest recommendation. In: SIGIR, pp. 363–372. ACM (2013)
20. Zhao, S., Zhao, T., King, I., Lyu, M.R.: Geo-teaser: geo-temporal sequential embedding rank for point-of-interest recommendation. In: WWW, pp. 153–162 (2017)

Using Multi-objective Optimization to Solve the Long Tail Problem in Recommender System

Jiaona Pang[1], Jun Guo[2(✉)], and Wei Zhang[2]

[1] Department of Computer Science and Technology, East China Normal University,
3663 Zhong Shan Rd. N., Shanghai, China
[2] Computer Center, East China Normal University,
3663 Zhong Shan Rd. N., Shanghai, China
jguo@cc.ecnu.edu.cn

Abstract. An improved algorithm for recommender system is proposed in this paper where not only accuracy but also comprehensiveness of recommendation items is considered. We use a weighted similarity measure based on non-dominated sorting genetic algorithm II (NSGA-II). The solution of optimal weight vector is transformed into the multi-objective optimization problem. Both accuracy and coverage are taken as the objective functions simultaneously. Experimental results show that the proposed algorithm improves the coverage while the accuracy is kept.

Keywords: Recommender system · Weighted similarity measure · Multi-objective optimization

1 Introduction

Personalized recommender system can help users quickly obtain the useful information they may be interested in from massive information, so as to alleviate the negative impact of information overload. An essential step in the recommendation algorithm is to calculate the similarity between users or items. There are many common methods, such as cosine similarity, pearson correlation coefficient and adjusted cosine similarity. Bobadilla *et al.* [3] proposed a weighted similarity measure which consists of linear combination of rating difference vector and weight vector, and utilized genetic algorithm (GA) to obtain the optimal similarity function. Gupta *et al.* [9] added fuzzy logic to the former, that is, the fuzzy c-means clustering algorithm was used to cluster a data set, and then used the similarity measure based on GA to calculate the similarity between the clustering values.

The items in the recommender system can be divided into popular items and long tail items. Long tail items refer to items with low popularity and low probability of being recommended. Coverage reflects the ability of the recommender system to explore long tail items. So far, many algorithms have been presented

© Springer Nature Switzerland AG 2019
Q. Yang et al. (Eds.): PAKDD 2019, LNAI 11441, pp. 302–313, 2019.
https://doi.org/10.1007/978-3-030-16142-2_24

to enhance the accuracy of the recommender system, where the comprehensiveness of recommended items are usually not considered. Thus, the long tail items rarely have the opportunities to be recommended. Therefore, it is necessary to improve the coverage of the recommender system. Ge et al. [7] suggested that beyond accuracy, coverage and serendipity should be considered as criteria for evaluating the performance of recommender systems, and proposed several methods of measuring coverage and serendipity. Park and Tuzhilin [12] enhanced the utilization of long tail items by dividing the item sets into popular sets and long tail sets, and clustering long tail items into different groups to reduce rating errors. Yin et al. [17] proposed a suite of recommendation algorithms to address the problem of long tail item recommendation. Wang et al. [16] proposed a novel multi-objective framework, and two contradictory objective functions are utilized to suggest both popular and novel items for users.

In order to meet the demands of different users, multiple indicators of the performance should be considered as much as possible, so the multi-objective optimization algorithm based recommendation algorithms have been proposed. Ribeiro et al. [13] proposed a hybrid recommendation algorithm, which uses strength pareto evolutionary algorithm (SPEA) to acquire the optimal hybrids maximizing both accuracy and diversity of the recommender system. Geng et al. [8] put forward a framework based on multi-objective evolutionary algorithm, where non-dominated neighbor immune algorithm (NNIA) is used to generate a series of recommendation lists of different focuses. Cui et al. [5] introduced a new diversity index and proposed a probabilistic multi-objective evolutionary algorithm which can achieve an excellent compromise between accuracy and diversity.

In this paper, a weighted similarity measure based on NSGA-II is presented. The weighted similarity measure in [3] is used to calculate the similarity between users, which is a linear combination of weight vector and user rating difference vector. Instead of only considering the weight vector corresponding to the highest accuracy in [3], we transform the solution of weight vector into a multi-objective optimization problem. Both accuracy and coverage are taken as the objective functions simultaneously, and then the optimal solution is obtained by NSGA-II.

2 Similarity Measure

Let $U = \{u_1, ..., u_s\}$ be a user set, and $I = \{c_1, ..., c_t\}$ be an item set. $r_x = [r_x^{(1)}, r_x^{(2)}, ..., r_x^{(t)}]$ denotes user x's ratings on the items, where $r_x^{(i)}$ is user x's rating on the item i. The range of ratings is from m to M, which are integers. Here, m denotes the user's lowest rating on the item, which indicates that the user is most displeased with the item. M denotes the highest rating of the user on behalf of the user is perfectly satisfied with the item. Note that in some cases the user has not rated some items, which will be denoted by a dot mark •.

The similarity measure used in this paper was proposed by Bobadilla et al. [3], which is related to weight vector and user rating difference vector. The similarity between two users can be expressed as follows:

$$sim_w(x, y) = \frac{1}{M - m + 1} \sum_{i=0}^{M-m} w^{(i)} v_{x,y}^{(i)}, \tag{1}$$

where $-1 \leq w^{(i)} \leq 1$ denotes the significance of the corresponding element $v_{x,y}^{(i)}$ in calculating the similarity between user x and y. $\boldsymbol{v}_{x,y} = [v_{x,y}^{(0)}, ..., v_{x,y}^{(M-m)}]$ denotes the difference between the ratings of user x and y, the dimension of which is the number of possible rating differences between two users. The element $v_{x,y}^{(i)}$ in the vector $\boldsymbol{v}_{x,y}$ denotes the ratio of items rated by user x and y, and in which the absolute value of the rating difference is i($i = \left| r_x^{(i)} - r_y^{(i)} \right|$), to the total number of items that both users have rated together. The following example clearly illustrates the meaning of $\boldsymbol{v}_{x,y}$.

Suppose that there are eight items in the recommender system, the minimum rating m is 1, and the maximum rating M is 5. Thus, we can see that the range of rating difference is $\{0, 1, 2, 3, 4\}$, and the dimension of vector $\boldsymbol{v}_{x,y}$ is 5. The rating vectors of user x and y are known as follows:

$$r_x = [5, 3, \bullet, 1, 2, \bullet, 4, 4],$$
$$r_y = [4, 1, 2, 4, 3, \bullet, \bullet, 4].$$

The total number of items rated by two users is 5, and only item 8 was rated equally, so that $v_{x,y}^{(0)} = 1/5$. There is no item for which user x rated it as 1 and user y rated it as 5, so $v_{x,y}^{(M-m)} = v_{x,y}^{(4)} = 0$. The value of other components in the vector can be calculated in the same way, and the vector $\boldsymbol{v}_{x,y}$ can be obtained as: $\boldsymbol{v}_{x,y} = [1/5, 2/5, 1/5, 1/5, 0]$.

Each component $w^{(i)}$ in the vector $\boldsymbol{w}$ denotes the effect of rating difference on the similarity between users. Obviously, the smaller the rating difference, the larger the value of the corresponding $w^{(i)}$ should be. It means that users with similar ratings are more similar.

With the Movielens and Netflix data sets where the reasonable range of ratings is 1 to 5, the ranges of $w^{(i)}$ are as shown: $0.6 \leq w^{(0)} \leq 1, 0.2 \leq w^{(1)} \leq 0.6$, $-0.2 \leq w^{(2)} \leq 0.2, -0.6 \leq w^{(3)} \leq -0.2, -1 \leq w^{(4)} \leq -0.6$.

The critical issue now is to achieve the best weight vector so that the corresponding similarity function is optimal. In this paper, the optimal weighted similarity function is obtained by NSGA-II, which can improve the coverage of recommender system and preserve the accuracy simultaneously. At this point, solving the optimal weight vector turns into a multi-objective optimization problem.

3   The Proposed Algorithm

In this section, we describe the weighted similarity measure based on NSGA-II in detail, including the objective functions, the genetic operators and the procedure of our proposed algorithm.

3.1   Objective Function

To improve coverage of the recommender system on the premise of ensuring the accuracy, the accuracy function and the coverage function are selected as the objective function of the multi-objective optimization problem. The accuracy can be measured by the mean absolute error (MAE), which can be described as follows:

$$\text{MAE} = \frac{1}{|U|} \sum_{u \in U} \frac{\sum_{i \in I_u} \left| \hat{r}_u^i - r_u^i \right|}{|I_u|}, \tag{2}$$

where U is the user set, I_u is the item set that has been rated by user u, $\hat{r}_u^i$ is the predictive rating on item i rated by user u, and r_u^i is the practical rating on item i rated by user u.

Coverage usually is used to measure the ratio of the number of items recommended to all users to the total number of all items, which reflects the ability of the recommender system to exploit long tail items. If the coverage of a recommender system is 100%, all items are recommended to at least one user. The calculation of the coverage is as follows:

$$\text{Coverage} = \frac{|\cup_{u \in U} L(u)|}{|I|}, \tag{3}$$

where $L(u)$ is a specified length of recommendation list for user u, I is the item set.

We choose Eqs. (2) and (3) as the objective functions, which are expressed as follows:

$$\begin{cases} \min f_1, & f_1 = \text{MAE} \\ \min f_2, & f_2 = -\text{Coverage} \end{cases}$$

3.2   Design of the Proposed Algorithm

Many multi-objective optimization algorithms have been proposed. Among them, there are some well-known algorithms such as non-dominated sorting genetic algorithm (NSGA) [15], SPEA [18] and pareto archived evolution strategy (PAES) algorithm [10]. NSGA-II [6] is an improved version from NSGA. Compared with NSGA, NSGA-II reduces computational complexity, accelerates execution speed of the algorithm, introduces an elite strategy, expands the sampling space and better maintains the diversity, so it has been widely used. In this paper, we choose NSGA-II to obtain the optimal weight vector $\boldsymbol{w}$.

Definition. Multi-objective optimization algorithm can optimize various contradictory objective functions simultaneously, coordinate each objective function and finally obtain a set of Pareto optimal solutions. According to [11], the multi-objective optimization can be described as follows:

$$\text{Min} \quad \hat{F}(\hat{\boldsymbol{x}}) = (f_1(\hat{\boldsymbol{x}}), f_2(\hat{\boldsymbol{x}}), ..., f_k(\hat{\boldsymbol{x}}))^{\mathsf{T}},$$
$$\text{S.t.} \quad \hat{\boldsymbol{x}} = (\hat{x}_1, \hat{x}_2, ..., \hat{x}_d) \in \Omega,$$

where $f_i(\hat{x})(1 \leq i \leq k)$ is the objective function, k is the total number of the objective functions, $\hat{x}$ is a d-dimensional decision vector and Ω is the decision space.

Let us first introduce Pareto domination. Suppose that $\hat{x}_a, \hat{x}_b \in \Omega$ are two feasible solutions to minimize the multi-objective function, if $f_i(\hat{x}_a) \leq f_i(\hat{x}_b)$ for $i = 1, 2, ..., k$ and $f_j(\hat{x}_a) < f_j(\hat{x}_b)$ for one or more objective function j, where $j = 1, 2, .., k$, and it is called $\hat{x}_a$ dominate $\hat{x}_b$, denoted as $\hat{x}_a \succ \hat{x}_b$. If no feasible solution in decision space can dominate $\hat{x}^*$, then $\hat{x}^*$ is referred to as the Pareto optimal. The set of all Pareto optimal solutions is known as the Pareto optimal set, and the values of the objective function under the objective space corresponding to the feasible solutions in Pareto optimal set are called the Pareto optimal front.

Encoding. The coding method commonly used in genetic algorithm is the binary coding. In this paper, each element $w^{(i)}$ in w is represented by a 10 bits binary strings $b_i^9...b_i^1 b_i^0$, where $i = \{0, 1, ..., M - m\}$. Each component $(-1 \leq w^{(i)} \leq 1)$ is calculated by the following expression:

$$w^{(i)} = \frac{2\sum_{j=10}^{9} 2^j b_i^j}{2^{10} - 1} - 1.$$

Initialize Population. Generate a population of size N randomly, which is the initial parent population.

Fast Non-dominated Sorting Approach. In our proposed algorithm, the population W is the weight vector w of all possible values in a reasonable ranges. In this paper, we have two objective functions f_1 and f_2.

First, for each individual w_p in W, two parameters need to be calculated: one is the number of individuals $\hat{n}_{w_p}$ that dominates the individual w_p, and the other is the sum of individuals $\hat{s}_{w_p}$ dominated by the individual w_p. The steps of identifying the population W into different non-dominated fronts $F_1, F_2, ..., F_R$ are as follows:

1: *for each $w_p \in W$*
2:     *initialize $\hat{s}_{w_p} = \emptyset, \hat{n}_{w_p} = 0$*
3:         *for each $w_q \in W$*
4:             *if $w_p \succ w_q$*
5:                 *add w_q to the set $\hat{s}_{w_q}$*
6:             *else if $w_q \succ w_p$*
7:                 *execute $\hat{n}_{w_p} = \hat{n}_{w_p} + 1$*
8:         *if $\hat{n}_{w_p} = 0$*
9:             *let $w_{p_{rank}} = 1$ and put them in set F_1*
10: *initialize $i = 1$*
11: *While F_i is not empty*
12:     *initialize $Q = \emptyset$*

13:    *for each $\boldsymbol{w}_p \in F_i$ and each $\boldsymbol{w}_q \in \hat{s}_{\boldsymbol{w}_p}$*
14:       *do $\hat{n}_{\boldsymbol{w}_q} = \hat{n}_{\boldsymbol{w}_q} - 1$*
15:       *if $\hat{n}_{\boldsymbol{w}_q} = 0$*
16:          *let $\boldsymbol{w}_{q_{rank}} = i + 1$ and put them in Q*
17:    *Execute $i = i + 1$, $F_i = Q$*

Crowding Distance Computation. Crowding distance is used to measure the number of individuals around a given individual in the population that belong to the same non-dominated front. The crowding distance of individuals in the non-dominated fronts $F_1, F_2, ..., F_R$ are calculated as follows:

1: *for each rank $j = 1, 2, ..., R$*
2:    *let $l = |F_j|$, $\hat{W} = F_j$*
3:    *for each objective function f_k (where $k = 1, 2$)*
4:       *$\hat{W} = sort(\hat{W}, f_k)$*
       // sort $\boldsymbol{w}_i$ in $\hat{W}$ in the ascending order according to the value of f_k.
5:       *$\hat{W}[\boldsymbol{w}_1, k]_{distance} = \hat{W}[\boldsymbol{w}_l, k]_{distance} = \infty$*
       // specify the crowding distance of two boundary individuals $\boldsymbol{w}_1$ and $\boldsymbol{w}_l$ is infinite
6:       *for other individuals $\boldsymbol{w}_i$ (where $i = 2, ..., l - 1$)*
7:          *$\hat{W}[\boldsymbol{w}_i, k]_{distance} = \hat{W}[\boldsymbol{w}_i, k]_{distance} + (\hat{W}[\boldsymbol{w}_{i+1}].k - \hat{W}[\boldsymbol{w}_{i-1}].k)/(f_k^{max} - f_k^{min})$*
8:       *for $i = 2, ..., l - 1$*
9:          *$\hat{W}[\boldsymbol{w}_i]_{distance} = \sum_k \hat{W}[\boldsymbol{w}_i, k]_{distance}$*
       // the total crowding distance value of each individual is obtained by summing the crowding distance values corresponding to k normalized objective functions

In the above algorithm, l denotes the number of individuals in the set F_j, $\hat{W}[\boldsymbol{w}_i, k]_{distance}$ denotes the crowding distance value of the individual $\boldsymbol{w}_i$ on the objective function f_k, $\hat{W}[\boldsymbol{w}_i].k$ denotes the value of the objective function f_k of the individual $\boldsymbol{w}_i$ in the set $\hat{W}$, and f_k^{max} and f_k^{min} are the maximum and minimum values of the objective function f_k respectively.

Crowded-Comparison Operator. After fast non-dominated sort and crowding distance computation, each individual $\boldsymbol{w}$ in the population W has the following two attributes:

(1) Non-domination rank $\boldsymbol{w}_{rank}$,
(2) Crowding distance $W[\boldsymbol{w}]_{distance}$.

The definition of crowded-comparison operator $\prec_n$ is as follows:
Let $\boldsymbol{w}_p$ and $\boldsymbol{w}_q$ be two individuals in W, if $\boldsymbol{w}_{p_{rank}} < \boldsymbol{w}_{q_{rank}}$ or $\boldsymbol{w}_{p_{rank}} = \boldsymbol{w}_{q_{rank}}$ and $W[\boldsymbol{w}_p]_{distance} > W[\boldsymbol{w}_q]_{distance}$, then $\boldsymbol{w}_p \prec_n \boldsymbol{w}_q$, indicating that the individual $\boldsymbol{w}_p$ is better than the individual $\boldsymbol{w}_q$.

Genetic Operators. As usual, we use the selection, crossover and mutation operators in the genetic algorithm to create a new population.

Selection: The selection of individuals in the population is based on the crowded-comparison operator $\prec_n$.

Crossover: One-point crossover technique is adopted in this paper, and the probability of crossover is 0.8.

Mutation: We use the single-point mutation operator like other multi-objective optimization genetic algorithms, and mutation probability is 0.05.

Procedure of the Proposed Algorithm. The weighted similarity measure based on NSGA-II is as follows:

Step 1 Generate 10 groups of random weight vectors, which constitute the parent population P_0, and the size N of the population is 10, initialize $z = 0$.
Step 2 Apply crossover and mutation on P_0 to generate the first generation offspring population Q_0 of size N.
Step 3 Combine P_z and Q_z to create a population R_z of size $2N$, that is, $R_z = P_z \cup Q_z$.
Step 4 For each weight vector w in R_z, put them into Eq. (1), then $2N$ similarity functions can be obtained.
Step 5 Identify $2N$ solutions w in R_z into several non-dominated fronts $F_1, F_2, ..., F_R$ according to the value of f_1 and f_2 on the training set by using the fast non-dominated sorting approach, and initialize $P_{z+1} = \emptyset$, $i = 1$. When $|P_{z+1}| + |F_i| \leq N$, the following loop is performed:
    Step 5.1 Calculate the crowding distance of all solutions in the set F_i, then do $P_{z+1} = P_{z+1} \cup F_i$ and $i = i + 1$.
Step 6 When $|P_{z+1}| + |F_i| > N$, sort F_i in descending order according to the crowded-comparison operator and add the first $N - |P_{z+1}|$ solutions from F_i to P_{z+1}.
Step 7 Generate the offspring population Q_{z+1} by applying crossover and mutation operators on P_{z+1}, and do $z = z + 1$.
Step 8 If the required number of iterations is not reached, go to Step 3.

After NSGA-II runs the specified number of iterations, we get a set of Pareto optimal solutions, which contains 10 sets of weight vectors. In this paper, we obtain the weighted mean of the weight vectors, which is regarded as the final solution of the weight vector under different number of iterations.

4   Experiments

In order to verify the effectiveness of the proposed algorithm, we design two sets of experiments from the following aspects:

(1) To explore the impact of the number of iterations on the experimental results.
(2) To explore the impact of the number of nearest neighbors on the performance of the recommender system.

4.1   Experimental Settings

Experimental Data Sets. In this paper, we use Movielens 1M [1] and Netflix [2] data sets to evaluate the effectiveness of our algorithm. Movielens 1M data set is provided by GroupLens research group, which contains 6040 users rating of 3952 movies, and each user rated at least 20 movies with a rating varies from 1 to 5. Netflix data set is published by Netflix, which contains approximately 100 million ratings of 17770 movies by 480189 anonymous users, and the rating is also varies from 1 to 5. In our experiments, we choose 80% of the data as the training set, and the remaining 20% of the data is used as the test set.

Evaluation Metrics. MAE, coverage, recall and precision are used to evaluate the performance of the proposed algorithm. MAE evaluates the error between the predicted ratings by recommender systems and the users' practical ratings. Coverage measures the ability of the recommender system to exploit long tail items. The combination of precision and recall is used to evaluate the classification accuracy of the recommender system.

In order to evaluate the effectiveness of the proposed algorithm, we compare our algorithm with the pearson correlation coefficient based collaborative filtering (CF) [14], the matrix factorization (MF) [4] and the weighted similarity measure based on GA [3]. We design two groups of experiments, which are tested respectively on the two data sets to compare the performance of different algorithms. In the first set of experiments, we compare the impact of the recommendation quality by running different number of iterations. In the other set of experiments, we observe the effect on the experimental results by changing the number of nearest neighbors or the number of the latent factors.

4.2   Experimental Results

Iterations. In the first part of the experiment, we refer to the literature [4] to set the parameters for MF, and the regularization parameter is set to 0.05, and the learning rate is set to 0.05. Compared with other algorithms, MF runs more number of iterations when it reaches the optimal value. Therefore, we set the number of iterations varies from 15 to 120 at intervals of 15 in MF, and we set the number of iterations vary from 5 to 40 at intervals of 5 for other algorithms. Thus, the actual number of iterations of MF is three times the abscissa value. K represents the number of latent factors for MF, while K represents the number of neighbors for other algorithms. We specify K to 70 on Movielens data set, and specify K to 90 on Netflix data set. The comparison results are shown in Figs. 1 and 2.

Figures 1 and 2 describe the performance of the four algorithms at different number of iterations on the two data sets, respectively. Figures 1 and 2 show that the four indicators of all algorithms reach stable gradually with the number of iterations increases. After the performance of the algorithm is stable, the coverage of our algorithm is improved by about 4% compared with the GA, and the values of recall and precision are also improved, while the value of

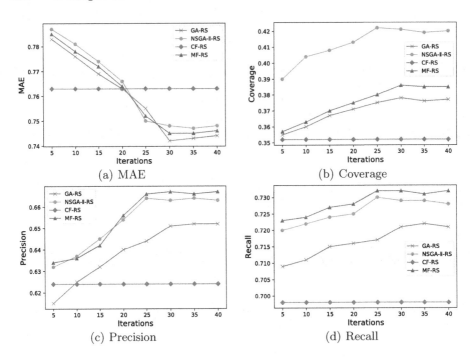

Fig. 1. Comparison results on Movielens under different iteration numbers.

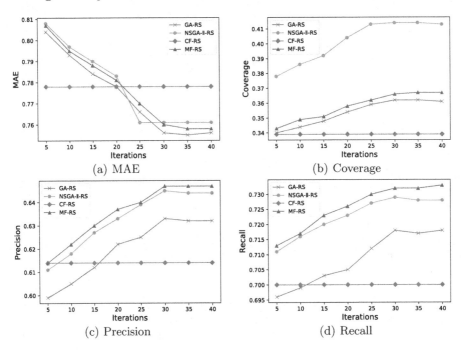

Fig. 2. Comparison results on Netflix under different iteration numbers.

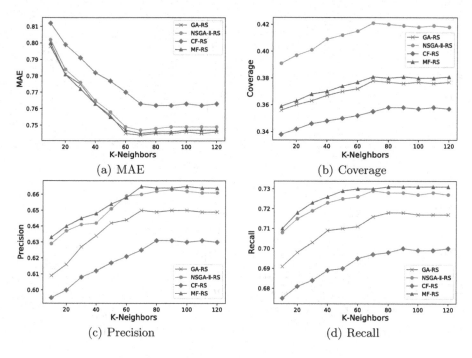

Fig. 3. Comparison results on Movielens under different values of K.

MAE is somewhat higher. The reason is that the GA focuses on improving accuracy, and our algorithm considers the coverage while keeping accuracy of the recommender system. In addition, GA and NSGA-II are superior to CF in the four indicators. MF is slightly better than both GA and NSGA-II in precision and recall, but the coverage of NSGA-II is improved 3.5% compared with the MF. The objective function of MF is to minimize the error between the predicted rating and the practical rating, and the proposed algorithm optimizes MAE and coverage simultaneously. Therefore, the predicted rating is not as accurate as MF to some extent, which shows that the proposed algorithm has played a certain role in mining long tail items in the recommender system. Figure 2 shows roughly the same trend of results as Fig. 1, except that it takes longer for the performance to stabilize, and the performance of the algorithms are worse, which is caused by the fact that Netflix data set is more sparse.

K-Neighbors. To explore the impact of the number of nearest neighbors, we specify the value of K vary from 10 to 120 with an interval of 10 on Movielens data set and specify the value of K vary from 10 to 150 on Netflix data set. We fix the number of iterations to 75 in MF, and set the number of iterations to 30 for the weighted similarity measure based on GA and NSGA-II. The experimental results are shown in Figs. 3 and 4.

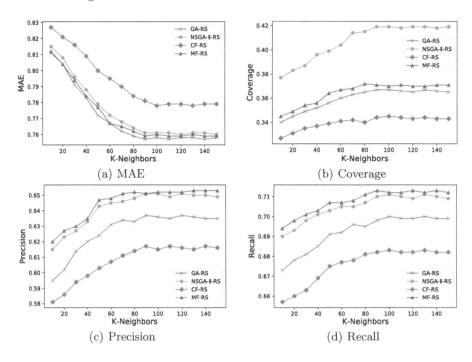

Fig. 4. Comparison results on Netflix under different values of K.

Figures 3 and 4 describe the performance of the four algorithms at different value of K on the two data sets, respectively. Figures 3 and 4 show that the four indicators of all algorithms tend to be stable with the increase of the value of K. Figure 4 shows a similar trend to Fig. 3, but the convergence is slower. The comparison results show that the predicted rating error of our proposed algorithm has a slight increase over the GA, but the precision and recall have been improved, meanwhile the coverage has also been greatly promoted. GA and NSGA-II are still superior to CF in the four indicators, while the performance of MF is slightly better than that of NSGA-II except for coverage. Figures 3 and 4 further illustrates the effectiveness of the proposed method.

5   Conclusion

In this paper, a weighted similarity measure based on NSGA-II is proposed, and achieving optimal weight vector is transformed into a multi-objective optimization problem. The accuracy function and coverage function are chosen as objective functions, and NSGA-II algorithm is used to obtain optimal solution. Compared with the CF algorithm, the MF algorithm and the weighted similarity measure based on GA, the proposed algorithm efficiently improves the coverage of recommender system while the accuracy is kept, and the precision and recall have also been improved to some extent. In the future work, we will choose more indicators as objective functions to further explore the effect on the performance of the recommender system. In practical applications, the objective functions can be chosen dynamically according to the recommendation emphasis.

References

1. Movielens dataset. https://grouplens.org/datasets/movielens/1M/
2. Netflix dataset. https://www.netflixprize.com/
3. Bobadilla, J., Ortega, F., Hernando, A., Alcalá, J.: Improving collaborative filtering recommender system results and performance using genetic algorithms. Knowl.-Based Syst. **24**(8), 1310–1316 (2011)
4. Chin, W.S., Zhuang, Y., Juan, Y.C., Lin, C.J.: A fast parallel stochastic gradient method for matrix factorization in shared memory systems. ACM Trans. Intell. Syst. Technol. **6**(1), 1–24 (2015)
5. Cui, L., Ou, P., Fu, X., Wen, Z., Lu, N.: A novel multi-objective evolutionary algorithm for recommendation systems. J. Parallel Distrib. Comput. **103**(C), 53–63 (2016)
6. Deb, K., Pratap, A., Agarwal, S., Meyarivan, T.: A fast and elitist multiobjective genetic algorithm: NSGA-II. IEEE Trans. Evol. Comput. **6**(2), 182–197 (2002)
7. Ge, M., Delgado-Battenfeld, C., Jannach, D.: Beyond accuracy: evaluating recommender systems by coverage and serendipity. In: ACM Conference on Recommender Systems, pp. 257–260 (2010)
8. Geng, B., Li, L., Jiao, L., Gong, M., Cai, Q., Wu, Y.: NNIA-RS: a multi-objective optimization based recommender system. Phys. A Stat. Mech. Appl. **424**, 383–397 (2015)
9. Gupta, A., Shivhare, H., Sharma, S.: Recommender system using fuzzy c-means clustering and genetic algorithm based weighted similarity measure. In: International Conference on Computer, Communication and Control, pp. 1–8 (2016)
10. Knowles, J.D., Corne, D.W.: Approximating the nondominated front using the Pareto archived evolution strategy. Evol. Comput. **8**(2), 149 (2000)
11. Konak, A., Coit, D.W., Smith, A.E.: Multi-objective optimization using genetic algorithms: a tutorial. Reliab. Eng. Syst. Saf. **91**(9), 992–1007 (2006)
12. Park, Y.J., Tuzhilin, A.: The long tail of recommender systems and how to leverage it. In: ACM Conference on Recommender Systems, Recsys 2008, Lausanne, Switzerland, pp. 11–18, October 2008
13. Ribeiro, M.T., Lacerda, A., Veloso, A., Ziviani, N.: Pareto-efficient hybridization for multi-objective recommender systems. In: ACM Conference on Recommender Systems, pp. 19–26 (2012)
14. Sarwar, B., Karypis, G., Konstan, J., Riedl, J.: Analysis of recommendation algorithms for e-commerce. Proceedings of ACM on E-Commerce (EC-2000) (2000)
15. Srinivas, N., Deb, K.: Muiltiobjective optimization using nondominated sorting in genetic algorithms. Evol. Comput. **2**(3), 221–248 (2014)
16. Wang, S., Gong, M., Li, H., Yang, J.: Multi-objective optimization for long tail recommendation. Knowl.-Based Syst. **104**(C), 145–155 (2016)
17. Yin, H., Cui, B., Li, J., Yao, J., Chen, C.: Challenging the long tail recommendation. Proc. VLDB Endow. **5**(9), 896–907 (2012)
18. Zitzler, E., Thiele, L.: Multiobjective evolutionary algorithms: a comparative case study and the strength pareto approach. IEEE Trans. Evol. Comput. **3**(4), 257–271 (1999)

Event2Vec: Learning Event Representations Using Spatial-Temporal Information for Recommendation

Yan Wang[(✉)] and Jie Tang

Department of Computer Science and Technology,
Tsinghua University, Beijing, China
yan-w16@mails.tsinghua.edu.cn, jietang@tsinghua.edu.cn

Abstract. Event-based social networks (EBSN), such as meetup.com and plancast.com, have witnessed increased popularity and rapid growth in recent years. In EBSN, a user can choose to join any events such as a conference, house party, or drinking event. In this paper, we present a novel model—Event2Vec, which explores how representation learning for events incorporating spatial-temporal information can help event recommendation in EBSN. The spatial-temporal information represents the physical location and the time where and when an event will take place. It typically has been modeled as a bias in conventional recommendation models. However, such an approach ignores the rich semantics associated with the spatial-temporal information. In Event2Vec, the spatial-temporal influences are naturally incorporated into the learning of latent representations for events, so that Event2Vec predicts user's preference on events more accurately. We evaluate the effectiveness of the proposed model on three real datasets; our experiments show that with a proper modeling of the spatial-temporal information, we can significantly improve event recommendation performance.

1 Introduction

Event-based social network (EBSN) is a new type of social network that has experienced increasing popularity and rapid growth. For instance, Meetup[1], one of the largest online social networks for facilitating offline group meetings, has attracted 30 million registered users who have created nearly 270,000 Meetup groups. Douban[2], a Chinese social networking service, has more than 200 million registered users and has hosted about 590,000 offline groups. These EBSN websites allow members to find and join groups unified by a common interest, such as politics, books, games, movies, health, careers or hobbies, and schedule a time to meet up together offline, which results in very interesting user behavior data combining both online and offline social interactions [9]. One challenging

---

[1] https://meetup.com.
[2] https://douban.com.

© Springer Nature Switzerland AG 2019
Q. Yang et al. (Eds.): PAKDD 2019, LNAI 11441, pp. 314–326, 2019.
https://doi.org/10.1007/978-3-030-16142-2_25

issue on these EBSN websites is how to keep users actively joining new events. Recommendation plays a critical role [11].

In contrast to conventional online social networks that mainly contain user's online interactions, users in EBSN can choose to join the event according to their interest in the event (based on the event content) and their availability (based on the event location and availability at the schedule time). Therefore, user's mobile behaviors presented in EBSN are explored typically in several important aspects, including event content, spatial influence [8] and temporal effect [5].

Many recent studies have exploited different factors to improve recommendation effectiveness. For instance, some efforts have been made to explicitly model the spatial information as in [15,20]. Some others exploit temporal cyclic effect to provide spatial or/and temporal novel recommendation like [18]. However, they lack an integrated analysis of the joint effect of all factors in a unified effective way and no previous work has explicitly modeled user's preference on both spatial and temporal factors to improve the recommendation performance.

In this work, we stand on the recent advances in embedding learning techniques and propose an embedding method—Event2Vec to encode events in a low-dimension latent space which integrates the spatial and temporal influence. In specific, we learn representations for three factors—the event, the location and the time simultaneously using the event sequential data attended by users. We propose to use multitask learning settings to model and predict user's preference on three factors naturally. The technique of shared embeddings are utilized in our proposed model to improve the efficiency.

In addition, our approach leverages the interactive influence between spatial and temporal factors presented in user's behaviors by modeling the combination of spatial-temporal information. In specific, events held at the same location could have very different topics at different time periods, thus attract varying groups of user. For instance, an urban park usually holds events like "picnic" in the afternoon while holds events like "jogging" at night. In the course of this paper, we will present how our embedding model exploits such joint and interactive influences of spatial and temporal factors in a natural way.

Finally, we propose a recommendation algorithm based on a similarity metric in the latent embedding space which is proved to be effective in our experiments. Compared with state-of-the-art recommendation frameworks, we can achieve a significant improvement.

2   Problem Definition

In this section, we will first clarify some terminology used in this paper, and then explicitly present our problem.

User behaviors are formulated as a set of four tuple $\{(u, e, l, \tau) : u \in U, e \in E, l \in L, t \in T\}$, where each means user u attended event e at location l, at time slot t. U is a set of users and E is a set of events, L^3 is a set of locations and T is

---

[3] The location l can be represented as a pair (longitude, latitude) or a specific address (e.g., "Wine Bar at MIST").

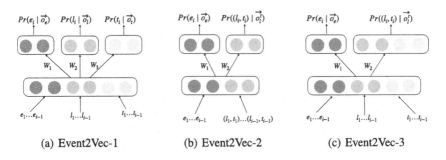

(a) Event2Vec-1   (b) Event2Vec-2   (c) Event2Vec-3

Fig. 1. Architectures of the three Event2Vec models

a set of time slots discretized from continuous timestamps. We use the notation $|\cdot|$ to denote the cardinality of a set — for example, $|L|$ indicates of the number of locations in set L.

For each user u, we create a user profile $D_u = \{(e_i, l_i, t_i), i = 1 \ldots n_u\}$, which is a sequence of events user u attended in chronological order.

Input: The input of our problem is an event-based social network $G = (U, E, L, T)$, and a set of user profiles $D = \{D_u : u \in U\}$.

Goal: Given a querying user u, our goal is to recommend upcoming events based on historical preferences of the user.

3   The Proposed Approach

In this section, we present the details of the proposed model—Event2Vec.

To incorporate different types of information, we learn latent representations for each event, location and time. Then, we model and predict user's preferences on the three factors explicitly to improve the recommendation accuracy.

In specific, the three factors are related to each other: for instance inferring user's preference on the location helps the inference of user's preference on the time. Predicting one helps in predicting the other one, and three factors altogether decides user's tendencies and behaviors. Therefore, we propose to take the perspective of multitask learning settings to naturally leverage the useful information contained in user's preferences on different factors which are related to each other. We set up three single tasks for predicting user's preference on the event, the location and the time respectively. We propose to use shared parameters (i.e., shared embeddings) in all three different tasks to learn latent representations which integrate different points of view. Shared embeddings are also important for the efficiency and generalization of low-dimensional representation learning in our proposed model.

We derive three different model architectures each with different target variables to implement the proposed model.

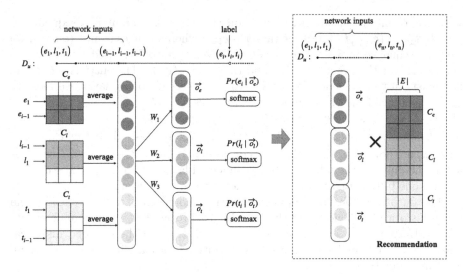

Fig. 2. The joint model with three target variables in which three similar networks are trained per each target variable. At serving, a nearest neighbor lookup is performed to generate a set of event recommendations.

In the first model (Event2Vec-1, Fig. 1(a)), we learn the embeddings for each event, location and time by learning to predict the next event user would attend, and the associated location and time simultaneously.

In the second model (Event2Vec-2, Fig. 1(b)), we learn the embeddings for each event and spatial-temporal pair (i.e., (l, t)) to further capture the interactive influences between spatial and temporal factors.

In the third model (Event2Vec-3, Fig. 1(c)), we propose a compromise between Event2Vec-1 and Event2Vec-2. We reserve distinct embeddings for each location and time but predict the spatial-temporal pair as a combination.

In the remainder of this section, we will describe the three models in more detail.

3.1   Our Models

Event2Vec-1. Event2Vec-1 learns low dimensional embeddings for each event, location and time in a fixed vocabulary and feeds these embeddings into a feed-forward neural network. The purpose of the neural network is to predict user's next behavior including the event to attend, the location and the time to go, using his/her historical behaviors. A user's history is represented by a variable-length sequence of sparse event, location and time IDs which are mapped to dense vector representations via the embeddings. However the network requires fixed-sized dense inputs. We find averaging the embeddings performed best among several strategies (sum, component-wise max, etc.).

More formally, we describe the proposed model starting with a single network of predicting the next event. Given a user profile $D_u = \{(e_j, l_j, t_j), j =$

$1 \ldots n_u\}$, to predict users' preferences on the event, the input is a sequence of $\{(e_1, l_1, t_1) \ldots (e_i, l_i, t_i)\}$ and the target is e_{i+1}, where i ranges from 1 to $n-1$. we feed the sequence into the neural network, which are represented by three one-hot vectors for the event, location and time respectively. The entry is set to one if it exists in the sequence, zero otherwise. In the embedding layer, we look up the embeddings from three embedding matrices, i.e., $C_e \in \mathbb{R}^{|E| \times d_e}$, $C_l \in \mathbb{R}^{|L| \times d_l}$ and $C_t \in \mathbb{R}^{|T| \times d_t}$, where d_e, d_l and d_t are the dimensions of the event, location and the time representations. By averaging, three fixed-sized vectors $\overrightarrow{e_{avg}}$, $\overrightarrow{l_{avg}}$ and $\overrightarrow{t_{avg}}$ are obtained. Then they are concatenated into a flat vector $\overrightarrow{v_{in}}$ which is fed as the input of the following fully-connected layers, with $|\overrightarrow{v_{in}}| = d_e + d_l + d_t$. We use one fully-connected layer parameterized by W_1 in our model.

The output of the fully-connected layer, denoted as $\overrightarrow{o_e} \in \mathbb{R}^{d_e}$, encodes the user's historical behaviors and thus can be used to predict the upcoming events user will attend. Let e_i denotes the target event, given the encoded historical behaviors $\overrightarrow{o_e}$, our model formulates the conditional probability $Pr(e_i \mid \overrightarrow{o_e})$ using a softmax function in Eq. 1.

$$Pr(e_i \mid \overrightarrow{o_e}) = \frac{\exp(\overrightarrow{e_i}^T \cdot \overrightarrow{o_e})}{\sum_{e' \in E} \exp(\overrightarrow{e'}^T \cdot \overrightarrow{o_e})} \tag{1}$$

where $\overrightarrow{e_i}$ and $\overrightarrow{e'}$ are row vectors of C_e. In order to make the model efficient for learning, the techniques of hierarchical softmax and negative sampling are used as proposed in Skip-Gram [6]. Similar to the single network of predicting the event, the other two neural networks with target variables of the location and time are built and output the probabilities of $Pr(l_i \mid \overrightarrow{o_l})$ and $Pr(t_i \mid \overrightarrow{o_t})$. Therefore, the objective of Event2Vec-1 is to minimize three cross entropy losses simultaneously. Figure 1(a) illustrates the architecture of Event2Vec-1 model.

At serving time we need to recommend top k events to the user. Our recommendation algorithm is based on the user-event cosine similarity in the embedding space. Since both spatial and temporal factors play important roles in event recommendation, so we utilize all output vectors of the neural networks to make recommendations.

In specific, we feed all user's historical behaviors into the neural networks and obtain the predicted vectors $\overrightarrow{o_e}$, $\overrightarrow{o_l}$ and $\overrightarrow{o_t}$ by forward propagation. We build user's preference $\overrightarrow{v_u}$ by concatenating them all together, i.e., $\overrightarrow{v_u} = \overrightarrow{o_e} \| \overrightarrow{o_l} \| \overrightarrow{o_t}$, where $\|$ is the concatenation operation. For each candidate event e_i associated with location l_i and time t_i, we get its final representation as $\overrightarrow{v_{e_i}} = \overrightarrow{e_i} \| \overrightarrow{l_i} \| \overrightarrow{t_i}$, where the embeddings are looked up in the embedding matrices—C_e, C_l and C_t.

Given a user u, for each event e_i which has not been attended by u, we compute its ranking score using Eq. 2, and select top k events with highest scores to recommend to the user.

$$S(u, e_i) = \overrightarrow{v_u}^T \cdot \overrightarrow{v_{e_i}} \tag{2}$$

Figure 2 demonstrates our proposed joint model with three target variables in which three similar networks are trained per each target variable. The trainable parameters include three embedding matrices, C_e, C_l and C_t, and the weight matrices of the fully connected layer, W_1, W_2 and W_3. Please note that parameters of the embedding matrices are shared and trainable in all three neural networks, while parameters of weight matrices are only updated through the associated neural network.

Event2Vec-2. The combination of the location and the time contain richer semantic information, however Event2Vec-1 doesn't consider such interactive influence between the spatial and temporal factors. A location usually holds different semantics at different time, and these semantics should have discriminative vectors. Therefore, Event2Vec-2 learns embeddings for each spatial-temporal pair. The spatial-temporal embedding matrice is denoted as $C_l^t \in \mathbb{R}^{|L \times T| \times d_l^t}$, where d_l^t means the dimension of spatial-temporal representation and $L \times T$ means the Cartesian product of L and T.

The architecture of Event2Vec-2 is illustrated in Fig. 1(b). There are two neural networks predicting the next event and the next spatial-temporal pair respectively. When making recommendations, the user preference is represented as $\overrightarrow{v_u} = \overrightarrow{o_e} \| \overrightarrow{o_l^t}$; and the candidate event e_j is represented as $\overrightarrow{v_{e_i}} = \overrightarrow{e_i} \| \overrightarrow{l_i^t}$, where $\overrightarrow{e_i}$ and $\overrightarrow{l_i^t}$ are row vectors of C_e and C_l^t.

Event2Vec-3. Since Event2Vec-2 divides the occurrences of each location into multiple time slots, the learning of embeddings suffer from the sparsity issue. In an attempt to alleviate the problem, we propose a new model—Event2Vec-3 to provide a trade-off between the discrimination and sparsity.

Event2Vec-3 reserves distinct embeddings for each event, location and time. However slightly different from Event2Vec-1, the location and the time are predicted as a combination. Each spatial-temporal pair (l, t) is represented by concatenating their distinct vectors $\overrightarrow{l}$ and $\overrightarrow{t}$ into a flat vector $(\overrightarrow{l} \| \overrightarrow{t}) \in \mathbb{R}^{d_l + d_t}$. The corresponding output $\overrightarrow{o_l^t}$ has the same length of $d_l + d_t$. The outputs of two neural networks are $Pr(e_i \mid \overrightarrow{o_e})$ and $Pr((l_i, t_i) \mid \overrightarrow{o_l^t})$ as shown in Fig. 1(c), where $Pr((l_i, t_i) \mid \overrightarrow{o_l^t})$ is calculated as in Eq. 3

$$Pr((l_i, t_i) \mid \overrightarrow{o_l^t}) = \frac{\exp((\overrightarrow{l_i} \| \overrightarrow{t_i})^T \cdot \overrightarrow{o_l^t})}{\sum_{(l', t') \in L \times T} \exp((\overrightarrow{l'} \| \overrightarrow{t'})^T \cdot \overrightarrow{o_l^t})} \tag{3}$$

It's worthy of noting that the embeddings of each location and time are shared among all spatial-temporal pairs (l, t).

4   Experiments

In this section, we evaluate the proposed model for the task of event recommendations. We first examine the performance of Event2Vec models compared with related models in Sect. 4.2. Then we examine the importance of spatial-temporal factors in Sect. 4.3; and finally different temporal patterns are compared and discussed in Sect. 4.4.

4.1 Experimental Setup

Datasets. We use three datasets in real-world domains, two from Douban and one from Meetup, for our experiments.

- **Meetup.** We collected the first dataset *Meetup* by crawling real events hosted in New York from meetup.com in 2016. For each event, we retrieved its geographic location, start time, and a list of users who attended. To reduce noise, we selected events that are attended by at least 20 users, and users who have attended at least 20 events. In the end, the *Meetup* dataset contains 4722 users and 5064 events.
- **Douban** [19]. We collected two datasets *Douban-bej* and *Douban-sha* by crawling events hosted in 2012 from douban.com located at Beijing and Shanghai respectively. For each event, we also retrieved its geographic location, start time, and a list of registered users who attended. Then we removed users who attended fewer than 20 events, and events attended by fewer than 20 users. We have 222795 attendances by 6513 users, 5326 events in the *Douban-bej* dataset; 6964 users, 4189 events and 241093 attendances in the *Douban-sha* dataset.

Data Preprocessing. To normalize the locations of events, we split the city into even grid cells according to coordinates, and each resultant location (gird) spans 0.13 km. The numbers of locations in the *Meetup*, *Douban-bej* and *Douban-sha* dataset are 1569, 813 and 626 respectively.

To capture the temporal characteristics in user's behaviors, we design a time discretizing scheme to smoothly map a continuous timestamp to a time slot. The preference variance exists in three time scales generally: hours of a day, different days in a week (or a month), and different months in a year, which is observed in [5] but not modeled. By experiments, we propose to divide the continuous time space into time slots using a weekday-hour pattern, such as "4 (day of the week), 1:00–2:00 (hour of the day)". Therefore, we can get at most $7 * 24$ discretized time slots on all three datasets. Other temporal patterns are compared and discussed in Sect. 4.4.

Comparison Methods. We compare our model with the following methods representing the state-of-the-art event-based recommendation techniques.

- **SVDFeature.** SVDFeature [3] is a machine learning toolkit designed to solve the feature-based matrix factorization. To compare with our model fairly, we implement it by incorporating more side information including the location and the time.

- **IRenMF.** IRenMF [10] is based on Weighted Matrix Factorization (WMF). IRenMF considers the influence of neighboring locations while modeling user's preferences.
- **Rank-GeoFM.** Rank-GeoFM [7] is a ranking based factorization method, which includes spatial influence in a latent model.
- **Event2Vec.** Our proposed methods for event recommendation, which incorporate spatial-temporal information using the embedding learning methods.

In summary, SVDFeature models the spatial-temporal information as simple bias, while both IRenMF and Rank-GeoFM model geographic influences as latent vectors using Matrix Factorization techniques.

For each individual user in the dataset, we sort his behaviors in time order and then mark off the last 10% events he attended for testing, while use the previous 90% historical events for training. In the experiments, we use a validation set to find the optimal hyper-parameters, and finally set d_e, d_l and d_t to 200, (we use the same dimension for simplicity, but they are not necessarily equal in practice). For implementation, we develope the model based on Tensorflow [1]. We use stochastic gradient descent (SGD) for optimization, and gradients are calculated using the back-propagation algorithm. We run each recommendation method for 5 times and report the average performances in Table 1.

Evaluation Metrics. We compare the performances through precision, recall, and f1-score as they are generally used in recommendation systems. We denote these metrics at top-k recommendation as p@k, r@k, f1@k respectively. Formally, if we define E_u^R as recommended events sorted by score in descending order and E_u^T as the true events attended by user u,

$$\text{p@}k = \frac{1}{|U|} \sum_{u \in U} \frac{|E_u^T \cap E_u^R[:k]|}{k}$$

$$\text{r@}k = \frac{1}{|U|} \sum_{u \in U} \frac{|E_u^T \cap E_u^R[:k]|}{|E_u^T|} \qquad (4)$$

$$\text{f1@}k = \frac{2 \cdot \text{p@}k \cdot \text{r@}k}{\text{p@}k + \text{r@}k}$$

4.2 Results

Table 1 shows the experimental results. We find Event2Vec models outperform other baselines significantly on all metrics, among which Event2Vec-2 achieves the best performance. The standard deviation of the performance from each method is less than 4×10^{-4}, confirming the reliability of our comparison results.

Baselines vs. Our Models. Several observations are made by comparing baselines and our models from the results. (1) Rank-GeoFM and IRenMF achieve a higher recommendation accuracy than SVDFeature on all metrics of performance, showing the benefits brought by factorizing the spatial-temporal influences into latent vectors instead of scalar bias used by SVDFeature. (2)

Table 1. Performance comparison

Dataset	Meetup						
Metric	p@1	p@5	p@10	r@1	r@5	r@10	f1@10
SVDFeature	0.0085	0.0131	0.013	0.0023	0.0188	0.0371	0.0192
IRenMF	0.0209	0.0234	0.0243	0.006	0.0335	0.0698	0.0360
Rank-GeoFM	0.0209	0.0278	0.0273	0.0058	0.0387	0.0763	0.0403
Event2Vec-1	0.1778	0.1237	0.0922	0.0463	0.1581	0.2312	0.1318
Event2Vec-2	**0.2006**	**0.1350**	**0.1014**	**0.0514**	**0.1715**	**0.2522**	**0.1447**
Event2Vec-3	0.1561	0.1099	0.0829	0.0403	0.1401	0.2097	0.1188
Dataset	Douban-bej						
Metric	p@1	p@5	p@10	r@1	r@5	r@10	f1@10
SVDFeature	0.0382	0.0296	0.026	0.0073	0.0267	0.0468	0.0334
IRenMF	0.0323	0.0311	0.0297	0.0069	0.0287	0.0502	0.0373
Rank-GeoMF	0.0344	0.0353	0.0326	0.007	0.0318	0.0543	0.0407
Event2Vec-1	0.244	0.1658	0.1275	0.0409	0.1284	0.1866	0.1515
Event2Vec-2	**0.2572**	**0.1748**	**0.1312**	**0.0451**	**0.1431**	**0.2055**	**0.1602**
Event2Vec-3	0.1154	0.0772	0.0571	0.0226	0.0726	0.1031	0.0735
Dataset	Douban-sha						
Metric	p@1	p@5	p@10	r@1	r@5	r@10	f1@10
SVDFeature	0.0456	0.0328	0.0269	0.0183	0.0631	0.1009	0.0425
IRenMF	0.0656	0.0533	0.0436	0.0284	0.1031	0.1568	0.0683
Rank-GeoFM	0.0692	0.0567	0.0452	0.0297	0.1063	0.1596	0.0704
Event2Vec-1	0.1721	0.0988	0.0718	0.054	0.1342	0.1825	0.1031
Event2Vec-2	**0.2245**	**0.1215**	**0.0884**	**0.0763**	**0.1825**	**0.2516**	**0.1308**
Event2Vec-3	0.1124	0.0653	0.0459	0.0419	0.1108	0.1479	0.07

Event2Vec models outperform other competitor methods by 4%–9% in terms of p@10 on three datasets. It shows the advantages of the proposed multitask learning framework and shared embeddings in modeling different related factors. Moreover, the proposed Event2Vec models explicitly predict user's preferences on three factors using the historical data. Therefore, we can see a significant improvement over other baseline methods in Table 1.

Event2Vecs. The performance of three Event2Vec models are very different and reflect their characteristics. (1) Event2Vec-2 achieves the best performance. Event2Vec-2 outperforms Event2Vec-1 by 0.9%–2.7% in terms of f1-score. The most possible reason is, Event2Vec-2 discriminates different location-time combinations and learn distinct representations for each of them to capture more accurate semantics. For example, the representation of "cafe-morning" learned by Event2Vec-2 could encode concrete and discriminative semantics probably like "breakfast", while in Event2Vec-1 it's represented by concatenating the vectors

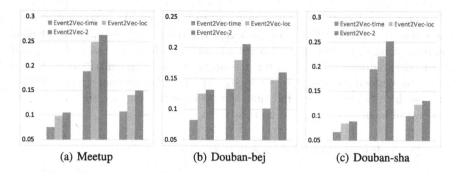

(a) Meetup   (b) Douban-bej   (c) Douban-sha

Fig. 3. The effect of different factors

Table 2. Comparison of temporal patterns

Dataset	Meetup			Douban-bej			Douban-sha		
Metric	p@10	r@10	f1@10	p@10	r@10	f1@10	p@10	r@10	f1@10
Weekday-hour	0.1045	0.2627	0.1495	0.1312	0.2055	0.1602	0.0884	0.2516	0.1308
Day-hour	0.0929	0.2355	0.1332	0.1037	0.1649	0.1274	0.0856	0.2575	0.1285
Month-weekday-hour	0.0893	0.2258	0.128	0.1208	0.19	0.1477	0.0901	0.2641	0.1343
Month-day-hour	0.0881	0.2239	0.1264	0.1129	0.1812	0.1391	0.086	0.2593	0.1292

of "cafe" and "morning" which may introduce the noises. From the results, we can conclude that in Event2Vec-2, the effectiveness of modeling interactive influence between the spatial and temporal factors is more significant than the issue caused by sparsity, thus Event2Vec-2 achieves the best performance. (2) The performance of Event2Vec-3 drops behind the other two Event2Vec methods, this is probably because during the back propagation, the updates on embeddings of the location and the time will influence each other, for that the boundary of the embeddings are blurred because of concatenating operation. Therefore it makes the representation learning of the location and the time less distinguishable and results in a worse performance than other Event2Vec models.

4.3   Impact of Different Factors

To explore the benefits of incorporating spatial and temporal influences into Event2Vec models respectively, we compare our Event2Vec model with two variants—Event2Vec-loc and Event2Vec-time. All three original Event2Vec models will reduce to the same architecture when only including one factor of the location or the time.

Event2Vec-time is the first simplified version where we ignore the spatial information in Event2Vec models.

Event2Vec-loc ignores the temporal information in Event2Vec models.

Event2Vec-2 is our best model by learning embeddings for spatial-temporal pairs.

We show the results on three datasets in Fig. 3. From Fig. 3, we first observe that Event2Vec-2 consistently outperforms the other two variants on all metrics, indicating that Event2Vec-2 takes advantage of both spatial and temporal influences simultaneously. Moreover, it's observed that the contributions of two factors to performance improvement are different. By comparing Event2Vec-time and Event2Vec-loc, we find that spatial influence is more significant than temporal influence for event recommendation.

4.4 Exploring Various Temporal Patterns

Our model recommends events to a user by taking advantage of the temporal influence. So far, we have evaluated its recommendation performance using a weekday-hour pattern, while its recommendation ability is not limited to one specific temporal pattern. By taking different definitions of temporal state, some other temporal patterns can be used for event recommendation with our model. For example, apart from the weekly pattern, we could also define the temporal state as daily pattern (day of the month); monthly pattern (month of the year); and their combinations. The only change made to our model is to divide time slots using different strategies. Table 2 shows the recommendation results of our model using different temporal patterns. The results show that the weekday-hour pattern achieves the best overall performance. By comparing the weekday-hour pattern and day-hour pattern, we observe that day of the week is more informative than day of the month, which indicates human behaviors exhibit stronger temporal cyclic patterns in a week than in a month (like working purpose on weekdays and entertainment purpose at weekends). However, the month-weekday-hour pattern and the month-day-hour pattern perform slightly worse than the weekday-hour pattern and the day-hour pattern on Meetup and Douban-bej dataset. Possible reasons could be that user's behaviors don't have strong patterns at month level and that adding monthly pattern additionally causes the sparsity issue to representation learning in time space.

5 Related Work

Event-based social networks (EBSN) have attracted much attention from research community. A great deal of research has been conducted on EBSNs. For example, Brown et al. [2] suggested that geographical closeness could influence the formation of online communities. Liu et al. [9] observed that 81.93% of event participations by a user are within 10 miles of his/her home location. Pham et al. [13] presented a graph-based model for event recommendation and Cheng et al. [4] developed a particular location recommendation method based on user preferences. Zhang et al. [20] used the location-based features for group recommendations in EBSN. Qiao et al. [15] proposes an approach to combine the heterogeneous social relationships, geographical features of events and implicit rating data from users to recommend events to users. However, most of these methods simply consider the spatial information as a bias factor and ignore the location-related semantic information.

From an algorithmic perspective, embedding techniques has been applied in a quantity of works such as network embedding [12], user profiling [16], social media prediction tasks [17], E-commerce product recommendation [14], and many other works. The embedding methods based on representing entries in low dimensional vector space, while preserving their properties, have been proved useful in multiple machine learning tasks such as classification, prediction and so on. However, no previous works have employed the representation learning methods in EBSN scenario where spatial and temporal factors have significant influences on user's behaviors.

6   Conclusion

In this paper, we study the recommendation problem in event-based social networks (EBSN). We proposed Event2Vec, a new embedding method that incorporates the spatial-temporal information jointly. We embed the event, location and time into low dimensional space based on event sequential data by taking advantages of the multitask learning and parameter sharing techniques. Different variants of Event2Vec are exploited to leverage the interactive influence between the spatial and temporal information.

We conducted extensive experiments to evaluate the performance of Event2Vec model on real-world datasets. The results showed superiority of our proposed model over other competitor methods. Moreover, we analyzed the effectiveness of spatial-temporal influences and compared different temporal patterns in user's behaviors in experiments.

References

1. Abadi, M., et al.: Tensorflow: a system for large-scale machine learning. In: OSDI 2016 (2016)
2. Brown, C., Nicosia, V., Scellato, S., Noulas, A., Mascolo, C.: Where online friends meet: social communities in location-based networks. In: ICWSM. Citeseer (2012)
3. Chen, T., Zhang, W., Lu, Q., Chen, K., Zheng, Z., Yu, Y.: SVDFeature: a toolkit for feature-based collaborative filtering. J. Mach. Learn. Res. **13**(Dec), 3619–3622 (2012)
4. Cheng, C., Yang, H., Lyu, M.R., King, I.: Where you like to go next: successive point-of-interest recommendation. In: IJCAI 2013, pp. 2605–2611 (2013)
5. Gao, H., Tang, J., Hu, X., Liu, H.: Exploring temporal effects for location recommendation on location-based social networks. In: Proceedings of the 7th ACM Conference on Recommender Systems, pp. 93–100. ACM (2013)
6. Goldberg, Y., Levy, O.: word2vec explained: deriving Mikolov et al'.s negative-sampling word-embedding method. CoRR abs/1402.3722 (2014). http://arxiv.org/abs/1402.3722
7. Li, X., Cong, G., Li, X.L., Pham, T.A.N., Krishnaswamy, S.: Rank-GeoFM: a ranking based geographical factorization method for point of interest recommendation. In: Proceedings of the 38th International ACM SIGIR Conference on Research and Development in Information Retrieval, pp. 433–442. ACM (2015)

8. Lian, D., Zhao, C., Xie, X., Sun, G., Chen, E., Rui, Y.: GeoMF: joint geographical modeling and matrix factorization for point-of-interest recommendation. In: Proceedings of the 20th ACM SIGKDD International Conference on Knowledge Discovery and Data Mining, pp. 831–840. ACM (2014)

9. Liu, X., He, Q., Tian, Y., Lee, W.C., McPherson, J., Han, J.: Event-based social networks: linking the online and offline social worlds. In: KDD 2012, pp. 1032–1040 (2012)

10. Liu, Y., Wei, W., Sun, A., Miao, C.: Exploiting geographical neighborhood characteristics for location recommendation. In: Proceedings of the 23rd ACM International Conference on Conference on Information and Knowledge Management, pp. 739–748. ACM (2014)

11. Nielsen, L.: Personas-User Focused Design, vol. 15. Springer, Heidelberg (2012)

12. Perozzi, B., Al-Rfou, R., Skiena, S.: Deepwalk: online learning of social representations. In: Proceedings of the 20th ACM SIGKDD International Conference on Knowledge Discovery and Data Mining, pp. 701–710. ACM (2014)

13. Pham, T.A.N., Li, X., Cong, G., Zhang, Z.: A general graph-based model for recommendation in event-based social networks. In: ICDE 2015, pp. 567–578 (2015)

14. Phi, V.T., Chen, L., Hirate, Y.: Distributed representation based recommender systems in e-commerce. In: DEIM Forum (2016)

15. Qiao, Z., Zhang, P., Cao, Y., Zhou, C., Guo, L., Fang, B.: Combining heterogenous social and geographical information for event recommendation. In: AAAI 2014, pp. 145–151 (2014)

16. Tang, D., Qin, B., Liu, T.: Learning semantic representations of users and products for document level sentiment classification. In: ACL, vol. 1, pp. 1014–1023 (2015)

17. Wijeratne, S., Balasuriya, L., Doran, D., Sheth, A.: Word embeddings to enhance twitter gang member profile identification. arXiv preprint arXiv:1610.08597 (2016)

18. Yuan, Q., Cong, G., Ma, Z., Sun, A., Thalmann, N.M.: Time-aware point-of-interest recommendation. In: Proceedings of the 36th International ACM SIGIR Conference on Research and Development in Information Retrieval, pp. 363–372. ACM (2013)

19. Zhang, W., Wang, J.: A collective bayesian poisson factorization model for cold-start local event recommendation. In: Proceedings of the 21th ACM SIGKDD International Conference on Knowledge Discovery and Data Mining, pp. 1455–1464. ACM (2015)

20. Zhang, W., Wang, J., Feng, W.: Combining latent factor model with location features for event-based group recommendation. In: KDD 2013, pp. 910–918 (2013)

Maximizing Gain over Flexible Attributes in Peer to Peer Marketplaces

Abolfazl Asudeh[1](✉) ⓘ, Azade Nazi[2], Nick Koudas[3], and Gautam Das[4]

[1] University of Michigan, Ann Arbor, USA
asudeh@umich.edu
[2] Google AI, Mountain View, USA
azade@google.com
[3] University of Toronto, Toronto, Canada
koudas@cs.toronto.edu
[4] University of Texas at Arlington, Arlington, USA
gdas@uta.edu

Abstract. Peer to peer marketplaces enable transactional exchange of services directly between people. In such platforms, those providing a service are faced with various choices. For example in travel peer to peer marketplaces, although some amenities (attributes) in a property are fixed, others are relatively *flexible* and can be provided without significant effort. Providing an attribute is usually associated with a cost. Naturally, different sets of attributes may have a different "gains" for a service provider. Consequently, given a limited budget, deciding which attributes to offer is challenging.

In this paper, we formally introduce and define the problem of *Gain Maximization over Flexible Attributes* (GMFA) and study its complexity. We provide a practically efficient exact algorithm to the GMFA problem that can handle any monotonic gain function. Since the users of the peer to peer marketplaces may not have access to any extra information other than existing tuples in the database, as the next part of our contribution, we introduce the notion of *frequent-item based count* (FBC), which utilizes nothing but the database itself. We conduct a comprehensive experimental evaluation on real data from AirBnB and a case study that confirm the efficiency and practicality of our proposal.

1 Introduction

Peer to peer marketplaces enable both "obtaining" and "providing" in a temporary or permanent fashion valuable services through direct interaction between people [6]. Travel peer to peer marketplaces such as *AirBnB, HouseTrip, HomeAway*, and *Vayable*[1], work and service peer to peer marketplaces such as *UpWork*,

A. Asudeh and A. Nazi—Work done at the University of Texas at Arlington.
This work was supported in part by NSF grant No. 1745925, grant W911NF-15-1-0020 from the Army Research Office, and a grant from AT&T.

---

[1] airbnb.com; housetrip.com; homeaway.com; vayable.com.

© Springer Nature Switzerland AG 2019
Q. Yang et al. (Eds.): PAKDD 2019, LNAI 11441, pp. 327–345, 2019.
https://doi.org/10.1007/978-3-030-16142-2_26

FreeLancer, *PivotDesk*, *ShareDesk*, and *Breather*[2], car sharing marketplaces such as *BlaBlaCar*, education peer to peer marketplaces such as *PopExpert*, and pet peer to peer marketplaces such as *DogVacay* are a few examples[3] of such marketplaces. In travel peer to peer marketplaces, for example, the service caters to accommodation rental; hosts are those providing the service (*service providers*), and guests, who are looking for temporary rentals, are receiving service (*service receivers*). Hosts list properties, along with a set of amenities for each, while guests utilize the search interface to identify suitable properties to rent. Figure 1 presents a sample set of rental accommodations. Each row corresponds to a property and each column represents an amenity. For instance, the first property offers `Breakfast`, `TV`, and `Internet` as amenities but does not offer `Washer`.

Although sizeable effort has been devoted to design user-friendly search tools assisting service receivers in the search process, little effort has been recorded to date to build tools to assist service providers. Consider for example a host in a travel peer to peer marketplace; while listing a property in the service for (temporary) rent, the host is faced with various choices. Although some amenities in the property are relatively *fixed*, such as number of rooms for rent, or existence of an elevator, others are relatively *flexible*; for example offering `Breakfast` or `TV` as an amenity. Flexible amenities can be added without a significant effort. Although amenities make sense in the context of travel peer to peer marketplaces (as part of the standard terminology used in the service), for a general peer to peer marketplace we use the term *attribute* and refer to the subsequent choice of attributes as *flexible attributes*.

Service providers participate in the service with specified objectives; for instance hosts may want to increase overall occupancy and/or optimize their anticipated revenue. Since there is a *cost* (e.g., monetary base cost to the host to offer internet) associated with each flexible attribute, it is challenging for service providers to choose the set of flexible attributes to offer given some budget limitations (constraints). An informed choice of attributes to offer should maximize the objectives of the service provider in each case subject to any constraints. Objectives may vary by application; for example an objective could be maximize the number of times a listing appears on search results, the position in the search result ranking or other. This necessitates the existence of functions that relate flexible attributes to such objectives in order to aid the service provider's decision. We refer to the service provider's objectives in a generic sense as *gain* and to the functions that relate attributes to gain as *gain functions* in what follows. In general, the gain function design may vary upon on the availability of information. However, in a peer to peer setting, a service provider may not have access to extra information other than the enlisted services.

In this paper, we aim to assist service providers in peer to peer marketplaces by suggesting those flexible attributes which maximize their gain. Given a service with known flexible attributes and budget limitation, our objective is to identify a set of attributes to suggest to service providers in order to maximize the gain. We

---

[2] upwork.com; freelancer.com; pivotdesk.com; sharedesk.net; breather.com.

[3] blablacar.com; popexpert.com; dogvacay.com.

ID	Breakfast	TV	Internet	Washer	ID	Breakfast	TV	Internet	Washer
Accom. 1	1	1	1	0	Accom. 6	1	0	1	0
Accom. 2	1	1	1	1	Accom. 7	1	0	0	0
Accom. 3	0	1	1	0	Accom. 8	1	1	0	1
Accom. 4	1	1	1	0	Accom. 9	0	1	1	1
Accom. 5	0	1	1	1	Accom. 10	1	0	0	1

Fig. 1. A sample set of rental accomodations

refer to this problem as *Gain Maximization over Flexible Attributes* (**GMFA**). Since the target applications involve mainly ordinal attributes, in this paper, we focus our attention on ordinal attributes and we assume that numeric attributes (if any) are suitably discretized. Without loss of generality, we first design our algorithms for binary attributes, and, due to the space limitations, provide the extension to ordinal attributes in the technical report [15].

Our contribution in this paper is twofold. First, we formally define the general problem of *Gain Maximization over Flexible Attributes* (**GMFA**) in peer to peer marketplaces and propose a general solution which is applicable to a general class of gain functions. Second, without making any assumption on the existence extra information other than the dataset itself, we introduce the notion of *frequent-item based count* as a simple yet compelling gain function in the absence of other sources of information. First, we prove that the general GMFA is NP-hard and there is no approximation algorithm with a fixed ratio for it unless P = NP. We provide a (practically) efficient exact algorithm to the GMFA problem for a general class of monotonic gain functions[4]. This generic proposal is due to the fact that gain function design is application specific and depends on the available information. Thus, instead of limiting the solution to a specific application, the proposed algorithm gives the freedom to easily apply any arbitrary gain function into it. In other words, it works for *any arbitrary monotonic gain function* no matter how and based on what data it is designed. In a rational setting in which attributes on offer add value, we expect that all gain functions will be monotonic.

The next part of our contribution focuses on the gain function design. It is evident that gain functions could vary depending on the underlying objectives. The gain function design, as discussed in technical report [15], is application specific and may vary upon on the availability of information such as query logs, reviews, or a weighting of attributes based on some criteria (e.g., importance) that can be naturally incorporated in our framework without changes to the algorithm. However, in a peer to peer setting, the users may usually have no extra information other than enlisted services. Thus, rather than assuming the existence of any specific extra information, we, alternatively, introduce the notion of *frequent-item based count (FBC)* that utilizes nothing but the existing tuples in the database to define the notion of gain for the absence of extra information. The motivation behind the definition of FBC is that (rational) service providers provide attributes based on demand. For example, in Fig. 1 the existence of TV

---

[4] Monotonicity of the gain function simply means that adding a new attribute does not reduce the gain.

and `Internet` together in more than half of the rows, indicates the demand for this combination of amenities. Also, as shown in the real case study provided in Sect. 5.3, popularity of `Breakfast` in the rentals located in Paris indicates the demand for this amenity there. Since counting the number of frequent itemsets is #P-complete [10], computing the FBC is challenging. In contrast with a simple algorithm that is an adaptation of Apriori [1] algorithm, we propose a practical output-sensitive algorithm for computing FBC that runs in the time linear in its output value. The algorithm uses an innovative approach that avoids iterating over the frequent attribute combinations by partitioning them into disjoint sets. In summary, we make the following contributions in this paper:

- We introduce the notion of flexible attributes and the novel problem of gain maximization over flexible attributes (GMFA) in peer to peer marketplaces.
- Studying the complexity of the problem, we propose an efficient algorithm that any arbitrary monotonic gain function can simply get plugged into it.
- While not promoting any specific gain function, we propose *frequent-item based count (FBC)* as a simple yet compelling gain function in the absence of extra information other than the dataset itself. We propose a practically efficient algorithm for assessing the gain.
- We conduct a comprehensive performance study on real dataset from AirBnB to evaluate the proposed algorithms. Also, in a real case study, we to illustrate the practicality of the approaches.

2 Preliminaries

Dataset Model: We model the entities under consideration in a peer to peer marketplace as a dataset $\mathcal{D}$ with n tuples and m attributes $\mathcal{A} = \{A_1, \ldots, A_m\}$. For a tuple $t \in \mathcal{D}$, we use $t[A_i]$ to denote the value of the attribute A_i in t. Figure 1 presents a sample set of rental accommodations with 10 tuples and 4 attributes. Each row corresponds to a tuple (property) and each column represents an attribute. For example, the first property offers `Breakfast`, `TV`, and `Internet` as amenities but does not offer `Washer`. Note that, since the target applications involve mainly ordinal attributes, we focus our attention on such attributes and we assume that numeric attributes (if any) are suitably discretized. Without loss of generality, throughout the paper, we consider the attributes to be binary and defer the extension of algorithms to ordinal attributes in the technical report [15]. We use $\mathcal{A}_t$ to refer to the set of attributes for which $t[A_i]$ is non zero; i.e. $\mathcal{A}_t = \{A_i \in \mathcal{A} \mid t[A_i] \neq 0\}$, and the size of $\mathcal{A}_t$ is k_t.

Query Model: Given the dataset $\mathcal{D}$ and set of binary attributes $\mathcal{A}' \subseteq \mathcal{A}$, the query $Q(\mathcal{A}', \mathcal{D})$ returns the set of tuples in $\mathcal{D}$ where contain $\mathcal{A}'$ as their attributes; formally:

$$Q(\mathcal{A}', \mathcal{D}) = \{t \in \mathcal{D} \mid \mathcal{A}' \subseteq \mathcal{A}_t\} \tag{1}$$

Table 1. Table of notations

Notation	Meaning	Notation	Meaning
$\mathcal{D}$	The dataset	$\mathcal{A}$	The set of the attributes in database $\mathcal{D}$
m	The size of $\mathcal{A}$	n	The number of tuples in database $\mathcal{D}$
$t[A_i]$	The value of attribute A_i in tuple t	$\mathcal{A}_t$	Set of non-zero attributes in tuple t
B	The budget	$cost[A_i]$	Cost to change the (binary) A_i to 1
$gain(.)$	The gain function	$\mathcal{L}_{\mathcal{A}_i}$	Lattice of attribute combinations $\mathcal{A}_i$
$V(\mathcal{L}_{\mathcal{A}_i})$	The set of nodes in $\mathcal{L}_{\mathcal{A}_i}$	$\mathcal{B}(v_i)$	The bit representative of the node v_i
$v(\mathcal{A}_i)$	The node with attribute combination $\mathcal{A}_i$	$v(\mathcal{B}_i)$	The node with the bit representative $\mathcal{B}_i$
$\ell(v_i)$	The level of the node v_i	$cost(v_i)$	The cost associated with the node v_i
$\rho(\mathcal{B}(v_i))$	The index of the right-most zero in $\mathcal{B}(v_i)$	$parent_T(v_i)$	Parent of v_i in the tree data structure

Similarly, the query model for the ordinal attributes is as following: given the dataset $\mathcal{D}$, the set of ordinal attributes $\mathcal{A}' \subseteq \mathcal{A}$, and values $\mathcal{V}$ where $V_i \in \mathcal{V}$ is a value in the domain of $A_i \in \mathcal{A}'$, $Q(\mathcal{A}', \mathcal{V}, \mathcal{D})$ returns the tuples in $\mathcal{D}$ that for attribute $A_i \in \mathcal{A}'$, $V_i \le t[A_i]$.

Flexible Attribute Model: In this paper, we assume an underlying $cost$[5] associated with each attribute A_i, i.e., a flexible attribute A_i can be added to a tuple t by incurring $cost[A_i]$. For example, the costs of providing attributes Breakfast, TV, Internet, and Washer, in Fig. 1, on an annual basis, are $cost = [1000, 300, 250, 700]$. For the ordinal attributes, $cost(A_i, V_1, V_2)$ represents the cost of changing the value of A_i from V_1 to V_2. Our approach places no restrictions on the number of flexible attributes in $\mathcal{A}$. For the ease of explanation, in the rest of paper we assume all the attributes in $\mathcal{A}$ are flexible.

We also assume the existence of a gain function $gain(.)$, that given the dataset $\mathcal{D}$, for a given attribute combination $\mathcal{A}_i \subseteq \mathcal{A}$ provides a score showing how desirable $\mathcal{A}_i$ is. For example in a travel peer to peer marketplace, given a set of m amenities, such a function could quantify the anticipated gain (e.g., visibility) for a host if a subset of these amenities are provided by the host on a certain property.

Table 1 presents a summary of the notation used in this paper. Next, we formally define the general *Gain Maximization over Flexible Attributes* (**GMFA**) in peer to peer marketplaces.

2.1   General Problem Definition

We define the general problem of *Gain Maximization over Flexible Attributes* (**GMFA**) in peer to peer marketplaces as a constrained optimization problem. The general problem is agnostic to the choice of the gain function. Given $gain(.)$, a service provider with a certain budget B strives to maximize $gain(.)$ by considering the addition of flexible attributes to the service. For example, in a travel

---
[5] Depending on the application it may represent a monetary value.

peer to peer marketplace, a host who owns an accommodation ($t \in \mathcal{D}$) and has a limited (monetary) budget B aims to determine which amenities should be offered in the property such that the costs to offer the amenities to the host are within the budget B, and the gain ($gain(.)$[6]) resulting from offering the amenities is maximized. Formally, our GMFA problem is defined as following:

Gain Maximization over Flexible Attributes (GMFA):

Given a dataset $\mathcal{D}$ with the set of binary flexible attributes $\mathcal{A}$ where each attribute $A_i \in \mathcal{A}$ is associated with cost $cost[A_i]$, a gain function $gain(.)$, a budget B, and a tuple $t \in \mathcal{D}$, identify a set of attributes $\mathcal{A}' \subseteq \mathcal{A} \backslash \mathcal{A}_t$, such that $\sum_{\forall A_i \in \mathcal{A}'} cost[A_i] \leq B$ while maximizing $gain(\mathcal{A}_t \cup \mathcal{A}', \mathcal{D})$.

2.2   Computational Complexity

In Theorem 1, we prove that GMFA is NP-hard[7] by reduction from quadratic knapsack (QKP) [8,19]. The reduction from the QPK shows that it can be modeled as an instance of GMFA; thus, a solution for QPK cannot not be used to solve GMFA. In addition to the complexity, this reduction shows the difficulty of designing an approximate algorithm for GMFA. Rader et al. [14] prove that QKP, in general, does not have a polynomial time approximation algorithm with fixed approximation ratio, unless P=NP. In Theorem 2, we prove that a polynomial approximate algorithm with a fixed approximate ratio for GMFA guarantees a fixed approximate ratio for QKP, and its existence contradicts the result of [14]. Furthermore, studies on the constrained set functions optimization, such as [7], also admits that maximizing a monotone set function up to an acceptable approximation, even subject to simple constraints is not possible. Due to the space limitations, further details on the problem complexity, as well as the proofs of Theorems 1 and 2, are provided in the technical report [15].

Theorem 1. *The problem of Gain maximization over flexible attributes (GMFA) is NP-hard.*

Theorem 2. *There is no polynomial-time approximate algorithm with a fixed approximate ratio for GMFA unless there is an approximate algorithm with a constant approximate ratio for QKP.*

---

[6] In addition to the input set of attributes, the function $gain(.)$ may depend to other variables such as the number of attribute (n); one such function is discussed in Sect. 4.

[7] Please note that the GMFA is NP-complete even for the polynomial time gain functions.

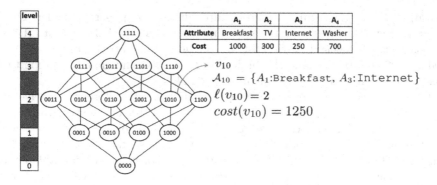

Fig. 2. Illustration of $\mathcal{L}_{\mathcal{A}}$ for Example 1

3   Solution

Considering the negative result of Theorem 2, we turn our attention to the design of an exact algorithm for the GMFA problem; even though this algorithm will be exponential in the worst case, we will demonstrate that is efficient in practice. In this section, our focus is on providing a solution for GMFA over any *monotonic* gain function. A gain function $gain(.)$ is monotonic, if given two set of attributes $\mathcal{A}_i$ and $\mathcal{A}_j$ where $\mathcal{A}_j \subset \mathcal{A}_i$, $gain(\mathcal{A}_j, \mathcal{D}) \leq gain(\mathcal{A}_i, \mathcal{D})$. As a result, *this section provides a general solution that works for any monotonic gain function*, no matter how and based on what data it is designed. In fact considering a non-monotonic function for gain is not reasonable here, because adding more attributes to a tuple (service) should not decrease the gain. For ease of explanation, we first provide the following definitions and notations.

Definition 1. Lattice of Attribute Combination: *Given an attribute combination $\mathcal{A}_i$, the lattice of $\mathcal{A}_i$ is defined as $\mathcal{L}_{\mathcal{A}_i} = (V, E)$, where the nodeset V, depicted as $V(\mathcal{L}_{\mathcal{A}_i})$, corresponds to the set of all subsets of $\mathcal{A}_i$; thus $\forall \mathcal{A}_j \subseteq \mathcal{A}_i$, there exists a one to one mapping between each $v_j \in V$ and each $\mathcal{A}_j$. Each node v_j is associated with a bit representative $\mathcal{B}(v_j)$ of length m in which bit k is 1 if $A_k \in \mathcal{A}_j$ and 0 otherwise. For consistency, for each node v_j in $V(\mathcal{L}_{\mathcal{A}_i})$, the index j is the decimal value of $\mathcal{B}(v_j)$. Given the bit representative $\mathcal{B}(v_j)$ we define function $v(\mathcal{B}(v_j))$ to return v_j. In the lattice an edge $\langle v_j, v_k \rangle \in E$ exists if $A_k \subset \mathcal{A}_j$ and $\mathcal{B}(v_k)$, $\mathcal{B}(v_j)$ differ in only one bit. Thus, v_j (resp. v_k) is parent (resp. child) of v_k (resp. v_j) in the lattice. For each node $v_j \in V$, level of v_j, denoted by $\ell(v_j)$, is defined as the number of 1's in the bit representative of v_j. In addition, every node v_j is associated with a cost defined as $cost(v_j) = \sum_{\forall A_k \in \mathcal{A}_j} cost[A_k]$.*

Definition 2. Maximal Affordable Node: *A node $v_i \in V(\mathcal{L}_{\mathcal{A}})$ is affordable iff $cost(v_j) \leq B$; otherwise it is unaffordable. An affordable node v_i is maximal affordable iff $\forall$ nodes v_j in parents of v_i, v_j is unaffordable.*

Example 1: As a running example throughout the paper, consider $\mathcal{D}$ as shown in Fig. 1, defined over the set of attributes $A = \{A_1$:Breakfast, A_2:TV, A_3:Internet, A_4:Washer$\}$ with cost to provide these attributes as $cost = [1000, 300, 250, 700]$. Assume the budget is $B = 1300$ and that the property t does not offer these attributes/amenities, i.e., $\mathcal{A}_t = \emptyset$.

Figure 2 presents $\mathcal{L}_\mathcal{A}$ over these four attributes. The bit representative for the highlighted node v_{10} in the figure is $\mathcal{B}(v_{10}) = 1010$ representing the set of attributes $\mathcal{A}_{10} = \{A_1$:Breakfast, A_3:Internet$\}$; The level of v_{10} is $\ell(v_{10}) = 2$, and it is the parent of nodes v_2 and v_8 with the bit representatives 0010 and 1000. Since $B = 1300$ and the cost of v_2 is $cost(v_2) = 250$, v_2 is an affordable node; however, since its parent v_{10} the cost $cost(v_{10}) = 1250$ and is affordable, v_2 is not a maximal affordable node. v_{11} and v_{14} with bit representatives $\mathcal{B}(v_{11}) = 1011$ and $\mathcal{B}(v_{14}) = 1110$, the parents of v_{10}, are unaffordable; thus v_{10} is a maximal affordable node.

A baseline approach for the GMFA problem is to examine all the 2^m nodes of $\mathcal{L}_\mathcal{A}$. Since for every node the algorithm needs to compute the gain, it's running time is in $\Omega(m2^m\mathcal{G})$, where $\mathcal{G}$ is the computation cost associated with the $gain(.)$ function. As the first improvement over the baseline, Improved GMFA (**I-GMFA**) improves upon this baseline by leveraging the monotonicity of the gain function, which enables it to prune some of the branches in the lattice while searching for the optimal solution. We provide further details about **I-GMFA** in the technical report [15]. **I-GMFA** has two efficiency issues that we resolve in General GMFA (**G-GMFA**) as following:

The Problem with Multiple Children Generation. The first efficiency issue is that, following the Definition 2, a node with multiple unaffordable parents, gets generated multiple times (by its parents). To address this, we adopt the set enumeration tree [16] and one-to-all broadcast in a hypercube [4] constructing a tree that guarantees to generate each node in $\mathcal{L}_\mathcal{A}$ only once.

Tree Construction: Considering the bit representation of a node v_i, let $\rho(\mathcal{B}(v_i))$ be the right-most 0 in $\mathcal{B}(v_i)$. The algorithm first identifies $\rho(\mathcal{B}(v_i))$; then it complements the bits in the right side of $\rho(\mathcal{B}(v_i))$ one by one to generate the children of v_i. Figure 3 demonstrates the resulting tree for the lattice of Fig. 2 for this algorithm. For example, consider the node v_3 ($\mathcal{B}(v_3) = 0011$) in the figure; $\rho(0011)$ is 2 (for attribute A_2). Thus, nodes v_1 and v_2 with the bit representatives $\mathcal{B}(v_1) = 0001$ and $\mathcal{B}(v_2) = 0010$ are generated as its children. As shown in Fig. 3, all the nodes of the lattice are generated once and only once. The only parent of each node is identified by flipping the bit $\rho(\mathcal{B}(v_i))$ in $\mathcal{B}(v_i)$ to one. We use parent$_T(v_i)$ to refer to the parent of the node v_i in the tree structure. Also, based on the way v_i is constructed, $\rho(\mathcal{B}(v_i))$ is the bit that has been flipped by its parent to generate it.

The Problem with Checking All Parents in the Lattice. In order to decide if an affordable node is maximal affordable, one has to check all its parents in the lattice (not the tree). If at least one of the parents is affordable, it is not maximal affordable. Thus, even though we construct a tree to avoid generating the children multiple times, the parents may get checked multiple times by their children. Therefore, in the following, we show how the monotonicity of the cost function can help to generate a node only if it does not have an affordable parent. In the lattice, each child has one less attribute than its parents. Thus, for a node v_i, one can simply determine the parent with the minimum cost (cheapest parent) by considering the cheapest attribute in $\mathcal{A}$ that does not belong to $\mathcal{A}_i$. The key observation is that, for a node v_i, if the parent with minimum cost is not affordable, none of the other parents is affordable; on the other hand, if the cheapest parent is affordable, there is no need to check the other parents as this node is not maximal affordable. Consequently, one only has to identify the missing attribute with the least cost and check if its cost plus the cost of attributes in the combination is at most B. Identifying the missing attribute with the smallest cost is in $O(m)$. For each node v_i, $\rho(\mathcal{B}(v_i))$ is the bit that has been flipped by $\text{parent}_T(v_i)$ to generate it. For example, consider $v_1 = v(0001)$ in Fig. 3; since $\rho(0001) = 3$, $\text{parent}_T(v_1) = v(0011)$. We can use this information to reorder the attributes and instantly get the cheapest missing attribute in $\mathcal{A}_i$. The key idea is that if we originally *order the attributes from the most expensive to the cheapest*, $\rho(\mathcal{B}(v_i))$ is the index of the cheapest attribute. Moreover, adding the cheapest missing attribute generates $\text{parent}_T(v_i)$. Hence, if the attributes are sorted on their cost in descending order, a node with an affordable parent in the lattice will never be generated in the tree data structure. Consequently, after presorting the attributes, there is no need to check if a node has an affordable parent.

Sorting the attributes is in $O(m \log(m))$. In addition, computing the cost of a node is thus performed in constant time, using the cost of its parent in the tree. For each node v_i, $cost(v_i)$ is $cost(\text{parent}_T(v_i)) - cost[A_{\rho(\mathcal{B}(v_i))}]$. Therefore, the algorithm of **G-GMFA** is in $\Omega\big(\max(m \log(m), \mathcal{G})\big)$ and $O(2^m \mathcal{G})$. Due to the space limitations, we provide pseudo-code of **G-GMFA** in the technical report [15].

4   Gain Function Design

In Sect. 3, we proposed a general solution that works for any arbitrary monotonic gain function. We conducted our presentation for a generic gain function because the design of the gain function is application specific and depends on the available information. The application specific nature of the gain function design, motivated the consideration of the generic gain function, instead of promoting a specific function. Consequently, any monotonic gain function can directly applied into the general algorithm **G-GMFA**. In our work, the focus is on understanding which subsets of attributes are attractive to users. Based on the application, in addition to the dataset $\mathcal{D}$, some extra information (such as query logs and user ratings) may be available that help in understanding the desire for combinations of attributes and could be a basis for the design of such a function.

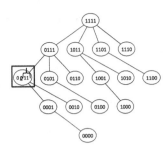

Fig. 3. Tree construction for Fig. 2

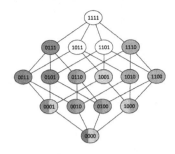

Fig. 4. Intersection of children of $C = \{A_1, A_2, A_4\}$ (Color figure online)

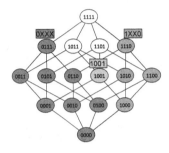

Fig. 5. Illustration of sublattice coverage by the tree (Color figure online)

id	v_i	$\mathcal{B}(v_i)$	Disjnt Ptrns	cnt
1	v_7	0111	$0XXX$	2^3
2	v_{14}	1110	$1XX0$	2^2
3	v_9	1001	$1\,0\,0\,1$	2^0

Fig. 6. $\text{FBC}(1111) = 8 + 4 + 1 = 13$

However, such comprehensive information that reflect user preferences are not always available. Consider a third party service for assisting the service providers. Such services have a limited view of the data [3] and may only have access to the dataset tuples. An example of such third party services is AirDNA[8] which is built on top of AirBnB. Therefore, instead on focusing on a specific application and assuming the existence of extra information, in the rest of this section, we focus on a simple, yet compelling variant of a practical gain function that only utilizes the existing tuples in the dataset to define the notion of gain in the absence of other sources of information. We provide a general discussion of gain functions with extra information in the technical report [15].

4.1 Frequent-Item Based Count (FBC)

Here, we propose a practical gain function that only utilizes the existing tuples in the dataset. It hinges on the observations that the bulk of market participants are expected to behave rationally. Thus, goods on offer are expected to follow a basic supply and demand principle. For example, based on the case study provided in Sect. 5.3, while many of the properties in Paris offer Breakfast, offering it is not popular in New York City. This indicates a higher demand for such an amenity in

---

[8] www.airdna.co.

Paris than in New York City. As another example, while many accommodations provide washer, dryer, and iron together, providing dryer without a washer and iron is rare. This reveals a need for the combination of these attributes. Utilizing this intuition, we define a frequent node in $\mathcal{L}_{\mathcal{A}}$ as follows:

Definition 3. Frequent Node: *Given a dataset $\mathcal{D}$, and a threshold $\tau \in (0, 1]$, a node $v_i \in V(\mathcal{L}_{\mathcal{A}})$ is frequent if and only if the number of tuples in $\mathcal{D}$ containing the attributes $\mathcal{A}_i$ is at least τ times n, i.e., $|Q(v_i, \mathcal{D})| \geq \tau n$.*[9]

For instance, in Example 1 let τ be 0.3. In Fig. 1, $v_3 = v(0011)$ is frequent because Accom. 2, Accom. 5, and Accom. 9 contain the attributes $\mathcal{A}_3 = \{A_3:\texttt{Internet}, A_4:\texttt{Washer}\}$; thus $|Q(v_3, \mathcal{D})| = 3 \geq 0.3 \times 10$. However, since $|Q(v_{11} = v(1011)), \mathcal{D})|$ is $1 < 0.3 \times 10$, v_{11} is not frequent.

Consider a tuple t and a set of attributes $A' \subseteq \mathcal{A} \backslash \mathcal{A}_t$ to be added to t. Let $\mathcal{A}_i$ be $\mathcal{A}_t \cup \mathcal{A}'$ and v_i be $v(\mathcal{A}_i)$. After adding $\mathcal{A}'$ to t, for any node v_j in $\mathcal{L}_{\mathcal{A}_i}$, t belongs to $Q(v_j, \mathcal{D})$. However, according to Definition 3, only the frequent nodes in $\mathcal{L}_{\mathcal{A}_i}$ are desirable. Using this intuition, Definition 4 provides a practical gain function utilizing nothing but the tuples in the dataset.

Definition 4. Frequent-item Based Count (FBC): *Given a dataset $\mathcal{D}$, and a node $v_i \in V(\mathcal{L}_{\mathcal{A}})$, the Frequent-item Based Count (FBC) of v_i is the number of frequent nodes in $\mathcal{L}_{\mathcal{A}_i}$. Formally*

$$FBC_\tau(\mathcal{B}(v_i), \mathcal{D}) = |\{v_j \in \mathcal{L}_{\mathcal{A}_i} \,|\, |Q(v_j, \mathcal{D})| \geq \tau\}| \tag{2}$$

For simplicity, throughout the paper we use $FBC(\mathcal{B}(v_i))$ to refer to $FBC_\tau(\mathcal{B}(v_i), \mathcal{D})$. In Example 1, consider $v_{15} = v(1111)$. In Fig. 4, we have colored the frequent nodes in $\mathcal{L}_{\mathcal{A}_{15}}$. Counting the number of colored nodes, FBC(1111) is 13.

Such a definition of a gain function has several advantages, mainly (i) it requires knowledge only of the existing tuples in the dataset (ii) it naturally captures changes in the joint demand for certain attribute combinations (iii) it is robust and adaptive to the underlying data changes. However, it is known that [10], counting the number of frequent itemsets is #P-complete. Consequently, counting the number of frequent subsets of a subset of attributes (i.e., counting the number of frequent nodes in $\mathcal{L}_{\mathcal{A}_i}$) is exponential to the size of the subset (i.e., the size of $\mathcal{L}_{\mathcal{A}_i}$). Therefore, for this gain function, even the verification version of GMFA is likely not solvable in polynomial time. Thus, in the rest of this section, we design a practical output sensitive algorithm for computing $FBC(\mathcal{B}(v_i))$.

4.2 FBC Computation

Given a node v_i, to identify $FBC(\mathcal{B}(v_i))$, the baseline traverses the lattice under v_i, i.e., $\mathcal{L}_{\mathcal{A}_i}$ counting the number of nodes in which more than τ tuples in the

---

[9] For simplicity, we use $Q(v, \mathcal{D})$ to refer to $Q(\mathcal{A}(v), \mathcal{D})$.

dataset match the attributes corresponding to v_i. Thus, this baseline is always in $\theta(n2^{\ell(v_i)})$. An improved method to compute FBC of v_i, is to start from the bottom of the lattice $\mathcal{L}_{\mathcal{A}_i}$ and follow the well-known Apriori [1] algorithm discovering the number of frequent nodes. This algorithm utilizes the fact that any superset of an infrequent node is also infrequent. The algorithm combines pairs of frequent nodes at level k that share $k-1$ attributes, to generate the candidate nodes at level $k+1$. It then checks the frequency of candidate pairs at level $k+1$ to identify the frequent nodes of size $k + 1$ and continues until no more candidates are generated. Since generating the candidate nodes at level $k+1$ contains combining the frequent nodes at level k, this algorithms is in $O(n.\text{FBC}(\mathcal{B}(v_i))^2)$.

Consider a node v_i which is frequent. In this case, Apriori will generate all the $2^{\ell(v_i)}$ frequent nodes, i.e., in par with the baseline solution. One interesting observation is that if v_i itself is frequent, since all nodes in $\mathcal{L}_{\mathcal{A}_i}$ are also frequent, $\text{FBC}(\mathcal{B}(v_i))$ is $2^{\ell(v_i)}$. As a result, in such cases, FBC can be computed in constant time. In Example 1, since node v_7 with bit representative $\mathcal{B}(v_7) = 0111$ is frequent $\text{FBC}(0111) = 2^3 = 8$ ($\ell(v_7) = 3$). This motivates us to compute the number of frequent nodes in a lattice without generating all the nodes. First, let us define the set of maximal frequent nodes as follows:

Definition 5. Set of Maximal Frequent Nodes: *Given a node v_i, dataset $\mathcal{D}$, and a threshold $\tau \in (0,1]$, the set of maximal frequent nodes is the set of frequent nodes in $V(\mathcal{L}_{\mathcal{A}_i})$ that do not have a frequent parent. Formally,*

$$\mathcal{F}_{v_i}(\tau, \mathcal{D}) = \{v_j \in V(\mathcal{L}_{\mathcal{A}_i}) | |Q(v_j, \mathcal{D})| \geq \tau n \text{ and } \forall v_k \in parents(v_j, \mathcal{L}_{\mathcal{A}_i}) : |Q(v_k, \mathcal{D})| < \tau n\} \quad (3)$$

In the rest of the paper, we ease $\mathcal{F}_{v_i}(\tau, \mathcal{D})$ with $\mathcal{F}_{v_i}$. In Example 1, the set of maximal frequent nodes of v_{15} with bit representative $\mathcal{B}(v_{15}) = 1111$ is $\mathcal{F}_{v_{15}} = \{v_7, v_{10}, v_{14}\}$, where $\mathcal{B}(v_7) = 0111$, $\mathcal{B}(v_{10}) = 1001$, and $\mathcal{B}(v_{14}) = 1110$. Unfortunately, unlike the cases where v_i itself is frequent, calculating the FBC of infrequent nodes is challenging. That is because the intersections between the frequent nodes in the sublattices of $\mathcal{F}_{v_i}$ are not empty (further details about this are provided in the technical report [15]). Therefore, in the following, we propose an algorithm that breaks the frequent nodes in the sublattices of $\mathcal{F}_{v_i}$ into disjoint partitions.

Let $\mathcal{U}_{v_i}$ be the set of all frequent nodes in $\mathcal{L}_{\mathcal{A}_i}$ – i.e., $\mathcal{U}_{v_i} = \bigcup_{\forall v_j \in \mathcal{F}_{v_i}} V(\mathcal{L}_{\mathcal{A}_j})$. In Example 1, $\mathcal{U}_{v_{15}}$ is a set of all colored nodes in Fig. 4. Our goal is to *partition* $\mathcal{U}_{v_i}$ to a collection $\mathcal{S}$ of disjoint sets such that (i) $\bigcup_{\forall S_i \in \mathcal{S}}(S_i) = \mathcal{U}_{v_i}$ and (ii) the intersection of the partitions is empty, i.e., $\forall S_i, S_j \in \mathcal{S}, S_i \cap S_j = \emptyset$; given such a partition, $\text{FBC}(\mathcal{B}(v_i))$ is $\sum_{\forall S_i \in \mathcal{S}} |S_i|$. Such a partition for Example 1 is shown in Fig. 5, where each color represents a set of nodes which is disjoint from the other sets designated by different colors.

In order to identify the disjoint partitions, we first define a *"pattern"* P as a string of size m, where $\forall 1 \leq i \leq m$: $P[i] \in \{0, 1, X\}$. Specially, we refer to the pattern generated by replacing all 1s in $\mathcal{B}(v_i)$ with X as the *initial pattern* for v_i. For example, in Fig. 5, there are four attributes, $P = 0XXX$ is a pattern (which is the initial pattern for v_7). The pattern $P = 0XXX$ covers all nodes whose bit representatives start with 0 (the nodes with green color in Fig. 4). More formally, the coverage of a pattern P is defined as follows:

Definition 6. *Given the set of attributes $\mathcal{A}$ and a pattern P, the coverage of pattern P is*[10]

$$COV(P) = \{v_i \subseteq V(\mathcal{L}_\mathcal{A}) | \forall 1 \leq k \leq m \text{ if } P[k] = 1 \text{ then } A_k \in \mathcal{A}_i, \text{ and if } P[k] = 0 \text{ then } A_k \notin \mathcal{A}_i\} \tag{4}$$

In Fig. 4, all nodes with green color are in $COV(0XXX)$ and all nodes with blue color are in $COV(XXX0)$. Specifically, node v_3 with bit representative $\mathcal{B}(v_3) = 0011$ is in $COV(0XXX)$ because its first bit is 0. Note that a node may be covered by multiple patterns, e.g., node v_6 with bit representative $\mathcal{B}(v_6) = 0110$ is in $COV(0XXX)$ and $COV(XXX0)$. We refer to patterns with disjoint coverage as *disjoint patterns*. For example, $01XX$ and $11XX$ are disjoint patterns. Figure 5, provides a set of disjoint patterns (also presented in the 4-th column of Fig. 6) that partition $\mathcal{U}_{v_{15}}$ in Fig. 4. The nodes in the coverage of each pattern is colored with a different color.

Consider a set of disjoint patterns $\mathcal{P}$ that partition $\mathcal{U}_{v_i}$. For a pattern $P \in \mathcal{P}$ let $x_k(P)$ be the number of Xs in P; the number of nodes covered by the pattern P is $2^{k_x(P)}$. Thus, FBC$(\mathcal{B}(v_i))$ is simply $\sum_{\forall P_j \in \mathcal{P}} 2^{k_x(P_j)}$. For example, considering the set of disjoint patterns in Fig. 5, the last column of Fig. 6 presents the number of nodes in the coverage of each pattern P_j (i.e., $2^{k_x(P_j)}$); thus FBC(1111) in this example is the summation of the numbers in the last column (i.e., 13).

In order to find the set of disjoint patterns that partition $\mathcal{U}_{v_i}$, we define a bipartite graph that allows us to assign every frequent node in $\mathcal{U}_{v_i}$ to one and only one partition. Due to the space limitations, we provide the details in the technical report [15].

5   Experimental Evaluation

We now turn our attention to the experimental evaluation of the proposed algorithms, over real-world data. In addition to the performance evaluation, we also provide a case study in Sect. 5.3 to illustrate the practicality of the approaches.

5.1   Experimental Setup

Hardware and Platform: All the experiments were performed on a Core-I7 machine 8 GB of RAM. The algorithms were implemented in Python.

Dataset: The experiments were conducted over real-world data collected from *AirBnB*, a travel peer to peer marketplace. We collected the information of approximately 2 million *real* properties around the world, shared on this website.

---

[10] Note that if $P[k] = X$, A_k may or may not belong to $\mathcal{A}_i$.

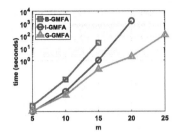

Fig. 7. Impact of varying m on GMFA algorithms

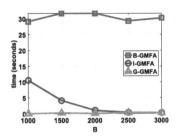

Fig. 8. Impact of varying B on GMFA algorithms

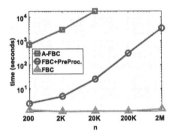

Fig. 9. Impact of varying n on FBC algorithms

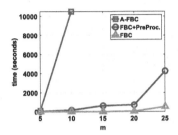

Fig. 10. Impact of varying m on FBC algorithms

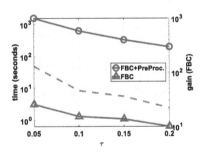

Fig. 11. Impact of varying τ on FBC algorithms

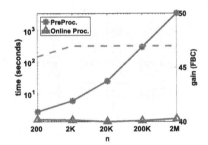

Fig. 12. Impact of varying n on pre-processing and online processing

AirBnB has a total number of 41 attributes for each property. Among all the attributes, 36 of them are boolean attributes, such as TV, Internet, Washer, and Dryer, while 5 are ordinal attributes, such as *Number of Bedrooms* and *Number of Beds*. We identified 26 (boolean) flexible attributes and for practical purposes we estimated their costs for a one year period. Notice that these costs are provided purely to facilitate the experiments; other values could be chosen and would not affect the relative performance and conclusions in what follows. In our estimate for attribute $cost[.]$, one attribute (Safety card) is less than $10, nine attributes

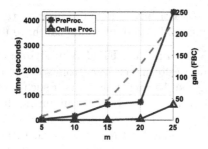

Fig. 13. Impact of varying m on preprocessing and online processing

(eg. `Iron`) are between $10 and $100, fourteen (eg. `TV`) are between $100 and $1000, and two (eg. `Pool`) are more than $1000.

Algorithms Evaluated: We evaluated the performance of our proposed algorithms: **B-GMFA** (Baseline GMFA), **I-GMFA** (Improved GMFA) and **G-GMFA** (General GMFA). According to Sect. 3, **B-GMFA** does not consider pruning the sublattices of the maximal affordable nodes and examines all the nodes in $\mathcal{L}_\mathcal{A}$. In addition, for Sect. 4, the performance of algorithms **A-FBC** (Apriori-FBC) and **FBC** is evaluated.

Default Values: n (number of tuples): 200,000; m (number of attributes): 15; B (budget): $2000; τ (frequency threshold): 0.1; $\mathcal{A}_t = \{\}$.

5.2   Experimental Results

Figure 7: Impact of Varying m on GMFA Algorithms. We first study the impact of the number of attributes (m) on the performance Sect. 3 algorithms where the GMFA problem is considered over the general class of monotonic gain functions. In this (and next) experiment we only consider the running time for the GMFA algorithms by deducting the gain function computation from the total time. Figure 7 presents the results for varying m from 5 to 25. The size of $\mathcal{L}_\mathcal{A}$ exponentially depends on the number of attributes (m). Thus, traversing the complete lattice, **B-GMFA** did not extend beyond 15 attributes (requiring more than 10K s to finish). Utilizing the monotonicity of the gain function, **I-GMFA** scaled up to 20 attributes. Still it took 1566 s for **I-GMFA** to finish with 20 attributes. Despite the exponential growth of $\mathcal{L}_\mathcal{A}$, transforming the lattice to a tree structure, reordering the attributes, and amortizing the computation costs, **G-GMFA** performed well for all the settings, requiring 116 s for 25 attributes.

Figure 8: Impact of Varying B on GMFA Algorithms. In this experiment, we vary the budget from $1K up to $3K Since **B-GMFA** traverses the complete lattice and its performance does not depend on B. For a small budget, the maximal affordable nodes are in the lower levels of the lattice and **I-GMFA** (as well as **G-GMFA**) require to traverse more nodes until they terminate their traversal. To resolve the issue when the budget is limited, one may identify the node v_i with the cheapest attribute combination and start the traversal from level $\ell(v_i)$ as no other maximal affordable node can have more attributes.

Figure 9: Impact of Varying n on FBC Algorithms. In next three experiments we compared the performance our proposed algorithm for computing **FBC** to that of adopting the *Apriori* algorithm for such computation (**A-FBC**). The algorithms are utilized inside **G-GMFA** computing the FBC (gain) of maximal frequent nodes. When comparing the performance of these algorithms, after running **G-GMFA**, we consider the total time of each run. The input to **FBC** is the set of maximal frequent nodes, which is computed offline. Thus, during the identification of the FBC of a node v_i there is no need to recompute them. Still, for a fair comparison, we include in the graphs a line that demonstrates total time (the time to identify maximal affordable nodes and the time required by **FBC**). First, we vary the number of tuples (n) from 200 to $2M$. As reflected in the figure, **A-FBC** does not extend beyond 20K tuples and even for $n = 20K$ it requires more than 17K s to complete. Even considering preprocessing in the total running time of **FBC** (blue line), it extends to 2M tuples with a total time less than 3K s. The running time of **FBC** itself does not depend on n. This is reflected in Fig. 9, as in all settings the time required by **FBC** is less than 2 s.

Figure 10: Impact of Varying m on FBC Algorithms. Next, we vary the number of attributes from 5 to 25. **A-FBC** requires 10K s for 10 attributes and does not extend beyond, in a reasonable time. On the other hand, **FBC** performs well for all the settings; the total time for preprocessing and running time for **FBC** on 25 attributes is around 4K s, while the time to run **FBC** itself is 510 s.

Figure 11: Impact of Varying τ on FBC Algorithms. We vary the frequency threshold (τ) from 0.05 to 0.2. **A-FBC** did not complete for any of the settings! Thus, in Fig. 11 we present the performance of **FBC**. To demonstrate the relationship between the gain and the time required by the algorithm, we add the FBC of the optimal solution in the right-y-axis and the dashed orange line. When the threshold is large, smaller number of nodes are frequent and the maximal frequent nodes appear at the lower levels of the lattice. Thus, preprocessing stops earlier. Also, having smaller number of nodes for larger thresholds, the gain (FBC) of the optimal solution decreases as the threshold increases. In this experiment, the gain of the optimal solution for $\tau = 0.05$ was 129 while it was 22 for $\tau = 0.2$. The performance of **FBC** linearly depends on its output value. The time required by **FBC** in this experiment is less than 3.2 s for all settings.

Figures 12 and 13: Impacts of Varying n and m in the Online and Offline Running Times. We evaluate the performance of **G-GMFA** with **FBC** and perform two experiments to study offline processing time (identifying the maximal affordable nodes) and the total online (query answering) time. We also include the right-x-axis and the dashed orange line to report the gain (FBC) of the optimal solution. First, we vary the number of tuples (n) between 200 and 2M (Fig. 12). While the preprocessing time increases from less than 3 s for $n = 200$ up to around $3,400$ s for $n = 2M$, the online processing time (i.e., **GMFA** and **FBC** times) does not depend on n, and for all the settings requires less than 2 s. This verifies the suitability of our proposal for online query answering at scale. Next, we vary m from 5 to 25. Despite the exponential increases in the size of $\mathcal{L}_A$, the total time to execute **G-GMFA** with **FBC** for 25 attributes is around 600 s. In practice, since $\mathcal{A}_t$ is probably not empty set, the number of remaining attributes is less than m and one may also select a subset of them as flexible attributes; we utilize $m = 25$ to demonstrate that even for the extreme cases, the proposed algorithms provide answers in reasonable time.

5.3   Case Study

We preformed a real case study on the actual AirBnB rental accommodations in two popular locations: **Paris** and **New York City** (NYC). We used the location information of the accommodations, i.e., latitude and longitude for the filtering, and found $42,470$ rental accommodations in Paris and $37,297$ ones in NYC. We considered two actual accommodations, one in each city, offering the same set of amenities. These accommodations lack providing the following amenities: Air Conditioning, Breakfast, Cable TV, Carbon Monoxide Detector, Doorman, Dryer, First- Aid Kit, Hair Dryer, Hot Tub, Indoor Fireplace, Internet, Iron, Laptop Friendly Workspace, Pool, TV, and Washer. We used the same cost estimation discussed in Sect. 5.1, assumed the budget $B = \$2000$ for both accommodations, and ran **GMFA** while considering **FBC** (with threshold $\tau = 0.1$) as the gain function. It took 0.195 s to finish the experiment for Paris 0.365 s for NYC. While the optimal solution suggests offering Breakfast, First Aid Kit, Internet, and Washer in Paris, it suggests adding Carbon Monoxide Detector, Dryer, First Aid Kit, Hair Dryer, Internet, Iron, TV, and Washer in NYC. Comparing the results for the two cases reveals the popularity of providing Breakfast in Paris, whereas the combination of Carbon Monoxide Detector, Dryer, Hair Dryer, Iron, and TV are preferred in NYC.

6   Related Work

Product Design: The problem of product design has been studied by many disciplines such as economics, industrial engineering, and computer science [2,17,18]. More specifically, manufacturers want to understand the preferences of their (potential) customers for products and services they are or may consider offering. Many factors like the cost and return on investment are currently considered. Work in this domain requires direct involvement of consumers, who choose preferences from a set of existing alternative products. While the existing work's focus is on identifying the set of attributes and information collection from various sources for a single product, our goal is to use the existing data for providing a tool that helps service providers as a part of peer to peer marketplace. In such marketplaces, service providers (e.g. hosts) are the customers of the website owner (e.g., AirBnB) who aim to list their service for other customers (e.g. guests) which makes the problem more challenging. The problem of item design in relation to social tagging is studied in [5], where the goal is to creates an opportunity for designers to build items that are likely to attract desirable tags when published. [13] also studied the problem of selecting the snippet for a product so that it stands out in the crowd of existing competitive products. Still, none studied our proposed problem.

Frequent Itemsets Count: Finding the number of frequent itemsets and number of maximal frequent itemsets has been shown to be #P-complete [10,11]. The authors in [9] provided an estimate for the number of frequent itemset candidates containing k elements rather than true frequent itemsets. Clearly, the set of candidate frequent itemsets can be much larger than the true frequent itemset. In [12] the authors theoretically estimate the average number of frequent itemsets under the assumption that the transactions matrix is subject to either simple Bernoulli or Markovian model. In contrast, we do not make any probabilistic assumptions about the set of transactions and we focus on providing a practical exact algorithm for the frequent-item based count.

7   Final Remarks

We proposed the problem of gain maximization over flexible attributes (GMFA) in the context of peer to peer marketplaces. Studying the complexity of the problem, we provided a practically efficient algorithm for solving GMFA that works for any arbitrary monotonic gain function. We presented frequent-item based count (FBC), as an alternative practical gain function in the absence of extra information, and proposed an efficient algorithm for computing it. The extensive experiment on a real dataset from AirBnB and the case study confirmed the efficiency and practicality of our proposal. The focus of this paper is such that it works for various application specific gain functions. Of course, which gain function performs well in practice depends on the application and the availability of the data. A comprehensive study of the wide range of possible gain functions and their applications left as future work.

References

1. Agrawal, R., Srikant, R., et al.: Fast algorithms for mining association rules. In: VLDB (1994)
2. Albers, S., Brockhoff, K.: Optimal product attributes in single choice models. JORS **37**, 647–655 (1980)
3. Asudeh, A., Zhang, N., Das, G.: Query reranking as a service. PVLDB **9**(11), 888–899 (2016)
4. Bertsekas, D.P., Özveren, C., Stamoulis, G.D., Tseng, P., Tsitsiklis, J.N.: Optimal communication algorithms for hypercubes. JPDC **11**(4), 263–275 (1991)
5. Das, M., Das, G., Hristidis, V.: Leveraging collaborative tagging for web item design. In: SIGKDD (2011)

6. Ertz, M., Durif, F.: Collaborative consumption or the rise of the two-sided consumer. J. Bus. Manage. **4**, 195–209 (2016)
7. Feldman, M., Izsak, R.: Constrained monotone function maximization and the supermodular degree. arXiv preprint arXiv:1407.6328 (2014)
8. Gary, M.R., Johnson, D.S.: Computers and Intractability: A Guide to the Theory of NP-Completeness. W.H. Freeman, New York (1979)
9. Geerts, F., Goethals, B., Bussche, J.: Tight upper bounds on the number of candidate patterns. TODS **30**, 333–363 (2005)
10. Gunopulos, D., Khardon, R., Mannila, H., Saluja, S., Toivonen, H., Sharma, R.S.: Discovering all most specific sentences. TODS **28**, 140–174 (2013)
11. Han, J., Pei, J., Yin, Y.: Mining frequent patterns without candidate generation. In: SIGMOD. ACM (2000)
12. Lhote, L., Rioult, F., Soulet, A.: Average number of frequent (closed) patterns in Bernoulli and Markovian databases. In: ICDM. IEEE (2005)
13. Miah, M., Das, G., Hristidis, V., Mannila, H.: Determining attributes to maximize visibility of objects. TKDE **21**, 959–973 (2009)
14. Rader Jr., D.J., Woeginger, G.J.: The quadratic 0–1 knapsack problem with series-parallel support. Oper. Res. Lett. **30**(3), 159–166 (2002)
15. Asudeh, A., Nazi, A., Koudas, N., Das, G.: Assisting service providers in peer-to-peer marketplaces: maximizing gain over flexible attributes. CoRR, abs/1705.03028 (2017)
16. Rymon, R.: Search through systematic set enumeration (1992)
17. Selker, T., Burleson, W.: Context-aware design and interaction in computer systems. IBM Syst. J. **39**, 880–891 (2000)
18. Shocker, A.D., Srinivasan, V.: A consumer-based methodology for the identification of new product ideas. Manage. Sci. **20**, 921–937 (1974)
19. Witzgall, C.: Mathematical methods of site selection for electronic message systems (EMS). NASA STI/Recon Tech, Report (1975)

An Attentive Spatio-Temporal Neural Model for Successive Point of Interest Recommendation

Khoa D. Doan[1]([✉]), Guolei Yang[2], and Chandan K. Reddy[1]

[1] Department of Computer Science, Virginia Tech, Arlington, VA, USA
khoadoan@vt.edu, reddy@cs.vt.edu
[2] Facebook Inc., Seattle, WA, USA
glyang@fb.com

Abstract. In a successive Point of Interest (POI) recommendation problem, analyzing user behaviors and contextual check-in information in past POI visits are essential in predicting, thus recommending, where they would likely want to visit next. Although several works, especially the Matrix Factorization and/or Markov chain based methods, are proposed to solve this problem, they have strong independence and conditioning assumptions. In this paper, we propose a deep Long Short Term Memory recurrent neural network model with a memory/attention mechanism, for the successive Point-of-Interest recommendation problem, that captures both the sequential, and temporal/spatial characteristics into its learned representations. Experimental results on two popular Location-Based Social Networks illustrate significant improvements of our method over the state-of-the-art methods. Our method is also robust to overfitting compared with popular methods for the recommendation tasks.

Keywords: Deep learning · Spatio-temporal data ·
Attention mechanism · Recurrent neural network ·
Long short term memory · Social networks

1 Introduction

Location-Based Social Networks (LBSNs) produce a huge amount of data, in both veracity and volume, thus providing opportunities for building personalized Point-of-Interest (POI) recommender systems. In a typical POI recommendation task, a user makes a sequence of check-ins at various POIs that are both geo-tagged and time-stamped, and the task is to recommend the next POI that the user is likely interested in visiting. Here a check-in comprises of which POI is visited, and additional contextual information such as the time or geotag of the visit. Finding an efficient way to represent the POI and its contextual information is essential because this can improve the performance of the model

© Springer Nature Switzerland AG 2019
Q. Yang et al. (Eds.): PAKDD 2019, LNAI 11441, pp. 346–358, 2019.
https://doi.org/10.1007/978-3-030-16142-2_27

and allow a better understanding of the seemingly complex inter-relationships of the heterogeneous properties of the POIs.

The recommendation task has been studied in numerous works [6,11,12,16, 20]. One of the most widely used technique is matrix factorization (MF), or a hybrid of MF and Markov Chain (MC). These methods, albeit having impressive performance, rely on strong independent assumption among different factors. Several attempts (e.g., Neighborhood-based MF methods [13,14,16]) have been made to overcome these limitations, but are unable to efficiently model the sequential, periodic check-in behaviors.

It has been shown that human movements usually demonstrate strong patterns in both spatial and temporal domain [3]. To take advantage of the spatio-temporal nature of check-ins, several recommendation systems have been proposed particularly for POIs (e.g., [18]). The state-of-the-art POI recommendation systems [5,15] use neural networks to learn the latent correlation between spatio-temporal features from historical check-ins and the next check-in location of a user. By mining spatio-temporal information from such correlations, these techniques are able to significantly outperform generic recommendation systems in the POI recommendation task.

In this paper, we will tackle this challenge and try to advance the state-of-the-art in POI recommendation systems. We propose a novel **Attentive Spatio-TEmporal Neural** (**ASTEN**) model that is able to recommend a POI by (1) extracting useful information from the most relevant POI visits reported by a user, and (2) minimizing the influence from non-relevant POI visits from the user. At the core of the proposed system is a Long-Short Term Memory (LSTM) Network structure, which employs the attention mechanism [2,4] to automatically select and extract information from the most relevant check-ins on a user's trajectory and make recommendations. ASTEN's network architecture overcomes the limitations of using a single hidden vector to represent a user's dynamical check-in behavior. As a result, our system is able to exploit long user trajectories without having to deal with the excessive noise. The main contributions of our work are:

- We propose a novel ASTEN model that addresses the challenge of noise handling in user trajectory data and advances the state-of-the-art of POI recommendation systems. This is achieved by combining the LSTM Network structure with a sophisticated attention mechanism specifically designed for spatio-temporal information present in LBSN datasets. To the best of our knowledge, this approach and the model design have not been studied for POI recommendation in the literature.
- We demonstrate the effectiveness of our method using three real-world LBSN datasets. Experiments show that our model outperforms existing POI recommendation systems. Our method is not only scalable but also robust to overfitting when the complexity increases.
- From our analysis of experimental results, we derive a set of practical implications that are useful for real-world applications.

2   Related Works

We describe the prior works that capture sequential, temporal and geographical influences in the context of LBSNs.

2.1   POI Recommendation

MF-based methods are arguably one of the best user-based collaborative filtering approaches [6,12]. Neighborhood-based MF methods attempt to incorporate temporal and spatial features. TimeSVD++ [11], for example, takes advantage of both the transition effect and the long-term transition pattern by modeling the user preference as a function of time. Similarly, recent works [13,16] model users' interest limited to the neighborhoods of the recently visited locations. In [6], PRME learns a personalized metric embedding and models the sequential POI transition. Another popular approach for modeling sequential data is MC, which learns a transition probability matrix over sequential events. In recent works [20], instead of estimating a single matrix for all users, each user can be mapped to a personalized transition probability matrix. For example, Factorizing Personalized MC (FPMC), which has the ability to model sequential data in an MF-based approach, is the state-of-the-art method [20].

Besides the cold-start problem, the common drawbacks of the MF based approaches are their strong independent assumptions among the factorized components and that their generalization strengths depend on designing a good feature space, which might not be a realistic assumption for many real-world problems.

2.2   Neural Models and Attention Mechanism

Progress in RNNs has shown impressive results in modeling sequential data [7]. Although RNN is theoretically capable of conditioning the model on all of the previous time-steps, the number of time-steps, in practice, what such a RNN model can remember is limited because of its difficulties in training.

Because RNN assumes discrete influence of the sequential events, it does not explain well real-world situations where the transition to a POI is continuously influenced by the historical spatial and temporal context. ST-RNN [15] models the continuous local temporal and spatial contexts with time-specific and distance-specific transition matrices and achieves a significant performance improvement in the recommendation task. RMTPP [5] jointly models the prediction of the time to next events and the event themselves. ST-RNN and RMTPP, however, suffer from the bottleneck problem in RNN where the use of the single hidden vector is insufficient to capture the complex characteristics of the sequences in a problem [2]. A recent success in training RNNs is a concept of attention [2,4]. For POI recommendation, however, it is not straightforward how the attention mechanism should be modeled.

In this paper, *we attempt to formalize the concepts of POI and check-in representations and describe how such representations can be embedded and learned within an efficient spatio-temporal attentive recurrent network structure.* The proposed model is able to **capture the sequential information, and spatio-temporal influence between check-ins in an end-to-end network that is robust to noisy check-in data.**

3   Data Description and Analysis

3.1   Data Description

We use three datasets collected from various activities of users on two popular LBSNs, namely, Foursquare (4SQ) and Gowalla. For 4SQ, we collect activities of users in the United States and in Europe separately and denote them as 4SQ-US and 4SQ-EU, respectively. For Gowalla, we use the dataset described in [3]. We pre-process the check-in data by filtering out POIs that were checked into by less than 10 users and users who checked-in less than 10 POIs. Table 1 summarizes the pre-processed datasets.

Table 1. Summary statistics of the LBSN datasets.

	4SQ-US	4SQ-EU	Gowalla
Number of users	21,878	15,387	52,484
Number of POIs	21,651	30,276	115,567
Number of check-ins	569,091	56,301	3,227,845
Average length	37	34	61

3.2   Check-In Data Exploration

Figures 1a–c show the temporal characteristics of the check-in activities. We observe that weekdays and weekend have different patterns of cumulative check-ins, defined as the total number of observed check-ins from all users at a specific hour of the day. Moreover, check-in activities form different patterns for different hours of the day inside the weekday or weekend group. Therefore, modeling POI and check-in representations should consider the temporal periodic variances and their interaction patterns.

Next, we perform analysis on the regularity of the check-in sequences. For each user, we calculate the transition distances between the sequentially visited POIs. We employ approximate entropy [19] as a measure of the regularity and unpredictability of local fluctuations in the resulting sequences of transition distances. We set a filtering level of 1 mile. Figures 1d–f show the histogram of approximate entropies of the sequences in the three datasets. We filter highly irregular series with infinite approximate entropies. In all datasets, the filter removes at most 25% of the users. We observe that the majority of the sequences

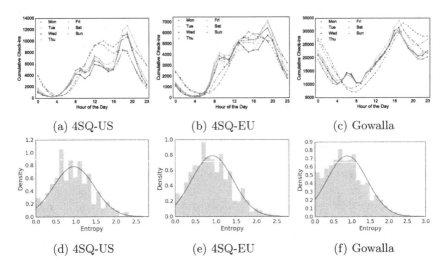

(a) 4SQ-US   (b) 4SQ-EU   (c) Gowalla

(d) 4SQ-US   (e) 4SQ-EU   (f) Gowalla

Fig. 1. Statistics of check-ins in the studied datasets. Figures (a–c) show the total number of check-ins on different days of the week. Solid lines correspond to weekdays. Dashed-lines correspond to weekends. Figures (d–f) show the distribution of approximate entropies of transition distances of user check-in sequences.

Table 2. Notations used in our paper.

Notation	Description
W_*	Weight matrices of the network architecture. A W_{ab} is a matrix of size $R^{m \times n}$, where m is the dimension of the input layer a and n is the dimension of the output layer b
b_*	Bias term associated with the corresponding W_*
l_t	One-hot encoding vector of POI at time t
p_t	Embedding of a POI at time t
s_{t_1,t_2}	Spatial distance between POIs at time t_1 and t_2. Note that t_1 and t_2 may not be consecutive time-steps
t_t	Temporal periodicity vector at time t

have low approximate entropies, which means that their transition distances exhibit not only regularity but also less fluctuation. This observation motivates us to model the distance transition behavior into a sequence's representation.

4   Proposed Methodology

4.1   Problem Definition

In this section, we formally define the successive recommendation task discussed in this paper.

Definition 1. (Check-in). *A check-in $C_u(t)$ is a tuple of $(u, l, t, s) \in U \times L \times T \times S$, where U is a set of unique users, L is a set of unique POIs, T is the continuous time domain, and S is continuous spatial domain, indexed by the latitude and longitude coordinates. $C_u(t)$ indicates that user u visited location l geo-tagged with coordinates s at time t.*

Definition 2. (User-historical check-ins). *A set of time-ordered, historical check-ins of a user u is defined as $C_u^{T_u} = \{C_u(t) : t \in [1, T_u]\}$, where T_u is the number of check-ins of user u.*

Definition 3. (Successive POI Recommendation). *Given a set of user-historical check-ins $C_u^{T_u}$, the successive point of interest recommendation task is to suggest the POI(s) that the user u will likely check-in after time T_u.*

In the following sections, we discuss our proposed method. Table 2 describes the notations used in our discussion.

4.2 POI Embedding and Check-In Representation

We propose to learn efficient representations of POIs and check-ins. Given a user u who performs a sequence of check-ins $C_u^{T_u} = (C_u(1), ..., C_u(T_u))$, where each check-in, as described, contains a POI $l^{(j)} \in L$, and the spatio-temporal information about the check-in, we learn two types of representations:

1. *POI embeddings*: we learn a function $f_{l^{(j)}} : L \mapsto R^m$ that maps every POI to a real-valued vector R^m where m is the dimension of the embedding.
2. *Check-in representations*: we learn a similar function $f_{C_u(t)} : C_u(t) \mapsto R^n$ that maps every check-in, which is a tuple of the checked POI, and its temporal information and spatial transition relationship to the previously checked POI, into a n-dimensional real-valued vector.

Given these objectives, we model a check-in x_t at time t as a function of the embedded visited-POI, its temporal context and its spatial transition distance shown in Eq. 1. p_t is the one-hot encoding of the checked POI at time step t. The temporal context is a set of one-hot vectors encoding the time periodicity and denoted by $t_t = concatenate(dom_t, dow_t, hr_t)$, where dom_t is the day of the month, dow_t is the day of the week and hr_t is the hour of the day. The spatial transition context $s_{t,t-1}$ is the great-circle distance between the checked POIs at time-points t and $t-1$. The spatio-temporal model is shown in Fig. 2a.

$$x_t = ReLU(W_{vx} * c_t + b_x)$$
$$c_t = concatenate(p_t, t_t, s_{t,t-1}) \qquad (1)$$

4.3   Recurrent Neural Networks and LSTMs

We model the hidden state as a latent representation of the past events. The predicted output, the ranked list of the recommended POIs, is a function of the hidden state:

$$h_t = \sigma(W_{hh}h_{t-1} + W_{xh}x_t + b_h)$$
$$\hat{y} = \text{softmax}(W_{hy}h_t + b_y) \tag{2}$$

where W_{hh} and W_{xh} are weight matrices of the hidden-hidden and input-hidden connections respectively, b_h is the hidden bias term, W_{hy} and b_y are the weights and bias of the hidden-output connections respectively, and σ is a Rectified Linear Unit (ReLU) [7] in our paper.

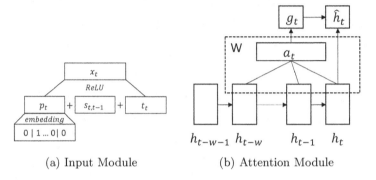

(a) Input Module   (b) Attention Module

Fig. 2. Network architecture of ASTEN. (a) Illustration of the recurrent input. (b) Illustration of the attention module: h_t in the vanilla RNN is replaced by $\hat{h}_t$.

In our model, we use a combination of the gradient-clipping technique to overcome the gradient exploding problem [7] and the LSTM units [9] to better capture long-term dependencies.

4.4   The Proposed ASTEN Model

Most of the existing RNNs rely on the last hidden activation vector as input into a feed-forward module, such as the softmax layer used in Sect. 4.3. Consequently, the last hidden state becomes the primary bottleneck of the neural model as discussed in Sect. 2.2, which often results in non-trivial model tuning and longer training time. In our paper, we introduce an attention or memory access mechanism that allows the recommendation task to pool a fixed set of the hidden states created in the previous time-steps in order to make the recommendation.

At a time step t, we learn the unit-length alignment vector $a_t \in R^W$. W is called the window and is a hyperparameter that determines how many of the previous hidden states in the previous timesteps should play a role in constructing the pooled hidden representation. An element at position w of a_t determines the amount of information from the previous hidden state h_{t-w} the model should retain and can be calculated by:

$$a_{t,w} = \frac{\text{score}(h_t, h_{t-w})}{\sum_{k=1}^{W} \text{score}(h_t, h_{t-k})} \tag{3}$$

where there are several options for the score function. A common version of the score function can be specified as:

$$\text{score}(h_t, h_k) = h_t W_{ah} h_k \tag{4}$$

Although there are other choices of the attention score function, we have seen better performance of the proposed score function, which is similar to the findings in [17]. Since spatial check-in characteristics and temporal transition distances could influence the check-in behaviors as discussed in Sect. 3, we propose modeling the score as a function of the relative relationship between the spatial and temporal properties of the check-ins at time $t - w$ and t. Specifically, score can be expressed as follows:

$$\text{score}(h_t, h_k) = h_t W_{ah} h_k + h_t W_{at} t_k + h_t w_{as} s_{t,k} \tag{5}$$

where t_k is the temporal periodicity vector at time k defined in Sect. 4.2, while $s_{t,k}$ – similar to the definition of $s_{t,t-1}$ also in Sect. 4.2 – is the great-circle spatial transition distance between check-in at time k and the current check-in at time t. Given a_t, the final attentive hidden state $\hat{h}_t$ can be calculated as follows:

$$\hat{h}_t = W_{ch} \text{concat}(h_t, g_t)$$

$$g_t = \sum_{w=1}^{W} a_{t,w} h_{t-w} \tag{6}$$

Our goal, therefore, is to learn parameters of the scoring function such that the scores reflect the similarity between the past hidden states and the current hidden state t based on their temporal and spatial similarities. The architecture of our proposed ASTEN model is shown in Fig. 2b.

4.5   Parameter Inference

We train an LSTM network that, given a sequence of check-ins, will recommend the next likely checked-in location. Given a sequential representation of check-ins, we minimize the cross entropy loss as follows:

$$W_*, w_*, b_* \frac{1}{T} \sum_{t=1}^{T} -y_{t+1} \log \hat{y}_t - (1 - y_{t+1}) \log (1 - \hat{y}_t) \tag{7}$$

where

$$\hat{y}_t = \frac{\exp(W_{hy} \hat{h}_t + b_y)}{\sum_{j=1}^{L} \exp(W_{hy}[j,:]\hat{h}_t + b_y[j])} \tag{8}$$

and W_{hy} and b_y are the weight matrix and bias vector of the softmax classifier to predict the next checked-in POI, respectively.

To train the proposed model, we adopt the gradient based backpropagation through time training technique [8] and Adam optimizer [10]. We also employ dropout technique [7] for learning all parameters. We set the dropout value to 0.2. We train our model using an initial learning rate of 0.01 and an exponential learning rate decay of 0.96 at every 100 train steps.

5   Experimental Results

In this section, we show the performance evaluation of our proposed model through empirical experiments.

5.1   Experimental Setup

We perform our experiments on real-world LBSN datasets, namely, Foursquare (Europe and US) and Gowalla, as described in Sect. 3.1. We employ the 5-fold cross validation technique. The performance metrics are reported from their averages across the folds.

To evaluate the performance, we employ two popular ranking metrics, **Recall@k** and **F1-score@k**, where k is the number of recommended POIs. We also report the Area under the ROC curve (AUC) in our experiments.

5.2   Comparison Methods

We compare the effectiveness of our ASTEN model with several representative recommendation methods:

1. *Most Popular Location* (**TOP**): recommend the most popular locations.
2. *Markov Chain* (**MC**): the popular MC model for sequential data. We choose the Markov order using its generalization error on the validation set.
3. *Spatio-temporal Analysis via Low Rank Tensor Learning* (**LRTL**) [1]: an extension of Matrix Factorization into three-dimensional user, spatial and temporal information.
4. *Factorizing Personalized Markov Chains* (**FPMC**) [20]: state-of-the-art Markov chain method based on matrix factorization.
5. *Personalized Ranking Metric Embedding* (**PRME**) [6]: state-of-the-art pairwise Metric Embedding method for POI recommendation that jointly models the sequential information, user preference and geographical influence.
6. *Recurrent Neural Network* (**RNN**): RNN model for discrete temporal data.
7. *Spatial Temporal RNN* (**ST-RNN**) [15]: state-of-the-art RNN-based POI recommender system that models both local temporal and spatial transition context via time-specific and distant-specific transition matrices respectively.

For the MF models, we perform grid-search to find the best hyperparameters using a validation set, which is 20% of the training data, before evaluating their performances on a hold-out test set. For RNN and ST-RNN, we use a similar learning rate, decay schedule, and batch sizes as those of ASTEN.

Table 3. Evaluation results of various methods on various LBSN datasets.

Dataset	Method	recall@1	recall@5	recall@10	F1@1	F1@5	F1@10	AUC
Foursquare-US	TOP	0.029	0.120	0.275	0.029	0.051	0.049	0.731
	MC	0.101	0.209	0.301	0.101	0.134	0.107	0.761
	LRTL	0.125	0.237	0.307	0.125	0.135	0.128	0.787
	FPMC	0.141	0.258	0.322	0.141	0.159	0.147	0.804
	PRME	0.148	0.265	0.343	0.148	0.161	0.153	0.820
	RNN	0.145	0.267	0.349	0.145	0.163	0.151	0.825
	ST-RNN	0.159	0.281	0.364	0.159	0.175	0.165	0.846
	ASTEN	**0.181**	**0.328**	**0.414**	**0.181**	**0.189**	**0.178**	**0.897**
Foursquare-EU	TOP	0.028	0.074	0.153	0.028	0.044	0.043	0.610
	MC	0.073	0.131	0.204	0.073	0.083	0.078	0.702
	LRTL	0.107	0.188	0.259	0.107	0.117	0.112	0.746
	FPMC	0.112	0.196	0.275	0.112	0.126	0.123	0.768
	PRME	0.120	0.208	0.291	0.120	0.131	0.125	0.780
	RNN	0.121	0.219	0.304	0.115	0.139	0.129	0.774
	ST-RNN	0.125	0.243	0.329	0.125	0.148	0.138	0.794
	ASTEN	**0.144**	**0.281**	**0.35**	**0.144**	**0.159**	**0.150**	**0.827**
Gowalla	TOP	0.009	0.025	0.061	0.009	0.013	0.012	0.566
	MC	0.019	0.054	0.097	0.019	0.065	0.062	0.601
	LRTL	0.026	0.063	0.132	0.026	0.077	0.071	0.608
	FPMC	0.044	0.083	0.174	0.044	0.091	0.089	0.652
	PRME	0.050	0.091	0.192	0.050	0.097	0.090	0.670
	RNN	0.048	0.098	0.189	0.048	0.121	0.095	0.673
	ST-RNN	0.061	0.120	0.223	0.061	0.138	0.120	0.695
	ASTEN	**0.081**	**0.152**	**0.266**	**0.081**	**0.165**	**0.158**	**0.735**

5.3   POI Recommendation Performance

Table 3 shows the averaged performance results across different metrics discussed in Sect. 5.1. TOP has the worst performance results, as expected. MC improves over TOP since it incorporates the sequential transitions into the model. However, MC's recall, F1-scores and AUC are worse than that of the three neural models that have a better memory capacity. Since FPMC combines the successes of MF-based models and MC-based models, in our experiments, FPMC outperforms MC by at least 2% in all metrics. FPMC also outperforms LRTL. PRME improves further upon FPMC and its performance is comparable to that of RNN. However, its performance is worse than that of ST-RNN and the proposed ASTEN model in our experiments.

Among the neural models, ST-RNN expectedly outperforms RNN by 2%−5% in our results. Nevertheless, ASTEN achieves a better performance improvement compared to ST-RNN. Moreover, when K increases, ASTEN experiences the highest recall improvement compared to the other methods, suggesting that the top-ranked POIs are more relevant to the recommendation. The results are consistent across all the datasets.

Table 4. Performance evaluation when adding spatial and temporal components.

Dataset	Method	recall@1	recall@5	recall@10
4SQ-US	LSTM	0.155	0.279	0.361
	ST-LSTM	0.161	0.303	0.378
	A-LSTM	0.159	0.309	0.371
	ASTEN	**0.181**	**0.328**	**0.414**
Gowalla	LSTM	0.049	0.105	0.192
	ST-LSTM	0.054	0.118	0.218
	A-LSTM	0.067	0.132	0.231
	ASTEN	**0.081**	**0.152**	**0.266**

5.4 ASTEN Performance Analysis

In this section, we present the performance improvements of our proposed model when various modeling components are being added in the 4SQ-US and Gowalla datasets. We look at the recall@k metric in the experiments on both 4SQ-US and Gowalla datasets in the following settings:

1. Discrete LSTM (**D-LSTM**), which is similar to the discrete (vanilla) RNN mentioned in Sect. 5.2 but using LSTM as hidden units.
2. Spatio-temporal LSTM (**ST-LSTM**), the ASTEN model without the spatio-temporal attention mechanism.
3. Attentive LSTM (**A-LSTM**), the ASTEN model without spatio-temporal embedding inputs as described in Sect. 4.2.
4. **ASTEN** model, which is our proposed model described in Sect. 4.4.

In Table 4, D-LSTM only slightly outperforms the previous RNN's recall discussed in Sect. 5.3. This result may be explained by the fact that although LSTM models are theoretically more robust to the gradient problems, in practice, this is not always the case. Both of the baseline neural models, however, have lower recall values compared to ST-LSTM and A-LSTM, both of which have comparable recall values in our experiments. We conjecture that the superior performances of these two models are due to the following reasons:

- Learning the spatio-temporal interaction of check-in sequences results in better recommendation quality.
- Our attentive mechanism captures better the check-in representation of a user, thus improving the recommendation quality.

Finally, our proposed model combines the spatio-temporal and attention mechanism and achieves the best performance in our experiments.

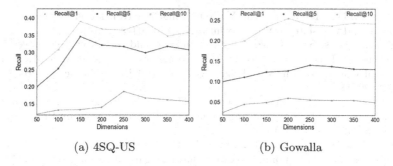

Fig. 3. Recall@1 when varying ASTEN's dimensions.

5.5 Varying Dimensions

To determine the effect of increasing the network complexity on its performance, we vary the dimensionality of POIs and check-in representations, and the hidden layer's LSTM size from 50 to 400, and compute the network's generalization using recall@1 for each case. As the dimension increases, the network's performance increases until an optimal value is achieved, after which the recall slowly decreases, though the decrease is not very significant. We notice that the generalization recall@1's, around and after the optimal values, are still better than that of the methods compared in Sect. 5.3, which indicates that increasing the network's performance is not sensitive to its capacity when the dimensionality is sufficient. We conjecture that this is probably because of the dropout technique employed in our model.

6   Conclusion

We proposed a novel end-to-end learning model that takes advantage of the sequential nature and spatial/temporal contextual information of user check-ins. We also proposed a novel attention/memory access mechanism that can effectively overcome the hidden layer bottleneck of RNNs. We have shown that the proposed ASTEN model outperforms various existing methods on real-world datasets. A primary goal of our work is to find efficient representations for a learning task and our results clearly illustrate that our method could achieve this goal. Our complexity analysis shows that ASTEN outperforms state-of-the-art methods even as the number of parameters increases.

Acknowledgments. This work was supported in part by the US National Science Foundation grants IIS-1619028, IIS-1707498 and IIS-1838730.

References

1. Bahadori, M.T., Yu, Q.R., Liu, Y.: Fast multivariate spatio-temporal analysis via low rank tensor learning. In: Advances in Neural Information Processing Systems, pp. 3491–3499 (2014)

2. Bahdanau, D., Cho, K., Bengio, Y.: Neural machine translation by jointly learning to align and translate. arXiv preprint arXiv:1409.0473 (2014)

3. Cho, E., Myers, S.A., Leskovec, J.: Friendship and mobility: user movement in location-based social networks. In: Proceedings of the 17th ACM SIGKDD International Conference on Knowledge Discovery and Data Mining, pp. 1082–1090. ACM (2011)

4. Chorowski, J.K., Bahdanau, D., Serdyuk, D., Cho, K., Bengio, Y.: Attention-based models for speech recognition. In: Advances in Neural Information Processing Systems, pp. 577–585 (2015)

5. Du, N., Dai, H., Trivedi, R., Upadhyay, U., Gomez-Rodriguez, M., Song, L.: Recurrent marked temporal point processes: embedding event history to vector. In: Proceedings of the 22nd ACM SIGKDD International Conference on Knowledge Discovery and Data Mining, pp. 1555–1564. ACM (2016)

6. Feng, S., Li, X., Zeng, Y., Cong, G., Chee, Y.M., Yuan, Q.: Personalized ranking metric embedding for next new poi recommendation. In: IJCAI, pp. 2069–2075 (2015)

7. Goodfellow, I., Bengio, Y., Courville, A.: Deep Learning, vol. 1. MIT Press, Cambridge (2016)

8. Guo, J.: Backpropagation through time. Harbin Institute of Technology, Unpubl. ms (2013)

9. Hochreiter, S., Schmidhuber, J.: Long short-term memory. Neural Comput. $9(8)$, 1735–1780 (1997)

10. Kingma, D., Ba, J.: Adam: a method for stochastic optimization. arXiv preprint arXiv:1412.6980 (2014)

11. Koren, Y.: Collaborative filtering with temporal dynamics. Commun. ACM $53(4)$, 89–97 (2010)

12. Koren, Y., Bell, R., Volinsky, C.: Matrix factorization techniques for recommender systems. Computer $42(8)$, 30–37 (2009)

13. Li, X., Cong, G., Li, X.L., Pham, T.A.N., Krishnaswamy, S.: Rank-GeoFM: a ranking based geographical factorization method for point of interest recommendation. In: Proceedings of the 38th International ACM SIGIR Conference on Research and Development in Information Retrieval, pp. 433–442. ACM (2015)

14. Liu, N.N., Zhao, M., Xiang, E., Yang, Q.: Online evolutionary collaborative filtering. In: Proceedings of the Fourth ACM Conference on Recommender Systems, pp. 95–102. ACM (2010)

15. Liu, Q., Wu, S., Wang, L., Tan, T.: Predicting the next location: A recurrent model with spatial and temporal contexts. In: AAAI, pp. 194–200 (2016)

16. Liu, Y., Wei, W., Sun, A., Miao, C.: Exploiting geographical neighborhood characteristics for location recommendation. In: Proceedings of the 23rd ACM International Conference on Conference on Information and Knowledge Management, pp. 739–748. ACM (2014)

17. Luong, M.T., Pham, H., Manning, C.D.: Effective approaches to attention-based neural machine translation. arXiv preprint arXiv:1508.04025 (2015)

18. Noulas, A., Scellato, S., Lathia, N., Mascolo, C.: Mining user mobility features for next place prediction in location-based services. In: IEEE 12th International Conference on Data Mining, pp. 1038–1043 (2012)

19. Pincus, S.M.: Approximate entropy as a measure of system complexity. Proc. Nat. Acad. Sci. $88(6)$, 2297–2301 (1991)

20. Rendle, S., Freudenthaler, C., Schmidt-Thieme, L.: Factorizing personalized markov chains for next-basket recommendation. In: Proceedings of the 19th International Conference on World Wide Web, pp. 811–820. ACM (2010)

Mentor Pattern Identification
from Product Usage Logs

Ankur Garg[1], Aman Kharb[2], Yash H. Malviya[3], J. P. Sagar[4], Atanu R. Sinha[5],
Iftikhar Ahamath Burhanuddin[5], and Sunav Choudhary[5(✉)]

[1] The University of Texas at Austin, Austin, USA
ankgarg@cs.utexas.edu
[2] Indian Institute of Technology Kharagpur, Kharagpur, India
kharb.aman00@gmail.com
[3] Apple Inc., Cupertino, USA
yashmalviya94@gmail.com
[4] Indian Institute of Technology Madras, Chennai, India
jpsagarm95@gmail.com
[5] Adobe Research, Bangalore, India
{atr,burhanud,schoudha}@adobe.com

Abstract. A typical software tool for solving complex problems tends to expose a rich set of features to its users. This creates challenges such as new users facing a steep onboarding experience and current users tending to use only a small fraction of the software's features. This paper describes and solves an unsupervised mentor pattern identification problem from product usage logs for softening both challenges. The problem is formulated as identifying a set of users (mentors) that satisfies three mentor qualification metrics: (a) the mentor set is small, (b) every user is close to some mentor as per usage pattern, and (c) every feature has been used by some mentor. The proposed solution models the task as a non-convex variant of an ℓ_1-norm regularized logistic regression problem and develops an alternating minimization style algorithm to solve it. Numerical experiments validate the necessity and effectiveness of mentor identification towards improving the performance of a k-NN based product feature recommendation system for a real-world dataset. Further, t-SNE visuals demonstrate that the proposed algorithm achieves a trade-off that is both quantitatively and qualitatively distinct from alternative approaches to mentor identification such as Maximum Marginal Relevance and K-means.

Keywords: Mentor identification · Unsupervised learning ·
Sparsity-coverage trade-off · L1-regularized logistic regression

All authors were affiliated with Adobe Research during the course of this work. An extended version of this work is available on https://arxiv.org.

Electronic supplementary material The online version of this chapter (https://doi.org/10.1007/978-3-030-16142-2_28) contains supplementary material, which is available to authorized users.

1   Introduction

Consider the set of software tools devised to assist in solving complex problems. Common examples include, but are not limited to, software that are classified as 3D modelling tools, image editing software or Integrated Development Environments (IDEs). These tools are often developed with numerous features, making them complicated to use and leading to multiple problems during adoption. For novice users, the onboarding process presents a steep learning curve, taking them months of practice and frustration to gain sufficient expertise [5] to use the tools productively. Furthermore, even after onboarding, most users don't use a large fraction of available features. For most users, this means that they use the software inefficiently [15] and are unable to guide or onboard new users. Figure 1 illustrates this issue with our real-world data.

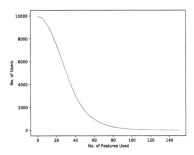

Fig. 1. Number of users vs Number of distinct features used for a software with 600 distinct features over a period of two months. Rapid decay suggests that most users use very few features.

Herein, we develop a method to soften the above challenges. We term the core problem as *mentor pattern identification* and propose an algorithm to solve it. A mentor is described by [6] as "someone who helps someone else learn something that he or she would have learned less well, more slowly, or not at all if left alone." Informally, our core problem translates to identifying a small set of existing users who collectively possess expertise (as per usage pattern) across all features of the software tool. If this is successful, then the usage patterns from these users may serve as *mentor patterns* for virtual/algorithmic mentoring of other users.

Nature of the Data. The characteristics of the mentor pattern identification problem are tied to information captured in the data. The difficulty and possible approaches to the problem can change with the presence of auxiliary information.

Our dataset consists of usage logs for a feature rich image editing software over an observation period of two months. The logs contain timestamped entries for each feature that was accessed by each registered user at each time point during the observation period. Approximately 10,000 users accessed over 600 features every month, generating a few million log entries per month. The features offered by the software are categorized as being useful to either experts or non-experts. Henceforth, we use the shorthands E and NE, respectively, to refer to expert and non-expert feature categories. In our data, this categorization is known at the outset for each feature and this auxiliary information makes our setting unique. Ground truth labels for mentors are not available. Hence, the problem is an *unsupervised* task.

We make two broad assumptions for the work presented here. First, patterns present in usage logs can capture candidacy for mentorship. Second, a feature recommendation system can be used to gauge whether mentor identification is beneficial. Subject to these assumptions, the paper makes the following contributions:

1. We formulate the mentor pattern identification problem from product usage logs in an unsupervised setting. We model the mentor set as achieving three desirable properties called mentor qualifications, *viz.* the mentor set is small, every user is close to some mentor *w.r.t.* usage pattern, and every feature has a mentor.
2. To the best of our knowledge, we are the first to consider a unique setting where features are categorized as being useful to either expert or non-expert users. Apart from this, we make no other non-standard assumptions in addressing the problem. We design a mentor identification algorithm that utilizes this feature categorization and solves a non-convex extension of an ℓ_1-norm regularized logistic regression.
3. We provide experimental evidence that mentor identification can improve k-NN feature recommendation performance for a real-world dataset. We further show that our approach outperforms other intuitive mentor identification attempts based on Maximum Marginal Relevance (MMR), K-means and frequent user selection.

2   Problem Setup and Feature Extraction

Let $\mathcal{U}$ denote the set of users and $\mathcal{F}$ denote the set of features in the software. We define the mentor pattern identification problem as the identification of a mentor set $\mathcal{M} \subseteq \mathcal{U}$ that achieves the following properties, which we refer to as *mentor qualifications*:

1. *Size Metric:* $\mathcal{M}$ should be small in size relative to $\mathcal{U}$. This is reasonable if mentors tend to possess expert knowledge.
2. *Closeness Metric:* Every user in $\mathcal{U}$ should be 'close' to some mentor in $\mathcal{M}$ *w.r.t.* feature usage pattern. This is a proxy for matching every user with a suitable mentor.
3. *Coverage Metric:* Every feature in $\mathcal{F}$ should have been used by some mentor in $\mathcal{M}$. This is a proxy for coverage provided by collective expertise of the mentors.

To capture 'closeness', we need a metric space representation of users. Let $\mathcal{U} = \{1, 2, \ldots, |\mathcal{U}|\}$ and $\mathcal{F} = \{1, 2, \ldots, |\mathcal{F}|\}$ be the index sets corresponding to the users and the features, where $|\cdot|$ denotes cardinality. We construct a feature space matrix $\boldsymbol{X} \in \mathbb{R}_+^{|\mathcal{U}| \times |\mathcal{F}|}$ where $\boldsymbol{X}(u, f)$ represents a proficiency score for user u w.r.t. feature f and the vector $\boldsymbol{X}(u, :)$ represents the usage pattern of user u. We use cosine similarity between $\boldsymbol{X}(u, :)$ and $\boldsymbol{X}(u', :)$ to capture closeness between users u and u'.

We construct $\boldsymbol{X}$ as follows. Let $\mathcal{L}$ represent the dataset. For every $(f, u) \in \mathcal{F} \times \mathcal{U}$, let the set $\mathcal{L}(f, u)$ represent all timestamps at which user u accessed feature f. If u never accessed f, then $\mathcal{L}(f, u)$ is empty. To gauge user proficiency, we use three usage-metrics from the literature [13]. For these metrics to be well-defined, we assume that all unused features have been removed from $\mathcal{F}$.

1. *Depth of Usage:* This metric captures frequency of usage. We collect this usage-metric in a matrix $\boldsymbol{D} \in [0, 1]^{|\mathcal{F}| \times |\mathcal{U}|}$ defined element-wise as

$$\boldsymbol{D}(f, u) = \frac{|\mathcal{L}(f, u)|}{\sum_{u' \in \mathcal{U}} |\mathcal{L}(f, u')|}. \tag{1}$$

2. *Timestamp:* This metric captures recency of usage. We collect this usage-metric in a matrix $\boldsymbol{T} \in [0, 1]^{|\mathcal{F}| \times |\mathcal{U}|}$ defined element-wise as

$$\boldsymbol{T}(f, u) = 1 - \frac{t_{\text{obs}} - \max \mathcal{L}(f, u)}{t_{\text{obs}} - \min \mathcal{L}(f, u)}, \tag{2}$$

where we have assumed $t_{\text{obs}} > \max_{f, u} \max \mathcal{L}(f, u)$, and that $\max \mathcal{L}(f, u) = \min \mathcal{L}(f, u) = 0$ if $\mathcal{L}(f, u) = \emptyset$.

3. *Niche Access:* This metric captures the importance of more esoteric features that do not get used enough across the user base. We collect this usage-metric in a matrix $\boldsymbol{A} \in [0, 1]^{|\mathcal{F}| \times |\mathcal{U}|}$ defined element-wise as

$$\boldsymbol{A}(f, u) = \begin{cases} \frac{1}{\log_2 (1 + k(f, u))}, & \mathcal{L}(f, u) \neq \emptyset, \\ 0, & \text{otherwise}, \end{cases} \tag{3}$$

where the quantity $k(f, u) \triangleq |\{u' \in \mathcal{U} \mid \mathcal{L}(f, u') \neq \emptyset\}|$ is the number of users who have accessed feature f at least once.

By construction, if u scores higher than u' on any of the usage metrics w.r.t. some feature f, it suggests that u has higher proficiency than u' w.r.t. feature f. We define $\boldsymbol{X}$ by $\boldsymbol{X}^{\mathrm{T}} = \boldsymbol{D} + \boldsymbol{T} + \boldsymbol{A}$. We further define for subsequent use, a matrix $\boldsymbol{X}_w$ parametrized by $\boldsymbol{w}^{\mathrm{T}} = (w_1, w_2, w_3) \in [0, 1]^3$ as the weighted linear combination $\boldsymbol{X}_w^{\mathrm{T}} = w_1 \boldsymbol{D} + w_2 \boldsymbol{T} + w_3 \boldsymbol{A}$. Here superscript T denotes transpose.

3   Mentor Identification Algorithm

We design an alternating minimization style algorithm [8] to select a mentor set that captures mentor qualifications well. The algorithm solves a non-convex extension of an ℓ_1-norm regularized logistic regression and utilizes the E/NE feature categorization described in Sect. 1. The pseudo-code is given as Algorithm 1. Salient assumptions behind the construction of the algorithm are described below.

Since each feature in $\mathcal{F}$ is categorizable as being useful to either experts (E) or non-experts (NE), $\mathcal{F}$ admits a binary labeling scheme. Let $y \in \{0,1\}^{|\mathcal{F}|}$ denote the vector of labels such that $y(f) = 1$ iff $f \in$ E. Each row of X_w^{T} represents the usage proficiencies (parametrized by w) for some feature across the set of all users. We assume that X_w^{T} can be used to explain the probability of each feature belonging to the E category, using the logistic regression model. Supposing that each label $y(f)$, $f \in \mathcal{F}$ is a realization of a Bernoulli random variable in $\{0,1\}$ and letting $(\beta_0, \boldsymbol{\beta}^{\mathrm{T}}) \in \mathbb{R}^{1+|\mathcal{U}|}$ denote the vector of regression coefficients, we have the label $y(f)$ is governed by the conditional probability

$$\Pr\big(y(f) = 1 \,|\, X_w(:, f); \beta, \beta_0\big) = \sigma\big(X_w(:, f)^{\mathrm{T}}\beta + \beta_0\big)$$
$$= \Big(1 + \exp\big(-X_w(:, f)^{\mathrm{T}}\beta - \beta_0\big)\Big)^{-1} \tag{4}$$

where $\sigma(t) \triangleq (1 + \exp(-t))^{-1}$ is the logistic function.

Since our goal is to identify the mentor set $\mathcal{M}$, we want to attach the following interpretation to the regression coefficients $\beta(u)$, $1 \le u \le |\mathcal{U}|$: *If $\beta(u)$ is non-zero and of significant magnitude, then user u belongs to the mentor set $\mathcal{M}$.* To encourage this interpretation of β in our learning algorithm, we introduce sparsity based regularization for β via the ℓ_1-norm penalty [17].

The problem at hand involves a joint estimation of all the unknowns w, β and β_0, resulting in a non-convex optimization problem. Hence we adopt an alternating minimization style algorithm. In Algorithm 1, $(\cdot)^{(k)}$ is used to specify values of iteratively calculated variables at iteration k, $\sigma(\cdot)$ denotes the logistic function, and $\|\beta\|_1 \triangleq \sum_{u \in \mathcal{U}} |\beta(u)|$ denotes the ℓ_1-norm of β. Of the three parameters ε, τ and C in Algorithm 1, the first two are tolerance parameters and we set them to $\tau = \varepsilon = 10^{-6}$ across all experiments. This is standard on many systems implementing optimization algorithms. C is the only effective parameter for which we have tested a wide range of values in our experiments.

Note that both subproblems (P$_1$) and (P$_2$) in Algorithm 1 are convex and are ℓ_1-norm regularized logistic regressions. We use efficient algorithms for the same to solve them [12]. Supplementary material provides some intuition as to why Algorithm 1 tends to select proficient users as mentors.

4   Experimental Results

A quantitative validation for unsupervised methods is challenging in general in the absence of ground truth. We evaluate the performance of algorithm in two ways. First, we analyze the mentor set, obtained using Algorithm 1, based on the mentor qualifications described in Sect. 2. Second, we show the effectiveness of our identified mentor set in improving the performance of a k-NN based

---

Algorithm 1. Mentor Identification Algorithm

Inputs: Factors $D, T, A \in [0,1]^{|\mathcal{F}| \times |\mathcal{U}|}$, feature labels $y(f) \in \{0,1\}$, $f \in \mathcal{F}$, ℓ_1-norm penalty multiplier $C > 0$, stopping tolerance $\varepsilon > 0$, and selection threshold $\tau > 0$.

Outputs: Estimated regression coefficients $\widehat{\beta} \in \mathbb{R}^{|\mathcal{U}|}$, $\widehat{\beta}_0 \in \mathbb{R}$, estimated weight vector $\widehat{w} = (\widehat{w}_1, \widehat{w}_2, \widehat{w}_3)^{\mathrm{T}} \in [0,1]^3_+$, and estimated mentor set $\mathcal{M}$.

Steps:

1. Initialize $w^{(0)} = 1$.
2. Let $t^{(k-1)} = X_{w^{(k-1)}}(:, f)^{\mathrm{T}} \beta + \beta_0$. At iteration $k \geq 1$, solve the ℓ_1-norm regularized logistic regression problem

$$\begin{aligned} \underset{\beta, \beta_0}{\text{minimize}} \quad & C\|\beta\|_1 + \sum_{f \in \mathcal{F}} \log \sigma\big((-1)^{y(f)} t^{(k-1)}\big) \\ \text{subject to} \quad & X^{\mathrm{T}}_{w^{(k-1)}} = w_1^{(k-1)} D + w_2^{(k-1)} T + w_3^{(k-1)} A, \end{aligned} \qquad (\mathrm{P}_1)$$

and let the solution be captured in $\beta^{(k)}$, $\beta_0^{(k)}$.
3. At iteration $k \geq 1$, solve the logistic regression problem

$$\begin{aligned} \underset{0 \leq w \leq 1}{\text{minimize}} \quad & \sum_{f \in \mathcal{F}} \log \sigma\big((-1)^{y(f)}(X_w(:, f)^{\mathrm{T}} \beta^{(k)} + \beta_0^{(k)})\big) \\ \text{subject to} \quad & X^{\mathrm{T}}_w = w_1 D + w_2 T + w_3 A, \end{aligned} \qquad (\mathrm{P}_2)$$

and let the solution be captured in $w^{(k)}$.
4. If $\frac{\|w^{(k-1)} - w^{(k)}\|_2}{\|w^{(k-1)}\|_2} \leq \varepsilon$ (*i.e.* relative change w.r.t. ℓ_2-norm in w across consecutive iterations is below the stopping tolerance) then set $K = k$ and go to next step, otherwise repeat steps 2 through 4 with $k \leftarrow k+1$ (*i.e.* next iteration).
5. Return $\widehat{\beta} = \beta^{(K)}$, $\widehat{\beta}_0 = \beta_0^{(K)}$, and $\widehat{w} = w^{(K)}$. Also return $\mathcal{M} = \{u \in \mathcal{U} \mid |\widehat{\beta}(u)| \geq \tau\}$, *i.e.* $\mathcal{M}$ consists of all indices at which elements of $\widehat{\beta}$ admit an absolute value of at least τ.

---

recommendation system. We also state the baseline approaches for comparative purposes and establish that our algorithm achieves a better trade-off among the evaluation metrics while improving the recommendation quality as well.

4.1   Dataset Details

The real-world dataset is described in Sect. 1. Table 1 provides some statistics of the dataset. For the purpose of validation through the recommendation algorithm, we let the test dataset $\mathcal{L}^{test}$ comprise of feature usage timestamps in the second month while timestamps of feature usage in the first month form the training dataset $\mathcal{L}$. Users whose feature usage timestamps are present in both $\mathcal{L}$ and $\mathcal{L}^{test}$ are about 6000 and denoted by $\mathcal{U}_{tt}$.

Table 1. Statistical characteristics of the real-world dataset

Statistic	Min	Max	Mean	Std. dev.
#E features used per user	0	40	3.87	4.00
#NE features used per user	1	152	30.55	16.63
#users per E feature	1	3216	227.48	473.32
#users per NE feature	1	7634	686.64	1222.05
E feature timestamps (days)	0	30	17.51	8.51
NE feature timestamps (days)	0	30	17.93	8.49

The real-world dataset mentioned above is owned by the software vendor and hence we cannot share it publicly. For this reason, in the spirit of reproducibility, we have also created a synthetic dataset that preserves empirical distributional characteristics (mean and variance of timestamps per user) and made it freely available under GNU GPL for download and reuse[1].

4.2 Evaluation Methodology

Mentor Qualifications Metrics. Quantitative comparison $w.r.t.$ mentor qualification metrics can be captured from the definitions in Sect. 2. The closeness metric can be captured as a scalar by the diversity value

$$div\_val_{\mathcal{M}} = 1.0 - \min_{u \in \mathcal{U} \setminus \mathcal{M}} \max_{m \in \mathcal{M}} \frac{X(u,:) \cdot X(m,:)}{\|X(u,:)\|_2 \|X(m,:)\|_2}. \tag{5}$$

If $div\_val_{\mathcal{M}}$ is low, for users there exist mentors who are similar to them in feature usage and thus, the mentor set is representative of all the different types of users in $\mathcal{U}$, a desirable characteristic. Further, we find support for this assertion by visualizing a geometrical representation of identified mentor and non-mentor sets from each baseline and our own approach, projected down to a 2-dimensional space, using the t-SNE plotting technique [14].

Two other qualification metrics are: the size metric, or, number of mentors $|\mathcal{M}|$, and the coverage metric represented by the number of features missed by the mentor set $\mathcal{M}$.

Evaluating Recommendations. We design a simple recommendation algorithm based on the popular k-NN collaborative filtering approach [10]. In this algorithm, recommendations are generated for a user based on the interpolated ratings of similar users. For our experiments, we find this set of similar users, for each user, from the mentor set. Cosine similarity between a user vector $X(u,:)$ and mentor vector $X(m,:)$ is used as a measure to find the top k similar mentors to a user in the test set. The vectors, $X(m,:)$, of the k most similar mentors are

---

[1] https://goo.gl/bEfE79.

then averaged along each feature to get the representative vector for a user u in the test set denoted by $\boldsymbol{X}_{avg}(u,:)$. We obtain the recommendation set for a user u, denoted by $\mathcal{R}_u$, by taking the top 10 highest valued features from $\boldsymbol{X}_{avg}(u,:)$. Now, $\{f \mid |\mathcal{L}^{test}(f,u)| > 0, f \in \mathcal{F}\}$ gives the set of features used by user u in the test set, and denoted by $\mathcal{T}_u$, then we can evaluate the performance of our recommendation using precision, recall and F_1-score.

Note that for this evaluation, we use the $\boldsymbol{X}_w$ matrix for mentor sets generated by our proposed approach and the $\boldsymbol{X}$ matrix for the baselines since it also depicts the benefit of assigning different weights to the three usage metrics.

4.3   Baselines for Comparison

Note that the *unsupervised setting* together with the nature of the usage log dataset and the absence of ground truth makes the problem unique. There are of course many natural baselines in the supervised setting [3] for similar problems that we cannot utilize in our setting. Under these restrictions, we compare our approach with the baselines mentioned below.

1. *Maximum Marginal Relevance (MMR):* This is a popular technique in text retrieval and summarization [4]. The algorithm is designed to reduce the redundancy in topics (present in each document) while retrieving a set of documents relevant to a given query. If we imagine topics as being product features and documents as users, then the importance of a topic in a document could be mapped to the proficiency score of a user u for the product feature f, *i.e.* $\boldsymbol{X}(u, f)$. With that hypothesis, the documents that would be found by the MMR algorithm will correspond to the mentors to be found in our problem setting. Since we want the selected mentor set to represent all the product features and be 'close' to the other users, there is an implied diversity that is necessary for our mentor set to have.

2. *K-means Clustering (with $K = 2$):* This is a popular algorithm for clustering (which is a form of unsupervised classification) [9]. Since our problem is about classifying each user as being either a mentor or a non-mentor in an unsupervised fashion, it can be trivial to think of using the K-means clustering algorithm with $K = 2$ clusters to arrive at a solution.

3. *Frequent Users:* Here we select users in the following manner: *For each feature $f \in \mathcal{F}$, select the user $u \in \mathcal{U}$, who uses that feature maximum number of times.* Mathematically, the set can be given by $\mathcal{M}_{freq} = \bigcup_{f \in \mathcal{F}} \{\arg\max_{u \in \mathcal{U}} |\mathcal{L}(u, f)|\}$. This follows from [13] and is a valid baseline in our case since it seems like a simple and intuitive way of generating the mentor set.

4. *Everyone:* We study the utility of using a mentor set to generate recommendations by using a baseline where no mentor identification algorithm is used prior to generating the recommendations and thus, the k-NN algorithm [16], finds the top k similar users from the whole user set $\mathcal{U}$. We denote this baseline by $\mathcal{M}_{\mathcal{U}}$.

Table 2. Performance of mentor identification algorithms on the real-world dataset

(a) Mentor Qualification metrics (lower is better)

Mentors	#Mentors	div_val	#Missed
$\mathcal{M}_{C=10^1}$	133	0.999	115
$\mathcal{M}_{C=10^0}$	404	0.934	2
$\mathcal{M}_{C=10^{-1}}$	324	0.962	0
$\mathcal{M}_{C=10^{-2}}$	387	0.934	0
$\mathcal{M}_{C=10^{-3}}$	399	0.897	0
$\mathcal{M}_{C=10^{-4}}$	481	0.876	0
$\mathcal{M}_{C=10^{-5}}$	845	0.862	0
$\mathcal{M}_{C=10^{-6}}$	2336	0.832	0
$\mathcal{M}_{mmr}$	9864	0.674	0
$\mathcal{M}_{2-means}$	2312	0.994	58
$\mathcal{M}_{freq}$	429	0.911	0

(b) Recommendation metrics on test set using $k = 30$ (higher is better)

Mentors	Prec (%)	Rec (%)	F_1 (%)
$\mathcal{M}_{C=10^1}$	44.55	19.85	27.46
$\mathcal{M}_{C=10^0}$	45.09	20.28	27.94
$\mathcal{M}_{C=10^{-1}}$	45.00	20.22	27.90
$\mathcal{M}_{C=10^{-2}}$	44.87	20.09	27.75
$\mathcal{M}_{C=10^{-3}}$	44.71	20.18	27.81
$\mathcal{M}_{C=10^{-4}}$	44.90	20.27	27.93
$\mathcal{M}_{C=10^{-5}}$	45.42	20.56	28.31
$\mathcal{M}_{C=10^{-6}}$	44.53	20.75	28.51
$\mathcal{M}_{mmr}$	44.36	20.45	27.99
$\mathcal{M}_{2-means}$	45.51	20.47	28.24
$\mathcal{M}_{freq}$	44.11	20.16	27.67
$\mathcal{M}_{\mathcal{U}}$	44.54	20.56	28.14

4.4   Comparison of Results

Since Algorithm 1 admits the tuning parameter C that controls the regularization *w.r.t.* the ℓ_1-norm penalty, we have evaluated the results of mentor qualifications across different values of C and then selected the best value of C for qualitative analysis via t-SNE plots. In all future references, $\mathcal{M}_{C=10^n}$, refers to the mentor set obtained using Algorithm 1 with $C = 10^n$. For the other baselines discussed in Sect. 4.3, we use the best parameter settings obtained by trial and error.

Analysis of Mentor Set Based on Mentor Qualifications. Table 2a shows the size of the mentor set, diversity value (div_val) and the number of features represented by mentor set *w.r.t.* the real-world dataset for each algorithm. Recall from Sect. 4.2 that lower values of div_val and number of features missed are desired. Further, lower number of mentors selected is preferred. We notice that while MMR does not miss any feature and has the lowest diversity value, it selects too many mentors. This is because our stopping criterion in MMR ensures that no features are missed. If instead we put a constraint on the size of the mentor set, then MMR misses many features. Thus, in both situations MMR performs worse than our approach. On the other hand, our algorithm with $C = 10^1$ selects very few mentors, but misses many features and has a high diversity value, both of which are undesirable. In general, there seems to be a trade-off between the size of the mentor set, the number of features missed and the diversity value. We pick the value $C = 10^{-5}$ as the operating point for our algorithm as it achieves a good balance among the three desirable qualification metrics. Figure 2 shows the t-SNE plots for the mentor sets identified by some of the algorithms.

We see that the $C = 10^{-5}$ and the frequency based ($\mathcal{M}_{freq}$) algorithm show a healthy spread in the set of mentors selected as per the t-SNE plots and correspondingly, they achieve lower div_val with a small mentor set in Table 2a. The 2-means algorithm selects mentors that cluster around one part of the t-SNE plot (see Fig. 2d) and correspondingly, achieves a poor trade-off in Table 2a by selecting too many mentors. Another observable characteristic of our algorithm as seen from Table 2a is that as value of C decreases, the number of mentors selected increases and more features are progressively covered until all features are represented.

We also point out that for this dataset our algorithm's run time is $3-4$ min whereas, the MMR baseline takes a few hours to identify a mentor set. This highlights that our approach is efficient and effective in identifying a mentor set which does well on mentor qualifications.

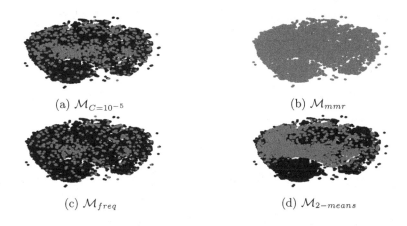

(a) $\mathcal{M}_{C=10^{-5}}$ (b) $\mathcal{M}_{mmr}$

(c) $\mathcal{M}_{freq}$ (d) $\mathcal{M}_{2-means}$

Fig. 2. Results on real-world dataset. Mentors indicated by red dots. (Color figure online)

Recommendation Performance. Table 2b shows the precision, recall and F_1-score value of recommendation evaluated on the test set, $\mathcal{L}^{test}$, of the real-world dataset using $k = 30$. First of all, we notice that without mentor identification, i.e. $\mathcal{M}_{\mathcal{U}}$, the recommendation quality is really poor. This indicates that the dataset and the scenario are particularly challenging for recommendation generation. We further notice that almost all of the mentor identification resultant sets lead to better downstream recommendation performance, suggesting that such techniques may be needed to boost recommendation quality, at least in the setup described, and warrant further investigation from the research community.

Based on the discussion in Sect. 4.4, we consider methods for which the number of features represented by the mentor set is equal to $|\mathcal{F}|$ in Table 2b. Finally, we note that $C = 10^{-5}$ is slightly better than $\mathcal{M}_{freq}$ at recommendation quality.

This effect is due to the lower value of div_val combined with more number of mentors for $C = 10^{-5}$. Other things being equal, lower div_val and higher $|\mathcal{M}|$ work as proxy for better representation which translates to better recommendations.

5 Related Work[2]

To the best of our knowledge, there does not exist a widely accepted and mathematically precise notion of a mentor. The absence of such a notion makes the problem challenging. Our experiments in Sect. 4.4 show that simple alternative approaches to mentor identification, such as MMR [4] and selecting frequent users for each feature, may not work well. Mentor identification is an important problem and some attempts have been made in the literature with application specific modeling assumptions under restrictive contexts [2,3]. The nature of our data is different and less restrictive from these prior works, warranting a fresh approach.

Usage data have been explored in the literature to help recommendation of features in a variety of scenarios. For example, [7,13] rely on usage of features in a software to make recommendation based on usage expertise. Unlike prior art in this area, we recognize that all features are not created equal and categorize them as E or NE. Further, we allow for a user to be a mentor for NE features, without being a mentor for E features, and vice versa. This distinction is not possible when applying the approach in [13] to our data and can adversely affect the mentor qualification metrics.

One of the early papers [11], explaining the Lumière project, relies on usage data although the interest is primarily to infer goals and needs of users, but not to identify mentors, nor expertise in usage of features. A distinction of our approach is that our end goal is not to build a recommendation system, but to identify mentors among users of a software. The identified mentor set can be used as input for a broad set of end goals, one of which can be making recommendations.

6 Conclusion

We formulate the problem of identifying a set of mentors from software usage logs in an unsupervised setting such that they satisfy three desirable properties termed as mentor qualifications. We present an algorithmic solution for the problem and showcase experimental results on a real-world dataset. We illustrate that our approach improves over baselines with regard to mentor qualifications. An application of our work is guided recommendation of features to a user of a software and show that the mentor set identified by our approach can be used to improve k-Nearest Neighbor based recommendations in an effective and efficient manner. Future work will focus on alternative ways to featurize and define patterns in sequential data ($e.g.$ [1]) to improve upon our current results.

---

[2] The authors thank Dr. Moumita Sinha for pointers to relevant literature.

References

1. Aggarwal, C.C., Bhuiyan, M.A., Hasan, M.A.: Frequent pattern mining algorithms: a survey. In: Aggarwal, C.C., Han, J. (eds.) Frequent Pattern Mining, pp. 19–64. Springer, Cham (2014). https://doi.org/10.1007/978-3-319-07821-2_2
2. Amatriain, X., Lathia, N., Pujol, J.M., Kwak, H., Oliver, N.: The wisdom of the few: a collaborative filtering approach based on expert opinions from the web. In: Proceedings of the 32nd International ACM SIGIR Conference on Research and Development in Information Retrieval, SIGIR 2009, pp. 532–539. ACM, New York (2009). https://doi.org/10.1145/1571941.1572033
3. Campbell, C.S., Maglio, P.P., Cozzi, A., Dom, B.: Expertise identification using email communications. In: Proceedings of the Twelfth International Conference on Information and Knowledge Management, CIKM 2003, pp. 528–531. ACM, New York (2003). https://doi.org/10.1145/956863.956965
4. Carbonell, J., Goldstein, J.: The use of MMR, diversity-based reranking for reordering documents and producing summaries. In: Proceedings of the 21st Annual International ACM SIGIR Conference on Research and Development in Information Retrieval, SIGIR 1998, pp. 335–336. ACM, New York (1998). https://doi.org/10.1145/290941.291025
5. Fernandes, J.F.N.: Softening the Learning Curve of Software Development Tools. Master's thesis, Universidade Técnica de Lisboa, Portugal, October 2011
6. Goldsmith, M., Lyons, L., Freas, A.: Coaching for Leadership: How the World's Greatest Coaches Help Leaders Learn. Jossey Bass/Pfeiffer, Hoboken (2000)
7. Grossman, T., Fitzmaurice, G.: An investigation of metrics for the in situ detection of software expertise. Hum.-Comput. Interact. **30**(1), 64–102 (2015)
8. Gupta, M.R., Chen, Y.: Theory and Use of the EM Algorithm, vol. 4. Now Publishers Inc., Delft (2011)
9. Hartigan, J.A., Wong, M.A.: Algorithm as 136: a k-means clustering algorithm. J. R. Stat. Soc. Ser. C (Appl. Stat.) **28**(1), 100–108 (1979)
10. Herlocker, J.L., Konstan, J.A., Borchers, A., Riedl, J.: An algorithmic framework for performing collaborative filtering. In: Proceedings of the 22nd Annual International ACM SIGIR Conference on Research and Development in Information Retrieval, SIGIR 1999, pp. 230–237. ACM, New York (1999). https://doi.org/10.1145/312624.312682
11. Horvitz, E., Breese, J., Heckerman, D., Hovel, D., Rommelse, K.: The lumiere project: bayesian user modeling for inferring the goals and needs of software users. In: Proceedings of the Fourteenth Conference on Uncertainty in Artificial Intelligence, UAI 1998, pp. 256–265. Morgan Kaufmann Publishers Inc., San Francisco (1998)
12. Lee, S.I., Lee, H., Abbeel, P., Ng, A.Y.: Efficient l1 regularized logistic regression. In: Proceedings of the Twenty-first National Conference on Artificial Intelligence (AAAI-06), pp. 1–9 (2006)
13. Ma, D., Schuler, D., Zimmermann, T., Sillito, J.: Expert recommendation with usage expertise. In: 25th IEEE International Conference on Software Maintenance (ICSM 2009), 20–26 September 2009, Edmonton, Alberta, Canada, pp. 535–538 (2009). https://doi.org/10.1109/ICSM.2009.5306386
14. van der Maaten, L., Hinton, G.: Visualizing data using t-SNE. J. Mach. Learn. Res. **9**(Nov), 2579–2605 (2008)

15. Matejka, J., Li, W., Grossman, T., Fitzmaurice, G.: CommunityCommands: command recommendations for software applications. In: Proceedings of the 22nd Annual ACM Symposium on User Interface Software and Technology, UIST 2009, pp. 193–202. ACM, New York (2009). https://doi.org/10.1145/1622176.1622214
16. Resnick, P., Iacovou, N., Suchak, M., Bergstrom, P., Riedl, J.: GroupLens: an open architecture for collaborative filtering of netnews. In: Proceedings of the 1994 ACM Conference on Computer Supported Cooperative Work, CSCW 1994, pp. 175–186. ACM, New York (1994)
17. Wu, T.T., Chen, Y.F., Hastie, T., Sobel, E., Lange, E.: Genome-wide association analysis by lasso penalized logistic regression. Bioinformatics **25**(6), 714–721 (2009). https://doi.org/10.1093/bioinformatics/btp041

Visual Data Mining

AggregationNet: Identifying Multiple Changes Based on Convolutional Neural Network in Bitemporal Optical Remote Sensing Images

Qiankun Ye, Xiankai Lu, Hong Huo, Lihong Wan, Yiyou Guo, and Tao Fang$^{(\boxtimes)}$

Department of Automation,
School of Electronic Information and Electrical Engineering,
Shanghai Jiao Tong University, Shanghai, China
{yqksjtu,carrierlxk,huohong,wanlihong917,guo,tfang}@sjtu.edu.cn

Abstract. The detection of multiple changes (i.e., different change types) in bitemporal remote sensing images is a challenging task. Numerous methods focus on detecting the changing location while the detailed "from-to" change types are neglected. This paper presents a supervised framework named AggregationNet to identify the specific "from-to" change types. This AggregationNet takes two image patches as input and directly output the change types. The AggregationNet comprises a feature extraction part and a feature aggregation part. Deep "from-to" features are extracted by the feature extraction part which is a two-branch convolutional neural network. The feature aggregation part is adopted to explore the temporal correlation of the bitemporal image patches. A one-hot label map is proposed to facilitate AggregationNet. One element in the label map is set to 1 and others are set to 0. Different change types are represented by different locations of 1 in the one-hot label map. To verify the effectiveness of the proposed framework, we perform experiments on general optical remote sensing image classification datasets as well as change detection dataset. Extensive experimental results demonstrate the effectiveness of the proposed method.

Keywords: Multiple change detection · Remote sensing · Aggregation network

1 Introduction

The aim of change detection is to compare two remote sensing images obtained from different times at the same location. With the development of earth observation, change detection has wide applications including ecosystem management, urban expansion and land cover monitoring [5,6,16]. Generally, change detection can be categorized into **multiple change detection** [15,19] (i.e., different change types) and **binary change detection** [11]. The former task describes

© Springer Nature Switzerland AG 2019
Q. Yang et al. (Eds.): PAKDD 2019, LNAI 11441, pp. 375–386, 2019.
https://doi.org/10.1007/978-3-030-16142-2_29

different kinds of changes, such as the change appearing among the land cover types. The latter task segments the input images into changed and unchanged regions.

In the literature, change detection algorithms can be distinguished to unsupervised [2–4,11,12,17,21] and supervised [17,20].

Traditional unsupervised methods are common in the domain of change detection. The change vector analysis (CVA) is used to compute the change information in [3]. Liu et al. [11] explore the local image descriptor to identify changes without image co-registration. Benedek et al. [2] investigate a multi-layer conditional mixed Markov (CXM) model for optical image change detection. Besides, the authors released a publicly available dataset called AirChange Benchmark set. These unsupervised methods greatly depend on the statistical property or the hand-crafted features that are easily affected by the image quality. Neural network can extract more robust features [13]. Thus the deep learned features are adopted. Gong et al. [7] proposed a method using The Restricted Boltzmann Machine (RBM) network to distinguish the changed and unchanged areas in synthetic aperture radar images. Liu et al. [12] proposed an unsupervised symmetric convolutional coupling network (SCCN) for change detection of optical and radar images. However, all the algorithms mentioned above can not discriminate different kinds of change. A few unsupervised methods are also applied to solve the multiple change detection. Bovolo et al. [4] proposed a compressed CVA framework (C^2VA) to detect multiple changes from bitemporal images. Considering the drawback of the C^2VA that the available spectral channels can be affected by noise, Zhang et al. [21] proposed a framework adopting the deep belief network (DBN) to transform the spectral channels into an abstract feature space. These two algorithms do not require any labeled training data and can be easily applied in some cases. Although these two algorithms can distinguish different kinds of changes, they can not identify the specific "from-to" change types (e.g. from farmland to building). However, this specific "from-to" change types is very important to some tasks such as land survey and so on. Furthermore, all these unsupervised methods do not use the ground truth and usually rely on some prior assumptions. However, irrelevant variations always exist in bitemporal optical remote sensing images, such as noise interference and illumination. Therefore, it may be challenging for the unsupervised methods in some situation.

In the literature, many efforts have been made to develop supervised change detection algorithms on account of the drawbacks of the unsupervised methods mentioned above. Zhan et al. [20] proposed a deep siamese convolutional network (DSCN) with contrastive loss [8] to determine whether the input images belong to unchanged regions or changed regions [8]. CNNs have many structures [14]. And the siamese convolutional network is one of them. Thus DSCN can extract better features for change detection. However, the application of DSCN is limited by defining the difficult pixel level labels. In addition, the DSCN can not detect multiple changes, let alone identify the specific "from-to" change types. Another intuitional method named post-classification comparison is classifying the two bitemporal images to create classification maps and then compare the classification maps pixel by pixel to estimate the multiple types change detection [17]. However, this approach did not consider temporal correlation of the two bitemporal images and critically depended on the accuracies of the single classification maps. And (under the assumption of independent errors in the maps) it is close to the product of the accuracies of the two images [4].

In this study, we focus on a much more challenging multiple change detection task, i.e., our model not only determines changed regions but also identifies the "from-to" change types. We decompose this "from-to" procedure into a multiple classification issue. In order to avoid creating difficult pixel level labels mentioned above, We use a patch level labels as a substitute. That is to say a dataset that containing all the classes of patches in the two bitemporal images will be set. Then a **one-hot label map** representing a specific change type (e.g., from farmland to building) will be set to label the two classes of patches (e.g., the first class is farmland and the second class is building). Then considering the temporal correlation of the two bitemporal image patches, we proposed a novel framework called **AggregationNet** to guarantee the effect of change detection by adopting the temporal correlation. This AggregationNet takes the image pairs as the input and outputs a probability map which can be used to identify the change types as the location of the maximum value in the probability map represents a specific change type. The AggregationNet consists of feature extraction part and a feature aggregation part. The feature extraction part used to extract the deep "from-to" features is a two-branch deep convolutional network which is composed of two backbone networks (e.g., Residual Network [9]). The feature aggregation part which is composed of a few convolution layers, two fully connected layers and a reshape layer is adopted to the mine temporal correlation of the two bitemporal image patches. The reshape layer is used to change the dimensions of features.

The remainder of this paper is organized as follows. The proposed method is described in Sect. 2. The experimental organization and results are presented in Sect. 3. Finally, we draw the conclusion of this paper in Sect. 4.

2   Proposed Method

We formulate the multiple change detection as a multiple classification task. An AggregationNet is built for this task and a one-hot label map is proposed for training this network. Figure 1 shows the detailed structure of the proposed AggregationNet and Fig. 2 shows the proposed one-hot label map.

2.1   Network Architecture

The proposed AggregationNet mainly consists of two parts: a two-branch feature extraction part and a feature aggregation part. The size of the output of the AggregationNet equals to the size of the one-hot label map. In the training phase,

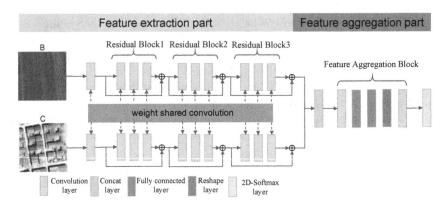

Fig. 1. Architecture of the proposed AggregationNet. Two images (belongs to class B and class C) are fed into the feature extraction part, and then feature aggregation part integrates the output feature for classification.

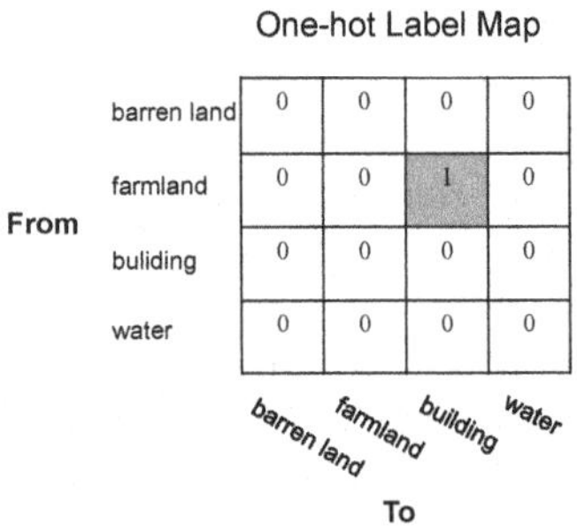

Fig. 2. The one-hot label map. The black region shown in the proposed one-hot label map indicates that the input image pairs belong to class *farmland* and *building*, respectively.

the cross entropy loss is applied to the output of the AggregationNet and the one-hot label map. In the testing phase, the output of AggregationNet is a probability map and the location of maximum value in the probability map represents the specific "from-to" change type of the two input patches. As shown in Fig. 1. Rather than comparing the outputs of two-branch convolution features directly, the proposed network aggregates the output features via a feature aggregation block.

Feature Extraction Part. The discriminating features of the input patches are extracted by the feature extraction part. The feature extraction part consists of a two-branch CNNs with weights shared. Specifically, Residual Network [9] is adopted as the backbone. ResNet is a variant of CNNs with shortcut connections. These shortcut connections and their corresponding convolutional layers constitute virtual residual blocks. An illustration of a residual block is shown in

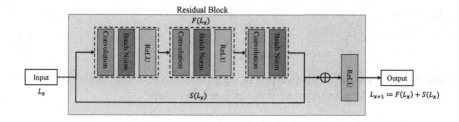

Fig. 3. A residual block in a ResNet.

Fig. 3. The residual block consists of three convolution layers. A batch normalization layer and a ReLU layer are implemented after every convolution layer. Each branch in this feature extraction part includes of a single convolution layer and subsequent three residual blocks. The kernel size of the first convolution layer is 7×7 and each residual block consists of three convolution layers with kernel sizes of 1×1, 3×3 and 1×1 as in [9].

Feature Aggregation Part. The feature aggregation part is composed of a few convolution layers, two fully connected layers and a reshape layer. The kernel sizes of the convolution layers are 1×1 because the sizes of the feature maps are small. The extracted features of two input images via the feature extraction part are concatenated by a *concat* layer in the feature aggregation part. After that, a convolution layer is added for information integrating before inputting to two fully connected (FC) layers. The FC layers work as a non-local operation which considers the global information for judging the changed area. It is noted that the nodes of the final fully connected layer are equal to the $2 * k^2$ where k is the number of the patch classes. As a result, we can employ a reshape layer to make the AggregationNet's output be two $k \times k$ maps. Then, we employ a convolution layer to perform high-level semantic feature aggregation [22] again with one output channel. A two-dimensional soft-max layer is added at the end of the network to normalize the output to the scale of [0, 1].

2.2 One-Hot Label Map

To train the proposed AggregationNet, a one-hot label map is designed for describing the "from-to" procedure. The size of the one-hot label map will increase with the growth of the classes of patches in a two-dimension style. If the number of the classes of the patches is k, the size of the label map is $k \times k$. That means the number of the possible change types is $k \times k$. Different cells of the label map represent different "from-to" change types. As shown in Fig. 2, if the input images belong to farmland and building respectively, then the corresponding label is a matrix in which only the *farmland-building* position is set to 1 and other elements are set to 0. Via this one hot label map, we transform the multiple change detection task into a multi-class classification task. Therefore, we can impose a cross entropy loss to train the AggregationNet. The cross entropy loss is:

$$\mathcal{L} = -\sum_{j=1}^{k \times k} y_j log(p_j) \tag{1}$$

where p_j is an element in the output of AggregationNet and y_j is the label value in the corresponding position of the one hot label map.

3   Experiment

Extensive experiments are conducted to demonstrate the effectiveness of the proposed AggregationNet. The reasonability of the proposed AggregationNet is validated on SAT-4 and SAT-6 datasets [1] compared with the post-classification method [17]. To highlight the effectiveness of the proposed method, AggregationNet is compared with other unsupervised and supervised methods on TISZADOB dataset. Furthermore, two extra newly built datasets are adopted to reflects how the proposed AggregationNet handles the "from-to" types in change detection.

3.1   Data and Experimental Settings

Dataset. Firstly, we perform the multiple change detection at the image level to validate the reasonability of the proposed AggregationNet on SAT-4 and SAT-6 datasets [1]. SAT-4 dataset contains 500000 image patches, including four classes: barren land, grassland, trees and others. SAT-6 contains 405000 image patches, covering 6 classes: barren land, trees, grassland, roads, building and water bodies. The image patches in these two datasets are with the size of 28 × 28 pixels. To validate the effectiveness of the proposed method, AggregationNet is compared with the state of art unsupervised and supervised methods on the change detection dataset named TISZADOB. TISZADOB is a subset in SZTAKI AirChange Benchmark Set [2]. TISZADOB consists of a pair of images and a label image with size 952 × 640 pixels. The resolution of each image is 1.5 m/pixel. Finally, we build two extra image pairs and evaluate the proposed AggregationNet on this new collected dataset named village and suburban. The resolutions of the village and suburban images are 0.5 m/pixel and 0.1m/pixel. The sizes of these two image pairs are 890 × 1024 pixels and 2304 × 1024 pixels separately.

Training and Testing Setting. To train AggregationNet, we first build the training pairs as [18]. Each class combines with all classes in the training set to compose different training pairs of different classes and each class of training pairs will be labeled with one one-hot label map. For SAT-4 and SAT-6 datasets, 70% of the total samples are randomly selected for training and the remaining are used for testing. For the change detection dataset, we adopt the training strategy proposed by [20].We crop the top-left corner of each image pair to 784×448 as the test set. And the rest constitute the training set. The regions in training set are cut to 28 × 28 patches overlapped. Then these patches are treated in the same way as the SAT datasets. We implement the proposed AggregationNet using the *caffe* [10], the training and testing are performed on an NVIDIA GTX 1080TI GPU. Stochastic gradient descent (SGD) is employed to train the network. We set the network learning rate to 0.001 and weight decay to 0.0005. The batch size is 64. The trained network is obtained after about 100 epoches.

Table 1. Comparison with three methods on the TISZADOB dataset. Best results viewed in bold.

Dataset	Methods	Measures		
		Precision (%)	Recall (%)	F-rate (%)
TISZADOB	CXM [2]	61.7	93.4	74.3
	SCCN [7]	**92.7**	79.8	85.8
	DSCN [20]	88.3	85.1	86.7
	AggregationNet	89.4	**95.6**	**92.4**

	barren land	grassland	others	trees
barren land	0.96	0.97	0.98	0.97
grassland	0.98	0.98	0.99	0.99
others	0.97	0.98	0.99	1.00
trees	0.98	0.99	1.00	0.99
OA	0.9816			

	barren land	building	grass land	road	trees	water
barren land	0.96	0.98	0.96	0.98	0.98	0.98
building	0.97	0.99	0.98	1.00	0.99	1.00
grass land	0.97	0.98	0.98	0.98	0.98	0.99
road	0.98	0.99	0.99	1.00	0.99	1.00
trees	0.97	1.00	0.98	1.00	0.99	0.99
water	0.98	1.00	0.99	1.00	1.00	1.00
OA	0.9857					

Fig. 4. Multiple change detection with the proposed AggregationNet on SAT-4 (left) and SAT-6 (right) datasets.

3.2   Results and Analysis

SAT Datasets. For SAT-4 and SAT-6, our proposed AggregationNet achieves overall accuracy (OA) of 98.16% and 98.57%, respectively. The detailed classification precision matrices for all classes are shown in Fig. 4. In order to verify the effectiveness of AggregationNet, we implement experiments to post-classification

	barren land	grassland	others	trees
barren land	0.94	0.95	0.96	0.96
grassland	0.95	0.96	0.97	0.97
others	0.96	0.97	0.98	0.99
trees	0.96	0.97	0.99	0.99
OA	0.9669			

	barren land	building	grass land	road	trees	water
barren land	0.93	0.96	0.94	0.96	0.97	0,97
building	0.96	0.98	0.96	0.98	0.99	0.99
grass land	0.94	0.96	0.94	0.97	0.97	0.97
road	0.96	0.98	0.97	0.99	0.99	0.99
trees	0.97	0.99	0.97	0.99	1.00	1.00
water	0.97	0.99	0.97	0.99	1.00	1.00
OA	0.9737					

Fig. 5. Multiple change detection with the backbone network on SAT-4 (left) and SAT-6 (right) datasets.

method. Thus, the backbone network of AggregationNet(the ResNet of the feature extraction part) is singly applied to all classes of SAT-4 and SAT-6 dataset twice. The backbone network achieves OAs of 96.69% and 97.37%, respectively. The detailed classification precision matrices for all classes are shown in Fig. 5.

In detail, there are 4^2 and 6^2 "from-to" change types for SAT-4 and SAT-6 dataset, and our method performs well on both changed pairs and unchanged pairs. For all unchanged types (i.e., the diagonal of the precision matrix) and the changed types, the corresponding OAs are higher than 96% and more competitive with the ones of the backbone network. In the results, the first row and the first column have inferior performance. The reason is that some images belonging to the "barren land" class are similar to the "grassland" class. This case also appears in the experiments with the backbone network. Overall, the proposed AggregationNet improves the overall accuracy by 1.47% and 1.20% on SAT-4 and SAT-6 dataset respectively. We speculate that the reason is AggregationNet not only extract the discriminating features of each input patch but also exploit the temporal correlation of the two input patches. The results verify the validity of the proposed AggregationNet and the proposed one-hot label map.

AirChange and Our Collected Dataset. To further evaluate the proposed AggregationNet, we compare it to CXM [2] as well as recent deep learning based methods: SCCN [7] and DSCN [20] on change detection task. A pair of images of TISZADOB is used for evaluation. To construct training and testing batches, we crop the images into patches with size of 28 × 28 and stride to 20 pixels. As a result, our method is based on patch level instead of pixel level (SCCN [7] and DSCN [20]). Other experimental settings follow the protocol in [20].

There are three evaluation criteria: precision, recall and F-measure rate which is the harmonic mean of precision and recall. The overall quantitative results are listed in Table 1. The proposed AggregationNet ranks first among all compared

Fig. 6. Visualization results of different methods for change detection on the TISZADOB dataset. The changed areas are with white color. (a) and (b) are image pairs; (c) Ground-truth; (d) CXM; (e) SCCN; (f) DSCN; (g) AggregationNet.

methods according to the recall (95.6%) and F-rate (92.4%). We visualize the change detection qualitative example in Fig. 6. The reason of the excellent performance is AggregationNet not only extracts the discriminating features of each input patch but also explore the temporal correlation of the input pairs. Furthermore, AggregationNet is a patch level method. Compared with pixel level change detection methods, AggregationNet can suppress the false positive samples effectively. Thus AggregationNet achieve a more excellent performance in term of recall than other methods. Moreover, without extra post-processing (e.g., k-NN in [20]), our methods generate low-noise change map when compared with Fig. 6(e) and (f). However, the patch level method can not distinguish the details in a patch. For example, AggregationNet can not precisely distinguish the edges of the changed regions. This drawback affects the performance in term of precision.

We further analyze the benefit of the proposed AggregationNet on a newly collected change detection dataset: village and suburban. We crop the images of the two datasets into patches with sizes of 28×28 pixels and 64×64 pixels. We choose a stride of 28 pixels and a stride of 64 pixels, separately. This dataset better reflects how the proposed AggregationNet handles "from-to" types in change detection. There are two image pairs in our collected dataset, as shown in Figs. 7 and 8. Figure 7 contains four classes: building, farm, greenhouse and vinylhouse, thus there are 16 potential change types. Figure 8 contains three classes: barren land, building, grassland, thus there are 9 potential change types.

Among these two datasets, (a) and (b) denote the input image pairs. (c) is the ground truth image. (d) visualizes the changed region and the corresponding change types with different colors. Considering all compared methods introduced on AirChange dataset can not deal with multiple change detection task, Figs. 7(d) and 8(d) only show the multiple change detection result based on the proposed AggregationNet. Based on patch level, the proposed AggregationNet can simultaneously detect the changed regions and identify the specific "from-to" change types. Apart from the qualitative evaluation result, we also report the quantitative results in terms of precision and recall for completeness in Table 2.

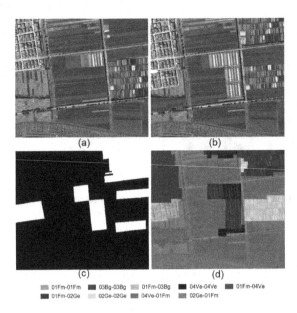

■ 01Fm-01Fm   ■ 03Bg-03Bg   ▨ 01Fm-03Bg   ■ 04Ve-04Ve   ■ 01Fm-04Ve
■ 01Fm-02Ge   ▨ 02Ge-02Ge   ■ 04Ve-01Fm   ▨ 02Ge-01Fm

Fig. 7. Our new collected village dataset and change detection results for the "form-to" change types using the proposed AggregationNet, where different colors indicate different change types. *Fm*, *Bg*, *Ve* and *Ge* indicate farmland, building, vinyl house and greenhouse, respectively.

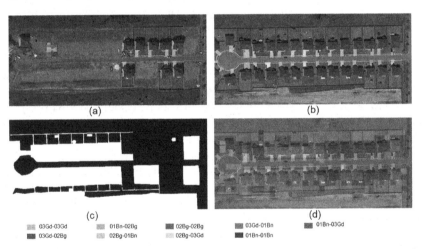

▨ 03Gd-03Gd   ▨ 01Bn-02Bg   ■ 02Bg-02Bg   ■ 03Gd-01Bn   ▨ 01Bn-03Gd
■ 03Gd-02Bg   ▨ 02Bg-01Bn   ▨ 02Bg-03Gd   ■ 01Bn-01Bn

Fig. 8. Our new collected suburban dataset and change detection results for the "form-to" change types using the proposed AggregationNet, where different colors indicate different change types. *Gd*, *Bn* and *Bg* mean grassland, barren land and building, respectively.

Table 2. Change detection results on new collected dataset with our proposed method.

Dataset	Measures	
	Precision (%)	Recall (%)
Village dataset	90.41	86.83
Suburban dataset	81.83	76.09

4 Conclusion

In this paper, we focus on the multiple change detection task based on deep learning. A novel CNN architecture called AggregationNet with one-hot label map is proposed. AggregationNet works well on optical remote sensing dataset (e.g., SAT-4 and SAT-6). Furthermore, the experimental results on change detection dataset illustrate that our AggregationNet can not only determine the changes region but also identify the specific "from-to" change types. Overall, this paper provides the first step to perform multiple change detection task based on deep learning. In the future, we plan to improve the detection results through reusing the features extracted by AggregationNet.

Acknowledgment. This study was partly supported by the National Science and Technology Major Project (21-Y20A06-9001-17/18), the National Key Research and Development Program of China (No. 2018YFB0505000), the National Natural Science Foundation of China (No. 41571402), the Science Fund for Creative Research Groups of the National Natural Science Foundation of China (No. 61221003).

References

1. Basu, S., Ganguly, S., Mukhopadhyay, S., DiBiano, R., Karki, M., Nemani, R.R.: DeepSat - a learning framework for satellite imagery. CoRR abs/1509.03602 (2015)
2. Benedek, C., Sziranyi, T.: Change detection in optical aerial images by a multilayer conditional mixed markov model. IEEE Trans. Geosci. Remote Sens. **47**(10), 3416–3430 (2009)
3. Bovolo, F., Bruzzone, L.: A theoretical framework for unsupervised change detection based on change vector analysis in the polar domain. IEEE Trans. Geosci. Remote Sens. **45**(1), 218–236 (2007)
4. Bovolo, F., Marchesi, S., Bruzzone, L.: A framework for automatic and unsupervised detection of multiple changes in multitemporal images. IEEE Trans. Geosci. Remote Sens. **50**(6), 2196–2212 (2012)
5. Che, M., Du, P., Gamba, P.: 2- and 3-D urban change detection with Quad-PolSAR data. IEEE Geosci. Remote Sens. Lett. **15**(1), 68–72 (2018)
6. Gong, M., Niu, X., Zhang, P., Li, Z.: Generative adversarial networks for change detection in multispectral imagery. IEEE Geosci. Remote Sens. Lett. **14**(12), 2310–2314 (2017)
7. Gong, M., Zhao, J., Liu, J., Miao, Q., Jiao, L.: Change detection in synthetic aperture radar images based on deep neural networks. IEEE Trans. Neural Netw. Learn. Syst. **27**(1), 125–138 (2016)

8. Hadsell, R., Chopra, S., LeCun, Y.: Dimensionality reduction by learning an invariant mapping. In: Proceedings of the IEEE Conference on Computer Vision and Pattern Recognition, pp. 1735–1742 (2006)

9. He, K., Zhang, X., Ren, S., Sun, J.: Deep residual learning for image recognition. In: Proceedings of the IEEE Conference on Computer Vision and Pattern Recognition, pp. 770–778 (2016)

10. Jia, Y., et al.: Caffe: convolutional architecture for fast feature embedding. In: Proceedings of ACM International Conference on Multimedia, pp. 675–678 (2014)

11. Liu, G., Delon, J., Gousseau, Y., Tupin, F.: Unsupervised change detection between multi-sensor high resolution satellite images. In: European Signal Processing Conference, pp. 2435–2439 (2016)

12. Liu, J., Gong, M., Qin, K., Zhang, P.: A deep convolutional coupling network for change detection based on heterogeneous optical and radar images. IEEE Trans. Neural Netw. Learn. Syst. **29**(3), 545–559 (2018)

13. Lu, X., Guo, Y., Liu, N., Wan, L., Fang, T.: Non-convex joint bilateral guided depth upsampling. Multimed. Tools Appl. **10**, 1–24 (2017)

14. Lu, X., Ma, C., Ni, B., Yang, X., Reid, I., Yang, M.: Deep regression tracking with shrinkage loss. In: ECCV, pp. 369–386 (2018)

15. Lv, P., Zhong, Y., Zhao, J., Zhang, L.: Unsupervised change detection model based on hybrid conditional random field for high spatial resolution remote sensing imagery. In: IEEE International Geoscience and Remote Sensing Symposium, pp. 1863–1866 (2016)

16. Ran, Q., Li, W., Du, Q.: Kernel one-class weighted sparse representation classification for change detection. Remote Sens. Lett. **9**(6), 597–606 (2018)

17. Singh, A.: Digital change detection techniques using remotely-sensed data. Int. J. Remote Sens. **10**(6), 989–1003 (1989)

18. Sun, Y., Wang, X., Tang, X.: Deep learning face representation from predicting 10, 000 classes. In: Proceedings of the IEEE Conference on Computer Vision and Pattern Recognition, pp. 1891–1898 (2014)

19. Wu, C., Zhang, L., Du, B.: Kernel slow feature analysis for scene change detection. IEEE Trans. Geosci. Remote Sens. **55**(4), 2367–2384 (2017)

20. Zhan, Y., Fu, K., Yan, M., Sun, X., Wang, H., Qiu, X.: Change detection based on deep siamese convolutional network for optical aerial images. IEEE Geosci. Remote Sens. Lett. **14**(10), 1845–1849 (2017)

21. Zhang, H., Gong, M., Zhang, P., Su, L., Shi, J.: Feature-level change detection using deep representation and feature change analysis for multispectral imagery. IEEE Geosci. Remote Sens. Lett. **13**(11), 1666–1670 (2016)

22. Zhang, P., Wang, D., Lu, H., Wang, H., Ruan, X.: Amulet: aggregating multi-level convolutional features for salient object detection. In: IEEE International Conference on Computer Vision (2017)

Detecting Micro-expression Intensity Changes from Videos Based on Hybrid Deep CNN

Selvarajah Thuseethan[✉], Sutharshan Rajasegarar, and John Yearwood

Deakin University, Geelong, VIC 3220, Australia
{tselvarajah,srajas,john.yearwood}@deakin.edu.au

Abstract. Facial micro-expressions, which usually last only for a fraction of a second, are challenging to detect by the human eye or machine. They are useful for understanding the genuine emotional state of a human face, and have various applications in education, medical, surveillance and legal sectors. Existing works on micro-expressions are focused on binary classification of the micro-expressions. However, detecting the micro-expression intensity changes over the spanning time, i.e., the micro-expression profiling, is not addressed in the literature. In this paper, we present a novel deep Convolutional Neural Network (CNN) based hybrid framework for micro-expression intensity change detection together with an image pre-processing technique. The two components of our hybrid framework, namely a micro-expression stage classifier, and an intensity estimator, are designed using a 3D and 2D shallow deep CNNs respectively. Moreover, we propose a fusion mechanism to improve the micro-expression intensity classification accuracy. Evaluation using the recent benchmark micro-expression datasets; CASME, CASME II and SAMM, demonstrates that our hybrid framework can accurately classify the various intensity levels of each micro-expression. Further, comparison with the state-of-the-art methods reveals the superiority of our hybrid approach in classifying the micro-expressions accurately.

Keywords: Micro-expression intensity ·
Convolutional Neural Networks · Hybrid framework ·
Fusion mechanism

1 Introduction

Facial expressions provide affluent information in understanding the emotional states of the human face. Facial expressions are broadly categorised into macro and micro expressions based on the spanning time and the strength of muscle movements. The macro-expression, also known as the regular expression, is easy to recognise under real-time settings through naked eyes since it continues for a considerable sprint of time. Unlike macro-expression, the micro-expression

ⓒ Springer Nature Switzerland AG 2019
Q. Yang et al. (Eds.): PAKDD 2019, LNAI 11441, pp. 387–399, 2019.
https://doi.org/10.1007/978-3-030-16142-2_30

provokes involuntarily as a rapid and brief facial expression. Hence, it is difficult to detect micro-expressions accurately and spontaneously by observers (e.g., another human). Presently, only the highly trained professionals can characterise the micro-expressions spontaneously, albeit with low accuracies. However, micro-expression is one of the most significant features for revealing the genuine emotions of a person, and therefore it is important to be analysed conscientiously. In the past, a widely used approach for analysing the micro expressions is through video processing techniques. Many micro-expression detection methods in the literature are confined, since they do not focus on systematic micro-expression profiling, such as detecting micro-expression intensity changes using spatial or temporal information.

Incorporating micro-expression is not only limited to traditional applications, such as lie or threatening behaviour detection, but also can be utilised in detecting positive behaviours, such as the level of confidence [2]. Micro-expressions are useful in many domains, such as educational, medical and legal sectors. For example, teachers can use the detected changes in micro-expressions to identify the students' stress, confidence level and other significant behavioural changes that happen in a classroom environment. In the medical field, doctors and psychologists can reveal the concealed emotions of their patients to provide further assistance. Micro-expression profiling plays a vital role during police investigations to recognise the abnormal behaviours, such as lies. For example, the intensity classification of micro-expressions can be used to analyse the onset of changes of a trouble maker in a public place. More importantly, the low level micro-expressions can also be detected by some intelligent algorithms, therefore, it is hard to conceal the actual emotion even for an expert human being. However, there are only a limited work exist in the literature that focus on comprehensive micro-expression profiling. This is due to the involuntariness and the shorter operating duration nature of the micro-expressions.

In this paper, we propose a novel hybrid micro-expression profiling framework that detects the *intensity changes* of micro-expression spontaneously from a video sequence. To the best of our knowledge, no previous work exist to detect the intensity changes of micro-expression on the fly. In our work, we use both spatial and temporal features, and a deep Convolutional Neural Network (CNN) based hybrid architecture, along with a fusion mechanism, to detect the intensity changes from videos. Further, we introduce a comprehensive image pre-processing technique to improve the micro-expression profiling, along with the major aspiration of this paper. In summary, the novelty and contributions of this research work are three-fold:

- We propose an efficient deep CNN based hybrid micro-expression intensity change detector to detect the fine-grained changes in a micro-expression. Our hybrid framework first classifies the micro-expression changes into three *stages*, namely *formation*, *peak* and *release*, using temporal information. Then, it detects the *frame level intensity changes* using spatial information.
- We introduce a fusion mechanism to enhance the automatic micro-expression intensity change detection accuracy.

– We demonstrate that our proposed hybrid framework is capable of accurately detecting the intensity changes through a cross-subject evaluation on three recent benchmark datasets. Moreover, we present a comparison results, before and after the fusion operation, to demonstrates the improvements achieved in the classification accuracy by the fusion process. Further, we compare our hybrid CNN based approach with the state-of-the-art algorithms and demonstrates its superiority in classifying the micro-expressions accurately.

The rest of this paper is organized as follows. Section 2 presents a review of the related work on micro-expression detection. We explain our proposed hybrid framework in Sect. 3. Experiment results and discussions are provided in Sect. 4. Finally, we conclude with suggestions for future work in Sect. 5.

2   Related Work

Here, we discuss the general history of micro-expressions, and provide a survey on the different micro-expression detection approaches used in the past. A recent review by Merghani et al. [8] provides an extensive summary of the literature on micro-expressions analysis. However, no work has been identified on the micro-expression intensity detection in the recent literature.

Two psychologists, namely Ekman and Friesen, have first discovered the micro-expression in 1969 while investigating a medical case, in which a subject tried to conceal the emotion sad nicely masked with a smile on his face [2]. Since then, numerous social psychologists, such as Ekman [2] and Gottman [3], have studied the micro-expressions extensively. Most of the techniques associated with the micro-expression are primarily focussed on detecting basic emotions, also known as prototypical or universal emotions, to provide Facial Action Coding System (FACS) based facial action unit annotations. Although the micro-expressions are hard to observe spontaneously, they are empirically inspected through psychological studies. Further, training programs (e.g., Ekman's Micro Expressions Training Tools[1]) have been created to enhance the support for micro-expression analysis.

In computer vision, only a handful of work exists on micro-expression analysis due to the high complexity of micro-expressions. Many current research focused on hand engineered features for analysis, such as 3D Histograms of Oriented Gradients (3DHOG) [1], Local Binary Pattern-Three Orthogonal Planes (LBP-TOP) and variations [1,11,14], and Histogram of Oriented Optical Flow (HOOF) [14]. For computer vision tasks, nowadays, deep CNN has become a mainstream feature extractor. As a reflection of this, deep CNN based micro-expression detection research have emerged over the past few years, and two noteworthy works on this are [5] and [9]. In the past, researchers have used standard classifiers to classify the micro-expressions, such as support vector machines [15], long short-term memory (LSTM) [5] and dual temporal scale CNN (DTSCNN) [9]. In this

---

[1] https://www.paulekman.com/micro-expressions-training-tools/.

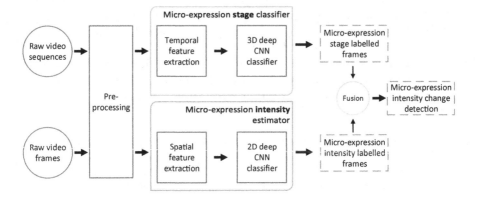

Fig. 1. Micro-expression intensity change detection framework.

paper, we focus on deep learning based feature extractors and classifiers for micro expression intensity detection.

In 2016, Kim et al. [5] adopted CNN for spontaneous micro-expression recognition from video sequences using spatiotemporal features. They encoded the spatial features from representative expression-state frames (i.e., onset, offset and apex) using CNN. Next, the temporal features are analysed after the learned spatial features are passed through an LSTM recurrent network. The overall accuracy achieved was 60.98% on CASME II [11], which is slightly better than the existing state-of-the-art approaches. Peng et al. [9] presented another significant work in micro-expression recognition using a two stream shallow network (4 layers only), namely DTSCNN. They used a selected set of images from the merged CASME I [12] and CASME II [11] datasets to train their model, which comprised four classes: positive, negative, surprise and others. Optical flow sequences, instead of raw sequences, are used to reduce the complexity of the classifier, and they achieved an overall accuracy of 66.67%.

In summary, works in [1,4–6,9,11,14,15] have only addressed the recognition (binary classification) problem of the micro-expressions. None of the existing works focused on detecting the *intensity changes* of micro-expressions. In this paper, we propose a novel hybrid framework for detecting the micro-expression intensity changes from video. Moreover, a significant problem faced by the deep learning based methods is the lack of larger datasets for training, contributing to the lower overall detection accuracy. Hence, we also present a carefully crafted hybrid framework, which includes a pre-processing mechanism, that prepares a comprehensive training set for use with the deep learning based micro-expression analysis. Next, we present our proposed framework.

3   The Proposed Hybrid Framework

Figure 1 illustrates the overall architecture of our proposed hybrid micro-expression intensity change detection framework. The framework primarily comprises a pre-processing unit, two separate deep CNN components and a fusion component. In the pre-processing phase, we augment and normalise the training set to use with the subsequent deep learning hybrid framework. In the deep CNN component, a *micro-expression stage classifier* classifies the three stages (formation, peak and release) of micro-expression changes using temporal feature information. The next component, namely the *micro-expression intensity estimator*, estimates the frame-wise micro-emotion intensities using spatial feature information. Finally, the fusion mechanism updates the intensity predictions incorporating the stage classifier predictions. Below, we explain each of the components of our hybrid framework in detail.

3.1 Pre-processing

In the pre-processing process, we convert the input video frames to grayscale, as the first step, in order to reduce the cross-database discrepancy between the video frames. Our descriptor comprises two significant pre-processing phases namely (a) data augmentation or synthetic sample generation and (b) normalisation. In the data augmentation step, we generate a set of synthesised frames in large amounts to increase the number of video frames, especially for training purposes using a deep learning model, which often require larger dataset. Adopting the approach of [10], we added random noise, using a 2D Gaussian distribution, in the eye centre and nose regions of the face to produce the synthetic frames. We used the individual frames to train the micro-expression intensity estimator, and the whole sequence to train the micro-expression stage classifier, which classifies the frames into three stages; *formation* (start of the micro-expression), *peak* (highest intensity level of the micro-expression) and *release* (end of the micro-expression).

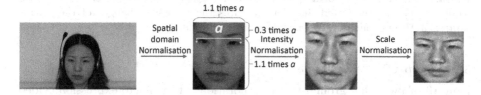

Fig. 2. Pre-processing steps: *Spatial domain normalisation* is performed based on the distance "a" between the active appearance model (AAM) facial feature points 37 and 46. The *intensity* and *scale* normalisations are applied subsequently.

We then perform a series of normalisation process in a sequel, as illustrated in Fig. 2. First, in the spatial normalisation step, a region of interest (ROI) is selected to eliminate the insignificant areas of the video frames for feature extraction, and each video frame is cropped accordingly. In this work, we eliminate not only the background information, but also some parts of the face, such as ears, chin and forehead, which do not reflect any micro-expression related information. The cropping of the facial region is performed based on the distance, indicated as a, between the active appearance model (AAM) facial feature points 37 and 46, as shown in the second image of Fig. 2. Second, we apply an intensity normalisation process, using the Contrast Limited Adaptive Equalization (CLAHE) [16] method, on each video frame to reduce the variation in the feature vector. An advantage of the CLAHE is that it redistributes the histogram part which exceeds the clip limit between all histogram bins, rather than just eliminating it. In this work, we use a Rayleigh distribution with clip limit of 0.01 and α value of 1. Third, in the scale normalisation step, we down-sampled the video frames to reduce the size to 128×128 pixels, using linear interpolation. Scale normalisation enables the same facial feature points of different video frames to co-occur approximately in the same location. Fourth, as the final step of the pre-processing, we performed the temporal normalisation to the input dataset (both training and testing datasets) of the micro-expression stage classifier. The input dataset is normalised to eight frames, and used as input to classify them as one of the micro-expression stages, i.e., formation, release or peak.

The pre-processed frames are used as input to our deep CNN components, as illustrated in Fig. 1, which we discuss in detail next.

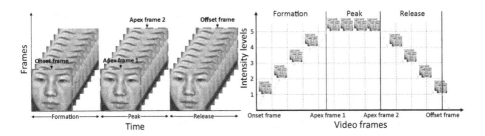

Fig. 3. The three stages of a spontaneously recognised micro-expression in a video sequence. The green dotted line demonstrates the intensity changes of the micro-expression throughout the sequence. (Color figure online)

3.2   Deep CNN Components

Figure 3 shows the ground-truth annotation of a micro-expression video sequence. In here, the onset and offset frames are the first and the last recognised frames of a particular micro-expression. The maximum intensity of the micro-expression (i.e., micro-expression intensity level 5 in our work) is identified from the frames between the apex frame 1 and the apex frame 2. The three stages, from onset frame to apex frame 1, apex frame 1 to apex frame 2 and apex frame 2 to offset frame, are named as *formation, peak* and *release* respectively. The micro-expression intensity shows a linearly increasing trend (levels 1 to 4) in the formation stage, and reaches the maximum (level 5) in the peak stage. After remaining unchanged for a while during the peak stage, the intensity drops linearly, during the release stage, to reach the minimum (level 1) again. Using the behaviour observed here, we perform the micro-expression stage detection, the frame-wise intensity estimation and a fusion process as explained below.

Micro-expression Stage Classifier. Micro-expression stage classifier is an action classifier, which classifies three stages of a micro-expression, called formation, peak and release. We use a shallow (i.e., three-layer) 3-dimensional convolutional network as the stage classifier. Further, two fully-connected (FC) layers are assembled, as the FC layer is known for learning a non-linear function in a computationally efficient way from the high-level feature space. We use a large-size convolutional kernel (i.e., input frames) while maintaining a fixed temporal dimension, which is not fine-tuned during the experiment. Unlike 2-dimensional CNNs, where each video frame is considered as an object, 3-dimensional CNNs treat the whole video sequence as an object for classification. As indicated in the figure, the size of the input image is $128 \times 128 \times 1 \times 8$ (*width* $\times$ *height* $\times$ *channel* $\times$ *frames*). Each convolution layer is followed by a dropout layer, in which the dropout probability is set and tuned during the experiment. Subsequently, a pooling layer is assembled, after each dropout layer, which uses max-pooling with a kernel size of $1 \times 2 \times 2$.

We use the same values for learning parameters, such as momentum, learning rate and decay, to train the hybrid framework on CASME [12], CASME II [11] and SAMM [1] datasets. Stochastic gradient decent with momentum (SGDM) optimiser was applied with the parameters momentum and learning rate set to 0.95 and 0.001 respectively. The classification output of this stage classifier forms one of the inputs of the subsequent fusion process to estimate the micro-expression intensity changes, as shown in Algorithm 1.

Micro-expression Intensity Estimator. In the literature, the intensities of action units (AUs) were used to estimate the intensities of *macro* emotions [7]. However, unlike macro expressions, micro-expression intensity estimation using AU intensity is challenging due to its rapid and feeble nature. Here, we introduce a novel frame-wise micro-expression intensity estimator using *relative intensity differences* between the frames of a video sequence. Zhao et al. [13] used a relative intensity based method to estimate the intensities of the macro-emotions. Inspired by this work, to assign the frame-wise intensities of micro-expressions in a video sequence, we use the following Eq. (1) to obtain the intensity level of jth frame I_j in a video sequence. The maximum and minimum intensities of micro-expression are denoted as $I_h = 5$ and $I_l = 1$, where the maximum intensity is attained at the peak micro-expression stage, and the minimum is attained at the formation and release micro-expression stages.

$$I_j = \left\lfloor \frac{j}{a_1}(I_h - I_l) \right\rceil \psi_{1 \leq j < a_1} + I_h \psi_{a_1 \leq j \leq a_2} + \left\lfloor \frac{n-j}{n-a_2}(I_h - I_l) \right\rceil \psi_{a_2 \leq j \leq n} \quad (1)$$

In Eq. (1), ψ is the indicator function applied on the frame number j, where $j = 1 \cdots n$, and n is the number of frames in a video sequence. a_1 and a_2 are the apex frame 1 and apex frame 2, respectively. The acquired micro-expression intensity curve for a video sequence is shown in Fig. 3.

For micro-expression intensity estimation, we propose a 2-dimensional deep CNN architecture, and the structure is similar to the one proposed for the micro-expression stage classifier, but with the replacement of all the 3-dimensional convolution layers with 2-dimensional convolution layers. The input layer of the micro-expression intensity estimator takes images with size $128 \times 128 \times 1$ (*width* $\times$ *height* $\times$ *channel*) as input. The output size of the last fully connected layer is set to 5, representing the five-level micro-expression intensity values/levels. We set the learning environmental settings of the micro-expression intensity change detector to the same values as used for the micro-expression stage classifier.

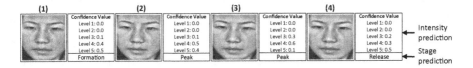

Fig. 4. For each frame, the predicted stage (formation, peak or release), and the confidence score for each intensity level, are shown. The red value indicate the highest confidence score and the corresponding intensity level of that frame. The green value indicates the confidence score and the corresponding intensity level that is assigned for that frame after the fusion process. (Color figure online)

3.3 Fusion Mechanism

The fusion process takes each frame and its intensity level obtained from the *micro-expression intensity estimator*, and adjusts its intensity level to match with the correct stage's (*obtained from the micro-expression stage classifier*; *formation, peak and release*) intensity level as described below. Figure 4 shows an example of this process.

– For the frames in the *formation and release stages*, we update the frames predicted with the highest intensity (i.e., level 5) by the intensity prediction of the second highest confidence score (e.g., frames (1) and (4) in Fig. 4). This adjustment is done because these frames are identified as belonging to either the formation stage or the release stage. Hence, their intensity levels are expected to be less than the maximum intensity level. We use the intensity level of the second highest confidence score to decide this.

– For the frames in the *peak stage*, we update the frames with predicted intensity levels 1, 2, 3 and 4 to intensity level 5 (e.g., frame (2)), if the second highest confidence value obtained is for level 5; otherwise we keep the intensity level of the highest confidence score (e.g., frame (3)). This adjustment is done because these frames are identified as belonging to the peak stage. Hence, their intensity levels are expected to be at the maximum level. We use the intensity level of the second highest confidence score to decide this.

Algorithm 1 explains the steps involved in the fusion process in detail.

Algorithm 1. Fusion process: S_i and F_i are the outputs of stage detector and intensity change detector components. Γ_i returns the intensity level for i^{th} confidence score rank. I_h indicates the highest intensity level. $\Omega(I)$ returns the confidence value of level I

```
1: procedure FUSION-PROCESS(S_i, F_i)
2:     For all i, S_i ∈ {'Formation','Release'}
3:     if Γ₁(F_i) ≠ I_h then
4:         I_i = Γ₁(F_i)
5:     else
6:         I_i = Γ₂(F_i)
7:     end if
8:     For all i, S_i ∈ {'Peak'}
9:     if Γ₁(F_i) = I_h or (Γ₁(F_i) ≠ I_h and Ω(I_h) ≥ Ω(Γ₂(F_i))) then
10:        I_i = I_h
11:    else
12:        I_i = Γ(I_i)
13:    end if
14:    Output: I
15: end procedure
```

---

4   Results and Analysis

In this section, we evaluate the proposed hybrid framework under divergent environments. The results are reported using classification accuracy measure obtained via leave-one-subject out cross validation procedure.

4.1   Micro-expression Intensity Change Detection

We present the results and analysis obtained using our proposed hybrid architecture, evaluated on three recent spontaneous micro-expression benchmark datasets CASME [12], CASME II [11] and SAMM [1]. CASME [12] dataset comprises 195 samples including eight micro-emotions namely, amusement, sadness, disgust, surprise, contempt, fear, repression and tense. The second dataset, CASME II [11], is an improved version of CASME [12], which comprises five distinct micro-expression classes: happiness, surprise, disgust, repression and others. SAMM [1] is another recent dataset with 159 micro-expression samples based on seven basic emotions; contempt, disgust, fear, anger, sadness, happiness, and surprise.

The confusion matrices shown in Fig. 5 presents the detection accuracies of the micro-expression stage classifier, where the average prediction results obtained are 89.37%, 93.2% and 93.6% on CASME [12], CASME II [11] and SAMM [1], respectively. We use this prediction result as the input for the fusion process to refine the micro-expression intensity change detection in videos. Furthermore, we evaluated the prediction accuracy of the micro-expression stage classifier *before* and *after* the *pre-processing* of datasets, and it clearly demonstrates a 10% increase in the classification rate on average. Although, the temporal normalisation is included in both of the experiments to maintain the uniformity of the fed CNN input, the above evaluation results are avoided for brevity.

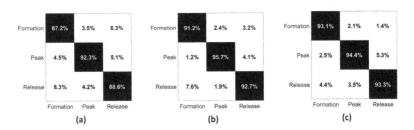

Fig. 5. Confusion matrix for the leave-one-subject-out classification of micro-expression stage classifier on (a) CASME [12], (b) CASME II [11] and (c) SAMM [1].

Table 1 shows the micro-expression intensity change classification results, adopting the entire validation methodology as described before. In order to validate the significance of the fusion process, we also carried out the experiment and recorded the prediction accuracies *before* and *after* the *fusion process*. In validation, before the fusion process, the intensity estimation is performed without the stage classifier. In validation, after the fusion process, the stage classifier is used to improve the output of the micro-expression intensity estimator. After the fusion process, the results demonstrate that the highest micro-expression intensity detection achieved is 76% with the SAMM [1] dataset, while the lowest is reported for the CASME [12] dataset, with 71%. As it can be seen, after the fusion process, we have a substantial increase (>7% better than the accuracy of before fusion) in the intensity prediction accuracy of any emotion, regardless of the micro-expression dataset. More precisely, the micro-expression intensity detection prediction results are more accurate when in combination with a stage classifier and the fusion process. For example, the Fig. 6 illustrates the intensity updation operation for a micro-expression video sequence after applying the fusion process, using confident scores (in this case, the micro-expression stages are correctly classified). In this, frames 9, 10 and 15 are adjusted to the correct intensity levels by the fusion process. This explicitly improves the classification accuracy from 62.5% to 75.0%.

Table 1. Micro-expression intensity change detection results before and after the fusion process. The pair of numbers represent the classification accuracies in the form of (before fusion, after fusion).

Micro-expressions	Datasets		
	CASME [12]	CASME II [11]	SAMM [1]
Amusement	(0.61, 0.68)	–	–
Anger	–	–	(0.69, 0.74)
Contempt	(0.57, 0.59)	–	(0.55, 0.60)
Disgust	(0.72, 0.79)	(0.79, 0.84)	(0.77, 0.81)
Fear	(0.77, 0.81)	–	(0.71, 0.79)
Happy	–	(0.73, 0.81)	(0.66, 0.71)
Others	–	(0.47, 0.52)	–
Repression	(0.52, 0.59)	(0.55, 0.59)	–
Sad	(0.81, 0.85)	–	(0.79, 0.84)
Surprise	(0.74, 0.77)	(0.81, 0.87)	(0.76, 0.83)
Tense	(0.58, 0.61)	–	–
Average rate	(0.67, 0.71)	(0.67, 0.73)	(0.70, 0.76)

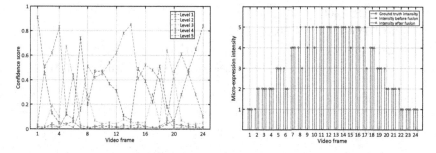

Fig. 6. Fusion process for a sequence in CASME [12] with emotion *happy*. Left figure illustrates frame-wise confidence scores and the right figure presents the frame-wise intensity labelling before and after the fusion process

4.2 Comparison with Existing Works

As discussed in the related work section, the closely related works to ours either targeted intensity estimation of the *macro*-expressions [13] or a binary classification (i.e., existence or not) of the micro-expression [1,4–6,11,14]. In order to perform a fair comparison of our framework with the existing works, we need to convert the intensity level (five levels) output of our framework to a binary form. Hence, we combined all the micro-expression intensity levels from 1 to 5 and considered them as the indicator of the existence of the micro-expression, while the level 0 is considered as the indicator of non-existence of the micro-expression. We compared these works using the CASME [12], CASME II [11] and SAMM

[1] datasets. Table 2 provides the comparison results. It can be observed that the deep learning based approaches, both CNN based approach of [5] and our approach have outperformed the hand-crafted feature-based approaches. Further, our scheme showed the highest accuracy among all; 77%, 82% and 91% on CASME [12], CASME II [11] and SAMM [1] respectively. This demonstrates that our hybrid architecture and the fusion mechanism helped to attain higher accuracy. Moreover, our proposed framework is capable of further classifying the various *intensity* levels within the micro-expression as demonstrated in the previous section.

Table 2. Comparison of micro-expression recognition methods against the proposed hybrid framework with leave-one-subject-out cross-validation on CASME [12], CASME II [11] and SAMM [1]. ‡ and § indicate the performance measures recognition accuracy and F-measure respectively.

Methods	Datasets		
	CASME [12]	CASME II [11]	SAMM [1]
LBP-TOP [1,11,14]	0.69‡	0.39§	0.55§
HOOF [14]	–	0.59‡	–
3DHOG [1]	–	–	0.89‡
Bi-WOOF [6]	–	0.61§	–
FHOFO [4]	0.55§	0.52§	–
CNN [5]	–	0.61‡	–
Ours	(0.77‡, 0.61§)	(0.82‡, 0.68§)	(0.91‡, 0.70§)

5 Conclusion

In this paper, we proposed a novel hybrid micro-expression intensity change detection framework, consisting of three components, namely, micro-expression stage classifier, micro-expression intensity estimator and a fusion mechanism. The hybrid framework is built on a combination of 3D and 2D deep CNN based architectures. The fusion mechanism is integrated to further enhance the classification efficiency. Moreover, we proposed a comprehensive preprocessing technique to cater the requirements of deep networks; a large amount of training data and reduced complexity. Experiments were carried out on three benchmark micro-expression datasets CASME [12], CASME II [11] and SAMM [1]. We demonstrated that our hybrid framework is capable of accurately detecting the intensity changes in the micro-expression. Further, we compared the accuracy before and after the fusion mechanism, and shown the improvement achieved in the accuracy of intensity change detection. In the future, we aim to apply this in combination with the macro-expression intensity changes for fast detection of onset of emotions.

References

1. Davison, A.K., Lansley, C., Costen, N., Tan, K., Yap, M.H.: SAMM: a spontaneous micro-facial movement dataset. IEEE Trans. Affect. Comput. **9**(1), 116–129 (2018)
2. Ekman, P.: Lie catching and microexpressions. Philos. Decept. **1**, 5 (2009)
3. Gottman, J.M., Levenson, R.W.: A two-factor model for predicting when a couple will divorce: exploratory analyses using 14-year longitudinal data. Fam. Process **41**(1), 83–96 (2002)
4. Happy, S.L., Routray, A.: Fuzzy histogram of optical flow orientations for micro-expression recognition. IEEE Trans. Affect. Comput. (2017). https://doi.org/10.1109/TAFFC.2017.2723386
5. Kim, D.H., Baddar, W.J., Ro, Y.M.: Micro-expression recognition with expression-state constrained spatio-temporal feature representations. In: Proceedings of the 2016 ACM on Multimedia Conference, pp. 382–386. ACM (2016)
6. Liong, S.T., See, J., Wong, K., Phan, R.C.W.: Less is more: micro-expression recognition from video using apex frame. Signal Process.: Image Commun. **62**, 82–92 (2018)
7. Lucey, P., Cohn, J.F., Prkachin, K.M., Solomon, P.E., Matthews, I.: Painful data: the UNBC-McMaster shoulder pain expression archive database. In: 2011 IEEE International Conference on Automatic Face and Gesture Recognition and Workshops (FG 2011), pp. 57–64. IEEE (2011)
8. Merghani, W., Davison, A.K., Yap, M.H.: A review on facial micro-expressions analysis: datasets, features and metrics. arXiv preprint arXiv:1805.02397 (2018)
9. Peng, M., Wang, C., Chen, T., Liu, G., Fu, X.: Dual temporal scale convolutional neural network for micro-expression recognition. Front. Psychol. **8**, 1745 (2017)
10. Simard, P.Y., Steinkraus, D., Platt, J.C.: Best practices for convolutional neural networks applied to visual document analysis. In: Null, p. 958. IEEE (2003)
11. Yan, W.J., et al.: CASME II: an improved spontaneous micro-expression database and the baseline evaluation. PloS One **9**(1), e86041 (2014)
12. Yan, W.J., Wu, Q., Liu, Y.J., Wang, S.J., Fu, X.: CASME database: a dataset of spontaneous micro-expressions collected from neutralized faces. In: 2013 10th IEEE International Conference and Workshops on Automatic Face and Gesture Recognition (FG), pp. 1–7. IEEE (2013)
13. Zhao, R., Gan, Q., Wang, S., Ji, Q.: Facial expression intensity estimation using ordinal information. In: Proceedings of the IEEE Conference on Computer Vision and Pattern Recognition, pp. 3466–3474 (2016)
14. Zheng, H., Geng, X., Yang, Z.: A relaxed K-SVD algorithm for spontaneous micro-expression recognition. In: Booth, R., Zhang, M.-L. (eds.) PRICAI 2016. LNCS (LNAI), vol. 9810, pp. 692–699. Springer, Cham (2016). https://doi.org/10.1007/978-3-319-42911-3_58
15. Zhu, X., Ben, X., Liu, S., Yan, R., Meng, W.: Coupled source domain targetized with updating tag vectors for micro-expression recognition. Multimed. Tools Appl. **77**(3), 3105–3124 (2018)
16. Zuiderveld, K.: Contrast limited adaptive histogram equalization. In: Graphics Gems, pp. 474–485 (1994)

A Multi-scale Recalibrated Approach for 3D Human Pose Estimation

Ziwei Xie[1] , Hailun Xia[1,2(✉)], and Chunyan Feng[1,2]

[1] Beijing Key Laboratory of Networks System Architecture and Convergence,
School of Information and Communication Engineering,
Beijing University of Posts and Telecommunications, Beijing 100876, China
{tomxie,xiahailun,cyfeng}@bupt.edu.cn
[2] Beijing Laboratory of Advanced Information, Beijing 100876, China

Abstract. The major challenge for 3D human pose estimation is the ambiguity in the process of regressing 3D poses from 2D. The ambiguity is introduced by the poor exploiting of the image cues especially the spatial relations. Previous works try to use a weakly-supervised method to constrain illegal spatial relations instead of leverage image cues directly. We follow the weakly-supervised method to train an end-to-end network by first detecting 2D body joints heatmaps, and then constraining 3D regression through 2D heatmaps. To further utilize the inherent spatial relations, we propose to use a multi-scale recalibrated approach to regress 3D pose. The recalibrated approach is integrated into the network as an independent module, and the scale factor is altered to capture information in different resolutions. With the additional multi-scale recalibration modules, the spatial information in pose is better exploited in the regression process. The whole network is fine-tuned for the extra parameters. The quantitative result on Human3.6m dataset demonstrates the performance surpasses the state-of-the-art. Qualitative evaluation results on the Human3.6m and in-the-wild MPII datasets show the effectiveness and robustness of our approach which can handle some complex situations such as self-occlusions.

Keywords: 3D human pose estimation · Recalibration module · Deep learning

1 Introduction

Estimating human pose from images is a hot spot in computer vision. It can be used in numerous applications, including human-computer interaction, medical rehabilitation, surveillance, action recognition, etc. The research on the human pose estimation is active but challenging for the ambiguity in human depiction. This can be attributed to the variant background, self-occlusions, limited viewpoint and so on [1]. In recent years, deep neural networks are utilized to mitigate the issues, and the tremendous network architectures are supported by large datasets [2–4]. In this paper, we aim at estimating 3D human pose from a single monocular RGB image.

In the context of 3D human pose estimation, lots of creative methods are proposed. In general, they can be divided into two categories: inferring 3D configuration from off-the-shelf 2D joints detector [5–8] and performing an end-to-end model to regress 3D

© Springer Nature Switzerland AG 2019
Q. Yang et al. (Eds.): PAKDD 2019, LNAI 11441, pp. 400–411, 2019.
https://doi.org/10.1007/978-3-030-16142-2_31

joint coordinates directly [9–12]. Restricted by the limits of 3D annotated human pose datasets, the shortage of scale and the lack of variance in natural settings, the model is not generalized enough to handle in-the-wild situations. To tackle the problem, some methods are proposed: using synthetic data [13], multi-view camera merging [14], and the hybrid learning for both 2D and 3D pose [15, 16].

In the hybrid learning, the cues in the original images and intermediate heat maps are exploited to introduce the spatial information and lower level features. With the use of large-scale in-the-wild 2D annotated datasets, both the accuracy and generalizability are promoted. Zhou et al. [16] proposed a weakly-supervised approach to introduce 3D geometric constraint. We reuse their work and make some change to further exploit image cues in the regression process.

The image classification is the foundation of other computer vision tasks. At the same time, the pre-trained networks can be used and fine-tuned to extract other high-level image features for other tasks. To a great extent, the work we did is inspired by some novel network designs for the task of image classification. For the most of the deep networks, they lie in the spatial domain and are composed of convolution layers. Convolution operation is spatial-symmetric naturally, but the saliencies are not uniformly distributed in the spatial domain for some highly structural tasks such as human pose estimation. For instance, when all the joints except head are settled, the distribution of head position is closer to the 'upper' position instead of uniformly. So, in the regression process of 3D pose estimation, we propose to introduce a recalibration approach [17, 18] to revise the response of each neuron by a spatial weight in residual modules [19].

Our contribution is introducing the spatial enhanced multi-scale recalibration to the base residual modules in 3D pose regression aiming to exploit the structural information in some human poses. This approach slightly outperforms than state-of-the-art baseline systems after fine-tuning. The experiments show that the Mean Per Joint Position Error (MPJPE [4]) is 4% lower than the state-of-the-art methods on the largest annotated 3D pose dataset. Moreover, the network possesses a high-performance in inferencing, so the system which is based on the approach has the ability to be equipped in some real-time applications.

This paper is organized as follow: After reviewing the related work in Sect. 2, Sect. 3 introduces the proposed approach, and the evaluation results are presented in Sect. 4. At last, Sect. 5 concludes the paper and previews the future work.

2   Related Work

Human pose estimation has been extensively studied in the past years [1, 20]. Providing a detailed overview is far beyond the scope of this work. In the context of this paper, we focus on related works on 3D human pose estimation which is one of the most challenging studies in computer vision. Simultaneously, we will review the spatial enhancement, especially the recalibration, for deep networks.

2.1   3D Human Pose Estimation

The 3D human pose can be simplified into a set of joints in 3D coordinate with some volumeless connections between joints. So, 3D human pose estimation can be formulated as a supervised regression problem which has the 3D coordinate as the output and the original images or predicted 2D pose as the inputs. With the alternative of input, most methods of 3D human pose estimation can be separated into two types.

The first type of methods is to train an end-to-end regression model or a data-based encoder directly predicting coordinates of each joint. Li et al. [11] regress and detect joints with a deep convolutional neural network. Pavlakos et al. [9] utilize volumetric representation in output space and proposes a coarse-to-fine approach to deal with the large dimensionality. Tekin et al. [10] introduce a deep learning regression architecture for structured prediction. An overcomplete auto-encoder is used to learn latent pose representation and joint dependencies. Li et al. [12] use two sub-networks for image and pose embedding. [21] proposes a new CNN as pose regressor to predict location maps for the 3D poses.

The other is separated into two subtasks [5–8]. For the first subtask, an off-the-shelf 2D human pose estimation method [22, 23] is used to detect different joints in a human pose, and it can be fine-tuned through large-scale in-the-wild 2D pose datasets. And the second subtask is regressing the 3D location of the corresponding 2D joint. Ramakrishna et al. [8] present a matching pursuit algorithm to estimate 3D pose from 2D projections. Moreno-Noguer et al. [5] use distance matrices to represent 3D and 2D pose and complete 2D to 3D lifting by a neural network architecture. Chen et al. [6] match the 2D pose in a bigdata 3D pose library through the nearest neighbors algorithm. Martinez et al. [7] create a simple network to lift 2D joint locations to 3D space. The state-of-the-art stacked hourglass network is adopted as the 2D joint detector.

Recently, some prior knowledge including structural or geometric constraints is utilized. Akhter and Black [24] set pose-dependent joint angle limits for lifting 2D location to 3D. Sun et al. [15] propose a bone-based pose representation to a structure-aware loss function. Zhou et al. [25] use a generative forward-kinematic layer to enforce the bone-length constraints in the prediction. In [16], they further introduce geometric constraints into their network to maintain consistency of bone-length ratios in the pose prediction.

2.2 Spatial Enhancement for Deep Networks

All the above methods did not exploit the spatial relations in the 3D pose regression. But for most of the image comprehension task, the relationship between two features is tighter when their spatial location is closer to each other. Some approaches are proposed to introduce this spatial priority [17, 26–29]. Spatial pooling is performed by He et al. [26], which explicitly splits the image lattice into several groups, and ignores the diversity of features in the same group. In [27], a set of regional proposals is utilized to summarize features. Another perspective is discriminate the salient features from others. Some approaches find the neurons that contribute most to the result through gradient back-propagation [28] or attention [29]. For the purpose of capturing nonlinear properties an reducing the number of parameters, a recalibration approach [17] is proposed to revise the response of neurons by a spatial level weight set. Recently, recalibration is enhanced by multi-scale strategy.

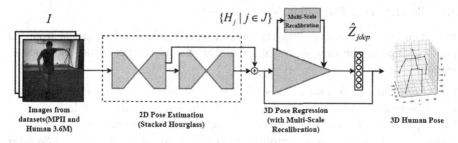

Fig. 1. Overview. A batch of images I are first fed into the 2D pose estimation module which consisted of two stacked hourglass network and the heatmaps for each joint are the outputs as $\{H_j|j \in J\}$. Every stack network produces an interim heatmap which contain the massive hidden features from multi layers, and all these heatmaps are coalesced as the input of the cascaded regression module. In regression module, multi-scale recalibration modules are added as the extra branches in residual structures. The regression module predicts the depth of each joint, as $\hat{Z}_{jdep}$, and the 3D pose estimations are constituted of predicted 2D joints positions and regressed joints depth from camera.

3 The Proposed Approach

The 3D human pose can be abstractly represented into a set of human joints coordinates with the volumeless connections between spatially associated joints. The set of joints is denoted as $P^{3D} \in R^{3 \times J}$, where J is the number of crucial joints in a human skeleton. Each joint in the set is a 3-dimension vector $P_j^{3D} = \left[X_j, Y_j, Z_j\right]^T, j \in 1, \ldots, J$. The first two coordinates are rescaled to the pixel coordinate which is consistent with the input RGB image I, and the third coordinate is the vertical depth from joint j to the camera. Our goal is to estimate the 3D human pose P^{3D} from a single RGB image I.

The overview of our approach is illustrated in Fig. 1. It consists of two substructures: 2D pose estimation and 3D pose regression with multi-scale recalibration modules. 2D joints set P^{2D} is detected by the 2D pose estimation module which predicts the joints locations with the highest probability. The vertical depth Z_{dep} is predicted by the regression module. The final output is the constitution of 2D joints set P^{2D} and joints vertical depth Z_{dep}.

The network we use is based on a weakly-supervised approach [16] which is trained from both images with 3D pose annotations and images with 2D pose annotations. As the main process of 3D human pose estimation, the regression module is improved by utilizing the multi-scale recalibration branches in residual modules [19]. Each multi-scale recalibration branch has its own weights, and these weights are fine-tuned separately during the training process. We describe the baseline pose estimation network and multi-scale recalibration regression below.

3.1   Baseline Pose Estimation Network

We adopt an end-to-end network [16] with a weakly-supervised constraint from 2D pose to 3D pose. The network contains two main modules. Brief introductions are as below.

2D Pose Estimation Module. The state-of-the-art network structure in [23] is utilized and reduces the stacked number from 6 to 2. The output of the module is a set of heatmaps $\{H_j | j \in J\}$ where for a given heatmap the module predicts the probability of a joint's coordinate on pixels coordinate system. To train the module, a Mean-Squared Error (MSE) loss is applied comparing the difference between predicted heatmap and ground-truth heatmap [23].

3D Pose Regression Module. The design of the regression module is inspired by the ResNet [19] which is constructed with residual modules. The number of residual modules of regression is also reduced to 8. A linear layer is applied as the output layer with outputting predicted joints depth $\hat{Z}_{jdep}$.

3.2   Multi-scale Recalibration Regression

Most 3D pose regression methods [5, 7] are aiming to lift the 2D joints positions to 3D directly by exploiting 2D information which is preparatory to the regression from a well-performed 2D joints detector. To reduce the ambiguity of 3D joint location, image cues [30] or geometric constraints [15, 16] are exploited in some methods. All the above methods neglect the spatial relations in images, which is often reflected in adjacent joints during the task of 3D pose regression. In 2D pose estimation, the problem is often offset through enlarging the local receptive field or combining features across scales, e.g. in [23], pooling and up-sampling are used repeatedly to gain high-level representations from multi-resolution. This method is proved to be effective (PCKh0.5 = 90.9%) in the task of 2D pose estimation.

Our approach is inspired by some novel networks [17, 18] designed for image classification. These networks adapt additional recalibration module to learn the spatial information which is benefits to some complex situations in classification. So, we adapt the multi-scale recalibration method in [18] to the 3D pose regression module for long-range features extraction.

As shown in Fig. 1, the 3D pose regression module utilizes the heatmaps H from 2D pose estimation module to predict the joints depth $\hat{Z}_{dep}$. The structure of the regression module is visualized in Fig. 2(a). It contains 8 cascaded residual modules which are totally the same. We define the residual module as a function,

$$Y = H + F(H) \tag{1}$$

The input of the residual module is H, and the output is Y, the convolution is equal to the function F(*). And the detail is showed in Fig. 2(b): 3 Convolution layers with batch normalization [31] and Rectified Linear Units (ReLUs) [32]. The output of the residual module with multi-scale recalibration Y* can be expressed as:

$$Y^* = H + F(H) \odot R(F(H)) \tag{2}$$

Note that $R(*)$ is the function of the multi-scale (scale factor $L = \{1, 2, 4\}$) recalibrations which are inserted after the convolution layers in the residual module (in Fig. 2(c)). '$\odot$' is element-wise multiply. The process of recalibration at scale factor L is constructed in Fig. 2(d). The structure of the recalibration module demonstrates that the input feature is squeezed to a narrow pipeline with the resolution of $L \times L$.

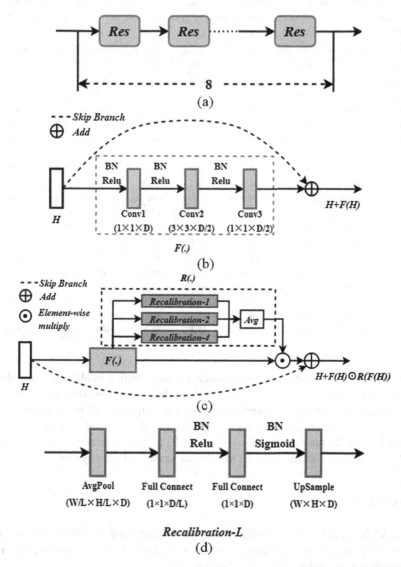

Fig. 2. (a) The regression module. 8 residual modules cascaded in line; (b) The details of residual module; (c) The residual module with multi-scale recalibration (d) L-Scale recalibration.

3.3   Training

We first pre-train the model follows the three-stage training schema in [16]. For each stage with an exponential decay learning rate from 0.001, and mini-batch of size 6.

For stage 1, the network is trained for 60 epochs, and for both stage 2&3, are 30 epochs.

Table 1. Quantitative results of our model on the Human3.6m dataset. The numbers are MPJPE (mm) results. BL means the baseline method which is trained without the Multi-SCALE Recalibration Modules, and Ours is the method we proposed with the Multi-Scale Recalibration Modules.

Method	Directions	Discussion	Eating	Greeting	Phoning	Photo	Posing	Purchases
Chen [6]	89.9	97.6	90.0	107.9	107.3	139.2	93.6	136.1
Tome [33]	65.0	73.5	76.8	86.4	96.3	110.7	68.9	74.8
Moreno [5]	69.5	80.2	78.2	87.0	100.8	102.7	76.0	69.7
Zhou [34]	87.4	109.3	87.1	103.2	116.2	143.3	106.9	99.8
Metha [35]	59.7	69.7	60.6	68.8	76.4	85.4	59.1	75.0
Pavakos [9]	58.6	64.6	63.7	62.4	66.9	70.7	57.7	62.5
Zhou [16] (BL)	54.8	60.7	58.2	71.4	62.0	**65.5**	**53.8**	55.6
Martinez [7]	51.8	56.2	58.1	**59.0**	69.5	78.4	55.2	58.1
Ours	56.0	61.3	**52.0**	63.2	**60.9**	69.2	60.9	**50.2**
Method	Sitting	Sitting down	Smoking	Waiting	walk dog	Walking	Walk pair	Average
Chen [6]	133.1	240.1	106.7	106.2	87.0	114.1	90.6	114.2
Tome [33]	110.2	172.9	85.0	85.8	86.3	71.4	73.1	88.4
Moreno [5]	104.2	113.9	89.7	98.7	82.4	79.2	77.2	87.3
Zhou [34]	124.5	199.2	107.4	118.1	114.2	79.4	97.7	79.9
Metha [35]	96.2	122.9	70.8	68.5	54.4	82.0	59.8	74.1
Pavakos [9]	76.8	103.5	65.7	61.6	67.6	56.4	59.5	66.9
Zhou [16] (BL)	75.2	111.6	64.2	66.1	51.4	63.2	55.3	64.9
Martinez [7]	**74.0**	**94.6**	**62.3**	59.1	65.1	49.5	52.4	62.9
Ours	76.1	103.6	62.5	**52.4**	**65.1**	**45.5**	**49.0**	**62.4**

After the training of the baseline network, the recalibration modules are introduced into the regression module. All the parameters of the baseline network are frozen, and the recalibration modules are optimized during the training, with the learning rate of 0.0025 and 30 epochs. Training for all the stages takes around 2 days in a Tesla M40 GPU, and the finetune process is done in 10 h. The inference spends 46 ms for each frame, so it is effective enough to be used to the real-time systems.

4   Evaluation

Our model is trained on Human3.6m dataset [4] and MPII dataset [2]. For the baseline network [16], we follow the training strategy of 3 stages (Sect. 3.3). And we fine-tune the recalibration modules on the baseline network.

Fig. 3. Qualitative results of our model on the Human3.6m dataset and MPII 2D pose dataset. Predicted poses are rotated and zoomed for the consistency of perspective with original image.

The result of the test data is shown in both quantity and quality. Quantitative evaluation (Table 1) is performed only on the Human3.6m dataset, and qualitative evaluation (Fig. 3) is performed on both Human3.6m and MPII.

We also perform an ablation study for our model, and the result is in Table 2.

4.1 Datasets

MPII Dataset. MPII dataset [2] is a large-scale in-the-wild 2D human pose dataset. It contains around 25k images over 40k people with annotated body joints. The training set of MPII is used to train the 2D pose estimation module in the baseline network. Despite the lack of 3D joint location, it can be exploited as the input of geometric constraints in the network.

Human3.6m Dataset. Human3.6m dataset [4] is used for both training and testing process. It is the largest public dataset for 3D human pose estimation. The dataset contains 3.6 million images in 11 subjects and 15 daily activities. All the data is captured by a motion capture system in an experimental indoor environment. We follow the standard protocol in [5, 9, 15, 16], which uses subjects S1, S5, S6, S7 and S8 for training and S9 and S11 for evaluating. The metric we use is Mean Per Joint Position Error (MPJPE) [4], which measures the average of millimeter Euclidean distance between the predicted and ground truth corresponding joints after aligning the root(pelvis) joint.

4.2   Quantitative Results

After the convergence of model around 30 epochs for the fine-tuning, we use the pre-trained model to the test dataset, and the result is summaries in Table 1 compared to most state-of-the-art methods in all the 15 activities under the standard protocol. Noted that in most activities, after the introducing and fine-tuning of the recalibration modules, we gain an enhancement from extra components. Our method reduces the MPJPE down to 62.4 in average. For most of the activities in Human3.6m, we exceed to baseline method (around 4% promotion) and achieve the state-of-the-art results.

4.3   Ablation Study and Analysis

We changed the scale factor set L in the recalibration modules. When the L = {1, 2, 4}, the recalibration modules capture coarse to fine grain, so relevance in different scale is exploited. As expected, the result when using most scale branches gains most profit to other 3 ablative scale factor sets. And we notice a quite interesting phenomenon: when L is ablated to only one element 1 (equivalent to Squeeze-and-excitation network [17]) and 4, deterioration appear when comparing to multi-scale one, but former is slightly better than later, on account of the relevancy between joints on full-range is much more influential to regional-range.

Table 2.   Result of ablation study.

L set	{1, 2, 4}	{1, 2}	{1}	{4}
MPJPE(avg)(mm)	**62.4**	63.1	63.4	64.3

4.4   Qualitative Results

Finally, we demonstrate some qualitative results on both Human3.6m and MPII dataset in Fig. 3. For the first line is the experiment result in Human3.6m which is in a limited controlled environment. Second to sixth lines is in-the-wild MPII dataset. As the results reveal, our model can dispose of most of the 3D human pose estimation problem in the wild and gain a precise prediction even in some self-occlusion situations. But our model has some limitation for predicting partial bodies, for the model is trained on full bodies, and these cases are not taken into consideration temporarily.

5   Conclusions and Future Work

In this paper, we first introduce the multi-scale recalibration approach in the 3D pose regression process for capturing more spatial information in the highly structural human pose. The evaluation results show the advancement of state-of-the-art baseline method. In the future, we plan to explore more geometric constraints in 3D human pose estimation to punishing some implausible predictions. We also expect this work can inspire some relevant work in 3D human pose estimation.

Acknowledgments. This work is supported by Chinese National Nature Science Foundation (61571062) and the 111 project (NO. B17007). We would like to thank Rui Zhang for helping with Fig. 3 and Dr. Pingyu Wang for instructive discussions. Also, we thank reviewers who gave us useful comments.

References

1. Sarafianos, N., Boteanu, B., Ionescu, B., Kakadiaris, I.A.: 3D human pose estimation: a review of the literature and analysis of covariates. Comput. Vis. Image Underst. **152**, 1–20 (2016)
2. Andriluka, M., Pishchulin, L., Gehler, P., Schiele, B.: 2D human pose estimation: new benchmark and state of the art analysis. In: Proceedings of the IEEE Conference on computer Vision and Pattern Recognition, pp. 3686–3693 (2014)
3. Sigal, L., Balan, A.O., Black, M.J.: HUMANEVA: synchronized video and motion capture dataset and baseline algorithm for evaluation of articulated human motion. Int. J. Comput. Vis. **87**(1–2), 4 (2010)
4. Ionescu, C., Papava, D., Olaru, V., Sminchisescu, C.: Human3.6M: large scale datasets and predictive methods for 3D human sensing in natural environments. IEEE Trans. Pattern Anal. Mach. Intell. **36**(7), 1325–1339 (2014)
5. Moreno-Noguer, F.: 3D human pose estimation from a single image via distance matrix regression. In: 2017 IEEE Conference on Computer Vision and Pattern Recognition (CVPR), pp. 1561–1570. IEEE (2017)
6. Chen, C.-H., Ramanan, D.: 3D human pose estimation = 2D pose estimation + matching. In: CVPR, p. 6 (2017)
7. Martinez, J., Hossain, R., Romero, J., Little, J.J.: A simple yet effective baseline for 3D human pose estimation. In: IEEE International Conference on Computer Vision, p. 3 (2017)
8. Ramakrishna, V., Kanade, T., Sheikh, Y.: Reconstructing 3D human pose from 2D image landmarks. In: Fitzgibbon, A., Lazebnik, S., Perona, P., Sato, Y., Schmid, C. (eds.) ECCV 2012. LNCS, vol. 7575, pp. 573–586. Springer, Heidelberg (2012). https://doi.org/10.1007/978-3-642-33765-9_41
9. Pavlakos, G., Zhou, X., Derpanis, K.G., Daniilidis, K.: Coarse-to-fine volumetric prediction for single-image 3D human pose. In: 2017 IEEE Conference on Computer Vision and Pattern Recognition (CVPR), pp. 1263–1272. IEEE (2017)
10. Tekin, B., Katircioglu, I., Salzmann, M., Lepetit, V., Fua, P.: Structured prediction of 3D human pose with deep neural networks. arXiv preprint: arXiv:1605.05180 (2016)
11. Li, S., Chan, A.B.: 3D human pose estimation from monocular images with deep convolutional neural network. In: Cremers, D., Reid, I., Saito, H., Yang, M.H. (eds.) ACCV 2014. LNCS, vol. 9004, pp. 332–347. Springer, Cham (2015). https://doi.org/10.1007/978-3-319-16808-1_23
12. Li, S., Zhang, W., Chan, A.B.: Maximum-margin structured learning with deep networks for 3D human pose estimation. In: Proceedings of the IEEE International Conference on Computer Vision, pp. 2848–2856 (2015)
13. Varol, G., et al.: Learning from synthetic humans. In: 2017 IEEE Conference on Computer Vision and Pattern Recognition (CVPR 2017) (2017)

14. Kadkhodamohammadi, A., Gangi, A., de Mathelin, M., Padoy, N.: A multi-view RGB-D approach for human pose estimation in operating rooms. In: 2017 IEEE Winter Conference on Applications of Computer Vision (WACV), pp. 363–372. IEEE (2017)

15. Sun, X., Shang, J., Liang, S., Wei, Y.: Compositional human pose regression. In: The IEEE International Conference on Computer Vision (ICCV) (2017)

16. Zhou, X., Huang, Q., Sun, X., Xue, X., Wei, Y.: Towards 3D human pose estimation in the wild: a weakly-supervised approach. In: IEEE International Conference on Computer Vision (2017)

17. Hu, J., Shen, L., Sun, G.: Squeeze-and-excitation networks. arXiv preprint: arXiv:1709.01507 (2017)

18. Wang, Y., Xie, L., Qiao, S., Zhang, Y., Zhang, W., Yuille, A.L.: Multi-scale spatially-asymmetric recalibration for image classification. arXiv preprint: arXiv:1804.00787 (2018)

19. He, K., Zhang, X., Ren, S., Sun, J.: Deep residual learning for image recognition. In: Proceedings of the IEEE Conference on Computer Vision and Pattern Recognition, pp. 770–778 (2016)

20. Moeslund, T.B., Granum, E.: A survey of computer vision-based human motion capture. Comput. Vis. Image Underst. **81**(3), 231–268 (2001)

21. Mehta, D., et al.: VNect: real-time 3D human pose estimation with a single RGB camera. ACM Trans. Graph. (TOG) **36**(4), 44 (2017)

22. Wei, S.-E., Ramakrishna, V., Kanade, T., Sheikh, Y.: Convolutional pose machines. In: Proceedings of the IEEE Conference on Computer Vision and Pattern Recognition, pp. 4724–4732 (2016)

23. Newell, A., Yang, K., Deng, J.: Stacked hourglass networks for human pose estimation. In: Leibe, B., Matas, J., Sebe, N., Welling, M. (eds.) ECCV 2016. LNCS, vol. 9912, pp. 483–499. Springer, Cham (2016). https://doi.org/10.1007/978-3-319-46484-8_29

24. Akhter, I., Black, M.J.: Pose-conditioned joint angle limits for 3D human pose reconstruction. In: Proceedings of the IEEE Conference on Computer Vision and Pattern Recognition, pp. 1446–1455 (2015)

25. Zhou, X., Sun, X., Zhang, W., Liang, S., Wei, Y.: Deep kinematic pose regression. In: Hua, G., Jégou, H. (eds.) ECCV 2016. LNCS, vol. 9915, pp. 186–201. Springer, Cham (2016). https://doi.org/10.1007/978-3-319-49409-8_17

26. He, K., Zhang, X., Ren, S., Sun, J.: Spatial pyramid pooling in deep convolutional networks for visual recognition. In: Fleet, D., Pajdla, T., Schiele, B., Tuytelaars, T. (eds.) ECCV 2014. LNCS, vol. 8691, pp. 346–361. Springer, Cham (2014). https://doi.org/10.1007/978-3-319-10578-9_23

27. Girshick, R.: Fast R-CNN. arXiv preprint: arXiv:1504.08083 (2015)

28. Xie, L., Zheng, L., Wang, J., Yuille, A.L., Tian, Q.: Interactive: inter-layer activeness propagation. In: Proceedings of the IEEE Conference on Computer Vision and Pattern Recognition, pp. 270–279 (2016)

29. Chen, L.-C., Yang, Y., Wang, J., Xu, W., Yuille, A.L.: Attention to scale: scale-aware semantic image segmentation. In: Proceedings of the IEEE Conference on Computer Vision and Pattern Recognition, pp. 3640–3649 (2016)

30. Simo-Serra, E., Quattoni, A., Torras, C., Moreno-Noguer, F.: A joint model for 2D and 3D pose estimation from a single image. In: 2013 IEEE Conference on Computer Vision and Pattern Recognition (CVPR), pp. 3634–3641. IEEE (2013)

31. Ioffe, S., Szegedy, C.: Batch normalization: accelerating deep network training by reducing internal covariate shift. arXiv preprint: arXiv:1502.03167 (2015)

32. Nair, V., Hinton, G.E.: Rectified linear units improve restricted boltzmann machines. In: Proceedings of the 27th International Conference on Machine Learning (ICML-10), pp. 807–814 (2010)

33. Tome, D., Russell, C., Agapito, L.: Lifting from the deep: convolutional 3D pose estimation from a single image. In: CVPR 2017 Proceedings, pp. 2500–2509 (2017)
34. Zhou, X., Zhu, M., Pavlakos, G., Leonardos, S., Derpanis, K.G., Daniilidis, K.: MonoCap: monocular human motion capture using a CNN coupled with a geometric prior. IEEE Trans. Pattern Anal. Mach. Intell. (2018)
35. Mehta, D., Rhodin, H., Casas, D., Sotnychenko, O., Xu, W., Theobalt, C.: Monocular 3D human pose estimation using transfer learning and improved CNN supervision. arXiv preprint: arXiv:1611.09813 (2016)

Gossiping the Videos:
An Embedding-Based Generative
Adversarial Framework for Time-Sync
Comments Generation

Guangyi Lv[1], Tong Xu[1], Qi Liu[1], Enhong Chen[1(✉)], Weidong He[1],
Mingxiao An[1], and Zhongming Chen[2]

[1] Anhui Province Key Laboratory of Big Data Analysis and Application,
School of Computer Science and Technology,
University of Science and Technology of China, Hefei, China
gylv@mail.ustc.edu.cn, cheneh@ustc.edu.cn
[2] Quantum Lab, Research Institute of OPPO, Shanghai, China

Abstract. Recent years have witnessed the successful rise of the time-sync *"gossiping comment"*, or so-called **"Danmu"** combined with online videos. Along this line, automatic generation of Danmus may attract users with better interactions. However, this task could be extremely challenging due to the difficulties of informal expressions and "semantic gap" between text and videos, as Danmus are usually not straightforward descriptions for the videos, but subjective and diverse expressions. To that end, in this paper, we propose a novel **E**mbedding-based **G**enerative **A**dversarial (**E-GA**) framework to generate time-sync video comments with "gossiping" behavior. Specifically, we first model the informal styles of comments via semantic embedding inspired by variational autoencoders (VAE), and then generate Danmus in a generatively adversarial way to deal with the gap between visual and textual content. Extensive experiments on a large-scale real-world dataset demonstrate the effectiveness of our E-GA framework.

1 Introduction

Recent years have witnessed the booming of the novel time-sync comments on online videos, or so-called **"Danmu"** [10,11], which describes the scene that massive comments flying across the screen just like bullets [14]. This new business mode could not only enrich the video with textual information but also attract viewers with better interactions. For instance, the report of iQiYi[1], a leading Danmu-enabled video-sharing platform in China, revealed that Danmus have improved the online user activities, such as views or comments, even by 100 times. Along this line, administrators are encouraged to improve the loyalty

---

[1] http://digi.163.com/14/0915/17/A66VE805001618JV.html.

© Springer Nature Switzerland AG 2019
Q. Yang et al. (Eds.): PAKDD 2019, LNAI 11441, pp. 412–424, 2019.
https://doi.org/10.1007/978-3-030-16142-2_32

of users with high-quality Danmus. However, due to the limitation of "grass-root" users, the quantity and quality of Danmu could be hardly ensured. Thus, solutions for automatic Danmu generation is urgently required.

Usually, prior arts conducted the short text generation mainly following the idea of tagging method [25], textual summarization [4,17] or Question-answering system [1]. Nevertheless, though large efforts have been made, these brilliant works may not be suitable for the Danmu generation task due to its unique characters. Indeed, Danmu is not just the objective statement of video content, more importantly, it could be the **"gossiping"** to the video. First, different from the image caption techniques, Danmu always indicates the **subjective** opinions, e.g., "*I like Penny*" and "*Sheldon is so cute*" (from the American TV sitcom "The Big Bang Theory"). Second, the content of Danmus could be more **diverse**, which is not limited to the current episode of video, e.g., we can see "*Bazinga*", the pet phrase of Sheldon, in Danmus at anywhere even without Sheldon. Besides, the expression of Danmus could be informal, as emotions (e.g., "O($\cap$_$\cap$)O") or slangs (e.g., "*lol*" which means laughing), which could be more **fluent** just like human talking, but cannot be interpreted by literal meanings and thus increase the difficulty of generation.

To that end, in this paper, we propose a novel **E**mbedding-based **G**enerative **A**dversarial framework (**E-GA**) to generate the gossiping Danmus of videos. Specifically, considering the informal expressions in Danmu, we represent both the video scenes and textual information as vectors. Then, to deal with the *semantic gap* between visual content and user opinions, a generative adversarial model is adapted to learn the latent mapping between visual space and semantic space. Along this line, the proper and diverse semantic vectors will be generated, and then decoded as sentences. To the best of our knowledge, we are among the first ones who attempt to generate Danmu-like comments with combining both embedding and adversarial approaches. Extensive experiments on a large-scale real-world dataset demonstrate the effectiveness of our E-GA framework, which validates the potential of our framework on generating "gossiping" text in Danmu-enable social media platforms.

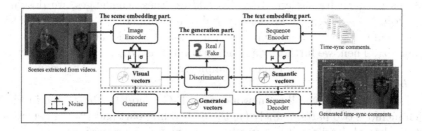

Fig. 1. The overall architecture of the generation framework.

2   Problem Definition and Technical Solution

In this paper, we target at generating Danmus for video frames. Formally, we first give the definition as follow:

Definition 1 (Danmu Generation). *Given the training set of video frames $\{v_i\}$, where $v_i \in V$ denotes the i-th frame in video, combined with related Danmus as $S_i = \{s_{ik}\}$. Our target is to learn a Danmu generator G, so that a series of Danmu-like comments $\{s'_{kj}\}$ could be produced for gossiping any given frame $v'_i \in V'$ in the test set.*

Specifically, as we mentioned above that we target at generating the "gossiping" Danmus for given video frames, we have to satisfy the following three requirements to ensure the gossiping characters:

1. **Relation.** The generated Danmus must be semantically related to the given frame.
2. **Diversity.** The generated Danmus should be more than only the description of the objective truth in the frame. They should be subjective and semantically diverse.
3. **Fluency.** The generated Danmu should be fluency, i.e., their style should be similar to the human-written comments.

Along this line, to satisfy all the three requirements above, we formulate our solution in the following way. First, according to the basic task, i.e., generating a sequence of comments given the video frame, we propose a generator G to model the probability distribution $P(s|v)$. Then, considering the requirements on *semantic relation*, we adopt the Generative Adversarial Networks (GANs) structure [16], and further introduce a noise vector τ, so that the requirements on *diversity* could also be satisfied.

Correspondingly, the generator G could be re-formulated as $\{s_k\} = G(\tau|v)$. However, here the generated Danmu, as the sequences of words, will be discrete but not continuous as prior arts. Thus, requirements of *fluency* could be unsatisfied with directly using the GAN [28]. Moreover, the informal expressions exist in Danmu may further increase the difficulty in understanding the relations between frames and text. To address these challenges, we design an **E**mbedding-based **G**enerative **A**dversarial framework (E-GA), where the frames V and comments S are first represented into low dimensional continuous spaces $\mathcal{H}_v$ and $\mathcal{H}_s$. Then, we further adapt our generator as $\{h_{sk}\} = G(\tau|h_v)$, in which $h_{sk} \in \mathcal{H}_s$ and $h_v \in \mathcal{H}_v$. Finally, Danmu sentences s_i will be reconstructed from h_{si}.

In summary, the overall framework of our E-GA model is illustrated in Fig. 1, which includes two parts, namely (1) the embedding part and (2) the generation part. Technical details will be introduced in the following sections.

2.1   The Embedding Part

First, we will introduce the detail of embedding part. In order to better model the internal relations for the frames and text, we choose to perform data representation via the Variational AutoEncoders (VAE) [13], which is based on a regularized standard autoencoder. It modifies the conventional ones by using a posterior distribution $q(z|x)$ instead of the deterministic embedding $\phi(x)$ for input x. Reconstruction of x is generated by sampling a vector z from $q(z|x)$ and then passing it through a decoder. In addition, to ensure that the embedding space is continuous where any point (vector) can be decoded to a valid sample, the posterior $q(z|x)$ is regularized with its KL-divergence from a prior distribution $p(z)$, which usually follows standard Gaussian $\mathcal{N}(0, 1)$. The objective function takes the following form:

$$L = -\mathbb{E}_{q(z|x)}[\log\ p(x|z)] + D_{KL}(q(z|x)\|p(z)), \tag{1}$$

where the expectation term is known as the reconstruction loss L_{rec}, while the other term denotes the KL-loss L_{KL}.

Though the VAE based model can achieve decoding vectors to human acceptable data, e.g., images or fluency sentences, its embedding ability has been largely weakened. Note that, the "embedding ability" here refers to how well the representations can reconstruct their original inputs. For example, if there are embedding vectors $h = \phi(x)$ that can be decoded to the inputs x with little loss, we normally say that ϕ have good embedding ability. In contrast, if a series of representations fail to reconstruct the original inputs, there is definitely a loss of information. At the same time, the associated reconstruction loss L_{rec} will be large. Thus, we are not going to use the VAE directly. Considering the KL term in Eq. 1, the KL divergence for diagonal Gaussian $\mathcal{N}(\mu, \sigma^2)$ can be formulated by:

$$L_{KL} = \sum_{i=1}^{N}(\mu_i^2 + \sigma_i^2 - \log(\sigma_i^2) - 1), \tag{2}$$

which is composed of the "μ-term" and the "σ-term". As we know, for a converged VAE, these two terms will ideally set μ and σ to 0 and 1 respectively, which will result in poor embedding effect. In our task, both of the ability of embedding and decoding are needed. On one hand, we need the proper representations μ to feed into the generator. On the other hand, we also need the decoder to generate new sentences from $h \sim \mathcal{N}(\mu, \sigma^2)$ rather than giving the existing sentences from the training set. To this end, we loose the KL constraint by replacing the μ-term with $\max(\mu_i^2 - \mu_0^2, 0)$:

$$L_{KL} = \sum_{i=1}^{N}(\max(\mu_i^2 - \mu_0^2, 0) + \sigma_i^2 - \log(\sigma_i^2) - 1), \tag{3}$$

so as to σ still converge to 1 while μ_i can be in the range of $[-\mu_0, \mu_0]$. Further, to measure the embedding capacity for the modified model, we define a metric as follows:

$$C = \mathbb{E}_{\mu_i \sim \mathcal{U}(-\mu_0, \mu_0)}\left[\frac{D_{KL}(q(z|x)\|p(z))}{H(q(z|x), p(z))}\right] = 1 - \frac{\sqrt{\ln 2\pi e}}{\mu_0}\arctan\frac{\mu_0}{\sqrt{\ln 2\pi e}}, \tag{4}$$

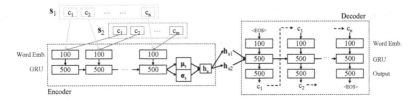

Fig. 2. The RNN structure of sentence encoder and decoder. Size of each layer is labeled on the box. Note that the encoder and decoder share the same parameters for word embedding layer.

where H denotes the cross entropy of the two distribution. C is valued in $[0, 1)$, and we could balance the effect of embedding and decoding by tuning μ_0 based on this. We will discuss more about this later in Sect. 3.4.

Next, to be specific, for video frames, we set up an encoder ϕ_v to encode an image $\boldsymbol{v} \in \mathcal{V}$ as a posterior distribution $q(\boldsymbol{h}_v|\boldsymbol{v})$. Typically, we use a diagonal Gaussian distribution $\mathcal{N}(\boldsymbol{\mu}_v, \boldsymbol{\sigma}_v^2)$ to present this posterior, where $(\boldsymbol{\mu}_v, \boldsymbol{\sigma}_v) = \phi_v(\boldsymbol{v})$. Then, to formulate the loss function and learn the model, a visual vector $\boldsymbol{h}_v$ is sampled from $q(\boldsymbol{h}_v|\boldsymbol{v})$ and then sent to a decoder ψ_v. The image is finally reconstructed as $\boldsymbol{v}' = \psi_v(\boldsymbol{h}_v)$. The reconstruction loss is in the form of Mean Squared Error (MSE):

$$L_{rec} = \frac{1}{N} \sum (\boldsymbol{v}' - \boldsymbol{v})^2. \tag{5}$$

Specially, the encoder ϕ_v and decoder ψ_v are implemented by deep convolutional networks with 4 layers as used in [19].

For Danmu sentences, the situation is a little different. We design character level Gated Recurrent Unit (GRU) [5] networks as encoder ϕ_s and decoder ψ_s, as shown in Fig. 2. At each time, a pair of sentences $(\boldsymbol{s}_1, \boldsymbol{s}_2)$ that are selected from the same frame are first put into the encoder by characters to get their posterior distributions $\mathcal{N}(\boldsymbol{\mu}_{s1}, \boldsymbol{\sigma}_{s1}^2)$ and $\mathcal{N}(\boldsymbol{\mu}_{s2}, \boldsymbol{\sigma}_{s2}^2)$. Like frame embedding, $\boldsymbol{h}_{s1}$ and $\boldsymbol{h}_{s2}$ which are sampled from the two distributions are put into the decoder. In the decoder, for every single sentence, the corresponding reconstruction loss is the sum of the negative log likelihood of the correct character at each step:

$$L_{rec}(\boldsymbol{s}) = -\log \ P(\boldsymbol{s}|\boldsymbol{h}_s) = -\sum_{t=1}^{N} \log \ P(c_t|\boldsymbol{h}_s, c_0, ..., c_{t-1}). \tag{6}$$

More importantly, to model the deeper semantic meaning of Danmus, we also involve a semantic loss formulated as:

$$L_{sem}(\boldsymbol{s}_1, \boldsymbol{s}_2) = \text{dist}(\boldsymbol{\mu}_{s1}, \boldsymbol{\mu}_{s2}), \tag{7}$$

in which we take the assumption of "temporal correlation" [14], i.e., comments appear in the same frame hold the similar topics (relevant to the frame, but semantically diverse). Here we choose cosine distance as the distance function dist(). Finally, the overall reconstruction loss for Danmu embedding is given by:

$$L_{rec} = L_{rec}(s_1) + L_{rec}(s_2) + L_{sem}(s_1, s_2). \tag{8}$$

2.2 The Generation Part

In the generation part, we set up Conditional Generative Adversarial Model which consists of two "adversarial" models: a generative model G that captures the data distribution, and a discriminative model D that estimates the probability that a sample came from the training data rather than G. Here, since we aim to produce semantic vectors from the visual vectors, both G and D are implemented by deep neural networks.

In detail, we choose to utilize our GAN as a Wasserstein GAN [2]. For G, the visual vector $\boldsymbol{h}_v$ and the noise vector $\boldsymbol{\tau}$ are first concatenated, and then put into the hidden layers with size 1000 and 500. Here, we perform the batch normalization [12] for every layer to reduce the internal-covariate-shift by normalizing its input distributions to the standard Gaussian distribution, and leaky ReLU with leak value 0.01 is used as the activation function. Then, a linear transformation is took place on the output to produce the "fake" semantic vector, i.e., $\boldsymbol{h}_s = G(\boldsymbol{\tau}|\boldsymbol{h}_v)$.

Similarly, for D, the input is the concatenation of a visual vector $\boldsymbol{h}_v$ and a (fake) semantic vector $\boldsymbol{h}_s$, while the hidden layers are sized as 2000 and 1000 with the same activation function. Please note that batch-norm should not be used for a discriminator since it can cause the model unable to converge. Finally, the critical output $y = D(\boldsymbol{h}_s|\boldsymbol{h}_v)$ is calculated by linearly mapping the hidden state to a scalar, which indicates whether the input semantic vector is fake or not. Furthermore, G and D are trained alternatively and the objective function of a two-player min-max game would be:

$$\min_{G} \max_{D} V(D, G) = \mathbb{E}_{p(\boldsymbol{h}_s|\boldsymbol{h}_v)}[D(\boldsymbol{h}_s|\boldsymbol{h}_v)] - \mathbb{E}_{p(\boldsymbol{\tau})}[D(G(\boldsymbol{\tau}|\boldsymbol{h}_v)|\boldsymbol{h}_v)].$$

2.3 Learning the Model

We then turn to introduce details about learning the model. With recalling the Fig. 1, the training process can be divided into two stages: (1) We separately learning the two autoencoders with the frames and comments from the videos. After the parameters are fine-tuned, we store the models including the image encoder ϕ_v, sequence encoder ϕ_s and the sequence decoder ψ_s for further use. (2) Based on the autoencoders, we train the generator $\mathbb{G}$ and the discriminator D in a generative adversarial way. Note that in this stage, parameters of ϕ_v, ϕ_s and ψ_s are kept unchanged, only $\mathbb{G}$ and D are updated.

To be specific, in both of the two stages, mini-batch gradient descent is used to optimize the models, where the batch size in our case is 32. For the autoencoders, we use SGD with momentum, where the learning rate and momentum are separately set as 0.1 and 0.6, and at the same time, gradient clipping is performed to constrain the L2 norm of the global gradients not larger than 1.0. To our pilot study, it is crucial to clip the gradients for most of the optimizing algorithms due to the exploding gradients problem even with a very small learning rate. For the GAN part, we take the RMSProp[2] algorithm with learning rate 10^{-5} and decay 0.9.

Another problem is the trade-off between reconstruction loss L_{rec} and KL-loss L_{KL} when training the embedding models. For a VAE-based model, directly minimizing $L_{rec} + L_{KL}$ may fail to encode useful information [3] in the embedding vector, since in most cases, L_{KL} is far more easy to be optimized, which will yield models that consistently set $Q(z|x)$ equal to $P(z)$. Thus, in our case, we design a simple annealing approach, in which $L_{rec} + \alpha L_{KL}$ is used to replace the original loss function, where α is initialized with 0 and then gradually increased to 1.

3   Experiments

3.1   Data Preparation

We choose to validate our work on a real-world dataset extracted from Bilibili, which is one of the largest video-sharing platforms in China. Specially, totally $2,716$ individual movies are extracted, which last for $232,485$ minutes and contain $9,661,369$ Danmus. To get scene images, we split the videos into frames for every one second.

Since the total number of the frames is too large, key frame extraction is carried out to eliminate the duplicated ones. First, we extract features for frames by constructing the scalable color descriptors (SCD) [15]. Then, based on these features, an affinity propagation algorithm is performed to cluster the frames, and the kernels are collected as our key frames. In our experiment, we got $214,953$ key frames with their corresponding Danmus. 80% of them are used as training data, while others for testing.

3.2   Experimental Setup

Baseline. As far as we know, few works about Danmu generation have been done before and there can be mainly three kinds of models for generation tasks. Thus, to evaluate our model, we consider the corresponding straightforward baseline models to compare with.

(1) Encoder-Decoder framework. We train a Convolutional Neural Network (CNN) as the encoder to get the representations of frames. The representations are then treated as inputs for a decoder implemented by a Recurrent Neural Network (RNN). The model is similar to the Neural Image Caption [22].

(2) Conditional Variational Autoencoders (CVAE). The CVAE [21] is based on traditional VAE which has an condition input y to both encoder and decoder. In our experiment, we take the representations of the frames as y.

---

(3) Simple Generative Adversarial Networks. Similar with CVAE, generative adversarial nets can be extended to a conditional model [16]. We can perform the conditioning by feeding extra information y (the representations of frames in our experiment) into both the discriminator and generator.

Artificial Judgement. Since heuristic rules could hardly judge whether a sentence should be "gossiping" of a given video, to evaluate the Danmu generation models, a human study is carried out, in which we have 40 experts who have years of experience in watching Danmu-enabled videos. While, as the amount of all generated Danmu is really huge for humans, we also developed a web-based GUI for online labeling. For each time a person logs in the system, 20 video frames are randomly sampled from the test set with their corresponding generated Danmus. Then he/she is asked to click the Danmus which are thought as fake. Our system will then label the clicked ones as "fake", and the others as "escaped". We evaluate the models based on the percentage of the "escaped" Danmus, which we call it *"Human Recall"*.

Metrics. To eliminate the errors caused by the human raters, we will take metrics which can be automatically computed as our alternative measurements. The *BLEU* score [18] which is a form of precision of word n-grams between generated and reference sentences has been commonly used in machine translation and image description. In this paper, we use the character level BLEU-4 score to measure the overall performance. The references set of BLEU are 3 sentences randomly selected from the existing comments of the corresponding frame. Additionally, we also define *Fluency* and *Diversity* metrics to measure the performances on multiple aspects. In detail, for each Danmu sentence s, we split it into n-gram tokens $t \in \mathcal{T}_s$. The Fluency and Diversity are separately defined in the form below:

$$Fluency = \frac{\sum_{t \in \mathcal{T}_s}[t \in \mathcal{T}]\mathrm{len}(t)}{\sum_{t \in \mathcal{T}_s}\mathrm{len}(t)}, \quad Diversity = 1 - \frac{1}{N}\sum_{s_i,s_j \in \mathcal{S}', i \neq j}\frac{2|\mathcal{T}_{s_i} \cap \mathcal{T}_{s_j}|}{|\mathcal{T}_{s_i}| + |\mathcal{T}_{s_j}|},$$

where $\mathcal{T}$ denotes the n-gram tokens for all human written sentences in the train set, $[t \in \mathcal{T}]$ is an indicator function whose value is 1 if $t \in \mathcal{T}$ otherwise is 0, and $\mathcal{S}'$ indicates all sentences generated for the same scene and N is the total number of the pair combinations in $\mathcal{S}'$.

Table 1. Performance of these models.

	Human Recall	BLEU-4	Fluency	Diversity
Encoder-Decoder	0.4572	0.168214	0.678827	0.904757
Conditional VAE	0.5580	0.174298	0.733117	0.948959
Simple-GAN	0.3454	0.129924	0.440087	0.705946
E-GA	**0.6274**	**0.177638**	**0.845072**	**0.964757**

3.3   Overall Results

The overall experimental results are summarized in Table 1. We can see that our proposed framework outperforms the other models in all of four metrics. Not surprisingly, all models except Simple-GAN achieve high performance on *Fluency*, since for those straightforward RNN-based models, it is easier to imitate human style languages, while simple implemented GAN fails due to the discreteness output in the task. However, all of these methods perform poor on *BLEU*, which we think is also reasonable since our task is quite different from those like translation or image description. As mentioned in Sect. 1, the Danmu senders do not aim to reveal the objective truth in most cases, so the existing Danmus cannot be considered as the only deterministic ground-truth in our experiment. Consequently, it is very difficult and sometimes no need to hit the existing Danmus precisely.

At the same time, we have observed that our model outperforms the others with significant margin on *Human-Recall* and *Diversity* due to the excellent generative ability of GAN. Thus, it is proved to be reasonable that the combination of embedding method and GAN is suitable for Danmu generation task. On one hand, the embedding technology simplifies the GAN structure into DNNs which are more easy to learn. On the other hand, it avoids the discrete problem when training a GAN in generating sequential data.

3.4   Balance for Embedding and Decoding Capacity

The performance of our framework can be affected by the embedding/decoding capacity of autoencoders, therefore, it is crucial for us to determine the associated parameter and also necessary to analyze the impacts of them. As mentioned in Sect. 2.1, the embedding effect of a VAE model is naturally opposite to its decoding ability, and thus we involved parameter μ_0 in making a trade-off. According to Eq. 4, there is a curve that the embedding capacity C changes along with μ_0. As shown in Fig. 3, C is zero at the beginning, which means the model is almost unable to perform sentence representation but perfect in generating. Then, as μ_0 increases, C grows rapidly, and at the same time, the embedding ability will become stronger. As μ_0 continues to become larger, the enhancement for embedding quality is getting less stark.

We examined this by setting up several autoencoders with different μ_0. Here, Table 2 gives some examples with μ_0 set to 0, 2, 4 and 8, and Fig. 3 shows the reconstruction loss changing with μ_0. For every case, three sentences are listed which separately indicate the "input", the "reconstruction" from μ and the "generation" from a sample from $\mathcal{N}(\mu, \sigma^2)$. Obviously, when μ_0 is zero, we got the best generation effect, however, we could hardly reconstruct the original sentence from its representation μ. Then, we can see for μ_0 valued 2 and 4, the reconstructed sentences are much better and the generated ones are still acceptable. At last, if μ_0 is much larger, the reconstruction quality reached the best, while the generated sentence became unreadable for humans. In summary, the results prove that our modification for VAE is reasonable, and in most cases, we can set μ_0 to around 2.

Table 2. Samples from trained autoencoders.

$\mu_0 = 0, C = 0.0$	$\mu_0 = 2, C = 0.2665$
INPUT 雪人保持的真好	卧槽 双击666啊！
(The snowman keeps well.)	(OMG Double click 666 AH!)
RECO. 好人的冲击力	卧槽 6666啊
(The impact of a good man.)	(OMG! 6666 AH!)
GEN. 高能预警QAQ	好可爱啊(oVⱼ)！！！
(Warning! High energy! QAQ)	(So cute it is (oVⱼ) !!!)

$\mu_0 = 4, C = 0.5063$	$\mu_0 = 8, C = 0.7129$
INPUT 绝对不能杀主角	他被亲了我好难受啊啊
(You can't kill the hero.)	(I feel so upset he was kissed.)
RECO. 绝对不是主角光环	他被亲了我好难受啊啊
(Definitely not the halo of the hero.)	(I feel so upset he was kissed.)
GEN. 好喜欢你，这首歌神的	黑暗示就好玩的
(Really like you, the god of the song's)	(The dark indicates good things to play.)

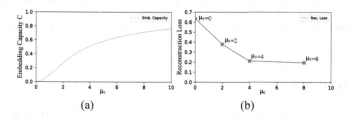

(a) (b)

Fig. 3. The embedding capacity (a) and reconstruction loss (b) w.r.t μ_0.

3.5 Case Study

At last, some typical scene images and the generated Danmus can be seen in Fig. 4. Row 1 and Row 2 are good and bad cases generated by our E-AG framework. Row 3 shows outputs from other baselines. For scenes in the first row, the generated Danmus are mainly focused on expressing viewers' different opinions on the frame, which have very high diversity. Especially, for the scene from row 1 column 2, we can easily recognize it as scared shot. Just like human viewers, our model not only generates Danmus to indicate "the ghost will come", but also send something like "BGM is lovely", "It is an interesting movie" to embolden themselves. Of course, we have to admit that there are also some Danmus do not fit the given scenes. While, to our further observation, we found that most of the miss-generated scenes are images with some strange content. Finally, for some results in the third row, we can hardly imagine the relationship between some of the comments and frames. In summary, the results are interesting, and furthermore, we could intuitively feel the diversity and the gossiping behavior of Danmu-enabled videos.

Fig. 4. Typical cases of generated Danmus. The Chinese sentences are translated into English.

4   Related Work

In this section, we will summarize the prior arts on three related topics, namely *Text Generation, Unsupervised Autoencoders* and the *Generative Adversarial Networks*.

Text Generation. Since we have witnessed only a few prior arts which focus on the Danmu analysis, especially for the Danmu generation, we will summarize related works on similar topics with Danmu-like Text Generation, i.e., the Image Caption which focus on extracts "meaningful" descriptions for images. Traditionally, early approaches rely on recognizing the visual elements, and then performing template model, n-gram model, or statistical machine translation to get sentences [8,20]. Recently, end-to-end methods [22,24] are proposed to combine deep convolutional networks and recurrent neural networks as autoregressive models. However, image caption techniques mainly focus on describing the objective facts, which is different from the task the Danmu generation who targets at expressing the subjective opinion of viewers.

Unsupervised Autoencoders. These NN-based techniques are designed for efficient embedding, with the aim of learning an encoder $\phi(x)$ by maximizing the likelihood of a probabilistic decoder $P(x|\phi(x))$. Though autoencoders have seen success in pre-training image [23] and sequence [6] models, they may not be effective at extracting for global semantic features, e.g., generating data from the continuous space. In contrast, recently, a variant method called Variational Autoencoder (VAE) [13] has become more widely used for learning generative models. The VAE learns representations not as single points, but as a distribution in the latent space, forcing them to fill the space rather than memorizing the training data as the isolated vectors. However, according to the features above, the VAE may not be suitable for embedding, due to the difficulty in reconstructing samples from the indeterministic representations.

Generative Adversarial Networks. GANs are methods to generate synthetic data with similar statistical properties as the real one [9]. Instead of explicitly defining a loss from a target distribution, GANs train a generator by receiving a loss from a discriminator which tries to differentiate between real and generated data. Though GANs and its variants have shown great success in Computer Vision domain [7,19], there are still challenges in applying them to the traditional NLP tasks [26–28].

5   Conclusion

In this paper, we proposed an embedding-based framework to generate Danmu-like comments for video scenes. In detail, we first represented key frames and comments into continuous spaces, and then learned the mapping between the two spaces via a generative adversarial approach. Along this line, the proper and diverse semantic vectors will be generated, and then decoded as sentences. Experiments on a real-world dataset showed the potential of our framework on generating "gossiping" text in Danmu-enable social media platforms. In the future, we will improve our framework with more comprehensive factors (e.g., positions, colors) which may help to better understand the meaning.

Acknowledgments. This research was partially supported by grants from the National Natural Science Foundation of China (Grant No. 61727809, U1605251, 61672483, and 61703386), the Anhui Provincial Natural Science Foundation (Grant No. 1708085QF140), and the Fundamental Research Funds for the Central Universities (Grant No. WK2150110006).

References

1. Alupului, M., Ames, A.L., Collopy, B.A.M., Pesot, J.F., Pierce, R., Steinmetz, D.C.: Question-answering system. US Patent App. 15/229,361, 5 August 2016
2. Arjovsky, M., Chintala, S., Bottou, L.: Wasserstein GAN. arXiv preprint arXiv:1701.07875 (2017)
3. Bowman, S.R., Vilnis, L., Vinyals, O., Dai, A.M., Jozefowicz, R., Bengio, S.: Generating sentences from a continuous space. arXiv preprint arXiv:1511.06349 (2015)
4. Chua, F.C.T., Asur, S.: Automatic summarization of events from social media. In: ICWSM (2013)
5. Chung, J., Gulcehre, C., Cho, K., Bengio, Y.: Empirical evaluation of gated recurrent neural networks on sequence modeling. CoRR (2014)
6. Dai, A.M., Le, Q.V.: Semi-supervised sequence learning. In: NIPS, pp. 3079–3087 (2015)
7. Denton, E.L., Chintala, S., Fergus, R., et al.: Deep generative image models using a Laplacian pyramid of adversarial networks. In: NIPS, pp. 1486–1494 (2015)

8. Farhadi, A., et al.: Every picture tells a story: generating sentences from images. In: Daniilidis, K., Maragos, P., Paragios, N. (eds.) ECCV 2010. LNCS, vol. 6314, pp. 15–29. Springer, Heidelberg (2010). https://doi.org/10.1007/978-3-642-15561-1_2

9. Goodfellow, I., et al.: Generative adversarial nets. In: NIPS, pp. 2672–2680 (2014)

10. He, M., Ge, Y., Chen, E., Liu, Q., Wang, X.: Exploring the emerging type of comment for online videos: Danmu. ACM Trans. Web (TWEB) **12**(1), 1 (2018)

11. He, M., Ge, Y., Wu, L., Chen, E., Tan, C.: Predicting the popularity of *DanMu*-enabled videos: a multi-factor view. In: Navathe, S.B., Wu, W., Shekhar, S., Du, X., Wang, X.S., Xiong, H. (eds.) DASFAA 2016. LNCS, vol. 9643, pp. 351–366. Springer, Cham (2016). https://doi.org/10.1007/978-3-319-32049-6_22

12. Ioffe, S., Szegedy, C.: Batch normalization: accelerating deep network training by reducing internal covariate shift. arXiv preprint arXiv:1502.03167 (2015)

13. Kingma, D.P., Welling, M.: Auto-encoding variational bayes. arXiv preprint arXiv:1312.6114 (2013)

14. Lv, G., Xu, T., Chen, E., Liu, Q., Zheng, Y.: Reading the videos: temporal labeling for crowdsourced time-sync videos based on semantic embedding. In: AAAI, pp. 3000–3006 (2016)

15. Manjunath, B.S., Ohm, J.R., Vasudevan, V.V., Yamada, A.: Color and texture descriptors. IEEE TCSVT **11**(6), 703–715 (2001)

16. Mirza, M., Osindero, S.: Conditional generative adversarial nets. arXiv preprint arXiv:1411.1784 (2014)

17. Neto, J.L., Freitas, A.A., Kaestner, C.A.A.: Automatic text summarization using a machine learning approach. In: Bittencourt, G., Ramalho, G.L. (eds.) SBIA 2002. LNCS (LNAI), vol. 2507, pp. 205–215. Springer, Heidelberg (2002). https://doi.org/10.1007/3-540-36127-8_20

18. Papineni, K., Roukos, S., Ward, T., Zhu, W.J.: BLEU: a method for automatic evaluation of machine translation. In: ACL, pp. 311–318 (2002)

19. Radford, A., Metz, L., Chintala, S.: Unsupervised representation learning with deep convolutional generative adversarial networks. arXiv preprint arXiv:1511.06434 (2015)

20. Rohrbach, M., Qiu, W., Titov, I., Thater, S., Pinkal, M., Schiele, B.: Translating video content to natural language descriptions. In: ICCV, pp. 433–440 (2013)

21. Sohn, K., Yan, X., Lee, H.: Learning structured output representation using deep conditional generative models. In: NIPS, pp. 3483–3491 (2015)

22. Vinyals, O., Toshev, A., Bengio, S., Erhan, D.: Show and tell: a neural image caption generator. In: CVPR, pp. 3156–3164 (2015)

23. Wang, N., Yeung, D.Y.: Learning a deep compact image representation for visual tracking. In: NIPS, pp. 809–817 (2013)

24. Wang, Z., et al.: Chinese poetry generation with planning based neural network. COLING (2016)

25. Wu, B., Zhong, E., Tan, B., Horner, A., Yang, Q.: Crowdsourced time-sync video tagging using temporal and personalized topic modeling. In: SIGKDD, pp. 721–730. ACM (2014)

26. Yu, L., Zhang, W., Wang, J., Yu, Y.: SeqGAN: sequence generative adversarial nets with policy gradient. In: AAAI (2017)

27. Zhang, K., et al.: Image-enhanced multi-level sentence representation net for natural language inference. In: ICDM, pp. 747–756 (2018)

28. Zhang, Y., Gan, Z., Carin, L.: Generating text via adversarial training (2016)

Self-paced Robust Deep Face Recognition with Label Noise

Pengfei Zhu, Wenya Ma, and Qinghua Hu[✉]

College of Intelligence and Computing, Tianjin University, Tianjin, China
{zhupengfei,wyma,huqinghua}@tju.edu.cn

Abstract. Deep face recognition has achieved rapid development but still suffers from occlusions, illumination and pose variations, especially for face identification. The success of deep learning models in face recognition lies in large-scale high quality face data with accurate labels. However, in real-world applications, the collected data may be mixed with severe label noise, which significantly degrades the generalization ability of deep models. To alleviate the impact of label noise on face recognition, inspired by curriculum learning, we propose a self-paced deep learning model (SPDL) by introducing a negative l_1-norm regularizer for face recognition with label noise. During training, SPDL automatically evaluates the cleanness of samples in each batch and chooses cleaner samples for training while abandons the noisy samples. To demonstrate the effectiveness of SPDL, we use deep convolution neural network architectures for the task of robust face recognition. Experimental results show that our SPDL achieves superior performance on LFW, MegaFace and YTF when there are different levels of label noise.

Keywords: Face recognition · Label noise · Self-pace learning

1 Introduction

Deep learning has achieved consistent breakthroughs in different tasks, including face recognition [3], object detection [19] and visual tracking [25]. The superior performance of deep learning owns to the representations of data with multiple levels of abstraction and massive well-labelled training data. For face recognition, despite the success of deep learning in face verification, it is hard to achieve satisfactory recognition accuracy without sufficient training data, especially when there are a large number of subjects in face identification. CosFace [20] uses a large-scale face dataset that consist of 5 millions face images from more than 90 thousands identities. FaceNet [18] is learned on a much larger dataset with 200 millions face images of 8 millions identities. The large-scale face databases with accurate labels can dramatically improve the performance of face recognition in that the deep learning models can be well trained. However, the high expense of labelling data makes it hard to get massive face data with accurate identification information. In real-world applications, the collected data are mixed with severe

© Springer Nature Switzerland AG 2019
Q. Yang et al. (Eds.): PAKDD 2019, LNAI 11441, pp. 425–435, 2019.
https://doi.org/10.1007/978-3-030-16142-2_33

label noise, which significantly degrades the generalization ability of deep learning models. For face recognition, it is much challenging to utilize the massive face data with inaccurate decision information to train a robust face recognition model.

How to acquire correctly labelled face dataset is one of the key challenges in face recognition tasks. One intuitive way is to manually collect and label the face images. The other way is semi-automatic annotation by online image searching. Searching results contain massive label noise, which should be manually corrected. Both strategies are in bad need of manual annotation, which suffers from high time consumption and labelling expense. In most cases, we get a small-scale dataset with accurate labels and a large-scale dataset with label noise. DCNNs such as Resnet [6] and ResNeXt [22] have hundreds of layers and millions of parameters, and therefore need a large number of accurately labelled samples for training. Although they can achieve the state-of-the-art result in many tasks, they perform poorly when they meet label noise. How to train a model with label noise is still a difficult task to solve.

For DCNNs, they update the parameters of deep learning models by training on fixed-size mini-batches that consist of random samples With stochastic gradient descent (SGD) and back propagation, each sample in mini-batches reflects its own influence by gradients propagation. If the data is clean enough, all samples in the mini-batch guide the DCNNs to train a satisfactory model. However, when there exists label noise in the massive face data, noisy samples could do harm to DCNNs in the training period. Noisy samples slow down convergence speed of DCNNs, or even make model unable to converge. In other words, the cleanness of samples in the training dataset directly affect the performance of the deep model. This paper studies how to overcome the negative impact of label noise on DCNNs.

Inspired by the recent success of self-paced learning (SPL) [10], we propose a self-paced deep learning model (SPDL) for robust face recognition with corrupted labels and outliers. First, we train a deep face recognition model on a small clean dataset. Then a more robust model with good generalization ability is trained on a large-scale dataset that contains label noise. In each iteration of the training process, SPDL learns the cleanness of all samples in each mini-batches by introducing the sample weights. Large weights are assigned to the samples with low classification loss via a negative l_1-norm regularizer. Then cleaner samples, i.e., samples with larger weights are used for loss calculation and back propagation of gradients. To verify the effectiveness of the proposed method, experiments are conducted on LFW [7], MegaFace [12], and YTF [21] datasets. Results show that SPDL can significantly improve the performance of DCNNs trained on both corrupted labels and outliers. This paper makes three key contributions towards face recognition.

- A novel robust self-paced learning framework is proposed for DCNNs.
- The difference of the influence on DCNNs is discussed between corrupted labels and outliers.

– We verify our proposed SPDL on three face recognition benchmarks and validate the robustness of SPDL.

2   Related Work

There are two types of label noise, i.e., corrupted labels and outliers. If a sample is not accurately labelled but belongs to one class of the training data, this is called corrupted labels. If a sample is not only mislabelled but does not belong to any class of the training data, the sample is a outlier. To deal with label noise, there are mainly three types of methods: noise-robust, noise-removal and noise-tolerant. The noise-robust method is the most straightforward and ideal method to deal with label noise. Patrini et al. proposed to improve the robustness to label noise by loss factorization in weakly supervised leaning [17]. Gao et al. divided the loss function into two parts: one irrelevant to noise and the other related with noise, by risk minimization [4]. The second type of methods consider that the noisy face images can be relabelled or discarded by a filter. These methods need to manually set a threshold for noise removal [24]. Brodley et al. proposed to detect noisy samples by setting classification confidence scores [2]. The third type of methods model the noise distribution. Thus, the classification model and the noise model are directly separated. The most common noise modelling method is to estimate the noise distribution by the Bayesian methods. However, there are few methods to train DCNNs with label noise, especially the large ratio of label noise.

Recent studies show that selecting a subset of good samples for training a classifier can lead to better results than using all the samples [13,14]. Curriculum learning (CL) [1] is one of the most representative works. CL introduced a heuristic measure of easiness to determine the selections of samples from the training data. CL has successfully improved the performance in a variety of vision tasks. MentorNet [11] is a work that learns data-driven curriculums for DCNNs trained on corrupted labels. It connects curriculum learning and StudentNet and achieves a high performance on benchmarks. But curruculums in CL are usually predefined and remain fixed during training. By comparison, SPL quantifies the easiness by the current sample loss. In SPL, there is a threshold λ controlling the sample selecting during training. The training instances with loss values less than the threshold are selected as easy samples to train. The training instances with loss values larger than λ are neglected during training. λ dynamically increases in the training process to make more complex samples will be learned, until all training instances are considered. SPL has been widely applied to various problems, including visual tracking [8], person re-identification [26] and multi-label learning [15].

3   Self-paced Robust Face Recognition

3.1   Label Noise

Label noise contains two types: corrupted labels and outliers. Assume that we have a training set with N samples $\mathcal{D} = \{x_n, y_n\}_{n=1}^{N}$, where y_n is the label of sample x_n and the real label is $\hat{y}_n$. For noisy samples, $y_n \neq \hat{y}_n$ and for clean sample $y_n = \hat{y}_n$. Let C denote the number of classes, $C < N$.

For corrupted labels, assuming there is a noisy sample $x_i \in N$ in the training set with label y_i and its real label is $\hat{y}_i$, then $y_i \neq \hat{y}_i$, $y_i \in C$, and $\hat{y}_i \in C$, In other words, the real labels of the samples with corrupted labels belong to the classes of the training set but they are labelled to other classes of the training set by mistake.

For outliers, if there is a noisy sample $x_i \in N$ in the training set with label y_i and its real label is $\hat{y}_i$, then $y_i \neq \hat{y}_i$, $y_i \in C$, and $\hat{y}_i \notin C$. In other words, the real labels of outliers do not belong to the classes of the training set but they are labelled to the known class of the training set by mistake.

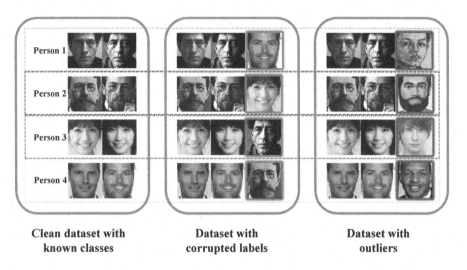

Clean dataset with known classes   **Dataset with corrupted labels**   **Dataset with outliers**

Fig. 1. The example of two kinds of label noise as there are four known classes. The images with green boxes are samples with corrupted labels, which are mislabeled but belong to the classes of the training set. The images with red boxes are outliers, whose real labels do not belong to the classes of the training set. (Color figure online)

As shown in Fig. 1, the dataset could be mixed with two types of label noise, i.e., corrupted labels and outliers. Both of them seriously impact our model training. The difference of the impact on DCNNs between corrupted labels and outliers will be shown in our experimental section (Fig. 2).

3.2 SPDL Framework

For clean dataset or other standard tasks, the loss function can be formulated as:

$$L(w) = \sum_{i=1}^{n} L(y_i, f(\mathbf{x}_i, \mathbf{w})), \tag{1}$$

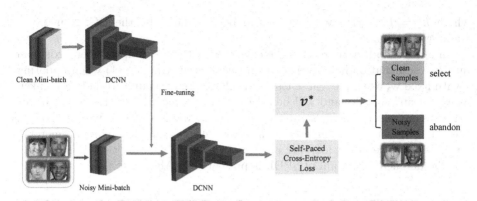

Fig. 2. The flowchart of self-paced robust deep face recognition. We first pre-train a deep model with a small clean dataset. Then we fine-tune on a large-scale noisy dataset with SPDL. v^* represents the cleanness of samples in each mini-batch. Images with red boxes represent noisy samples while images with green boxes represent clean samples. SPDL automatically chooses clean samples for training and abandons noisy samples. (Color figure online)

where $L(y_i, f(x_i, w))$ is the loss function of the sample x_i which measures the cost between the ground truth label y_i and the estimated label $f(x_i, w)$. Here w represents the model parameter inside the decision function f. Our work is to learn the network parameters w by minimizing L. For SPL, its strategy is to generate a curriculum which includes easy and diverse samples during learning. The training instances with loss values larger than a threshold, λ, are neglected during training and λ dynamically increases in the training process to include more complex samples, until all training instances are considered. We first train a deep face recognition model with a small clean dataset. Faced with label noise, we utilize the self-paced strategy to estimate the cleanness of every sample in mini-batches. The loss function can be written as:

$$L(w) = \sum_{i=1}^{n} v_i L(y_i, f(\mathbf{x_i}, \mathbf{w})),\tag{2}$$

where v_i represents the cleanness of the sample x_i and it also represents the weight of x_i to DCNNs in the training process. It acts on back propagation to supervise the model training. In this case, the objective function can be further formulated as

$$\min_{\mathbf{w},v} \mathbf{E}(\mathbf{w}, v) = \sum_{i=1}^{n} v_i L(y_i, f(\mathbf{x}_i, \mathbf{w})) - \lambda \sum_{i=1}^{n} v_i,\tag{3}$$

With $\mathbf{w}$ fixed, the goal optimum $v^* = [v_1^*, \ldots, v_n^*]$ can be easily obtained. v^* can be solved as follows:

$$v_i^* = \begin{cases} 1 - \frac{L_i}{\lambda}, & L_i < \lambda, \\ 0, & otherwise. \end{cases}\tag{4}$$

where $L_i = L(y_i, f(\mathbf{x}_i, \mathbf{w}))$, which can be calculated by the DCNN in every training iteration.

In our method, v_i represents the cleanness of the ith sample. There exists an intuitive explanation behind this alternative search strategy: (1) when updating v with fixed $\mathbf{w}$, a sample whose loss is smaller than a certain threshold λ is taken as an "clean" sample, and will be selected in training $(v_i^* = (1 - \frac{L_i}{\lambda}) > 0)$, or otherwise unselected noisy sample $(v_i^* = 0)$. (2) when updating $\mathbf{w}$ with fixed v, the classifier is trained only on the selected "clean" samples. The parameter λ controls the pace of the learning process.

The detailed algorithm of SPDL is presented in Algorithm 1.

Algorithm 1. SPDL

Require: Input: D(training data), B(batch size), C(number of classes), M(pre-trained model parameters), E(total iteration)
Ensure: Output: The model parameters $\mathbf{W}$
1: Initialize parameters M;
2: **for** $j = 0 \rightarrow E$ **do**
3:     **for** $i = 0 \rightarrow B$ **do**
4:         Calculate the loss of each sample $L_i = L(y_i, f(\mathbf{x}_i, \mathbf{w}))$;
5:         **if** $L(y_i, f(\mathbf{x}_i, \mathbf{w})) > \lambda$ **then**
6:             $v_i = 0$;
7:         **else**
8:             $v_i = 1 - \frac{L_i}{\lambda}$;
9:         **end if**
10:         $L_i = v_i * L(y_i, f(\mathbf{x}_i, \mathbf{w}))$
11:     **end for**
12:     Update $\mathbf{w}$ via BP;
13: **end for**

But we set $\lambda = 12.0$ fixed in our experiments. In early training process, limited by the performance of the trained model, the loss of every sample may be very big. Then the weight of all samples may be small. For noisy samples and hard samples their weight is 0. During the first few iterations, clean samples will be used to train DCNN and noisy samples will be abandoned. Our purpose is to ensure the DCNN can learn satisfactory parameters to accelerate the convergence speed. In the later training process, with the improvement of model's accuracy, the value of λ will not change to ensure our model the ability of learning hard samples. With this strategy, all the clean samples in mini-batches will be selected for training and some *easy noisy* samples with low weights will be selected, which can improve the generalization of the deep models.

4   Experiments

This section empirically verifies the proposed method on different benchmarks in terms of both corrupted labels and outliers. We first train a model with a small clean datasets and then fine-tune on a large-scale noisy datasets. All experiments are conducted using the Caffe [9] framework with four NVIDIA Titan xp GPUs. Stochastic gradient descent (SGD) is utilized to train the Resnet-80, a kind of residual networks. Face images are all cropped to 256 × 256 pixels. We compared our SPDL method against DCNN with cross-entropy loss. λ are kept consistent between the strategy in our SPDL, In our experiments, we use different datasets to train our models and evaluate them on different datasets. The following sections discuss the experimental setups and results in more detail.

4.1 Datasets

The datasets in our experiments are shown in Table 1. CASIA-WebFace [23] was used as a small clean dataset to pre-train a model firstly. Labeleb Faces in the Wild (LFW) [7] and YouTube Faces (YTF) [21] are two benchmarks in face recognition. MegaFace [12] and FaceScrub [16] are two parts of MegaFace Challenge 1. MegaFace dataset is the gallery dataset comprised of photos from Flickr users and FaceScrub dataset is comprised of celebrities. MS is a clean subset of MS-Celeb-1M [5]. All the training and test operations are based on those datasets.

Table 1. The information of face datasets.

Datasets	Subjects	Images	Role
LFW	5,749	13,233	Test
CASIA-webface	10,575	494,414	Train
MS	41,857	3,095,536	Train
MegaFace	530	100,000	Test
FaceScrub	80	3,530	Test

In our experiments, because there are few open-source face datasets with different types of label noise, we utilize MS to create our noisy datasets as shown in Tabel 2. For corrupted label noise, we separately select 30%, 50%, 70% face images from the MS dataset to establish noisy data. The label of the noisy samples is randomly set as other classes in the training data. For examples, for the dataset with 30% noisy data, about 928,661 images should be set as noisy samples. We select 928,642 images from 12,654 classes and set their label randomly among the other 29,203 classes. For outliers, we shuffled the data of all datasets first. Then we select 30%, 50%, 70% data from the datasets separately and set this label randomly. Here we need to address that in our experiments we keep the total number of face images in every noisy datasets the same as original MS with 3,095,536 images.

Table 2. The comparison of different types and ratios of MS in our experiments.

Original dataset	Noisy type	Noise ratio	Subjects	Clean images
MS	No	0	41,857	3,095,536
MS	Corrupted labels	30%	41,857	2,166,875
MS	Corrupted labels	50%	41,857	1,547,768
MS	Corrupted labels	70%	41,857	928,661
MS	Outliers	30%	29,203	2,166,894
MS	Outliers	50%	20,855	1,547,798
MS	Outliers	70%	12,493	928,658

4.2   Results and Discussion

This subsection shows the experimental results on different benchmarks of corrupted label noise and outlier noise. All the settings of experiments are fixed for fair comparison. Table 3 shows the face recognition performance when the deep model is trained on the MS dataset without label noise. Raw means the original cross entropy loss is used for traning DCNN directly. CASIA-WebFace is used to pre-train a deep model. For the MegaFace dataset, the accuracy is the identification rate using uncropped FaceScrub set with 1,000,000 distractors of MegaFace Challenge 1.

Table 3. The face recognition rate with clean datasets.

Datasets	Method	Accuracy (%)		
		LFW	MegaFace	YTF
MS	Raw	99.48	63.57	94.72

Table 4 shows the comparison between using the original cross entropy loss and SPDL under different ratios of corrupted labels noise. With direct training, DCNN is robust at lower ratio noise. But with noisy samples increasing, DCNN is unable to learn a satisfactory model. By comparison, our SPDL method achieves significant improvement. The result of SPDL with 30% noise is better than directly learning on clean MS dataset. The result of SPDL with 50% noise is not bad than directly learning on clean MS dataset. When the noise ratio is 70%, the original cross-entropy loss is unable to converge, but our SPDL achieves not bad results. To sum, the proposed SPDL is robust for corrupted labels noise compared with the original DCNN.

Table 5 shows the comparison between the original cross entropy loss and SPDL under different ratios of outlier noise. The superiority of the proposed SPDL to the original DCNN is much significant. Under 50% noise ratio, SPDL achieves much better results on three benchmarks compared with the baseline.

Table 4. The comparison between the original cross entropy loss and SPDL under different ratios of corrupted labels noise. "-" represents the DCNN can not converge.

Dataset	Noise ratio	Method	Accuracy (%)		
			LFW	MegaFace	YTF
MS	30%	Raw	99.18	55.74	94.14
MS	30%	SPDL	**99.52**	**65.28**	**95.36**
MS	50%	Raw	94.97	21.83	88.10
MS	50%	SPDL	**99.63**	**62.90**	**95.26**
MS	70%	Raw	-	-	-
MS	70%	SPDL	**97.43**	**29.67**	**70.08**

When the noise ratio is 70%, it also has the ability to learn a good model, which validates the robustness of the proposed method to outliers. The experiments shows that our SPDL method is also robust for outliers noise, especially for high ratio noise.

Table 5. The comparison between the original cross entropy loss and SPDL under different ratios of outliers noise.

Dataset	Noise ratio	Method	Accuracy (%)		
			LFW	MegaFace	YTF
MS	30%	Raw	99.03	55.89	93.36
MS	30%	SPDL	**99.58**	**65.64**	**95.52**
MS	50%	Raw	98.36	33.08	93.07
MS	50%	SPDL	**99.53**	**66.44**	**95.66**
MS	70%	Raw	87.83	1.67	91.02
MS	70%	SPDL	**98.93**	**37.40**	**94.82**

By comparison, DCNN is more robust for outliers noise than corrupted labels noise when the ratio of noise very high. The proposed SPDL significantly improves the robustness of DCNN. Our SPDL method is robust to both corrupted labels noise and outliers noise, especially for the high ratio label noise.

5 Conclusions

In this paper, we proposed a self-paced deep learning model (SPDL) for robust face recognition with label noise, including corrupted labels and outliers. Sample weights are embedded to evaluate the cleanness of the face images in each mini-batch. A negative l_1-norm regularizer is introduced to the loss function of deep convolutional neural network. During each iteration in the training process, only the clean samples with large weights contributes to back propagation of gradients and therefore alleviate the impact of noisy samples. Experimental results on LFW, Meageface and YTF validate the effectiveness of the proposed SPDL.

Acknowledgements. This work was supported by the National Natural Science Foundation of China under Grants 61502332, 61876127 and 61732011, Natural Science Foundation of Tianjin Under Grants 17JCZDJC30800.

References

1. Bengio, Y., Louradour, J., Collobert, R., Weston, J.: Curriculum learning. In: Proceedings of the 26th Annual International Conference on Machine Learning, pp. 41–48. ACM (2009)
2. Brodley, C.E., Friedl, M.A., et al.: Identifying and eliminating mislabeled training instances. In: Proceedings of the National Conference on Artificial Intelligence, pp. 799–805 (1996)
3. Deng, J., Guo, J., Zafeiriou, S.: ArcFace: additive angular margin loss for deep face recognition. arXiv preprint arXiv:1801.07698 (2018)
4. Gao, W., Wang, L., Li, Y.F., Zhou, Z.H.: Risk minimization in the presence of label noise. In: AAAI, pp. 1575–1581 (2016)
5. Guo, Y., Zhang, L., Hu, Y., He, X., Gao, J.: MS-Celeb-1M: a dataset and benchmark for large-scale face recognition. In: Leibe, B., Matas, J., Sebe, N., Welling, M. (eds.) ECCV 2016. LNCS, vol. 9907, pp. 87–102. Springer, Cham (2016). https://doi.org/10.1007/978-3-319-46487-9_6
6. He, K., Zhang, X., Ren, S., Sun, J.: Deep residual learning for image recognition. In: Proceedings of the IEEE Conference on Computer Vision and Pattern Recognition, pp. 770–778 (2016)
7. Huang, G.B., Jain, V., Learned-Miller, E.: Unsupervised joint alignment of complex images. In: ICCV (2007)
8. Huang, W., Gu, J.J., Ma, X., Li, Y.: Self-paced model learning for robust visual tracking. J. Electron. Imaging **26**(1), 013016 (2017)
9. Jia, Y., et al.: Caffe: convolutional architecture for fast feature embedding. arXiv preprint arXiv:1408.5093 (2014)
10. Jiang, L., Meng, D., Zhao, Q., Shan, S., Hauptmann, A.G.: Self-paced curriculum learning. In: AAAI, vol. 2, p. 6 (2015)
11. Jiang, L., Zhou, Z., Leung, T., Li, L.J., Fei-Fei, L.: MentorNet: learning data-driven curriculum for very deep neural networks on corrupted labels. In: International Conference on Machine Learning, pp. 2309–2318 (2018)
12. Kemelmacher-Shlizerman, I., Seitz, S.M., Miller, D., Brossard, E.: The MegaFace benchmark: 1 million faces for recognition at scale. In: Proceedings of the IEEE Conference on Computer Vision and Pattern Recognition, pp. 4873–4882 (2016)
13. Lapedriza, A., Pirsiavash, H., Bylinskii, Z., Torralba, A.: Are all training examples equally valuable? arXiv preprint arXiv:1311.6510 (2013)
14. Lee, Y.J., Grauman, K.: Learning the easy things first: self-paced visual category discovery. In: 2011 IEEE Conference on Computer Vision and Pattern Recognition (CVPR), pp. 1721–1728. IEEE (2011)
15. Li, C., Wei, F., Yan, J., Zhang, X., Liu, Q., Zha, H.: A self-paced regularization framework for multilabel learning. IEEE Trans. Neural Netw. Learn. Syst. **29**(6), 2660–2666 (2018)

16. Ng, H.W., Winkler, S.: A data-driven approach to cleaning large face datasets. In: 2014 IEEE International Conference on Image Processing (ICIP), pp. 343–347. IEEE (2014)

17. Patrini, G., Nielsen, F., Nock, R., Carioni, M.: Loss factorization, weakly supervised learning and label noise robustness. In: International Conference on Machine Learning, pp. 708–717 (2016)

18. Schroff, F., Kalenichenko, D., Philbin, J.: FaceNet: a unified embedding for face recognition and clustering. In: Proceedings of the IEEE Conference on Computer Vision and Pattern Recognition, pp. 815–823 (2015)

19. Wang, H., Wang, Q., Gao, M., Li, P., Zuo, W.: Multi-scale location-aware kernel representation for object detection. In: Proceedings of the IEEE Conference on Computer Vision and Pattern Recognition, pp. 1248–1257 (2018)

20. Wang, H., et al.: CosFace: large margin cosine loss for deep face recognition. arXiv preprint arXiv:1801.09414 (2018)

21. Wolf, L., Hassner, T., Maoz, I.: Face recognition in unconstrained videos with matched background similarity. In: 2011 IEEE Conference on Computer Vision and Pattern Recognition (CVPR), pp. 529–534. IEEE (2011)

22. Xie, S., Girshick, R., Dollár, P., Tu, Z., He, K.: Aggregated residual transformations for deep neural networks. In: 2017 IEEE Conference on Computer Vision and Pattern Recognition (CVPR), pp. 5987–5995. IEEE (2017)

23. Yi, D., Lei, Z., Liao, S., Li, S.Z.: Learning face representation from scratch. arXiv preprint arXiv:1411.7923 (2014)

24. Zhang, J., Sheng, V.S., Li, T., Wu, X.: Improving crowdsourced label quality using noise correction. IEEE Trans. Neural Netw. Learn. Syst. **29**(5), 1675–1688 (2018)

25. Zhang, Y., Wang, L., Qi, J., Wang, D., Feng, M., Lu, H.: Structured siamese network for real-time visual tracking. In: Ferrari, V., Hebert, M., Sminchisescu, C., Weiss, Y. (eds.) ECCV 2018. LNCS, vol. 11213, pp. 355–370. Springer, Cham (2018). https://doi.org/10.1007/978-3-030-01240-3_22

26. Zhou, S., et al.: Deep self-paced learning for person re-identification. Pattern Recogn. **76**, 739–751 (2018)

Multi-Constraints-Based Enhanced Class-Specific Dictionary Learning for Image Classification

Ze Tian and Ming Yang[✉]

School of Computer Science and Technology, Nanjing Normal University,
Nanjing 210023, China
zetian_edu@126.com, m.yang@njnu.edu.cn

Abstract. Sparse representation based on dictionary learning has been widely applied in recognition tasks. These methods only work well under the conditions that the training samples are uncontaminated or contaminated by a little noise. However, with increasing noise, these methods are not robust for image classification. To address the problem, we propose a novel multi-constraints-based enhanced class-specific dictionary learning (MECDL) approach for image classification, of which our dictionary learning framework is composed of shared dictionary and class-specific dictionaries. For the class-specific dictionaries, we apply Fisher discriminant criterion on them to get structured dictionary. And the sparse coefficients corresponding to the class-specific dictionaries are also introduced into Fisher-based idea, which could obtain discriminative coefficients. At the same time, we apply low-rank constraint into these dictionaries to remove the large noise. For the shared dictionary, we impose a low-rank constraint on it and the corresponding intra-class coefficients are encouraged to be as similar as possible. The experimental results on three well-known databases suggest that the proposed method could enhance discriminative ability of dictionary compared with state-of-art dictionary learning algorithms. Moreover, with the largest noise, our approach both achieves a high recognition rate of over 80%.

Keywords: Sparse representation · Dictionary learning ·
Low-rank matrix recovery · Discriminative coefficients ·
Image classification

1 Introduction

Sparse representation technique has achieved the great performance in signal processing applications consisting of compressed sensing [5], image denoising [6]. As the extension of signal processing field, Wright et al. exploit sparse representation-based classification (SRC) [23] for face recognition. SRC assumes that test samples could be a linear combinations of dictionary atoms composed of original training samples and the size of atoms is generally fixed, which could

© Springer Nature Switzerland AG 2019
Q. Yang et al. (Eds.): PAKDD 2019, LNAI 11441, pp. 436–448, 2019.
https://doi.org/10.1007/978-3-030-16142-2_34

decrease the performance of image classification since the training samples generally include pixel corruptions or occlusion corruptions. Therefore, lots of dictionary learning algorithms are proposed and have led to promising results.

Dictionary learning is aimed at designing dictionary to better fit the data. To get the suitable and appropriate dictionary, Aharon et al. [1] with K-SVD use the generalizing k-means clustering process to optimize the dictionary atoms. To scales up gracefully to large databases with millions of samples, Mairal et al. [16] propose online optimization algorithm based on stochastic approximations. In addition, Yang et al. [26] use the inconsistency of atoms to remove the redundancy of atoms. However, these methods cannot handle identification tasks effectively.

Recently, dictionary learning has successfully been applied to face recognition [14,25], object recognition [11,21], digit recognition [11,19] and gender classification [14,25]. The discriminative dictionary learning with sparse representation can be roughly divided into two categories. The first category of approaches learn class-specific (particular) sub-dictionary from each class of samples. To obtain compact between-class dictionaries, Ramirez et al. [19] with DLSI introduce the structural inconsistency into between-class dictionaries. Yang et al. [25] with FDDL incorporate Fisher discrimination criterion into dictionaries and coefficients to improve the recognition performance. Liu et al. [14] replace simple linear classifier with bilinear discriminative classifier. In fact, the above methods ignore common patterns in training samples, which may decrease the dictionary's discriminative power. Therefore, Kong et al. [11] divide dictionary learning framework into particular dictionaries and a shared dictionary (COPAR). However, Vu et al. [21] point out particular features may get represented by shared dictionary in COPAR so that it greatly decrease the classification ability. Therefore, Vu et al. with LRSDL impose low-rank constraint on shared dictionary, but could not consider these condition that dictionary from same class should be as correlated as possible and samples corrupted by large noise.

The second set of approaches learn a dictionary shared by all classes. To promote K-SVD to obtain classification ability, Zhang et al. [27] with D-KSVD embed a simple linear classifier into it. To enhance discriminative sparse codes of D-KSVD, Jiang et al. [10] with LC-KSVD introduce a new label consistency constraint on D-KSVD but LC-KSVD's label consistency isn't smooth enough. Therefore, Zhang et al. [28] improve label consistency by cosine similarity among signals and introduce sparse codes auto-extractor into LC-KSVD. Based on LC-KSVD, Xu et al. [24] add within-class-similar representation coefficients into D-KSVD to enhance the classification ability of linear classifier.

However, the above methods only work well for clean signals or signals with a little noise. Inspired by low-rank matrix recovery [13], most of researchers try to learn low-rank class-specific dictionaries to remove noise in training samples. Chen et al. [3] combine low-rank constraint and structural incoherence on dictionary learning, which has been proven to be an effective method for dealing with noise in images. Ma et al. [15] with DLRD integrate low-rank minimization and Fisher-based idea into dictionary learning. Li et al. [12] with $D^2L^2R^2$

embed Fisher-based ideas into particularity dictionaries and coefficients. These methods both take advantage of global structure of the data. However they can hardly handle the locality information of data. Therefore, Wang et al. [22] exploit locality constraint on coding schemes to explore data's manifold structure. The above methods only consider class-specific features in images, but not introduce common pattern from between-class data. Therefore, Jiang et al. [9] impose low-rank constraint on the class-specific dictionary and common dictionary. It is inappropriate to obtain a low-rank class-specific dictionary since dictionary from the same class should be as similar or correlated as possible not from all classes. Then, Rong et al. [20] with LRD^{2}L introduce low-rank constraint into a shared dictionary and particular dictionaries, but fails to consider to minimize the distance of intra-class subspace (coding coefficients) corresponding to particular dictionaries and maximize the distance of inter-class subspace corresponding to particular dictionaries, which may makes intra-class subspace dispersed and inter-class subspace overlapped. Moreover, Rong et al. ignore the intra-class shared coefficients to be as similar as possible.

In order to alleviate the problems of LRD^{2}L and LRSDL model, we aim at designing a multi-constraints-based enhanced class-specific dictionary learning method for image classification. Multi-constraints in our model include low-rank, sparse, minimizing within-class scatter and maximizing between-class scatter constraints. Next, we highlight some characteristics of our approach below:

(1) Dictionary learning framework based on sparse representation is composed of class-specific dictionaries and shared dictionary.

(2) MECDL makes the intra-class subspace compact and the inter-class subspace dispersal by minimizing the intra-class scatter of subspace and maximizing the inter-class scatter of subspace. At the same time, the intra-class shared coefficients are urged to be as similar as possible.

(3) To increase the discriminability in class-specific dictionaries, we apply Fisher discrimination criterion on it, which means the c-th samples can well be represented by c-th sub-dictionary not by j-th sub-dictionary, $j \neq c$. To remove the noise and increase the dictionary's representation ability, the class-specific dictionaries and shared dictionary are introduced into low-rank recovery technology. Then, the corrupted samples are separated three types of information by our method: class-specific features, shared features and noise.

The rest of the paper is organized as follows. Section 2 introduces the details of multi-constraints-based enhanced class-specific dictionary learning (MECDL) method. Experiments are conducted and presented in Sect. 3, followed by conclusion in Sect. 4.

2   Multi-Constraints-Based Enhanced Class-Specific Dictionary Learning (MECDL)

2.1   Notation

Let $Y = [Y_1, ..., Y_c, ..., Y_C] \in \Re^{d \times n}$ be the n d-dimensional training samples from C classes where Y_c comprises those in class c and $n = \sum_{c=1}^{C} n_c$. X^i, X, and X^0 represent the coding coefficients of Y corresponding to the dictionary D_i, D, and D_0, respectively. Denote by X_c^0 X_c^i, X_c, and $\overline{X}_c = [(X_c)^T, (X_c^0)^T]^T$ the sparse coding coefficients of Y_c on D_0, D_i, D, and $\overline{D}$, respectively, where X^T is the transpose of the matrix X, $\overline{D} = [D, D_0] \in \Re^{d \times K}$ with $K = k + k_0$ is the total dictionary, $D = [D_1, ..., D_c, ..., D_C] \in \Re^{d \times k}$ with $k = \sum_{c=1}^{C} k_c$ is a structural dictionary, $D_0 \in \Re^{d \times k_0}$ is a shared dictionary. Let m_c, m^0 and m be the mean vector of X_c, X^0 and X columns, respectively. Next, M_c^0, M_c, M^0, and M are the mean matrices of X_c^0, X_c, X^0 and X, respectively.

2.2   MECDL Model

It is assumed that c-th training samples Y_c contaminated by noise are separated into three parts including particular features $\overline{Y}_c$, shared common patterns $\widehat{Y}_c$ and sparse noise E_c, i.e., $Y_c = \overline{Y}_c + \widehat{Y}_c + E_c$. According to sparse representation theory, $\overline{Y}_c$ can be linearly represented by particular dictionaries D and $\widehat{Y}_c$ can be linearly represented by a shared dictionary D_0, i.e., $Y_c = DX_c + D_0 X_c^0 + E_c$. Since D_c is related with Y_c, it is expected to well represent Y_c, i.e., $Y_c = D_c X_c^c + D_0 X_c^0 + E_c$, which enhances discriminability in the class-specific dictionaries. In addition, $r(D_c) = \sum_{j=1,j\neq c}^{C} \left\| D_c X_j^c \right\|_F^2$ is embedded into our model to prevent c-th dictionary to represent j-th samples to enhance the discriminative ability of class-specific sub-dictionary. In order to make full use of subspace corresponding to dictionary, $\overline{g}(\overline{X}_c) = g(X_c) + \left\| X_c^0 - M_c^0 \right\|_F^2$ is added into the objective function, where $g(X_c) = \left\| X_c - M_c \right\|_F^2 - \left\| M_c - M \right\|_F^2 + \eta \left\| X_c \right\|_F^2$ makes intra-class subspace compact and keeps inter-class subspace dispersed, the term $\left\| X_c^0 - M_c^0 \right\|_F^2$ encourages the shared coefficients from the same class to be as similar as possible. Based on these analyses, the objective function of MECDL is designed as follows:

$$\underset{D_c, X_c, D_0, X_c^0, E_c}{\arg \min} \quad \left\| D_c \right\|_* + \left\| D_0 \right\|_* + \lambda_1 \left\| \overline{X}_c \right\|_1 + \lambda_2 \overline{g}(\overline{X}_c) + \beta \left\| E_c \right\|_1 + r(D_c)$$

$$s.t. \; Y_c = DX_c + D_0 X_c^0 + E_c, Y_c = D_c X_c^c + D_0 X_c^0 + E_c, \tag{1}$$

where λ_1, λ_2, and β denote balance parameters to balance the minimization of the six terms. The model is composed of five parts: two constraint term, the Fisher-based coefficients term $g(X_c)$, the representation coefficients based on l_1-norm term, the nuclear norm $\left\| \right\|_*$ term which promotes dictionary to get the clear images from the corrupted training samples. When λ_2 is set as 0, our model degenerate into LRD^{2}L [20] model.

2.3 Optimization of MECDL

The objective function in Eq. (1) is not convex for D, D_0, X_c and X_c^0 simultaneously, but it is convex for one of them when the others are fixed. To solve the optimization problem in Eq. (1), we divide it into three sub-problems: (1) Updating the sparse coefficients X_c and X_c^0 by class by class by fixing D and D_0; (2) Updating the dictionary D by computing D_c class by class while the others fixed; (3) Updating the dictionary D_0 by fixing other variables. We describe the detailed implementations of solving these three sub-problems in this sub-section.

Updating Coding Coefficients X_c and X_c^0: When D and D_0 are fixed, X_c and X_c^0 are simultaneously updated class by class. The objective function (1) is introduced two auxiliary variables H and S. Similar as [20], problem (1) is reduced to the following optimization problem:

$$\underset{X_c, H, X_c^0, S, E_c}{\arg\min} \quad \lambda_1 \|H\|_1 + \lambda_1 \|S\|_1 + \beta_1 \|E_c\|_1 + \lambda_2 (\|X_c N_c\|_F^2$$

$$- \|X_c P_c - G\|_F^2 - \sum_{k=1, k\neq c}^{C} \left\| Z - X_c B_c^{\ k} \right\|_F^2 + \eta \|X_c\|_F^2 + \left\| X_c^0 N_c \right\|_F^2) \qquad (2)$$

$$s.t. \ Y_c = DX_c + D_0 X_c^0 + E_c, X_c = H, X_c^0 = S,$$

where $G = \sum_{k=1, k\neq c}^{C} X_k B_k^{\ c}$, $Z = X_k A_k^{\ k}/n_k - \sum_{j=1, j\neq c}^{C} X_j B_j^{\ k}$, $B_j^{\ k} = A_j^{\ k}/n$, $P_c = A_c^{\ c}/n_c - A_c^{\ c}/n$, $N_c = I_{n_c \times n_c} - A_c^{\ c}/n_c$, I is an identity matrix, $A_j^{\ k}$ is a matrix of size $n_j \times n_k$ with all entries being 1. For derivation details of discriminative $g(X_c)$, we refer to Ref. [25]. The above problem can be solved by ALM [13] method:

$$\underset{X_c, H, X_c^0, S, E_c}{\arg\min} \quad \lambda_1 \|H\|_1 + \lambda_1 \|S\|_1 + \beta_1 \|E_c\|_1 + \lambda_2 (\|X_c N_c\|_F^2$$

$$- \|X_c P_c - G\|_F^2 - \sum_{k=1, k\neq c}^{C} \left\| Z - X_c B_c^{\ k} \right\|_F^2 + \eta \|X_c\|_F^2 + \left\| X_c^0 N_c \right\|_F^2)$$

$$+ \langle T_1, Y_A - DX_c - E_c \rangle + \langle T_2, X_c - H \rangle + \langle T_3, X_c^0 - S \rangle \qquad (3)$$

$$+ \frac{u}{2} (\|Y_A - DX_c - E_c\|_F^2 + \|X_c - H\|_F^2 + \left\| X_c^0 - S \right\|_F^2),$$

where λ_1 controls sparsity of coding coefficients, β_1 balances the level of noise, $Y_A = Y_c - D_0 X_c^0$, T_1, T_2 and T_3 are Lagrange multipliers, and u is a positive parameter. $\langle D, D_0 \rangle = Tr(D^T D_0)$ is sum of the diagonal elements of the matrix $D^T D_0$.

Updating Class-Specific Dictionaries D: When variables X_c, X_c^0, and D_0 are fixed, we update D by computing D_c class by class. The sub-dictionary D_c are updated, and the corresponding coefficients X_c^c are updated simultaneously to meet the constraint. Similar to solving strategy of this work in [20], the objective function of Eq. (1) introduced two auxiliary variables B and Z is reduced to the following optimization problem:

$$\underset{D_c, X_c^c, E_c, B, Z}{\arg\min} \quad \|B\|_* + \lambda_1 \|Z\|_1 + r(D_c) + \beta_2 \|E_c\|_1$$

$$s.t. \ Y_c = D_c X_c^c + D_0 X_c^0 + E_c, D_c = B, X_c^c = Z, \tag{4}$$

where β_2 represents the balance parameter of noise. The problem (4) can be converted by ALM [13] method:

$$\underset{D_c, X_c^c, E_c, B, Z}{\arg\min} \quad \|B\|_* + \lambda_1 \|Z\|_1 + r(D_c) + \beta_2 \|E_c\|_1$$

$$+ \langle T_1, Y_A - D_c X_c^c - E_c \rangle + \langle T_2, D_c - B \rangle + \langle T_3, X_c^c - Z \rangle \tag{5}$$

$$+ \frac{u}{2}(\|Y_A - D_c X_c^c - E_c\|_F^2 + \|D_c - B\|_F^2 + \|X_c^c - Z\|_F^2),$$

where T_1, T_2 and T_3 are Lagrange multipliers and u is a balance parameter. Equation (5) can be solved by using ALM algorithm. The detailed implementation of ALM can be referred to the Ref. [20].

Updating Shared Dictionary D_0: When variables X_c, X_c^0, and D are fixed, we learn the dictionary D_0 and the corresponding coding coefficients X_c^0 are also updated to meet the constraint. Similar as [20], using all-classes samples update the shared dictionary D_0. Then the objective function Eq. (1) introduced two auxiliary variables Q and L becomes the following equivalent optimization problem:

$$\underset{D_0, X^0, E, Q, L}{\arg\min} \quad \|Q\|_* + \lambda_1 \|L\|_1 + \beta_1 \|E\|_1$$

$$s.t. \ Y = DX + D_0 X^0 + E, D_0 = Q, X^0 = L, \tag{6}$$

where β_1 controls the balance parameter of noise. Equation (6) can be further reduced to the following ALM [13] method:

$$\underset{D_0, X^0, E, Q, L}{\arg\min} \quad \|Q\|_* + \lambda_1 \|L\|_1 + \beta_1 \|E\|_1$$

$$+ \langle T_1, Y_B - D_0 X^0 - E \rangle + \langle T_2, D_0 - Q \rangle + \langle T_3, X^0 - L \rangle \tag{7}$$

$$+ \frac{u}{2}(\|Y_B - D_0 X^0 - E\|_F^2 + \|D_0 - Q\|_F^2 + \|X^0 - L\|_F^2),$$

where T_1, T_2 and T_3 are Lagrange multipliers, u is a balance parameter, and Y_B denotes $Y - DX$. Equation (7) can be solved by using ALM algorithm. The detailed implementation of ALM can be referred to the Ref. [20].

Complete Algorithm of MECDL: To get a better representation power of Y_c and a good performance, similar to Refs. [9,20], we initialize each class-specific sub-dictionary by applying the singular value decomposition (SVD) on Y_c and normalize the columns in D_c to unit vectors. The specific initialization method can be written as follows:

$$Y_c = U_c S_c V_c^T, \quad D_c = U_c(1:d, 1) S_c(1, 1) V_c(1:k_c, 1)^T,$$

$$D_c(:, i) = D_c(:, i) / norm(D_c(:, i)) \quad for \quad i = 1, 2, ..., k_c. \tag{8}$$

Algorithm 1. The complete algorithm of MECDL.

Input:
Data matrix Y, initial shared dictionary D_0 and class-specific dictionaries D, parameters λ_1, λ_2, η, β_1, and β_2.
Output:
The class-specific dictionaries D, shared dictionary D, the coefficients matrix X and X^0, and the error matrix E.

1: **While** not converged and the maximal iteration number is not reached **do**
2: Fix D and D_0, simultaneously update X_c and X_c^0 by solving the problem (3) with the Inexact ALM algorithm [13];
3: Fix X_c, X_c^0, and D_0, update D by updating each D_c, $c = 1, ..., C$, by solving Eq. (5) with the Inexact ALM algorithm presented in Ref. [20];
4: Fix X_c, X_c^0, and D, update D_0 by solving Eq. (7) with the Inexact ALM algorithm presented in Ref. [20].
5: **end**

This initialization could capture the most significant class-specific information. For the shared dictionary D_0, similar to [25], we initialize the shared dictionary D_0 with eigenvectors of Y, which is related to the largest k_0 eigenvalues. Our experiments can lead to a desirable result by this initialization strategy.

Once the class-specific dictionaries D and shared dictionary D_0 are initialized, we can proceed by iteratively repeating the above process until a stopping criterion is reached. The complete algorithm of MECDL model is shown as Algorithm 1.

Complexity Analysis: In this section, we analyze the total complexity of MECDL method. The number of dictionary atoms from each class (shared class) are assumed to be same, which means $k_i = k_0 = k$, $i = 1, ..., C$. To calculate simply, the number of the samples from different classes are assumed to be same, i.e. $n_i = n, i = 1, ..., C$. For simplicity, we use the following facts to analyse complexity: (1) if $A \epsilon \Re^{m \times n}$, $B \epsilon \Re^{n \times p}$, then the complexity of AB is $O(mnp)$; (2) iterative number of each iterative algorithm is set to be q.

The main complexity cost of our method consists of updating coding coefficient matrix X_c and X_c^0, class-specific dictionaries D, and shared dictionary D_0 in Formulae (3), (5), and (7). In the solving process of problem (3), we can use the algorithm in the Refs. [2,8] to solve the lyapunov equation problem $XA + BX = C$. The time cost of the lyapunov equation to solve X_c and X_c^0 is respectively $O((2 + 4\sigma)((Ck)^3 + n^3) + \frac{5}{2}(Ckn^2 + n(Ck)^2))$ and $O((2 + 4\sigma)(k^3 + n^3) + \frac{5}{2}(kn^2 + n(k)^2))$ in the Ref. [2], where σ is the average number of QR steps required to make a sub-diagonal element negligible. Similar to the Ref. [20], the overall complexity of solving Eqs. (5) and (7) is $CqO(dnk + dk^2 + k^3)$ and $qO(d(Cn)k + dk^2 + k^3)$, respectively. Here we have supposed $C + 1 \approx C$ for large C. So the computational complexity of MECDL is $CqO((2 + 4\sigma)((Ck)^3 + n^3) + \frac{5}{2}(Ckn^2 + n(Ck)^2)) + CqO(dnk + dk^2 + k^3)$.

Classification Based on MECDL: After dictionary learning process, we obtain low-rank double dictionaries including shared dictionary D_0 and class-specific dictionaries D. For a corrupted test sample, its coding coefficient vector is obtained by solving:

$$\arg\min_{x,x^0,e} \|x\|_1 + \gamma\|x^0\|_1 + \omega\|e\|_1$$

$$s.t. \ y = Dx + D_0x^0 + e, \tag{9}$$

where γ and ω are positive-valued parameters, $\bar{x} = [x^T, (x^0)^T]^T$ denotes sparse vector of a new test sample y, e represents error matrix, $x = [x^1; ...; x^c; ...; x^C]$. Equation (9) can be further reduced to the following ALM [13] method. Introducing two relaxation variables H and S into Eq. (9), and the problem can be converted to minimize an unconstrained problem:

$$\arg\min_{x,H,x^0,S,e} \|H\|_1 + \gamma\|S\|_1 + \omega\|e\|_1 + \langle T_1, y - Dx - D_0x^0 - e\rangle + \langle T_2, x - H\rangle +$$

$$\langle T_3, x^0 - S\rangle + \frac{u}{2}(\|y - Dx - D_0x^0 - e\|_2^2 + \|x - H\|_2^2 + \|x^0 - S\|_2^2), \tag{10}$$

where T_1, T_2 and T_3 are Lagrange multipliers, u is a balance parameter. Same as [20], the above problem can be solved by inexact ALM algorithm [13]. Once x, x^0 and e are obtained, we use the following strategy for classification:

$$\text{identity}(y) = \arg\min_{1 \le c \le C} \|y - D_c x^c - D_0 x^0 - e\|_2^2. \tag{11}$$

3 Experiments

3.1 Databases and Experimental Settings

Databases: COIL-20 [18] consists 1440 images of 20 subjects, with each subject having 72 images. For each subject, we randomly select 10 images for training, and the rest as testing samples. Extended Yale B [7] consists 2414 images of 38 individuals, with each individual having around 60 images. Similar to Ref. [12], we select the first 15 individuals in this experiment. For each individual, we randomly select 20 images for training, and the rest as testing samples. AR database [17] includes 4,000 images of 126 subjects. These images are affected by lighting, expression change, and occlusion including scarfs and sunglasses. Similar to Refs. [4, 20], we select the subset containing 1400 images of 100 individuals with 50 females and 50 males, which means each individual has 14 images. For each individual, the first 7 images are used for training, and the others for testing.

Experimental Settings: In all experiments, all samples need to be prepro-cessed before training. Firstly, the sample images are uniformly cropped to 32×32 for COIL-20, 54×48 for Extended Yale B, 60×43 for AR. Next, a certain percentage of pixels of each image are manually corrupted by occlu-sion corruptions (i.e., a varying percentage (10%–50%) of image pixels manually corrupted by block image at the random location) or by pixel corruptions (i.e., randomly selected a percentage of pixels (10%–40%) in each image contaminated with noise uniformly distributed over $[0, V_{max}]$), where the largest value V_{max} in image. Figure 1 shows different database are corrupted by different percent-age of corruptions. Our method is compared with LCKSVD [10], FDDL [25], COPAR [11], LRSDL [21], DLRD [15], LRD^{2}L [20]. To verify the validity of the experiment, the size of the total dictionary should be same. The overall dictio-nary (class-specific dictionaries and shared dictionary) size of LCKSVD, FDDL, COPAR, LRSDL, DLRD, LRD^{2}L and MECDL are set to be 10×20, 10×20, $6 \times 20 + 80$, $6 \times 20 + 80$, 10×20, $6 \times 20 + 80$ and $6 \times 20 + 80$ for COIL-20 database, 20×15, 20×15, $15 \times 15 + 75$, $15 \times 15 + 75$, 20×15, $15 \times 15 + 75$, and $15 \times 15 + 75$ for Extended Yale B database, 7×100, 7×100, $5 \times 100 + 200$, $5 \times 100 + 200$, 7×100, $5 \times 100 + 200$, and $6 \times 100 + 100$ for AR database. Experiments are repeated 5 times to calculate the average recognition and the corresponding standard deviation.

There are five parameters in MECDL model: λ_1 for the sparse coefficients term, λ_2 and η for the discriminative coefficients term, β_1 and β_2 related with the sparse noise term. According to $\eta = 1$ in [25], we select the optimal solution of η from a small set $\{0.8, 0.9, 1, 1.1, 1.2\}$. Experimental results show that $\eta = 1.1$ makes the discriminative coefficient term $g(X_c)$ stable and convex. For the sake of fairness, the value of λ_1 is searched from 0.001 to 1, the value of λ_2, β_1 and β_2 is selected from 0.01 to 1, and the value of other parameters in all model is searched from 0.01 to 1. In our model, parameters are set to be $\eta = 1.1$ for all databases, $\lambda_1 = 0.1$, $\lambda_2 = 0.4$, $\beta_1 = 0.09$, and $\beta_2 = 0.055$ for Extended Yale

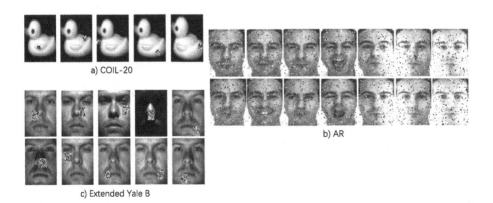

a) COIL-20

b) AR

c) Extended Yale B

Fig. 1. Examples with different percentage of corruptions from three databases.

B database, $\lambda_1 = 0.1$, $\lambda_2 = 0.05$, $\beta_1 = 0.1$ and $\beta_2 = 0.015$ for AR database, $\lambda_1 = 0.4$, $\lambda_2 = 0.6$, $\beta_1 = 0.3$, and $\beta_2 = 0.04$ for COIL-20 database.

3.2 Results

Table 1 shows the average recognition rates and standard deviation on Extended Yale B database. These images are corrupted by different percentage (10%–50%) of occlusion corruptions. Although LRSDL achieves the better performance without noise, the experimental performance of it with small or large noise is lower than our method, which indicates the proposed method is robust to image with

Table 1. Recognition accuracy (%) on Extended Yale B database with various occlusion corruptions percentage (%).

Corruptions	MECDL	LRD²L	DLRD	LRSDL	COPAR	FDDL	LCKSVD
0%	98.01 ± 0.32	96.80 ± 0.80	95.09 ± 0.93	$\mathbf{98.91 \pm 1.22}$	95.96 ± 0.40	96.93 ± 1.03	94.50 ± 1.20
10%	$\mathbf{97.89 \pm 0.81}$	95.68 ± 0.95	93.82 ± 0.72	93.72 ± 0.45	94.88 ± 0.84	94.53 ± 1.88	91.89 ± 0.97
20%	$\mathbf{95.09 \pm 1.03}$	93.60 ± 1.02	92.25 ± 1.24	91.49 ± 0.98	90.34 ± 0.94	91.40 ± 1.12	85.62 ± 1.57
30%	$\mathbf{90.68 \pm 1.34}$	88.29 ± 1.19	87.06 ± 1.58	85.40 ± 1.35	84.19 ± 1.19	85.81 ± 1.45	67.39 ± 3.13
40%	$\mathbf{84.94 \pm 0.68}$	83.23 ± 1.84	82.26 ± 1.09	81.55 ± 0.69	79.63 ± 1.96	80.59 ± 1.24	60.93 ± 1.71
50%	$\mathbf{80.51 \pm 1.35}$	79.81 ± 2.58	76.03 ± 1.66	74.97 ± 1.78	73.14 ± 2.12	73.91 ± 1.07	53.20 ± 2.73

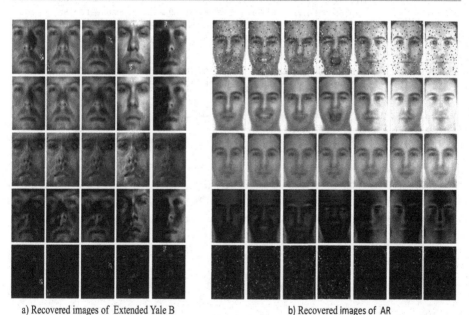

a) Recovered images of Extended Yale B   b) Recovered images of AR

Fig. 2. Examples of the proposed MECDL. First row: testing images with 10% occlusion corruptions or pixel corruptions; Second row: the recovered images $Dx + D_0 x^0$; Third row: class-specific images in Dx; Fourth row: shared images in $D_0 x^0$; Fifth row: error images in e.

increasing noise. When these images contain 40% or 50% block corruptions, our method is almost 0.7% higher than the second highest performance of LRD^{2}L. Figure 2(a) visualizes some results of image recovery by the proposed MECDL method.

From the Fig. 3, the proposed method achieves the highest performance on AR database with images corrupted by pixel corruptions (10%–40%). In addition, without noise, the classification accuracy of LRD^{2}L is slightly higher than our method, which may be because the size of the shared dictionary plays a important role without noise. However, with increasing corruptions, our method is higher than LRD^{2}L due to the discriminative coefficients. With 40% pixel corruptions, our method is almost 3% higher than other methods. Figure 2(b) visualizes some examples of recovery images generated by the MECDL method.

Recognition rates on COIL-20 database under different levels of corruptions are demonstrated on Table 2. Apparently, our method performs the best perfor-

Table 2. Recognition accuracy (%) on COIL-20 database with various occlusion corruptions percentage (%).

Corruptions	MECDL	LRD^{2}L	DLRD	LRSDL	COPAR	FDDL	LCKSVD
0%	**91.47 ± 0.96**	91.21 ± 1.06	89.27 ± 1.15	88.95 ± 1.79	87.19 ± 0.60	87.97 ± 2.07	90.77 ± 0.27
10%	**90.29 ± 1.80**	89.71 ± 1.94	88.76 ± 1.48	87.47 ± 1.27	85.53 ± 1.38	87.27 ± 1.54	89.68 ± 1.03
20%	**88.63 ± 1.75**	88.21 ± 1.42	86.34 ± 1.59	85.77 ± 0.89	82.13 ± 1.45	85.58 ± 1.34	82.74 ± 2.72
30%	**84.07 ± 1.71**	83.94 ± 1.35	83.32 ± 1.32	83.15 ± 1.85	78.73 ± 1.53	80.97 ± 0.78	79.24 ± 3.26
40%	**80.48 ± 1.36**	79.86 ± 1.11	78.53 ± 1.82	77.55 ± 1.38	73.48 ± 1.24	76.92 ± 0.67	76.13 ± 1.06

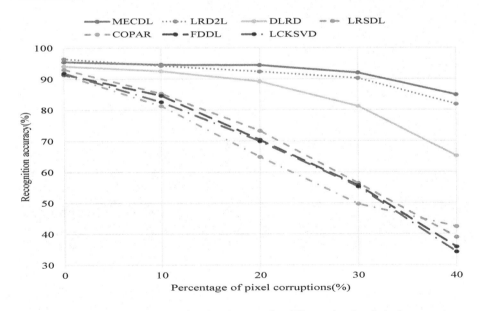

Fig. 3. Recognition accuracy on AR database under different levels of pixel corruptions.

mance compared with other methods. With increasing noise, the performance of low-rank dictionary learning methods is higher than dictionary learning without low-rank, which indicates that low-rank dictionary learning method is suitable for corrupted images.

4   Conclusion

In this paper, we propose a multi-constraints-based enhanced class-specific dictionary learning method for image classification. Our dictionary learning framework contains low-rank class-specific dictionaries and a low-rank shared dictionary. To enhance discriminablity in dictionary and coefficients, the class-specific dictionaries and it's coefficients are simultaneously introduced into Fisher-based ideas. Moreover, the intra-class shared coefficients corresponding to the shared dictionary should be as similar as possible. With increasing noise, our method could be robustness and effective compared with some state-of-art methods. In addition, with the largest noise, our approach both achieves a high recognition rate of over 80%.

Acknowledgments. This work was supported by the National Nature Science Foundation of China (61876087, 61272222), and the Key Program of the National Natural Science Foundation of China (61432008).

References

1. Aharon, M., Elad, M., Bruckstein, A.: K-SVD: an algorithm for designing over-complete dictionaries for sparse representation. IEEE Trans. Sig. Process. **54**(11), 4311–4322 (2006)
2. Bartels, R.H., Stewart, G.W.: Solution of the matrix equation AX + XB = C [F4] (algorithm 432). Commun. ACM **15**(9), 820–826 (1972)
3. Chen, C., Wei, C., Wang, Y.F.: Low-rank matrix recovery with structural incoherence for robust face recognition. In: CVPR, pp. 2618–2625 (2012)
4. Chen, Y., Su, J.: Sparse embedded dictionary learning on face recognition. Pattern Recogn. **64**, 51–59 (2017)
5. Donoho, D.L.: Compressed sensing. IEEE Trans. Inf. Theory **52**(4), 1289–1306 (2006)
6. Elad, M., Aharon, M.: Image denoising via sparse and redundant representations over learned dictionaries. IEEE Trans. Image Process. **15**(12), 3736–3745 (2006)
7. Georghiades, A.S., Belhumeur, P.N., Kriegman, D.J.: From few to many: illumination cone models for face recognition under variable lighting and pose. IEEE Trans. Pattern Anal. Mach. Intell. **23**(6), 643–660 (2001)
8. Golub, G.H., Nash, S., Loan, C.V.: A Hessenberg-Schur method for the problem AX + XB = C. IEEE Trans. Autom. Control. **24**(6), 909–913 (1978)
9. Jiang, X., Lai, J.: Sparse and dense hybrid representation via dictionary decomposition for face recognition. IEEE Trans. Pattern Anal. Mach. Intell. **37**(5), 1067–1079 (2015)
10. Jiang, Z., Lin, Z., Davis, L.S.: Label consistent K-SVD: learning a discriminative dictionary for recognition. IEEE Trans. Pattern Anal. Mach. Intell. **35**(11), 2651–2664 (2013)

11. Kong, S., Wang, D.: A dictionary learning approach for classification: separating the particularity and the commonality. In: Fitzgibbon, A., Lazebnik, S., Perona, P., Sato, Y., Schmid, C. (eds.) ECCV 2012. LNCS, vol. 7572, pp. 186–199. Springer, Heidelberg (2012). https://doi.org/10.1007/978-3-642-33718-5_14

12. Li, L., Li, S., Fu, Y.: Learning low-rank and discriminative dictionary for image classification. Image Vis. Comput. **32**(10), 814–823 (2014)

13. Lin, Z., Chen, M., Ma, Y.: The augmented lagrange multiplier method for exact recovery of corrupted low-rank matrices. CoRR. abs/1009.5055 (2010)

14. Liu, H., Yang, M., Gao, Y., Yin, Y., Chen, L.: Bilinear discriminative dictionary learning for face recognition. Pattern Recogn. **47**(5), 1835–1845 (2014)

15. Ma, L., Wang, C., Xiao, B., Zhou, W.: Sparse representation for face recognition based on discriminative low-rank dictionary learning. In: CVPR, pp. 2586–2593 (2012)

16. Mairal, J., Bach, F.R., Ponce, J., Sapiro, G.: Online dictionary learning for sparse coding. In: ICML, pp. 689–696 (2009)

17. Martinez, A.M.: The AR face database. CVC Technical report, 24 (1998)

18. Murase, H., Nayar, S.K.: Visual learning and recognition of 3-D objects from appearance. Int. J. Comput. Vis. **14**(1), 5–24 (1995)

19. Ramírez, I., Sprechmann, P., Sapiro, G.: Classification and clustering via dictionary learning with structured incoherence and shared features. In: CVPR, pp. 3501–3508 (2010)

20. Rong, Y., Xiong, S., Gao, Y.: Low-rank double dictionary learning from corrupted data for robust image classification. Pattern Recogn. **72**, 419–432 (2017)

21. Vu, T.H., Monga, V.: Fast low-rank shared dictionary learning for image classification. IEEE Trans. Image process. **26**(11), 5160–5175 (2017)

22. Wang, S., Fu, Y.: Locality-constrained discriminative learning and coding. In: CVPR, pp. 17–24 (2015)

23. Wright, J., Yang, A.Y., Ganesh, A., Sastry, S.S., Ma, Y.: Robust face recognition via sparse representation. IEEE Trans. Pattern Anal. Mach. Intell. **31**(2), 210–227 (2009)

24. Xu, L., Wu, X., Chen, K., Yao, L.: Supervised within-class-similar discriminative dictionary learning for face recognition. J. Vis. Commun. Image Represent. **38**, 561–572 (2016)

25. Yang, M., Zhang, L., Feng, X., Zhang, D.: Fisher discrimination dictionary learning for sparse representation. In: CVPR, pp. 543–550 (2011)

26. Yang, M., Zhang, L., Yang, J., Zhang, D.: Metaface learning for sparse representation based face recognition. In: ICIP, pp. 1601–1604 (2010)

27. Zhang, Q., Li, B.: Discriminative K-SVD for dictionary learning in face recognition. In: ICCV, pp. 2691–2698 (2010)

28. Zhang, Z., Li, F., Chow, T.W.S., Zhang, L., Yan, S.: Sparse codes auto-extractor for classification: a joint embedding and dictionary learning framework for representation. IEEE Trans. Sig. Process. **64**(14), 3790–3805 (2016)

Discovering Senile Dementia from Brain MRI Using Ra-DenseNet

Xiaobo Zhang, Yan Yang$^{(\boxtimes)}$, Tianrui Li, Hao Wang, and Ziqing He

School of Information Science and Technology,
Southwest Jiaotong University, Chengdu 611756, China
{xiaobo_zhang,yyang,trli}@swjtu.edu.cn, {hwang,zqhe}@my.swjtu.edu.cn

Abstract. With the rapid development of medical industry, there is a growing demand for disease diagnosis using machine learning technology. The recent success of deep learning brings it to a new height. This paper focuses on application of deep learning to discover senile dementia from brain magnetic resonance imaging (MRI) data. In this work, we propose a novel deep learning model based on Dense convolutional Network (DenseNet), denoted as *ResNeXt Adam DenseNet* (Ra-DenseNet), where each block of DenseNet is modified using ResNeXt and the adapter of DenseNet is optimized by Adam algorithm. It compresses the number of the layers in DenseNet from 121 to 40 by exploiting the key characters of ResNeXt, which reduces running complexity and inherits the advantages of Group Convolution technology. Experimental results on a real-world MRI data set show that our Ra-DenseNet achieves a classification accuracy with 97.1% and outperforms the existing state-of-the-art baselines (i.e., LeNet, AlexNet, VGGNet, ResNet and DenseNet) dramatically.

Keywords: Senile dementia · Deep learning ·
Magnetic resonance imaging (MRI) ·
ResNeXt Adam DenseNet (Ra-DenseNet)

1 Introduction

Senile dementia (e.g., Alzheimer's disease) is a chronic neurodegenerative disease that usually starts slowly and then worsens over time. The clinical manifestations of senile dementia are accompanied by memory disorder, aphasia, apraxia, agnosia, visuospatial impairment, executive dysfunction, and other performance of comprehensive dementia [3]. The exact pathogenesis of senile dementia is poorly understood so far. However, senile dementia appears constantly in real world.

Machine learning has been instrumental for the advances of data analysis and artificial intelligence. As current machine learning matures, many attempts applying machine learning models have been made to assist modern medical diagnosis. A certain number of biological diseases including Alzheimer's disease have been treated with diagnostics supportive by intelligent technologies, see [15, 21, 26, 27]. The diagnosis of senile dementia usually requires brain magnetic

© Springer Nature Switzerland AG 2019
Q. Yang et al. (Eds.): PAKDD 2019, LNAI 11441, pp. 449–460, 2019.
https://doi.org/10.1007/978-3-030-16142-2_35

resonance imaging (MRI) data collected from clinical medicine [12]. Although traditional machine learning methods have been exploited to discover senile dementia from MRI data, there are few attempts to use deep learning technology. In this work, we make attempt to discover Alzheimer's disease from MRI data using deep learning technology. In addition, we collect a brain MRI data set of Alzheimer's disease from the Open Access Series of Imaging Studies (OASIS), which are publicly available for Alzheimer's disease research [16].

In this paper, the task of discovering senile dementia (i.e., Alzheimer's disease) with brain MRI data is performed by an improved Dense convolutional Network (DenseNet). The DenseNet model has a character of efficient parameters, where each layer directly accesses to the gradients from the loss function data [5]. Moreover, DenseNet can reduce the risk of over-fitting when it is fed with a small size of training data. This is a crucial point as there are limited MRI Alzheimer's disease data. Therefore, we build our model upon DenseNet. In practice, we modify each block of DenseNet with ResNeXt [29]. Meanwhile, the adapter of DenseNet is optimized by Adam algorithm [7]. Thus, we call the proposed model _ResNeXt Adam DenseNet_ (Ra-DenseNet). In our experiments, we evaluate the classification performance of the proposed Ra-DenseNet by comparing with LeCun Network (LeNet) [10], Alex Network (AlexNet) [9], Visual Geometry Group Network (VGGNet) [22], Residual Networks (ResNet) [4] and DenseNet [5]. Experimental results show that our Ra-DenseNet model makes considerable improvement over the above-mentioned deep learning models.

The remaining part of this paper is organized as follows. Section 2 gives a brief review to the related work about the senile dementia prediction using brain MRI data. Section 3 presents the proposed Ra-DensNet model. Extensive experiments are conducted in Sect. 4. Finally, Sect. 5 concludes this paper.

2   Related Work

Several types of machine learning technologies have been studied to discover Alzheimer's disease from medical data in recent years. For example, Sorg et al. [24] analyzed functional and structural MRI data by a resting-state networks (RSNs) for Alzheimer's disease. It demonstrates that functional brain disorders can be characterized by functional-disconnectivity profiles of RSNs. Kong et al. [8] exploited independent component analysis and non-negative matrix factorization to identify significant genes and related pathways in the microarray gene expression data set of Alzheimer's disease.

Classification technologies are also employed on Alzheimer's disease research. Zhang et al. [31] proposed a multi-modal classification approach for Alzheimer's disease and mild cognitive impairment. This approach combines three modalities of biomarkers upon a kernel method. Liu et al. [11] discussed a local patch-based subspace ensemble method for the detection of Alzheimer's disease. It builds multiple individual classifiers with different subsets of local patches and then combines them to provide a more accurate and robust classification result. In addition, Zhang et al. [32] investigated a multi-modal multi-task learning approach to jointly predict multiple variables from multi-modal Alzheimer's disease data. Young et al. [30] used a Gaussian process classification to predict the probability of conversion for the Alzheimer's disease in patients with mild cognitive impairment. Gray et al. [2] discussed random forest-based similarity measures for multi-modal classification of Alzheimer's disease on neuroimaging and biological data. Recently, Tong et al. [28] presented a multi-modality classification framework with nonlinear graph fusion to exploit the complementary in the multi-modal data of Alzheimer's disease.

The brain MRI data of the Alzheimer's disease were also analyzed with clustering algorithms, e.g., K-Means, K-Medoids, Gaussian Mixture Model (GMM), Affinity Propagation (AP), Density Peaks (DP) and Cluster-based Similarity Partitioning Algorithm (CSPA) [33]. Meanwhile, the brain tumour was detected by using K-Means clustering algorithm [20]. In addition, non brain MRI data sets were studied by neuroimaging and deep learning methods [17,19].

For feature selection technology, Liu et al. [13] proposed a multi-task feature selection approach to preserve the complementary inter-modality information for Alzheimer's Disease and mild cognitive impairment identification. Zhu et al. [34] also employed a feature selection method by transferring the relational and inherent information in the data into a sparse multi-task learning framework, i.e., regression task and classification task of Alzheimer's disease diagnosis.

In terms of the early diagnosis of Alzheimer's disease, Moradi et al. [18] presented used Alzheimer's disease MRI data to predict the conversion of mild cognitive impairment from one to three years before clinical diagnosis. Fang et al. [1] investigated a Gaussian discriminant analysis method for the early diagnosis of mild cognitive impairment in Alzheimer's disease.

Apart for the above-mentioned methods, Hon et al. [6] performed two CNN architectures, i.e., VGGNet and GoogLeNet Inception, for the Alzheimer's disease diagnosis problem. Liu et al. [14] developed a multi-scale modeling variant-to-function-to-network framework to investigate the causal effect of rare non-coding variants on the Alzheimer's disease.

3   Methods

In this section, we first review LeNet, AlexNet, VGGNet, ResNet, ResNeXt, DenseNet and Adam algorithm. Then, we propose our Ra-DenseNet, which combines the key characters of ResNeXt and Adam algorithm with DenseNet architecture.

The LeNet Architecture. The LeNet [10] model is one of the most classic and basic model in convolutional neural network, which consists of three layers, i.e., a convolution layer and two pooling layers. Then the outputs are fed into a full connection layer to produce the final results.

The AlexNet Architecture. The AlexNet [9] model is a deeper and more effective model than LeNet [10]. The AlexNet model consists of five convolution layers containing pool layers and three fully connected layers. Besides, it uses a data enhancement technology.

The VGGNet Architecture. The VGGNet [22] is proposed by Oxford University. It has 16 layers of network structure, which includes 13 convolution layers and 3 fully connected layers.

The ResNet and ResNeXt Architecture. For ResNet [4], it not only makes the number of network layers deeper than VGGNet, but also proposes to use residual error which aims at solving the exploding gradient problem from deep networks. For ResNeXt [29], it combines the wide residual network and the group convolution in AlexNet [9] with the basis of ResNet [4]. Note that the number of channels in each layer of ResNeXt is doubled and the network is also widened. In such a way, it can improve the classification performance without increasing the complexity of parameters as well as reduce the number of super-parameters.

The DenseNet Architecture. DenseNet [5] is a deeper model which has fewer parameters. In DenseNet, each layer obtains additional inputs from all preceding layers and then transfers its own feature-maps to all subsequent layers, which can enhance the transmission of the feature map and reduce the vanishing gradient problem of deep learning.

The Adam Algorithm. The Adam algorithm [7] is a method for efficient stochastic optimization. It only requires the first-order gradients with a few memory requirements, which can compute an adaptive learning rate for each parameter generated from the estimates of the first and second moments of the gradients. The main function of Adam is shown as follows:

$$\theta_t = \theta_{t-1} - \frac{\alpha \hat{m}_t}{\sqrt{\hat{v}_t} + \varepsilon}, \tag{1}$$

where t denotes the number of times, α and ε are taken constant values, m_t is the exponential moving mean obtaining from the first moment of gradient, and ν_t is a square gradient achieving from the two moment of the gradient [7].

Ra-DenseNet (the Proposed Model). Our Ra-DenseNet builds upon DenseNet, meanwhile, exploits the key characteristics of ResNeXt and Adam. The structure of the proposed Ra-DenseNet is shown in Fig. 1. It consists of 40 layers and 4 blocks. In order to avoid over-fitting caused by excessive layers, we choose 3 identical sub-structures for each block. Each substructure contains three convolution layers with the convolution kernel of $1 * 1$, $3 * 3$, $1 * 1$, respectively. In order to reduce the number of input feature maps, we perform the preceding layer with the $3 * 3$ convolution in each dense block using a $1 * 1$ convolution operation, which can not only reduce the dimensionality and computation, but also fuse the features of each channel. Combining the block method used in ResNeXt, the following layer with the $3 * 3$ convolution of each dense block also contains a $1 * 1$ convolution operation in order to expand the number of channels. In such a way, the width can be doubled as introduced in [29]. At the same time, we use the technology of group convolution such that the correlation between feature mapping has been increased, which is similar to the regularization effect. Besides, we choose Adam optimizer [7] with better adaptation and convergence, unlike the Momentum optimizer [25] used in the original DenseNet network.

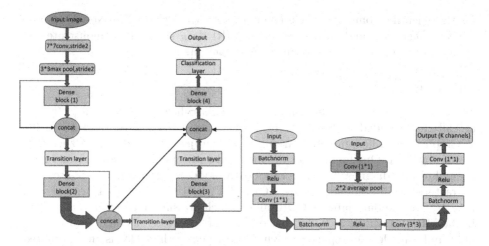

Fig. 1. Main structure of Ra-DenseNet   **Fig. 2.** Dense block and transition layer

The input of the Ra-DenseNe network is a tensor of [224, 224, 3]. First, we feed the input tensor into the convolution layer with a convolution kernel of 7 * 7 and a step size of 2. The tensor of [112, 112, 16] is obtained, and the tensor of size [56, 56, 16] is captured by a max pooling layer with a convolution kernel of 3 * 3 and a step size of 2. Then through the first dense block, the tensor of [56, 56, 52] is gotten, and the tensor of [28, 28, 52] is received through the transition layer. Next, through the same three dense blocks and two transition layers, the tensor of [7, 7, 160] is captured, and through the max pooling layer with a convolution kernel of 7 * 7, the tensor of [1, 1, 160] is obtained, which is then reshaped into a 160-dimensional eigenvector. And through a fully connected layer a 1000-dimensional eigenvector is gotten. Finally, with the softmax classifier, the loss function is formed by the cross-entropy of the label y and the output. The structures of Dense Block and Transition Layer for Ra-DenseNet are illustrated by Fig. 2.

Compared with DenseNet, our Ra-DenseNet has the following features: (1) It's number of layers is much smaller and less than one third of DenseNet. (2) It's channels of each block is wider than DenseNet because of adding convolution operation, and it's structure is updated by the technology of group convolution. (3) The adaptability and convergence of it's optimizer are better than DenseNet. In this work, we aim to create a better neural network structure and propose a suitable model for brain MRI data. Ra-DenseNet may be not suitable for other data sets compared with DenseNet, but its identification results for brain MRI data is promising (see experimental results).

4   Experiments

In this section, we evaluate the classification performance of the proposed model. We use a brain MRI data set collected from OASIS, which can be accessed freely for the scientific community. We compare with the LeNet, AlexNet, VGGNet, ResNet, DenseNet and Ra-DenseNet on MRI data sets, and demonstrate the classification performance in terms of Accuracy metric.

4.1 Data Set

The initial data set consists of a cross-sectional collection of 416 subjects that are all right-handed and include both men and women aged from 18 to 96. We applied the 235 complete marking of 416 subjects to study the Alzheimer's disease by MRI. One hundred of the included subjects older than 60 years have been clinically diagnosed with very mild to moderate Alzheimer's disease. The characteristics of each subject are AGE, M/F, EDUC, MMSE, eTIV, ASF and nWBV, and the data label is CDR, all of which are summarized in Table 1 [24]. In the MRI, the values as 0, 0.5, 1, 2 of CDR are expressed as no dementia, very mild, mild, moderate, separately, while the value of the eTIV is the estimated total intracranial volume (cm^3) and the nWBV is the percent of all voxels in the atlas-masked. Further information on eTIV and nWBV are drawn with ages in Figs. 3 and 4 for the data included in the samples [33].

Table 1. Imaging measures in the brain MRI data sets

Age	Age at time of image acquisition (years)
Sex	Male or female
Edu	Years of education
MMSE	Ranges from 0 (worst) to 30 (best)
ASF	Atlas scaling factor
eTIV	Estimated total intracranial volume (cm^3)
nWBV	Expressed as the percent of all voxels in the atlas-masked image
CDR	0 = no dementia, 0.5 = very mild, 1 = mild, 2 = moderate

4.2 Experimental Setup

All experiments are performed by a PC Server (Intel(R) Core(TM) i5-3337U CPU @ 1.80 GHz, memory 4 GB). Firstly, we implement Python script to process the initial MRI data sets for one MRI by one sample. Secondly, we select the 235 MRI samples with data labels and use eighty percent of all samples as a training set with ten percent as a verification set and ten percent as a test set, where the four different classifications with the values of CDR in initial MRI data sets are shown in Fig. 5. Thirdly, we choose the LeNet, AlexNet, VGGNet, ResNet, DenseNet and Ra-DenseNet architectures on the brain MRI data sets applying the Accuracy value as the evaluation of the performance in service of the TensorFlow software platform. Finally, we compare the evaluation results of the six neural network architectures.

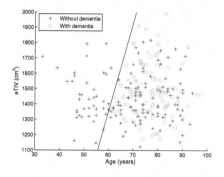

Fig. 3. Plots of automated anatomical measures with eTIV and age

Fig. 4. Plots of automated anatomical measures with nWBV and age

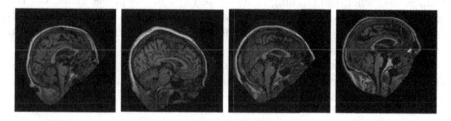

Fig. 5. Four images with different classifications of brain MRI data sets, that is, the CDR values are 0, 0.5, 1, 2 from left to right for MRI

4.3 Training

In this subsection, we give the detailed training processes of the LeNet, AlexNet, VGGNet, ResNet, DenseNet and Ra-DenseNet architectures with the MRI data sets and performance comparisons of the training results of each network.

In the experiments, the brain MRI data sets are divided into four categories, that is, disease-free, mild, moderate and severe. The training set, the verification set and the test set are distributed according to the ratio of 8:1:1. Hence 188 training sets and 23 verification sets with 24 test sets are gotten. We choose the smaller growth rate k = 12 to avoid that the network will become too wide. In addition, there is a transition layer between each dense block to control the output size between blocks so that unstable training caused by excessive channels can be avoided. The parameter here is set to be 0.5, which indicates the times that the output of each block will be reduced. Since there are many connections in the entire network, we use dropout technology [23] to randomly reduce branches to avoid over-fitting. The basic attributes of each network in training process are shown in Table 2, respectively, which also contains the layers, params and running time of each epoch.

With the increase of training times, the changes of the parameter of the neural network will be smaller and smaller. Therefore, we use the learning rate that changes with training times, which is initially 0.0005. When the times of

Table 2. The architecture of each network

Method	Depth	Params	Running time of each epoch
LeNet	5	0.06M	0.05 ms
AlexNet	8	60M	14.56 ms
VGGNet	16	138M	128.62 ms
ResNet	34	0.46M	51.59 ms
DenseNet	121	27.2M	80.65 ms
Ra-DenseNet	40	1M	60.16 ms

training exceed half of the preset number, the learning rate will be reduced by 10 times, and when the training times exceed 4/5 of the preset value, the rate will be reduced by another 10 times. Each batch is set to consist of 8 and 16 samples as batch size separately, and the training times, i.e. epochs is reset to be 500, 1000, 1500 and 2000.

More intuitively, when the batch size is 8 and epochs reach to 2000, the loss values comparison results and the accuracy performance are shown in Figs. 6 and 7, respectively.

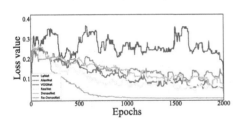
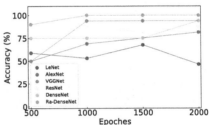

Fig. 6. Loss value comparison in training process with batch size of 8

Fig. 7. Accuracy performance comparison in training process with batch size of 8

And when batch size is 16 and epochs reach to 2000, the loss values comparison results are shown in Fig. 8 and the accuracy performance comparison is shown in Fig. 9, respectively.

A comparative analysis of Figs. 6 and 8 shows that the loss values of Ra-DenseNet are much less than other networks in the whole training process with different batch sizes. Figures 7 and 9 present that the accuracies of Ra-DenseNet are higher than other networks during training when the batch size is 8 or 16.

4.4   Results and Discussions

In this subsection, we give the testing results of the comparisons after the experiments with fifty running times for the LeNet, AlexNet, VGGNet, ResNet,

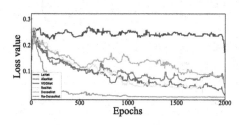
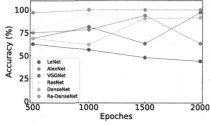

Fig. 8. Loss value comparison in training process with batch size of 16

Fig. 9. Accuracy performance comparison in training process with batch size of 16

DenseNet and Ra-DenseNet architecture on the brain MRI data sets. In case of the batch size of 8, the average values of accuracy of each network for the brain MRI data set are shown in Table 3, While the average accuracies of each network are given in Table 4 when the batch size is 16.

Table 3. The performances of each network with batch size of 8

Method	500 epochs	1000 epochs	1500 epochs	2000 epochs
LeNet	52.5%	56.7%	50.9%	63.4%
AlexNet	69.6%	73.3%	78.3%	81.5%
VGGNet	80.0%	83.6%	86.2%	90.4%
ResNet	82.0%	87.9%	90.3%	91.3%
DenseNet	79.1%	81.6%	83.2%	81.7%
Ra-DenseNet	**89.6%**	**93.8%**	**96.2%**	**96.3%**

Table 4. The performances of each network with batch size of 16

Method	500 epochs	1000 epochs	1500 epochs	2000 epochs
LeNet	52.1%	56.7%	58.3%	62.5%
AlexNet	62.1%	70.7%	76.3%	82.3%
VGGNet	78.7%	82.9%	86.7%	91.2%
ResNet	81.6%	89.3%	92.5%	92.5%
DenseNet	71.7%	79.1%	85.3%	85.4%
Ra-DenseNet	**87.9%**	**90.4%**	**95.8%**	**97.1%**

In Tables 3 and 4, the highest accuracies have been highlighted. It is noted that the Ra-DenseNet has the highest accuracy value with each different epochs for the recognition of the Alzheimer's Disease by the brain MRI data sets. More intuitively, in case of the batch size of 8, the comparison of accuracies for each network is shown in Fig. 10, while the comparison of accuracies is presented in Fig. 11 with the batch size of 16.

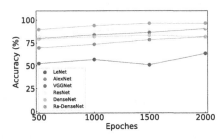

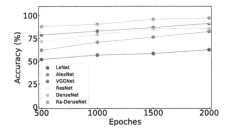

Fig. 10. Accuracy comparison results in testing process when batch size is 8

Fig. 11. Accuracy comparison results in testing process when batch size is 16

A comparative analysis of Figs. 10 and 11 shows that the accuracy values of Ra-DenseNet are much higher than other networks in the testing process with different epochs, i.e. 500, 1000, 1500 and 2000. Figures 10 and 11 also present that the accuracies of Ra-DenseNet appear an upward trend and the highest leveling over the course of the testing when the batch sizes are taken 8 and 16. Overall, the Ra-DenseNet model we proposed performs the best in the recognition of Alzheimer's disease on the MRI data sets. Ra-DenseNet may be not suitable for other data sets compared with others' deep learning network, but its recognize results for brain MRI data is excellent.

5 Conclusion

It is instrumental to use MRI data (or other types of medical data) to assist physicians to make a quick and exact diagnosis and treatment for diseases in modern medical industry. With the increase of patients and the waiting time of disease treating in hospital, it also calls for a large amount of demand for computer-aided diagnosis. In this paper, we proposed a new deep learning model (i.e., Ra-DenseNet), which utilizes the key characters of DenseNet, ResNeXt and Adam algorithm. After training the proposed Ra-DenseNet model on brain MRI data, Ra-DenseNet can effectively discover Alzheimer's disease in test. In this work, we evaluated our model on a brain MRI data set. Our future work includes collecting more MRI data sets, and then performing our Ra-DenseNet on these data sets. Meanwhile, we will study a new network architecture based on deep learning technology for others' diseases.

Acknowledgment. This work was supported by the National Natural Science Foundation of China (No. 61572407) and the Seeding Project of Scientific and Technological Innovation in Sichuan Province of China (No. 2018102).

References

1. Fang, C., Li, C., Cabrerizo, M., et al.: A Gaussian discriminant analysis-based generative learning algorithm for the early diagnosis of mild cognitive impairment in Alzheimer's disease. In: Proceedings of the IEEE International Conference on Bioinformatics and Biomedicine, pp. 538–542 (2017)
2. Gray, K.R., Aljabar, P., Heckemann, R.A., et al.: Random forest-based similarity measures for multi-modal classification of Alzheimer's disease. NeuroImage **65**, 167–175 (2013)
3. Harman, D.: Alzheimer's disease pathogenesis. Ann. N. Y. Acad. Sci. **1067**, 454–560 (2007)
4. He, K., Zhang, X., Ren, S., et al.: Deep residual learning for image recognition. In: Proceedings of the IEEE Conference on Computer Vision and Pattern Recognition, pp. 770–778 (2016)
5. Huang, G., Liu, Z., Weinberger, K.Q., et al.: Densely connected convolutional networks. In: Proceedings of the IEEE Conference on Computer Vision and Pattern Recognition, vol. 1, pp. 4700–4708 (2017)
6. Hon, M., Khan, N.M: Towards Alzheimer's disease classification through transfer learning. In: Proceedings of the IEEE International Conference on Bioinformatics and Biomedicine, pp. 1166–1169 (2017)
7. Kingma, D.P., Ba, J.: Adam: a method for stochastic optimization. Comput. Sci. (2014)
8. Kong, W., Mou, X., Hu, X.: Exploring matrix factorization techniques for significant genes identification of Alzheimers disease microarray gene expression data. In: Proceedings of the IEEE International Conference on Bioinformatics and Biomedicine, vol. 12, no. 5, p. S7 (2011)
9. Krizhevsky, A., Sutskever, I., Hinton, G.E.: ImageNet classification with deep convolutional neural networks. In: Proceedings of the 25th International Conference on Neural Information Processing Systems, vol. 1, pp. 1097–1105 (2012)
10. LeCun, Y., Bottou, L., Bengio, Y., et al.: Gradient-based learning applied to document recognition. Proc. IEEE **86**(11), 2278–2324 (1998)
11. Liu, M., Zhang, D., Shen, D.: Ensemble sparse classification of Alzheimer's disease. Neuroimage **60**(2), 1106–1116 (2012)
12. Liu, Y.: Magnetic resonance imaging. In: Current Laboratory Methods in Neuroscience Research, pp. 249–270 (2013)
13. Liu, F., Wee, C.Y., Chen, H., et al.: Inter-modality relationship constrained multi-modality multi-task feature selection for Alzheimer's disease and mild cognitive impairment identification. NeuroImage **84**, 466–475 (2014)
14. Liu, Q., Chen, C., Gao, A., et al.: VariFunNet, an integrated multiscale modeling framework to study the effects of rare non-coding variants in genome-wide association studies: applied to Alzheimer's disease. In: Proceedings of the IEEE International Conference on Bioinformatics and Biomedicine, pp. 2177–2182 (2017)
15. Luo, Y.M., Weng, H., Zhang, L., et al.: Salt restriction: recognition and treatment of chronic kidney disease related edema in ancient literature mining. In: Proceedings of the IEEE International Conference on Bioinformatics and Biomedicine, pp. 1369–1375 (2017)
16. Marcus, D., Wang, T., Parker, J., et al.: Open Access Series of Imaging Studies (OASIS): cross-sectional MRI data in young, middle aged, nondemented, and demented older adult. J. Cogn. Neurosci. **19**(9), 1498–1507 (2007)

17. Milletari, F., Ahmadi, S.-A., Kroll, C., et al.: Hough-CNN: deep learning for segmentation of deep brain regions in MRI and ultrasound. Comput. Vis. Image Underst. **164**, 92–102 (2017)

18. Moradi, E., Pepe, A., Gaser, C., et al.: Machine learning framework for early MRI-based Alzheimer's conversion prediction in MCI subjects. Neuroimage **104**, 398–412 (2015)

19. Nichols, T.E., Das, S., Eickhoff, S.B., et al.: Best practices in data analysis and sharing in neuroimaging using MRI. Nat. Neurosci. **20**(3), 299–303 (2017)

20. Panda, A.K., Kumar, M., Chaudhary, M.K., et al.: Brain tumour extraction from MRI images using k-means clustering. Int. J. Innov. Res. Electr. Electron. Instrum. Control Eng. **4**(4), 356–359 (2016)

21. Peng, Y., Tang, C., Chen, G., et al.: Multi-label learning by exploiting label correlations for TCM diagnosing Parkinson's disease. In: Proceedings of the IEEE International Conference on Bioinformatics and Biomedicine, pp. 590–594 (2017)

22. Russakovsky, O., Deng, J., Su, H., et al.: ImageNet large scale visual recognition challenge. Int. J. Comput. Vis. **115**(3), 211–252 (2015)

23. Srivastava, N., Hinton, G., Krizhevsky, A., et al.: Dropout: a simple way to prevent neural networks from overfitting. J. Mach. Learn. Res. **15**(1), 1929–1958 (2014)

24. Sorg, C., Riedl, V., Muhlau, M., et al.: Selective changes of resting-state networks in individuals at risk for Alzheimer's disease. Proc. Natl. Acad. Sci. **104**(47), 18760–18765 (2007)

25. Sutskever, I., Martens, J., Dahl, G., et al.: On the importance of initialization and momentum in deep learning. In: Proceedings of the International Conference on Machine Learning, pp. 1139–1147 (2013)

26. Tahmasian, M., Shao, J., Meng, C., et al.: Based on the network degeneration hypothesis: separating individual patients with different neurodegenerative syndromes in a preliminary hybrid PET/MR study. J. Nucl. Med. **57**, 410–415 (2016)

27. Tang, X., Hu, X., Yang, X., et al.: A algorithm for identifying disease genes by incorporating the subcellular localization information into the protein-protein interaction networks. In: Proceedings of the IEEE International Conference on Bioinformatics and Biomedicine, pp. 308–311 (2016)

28. Tong, T., Gray, K., Gao, Q., et al.: Multi-modal classification of Alzheimer's disease using nonlinear graph fusion. Pattern Recogn. **63**, 171–181 (2017)

29. Xie, S., Girshick, R., Dollár, P., et al.: Aggregated residual transformations for deep neural networks. In: Proceedings of the IEEE Conference on Computer Vision and Pattern Recognition, pp. 5987–5995 (2017)

30. Young, J., Modat, M., Cardoso, M.J., et al.: Accurate multimodal probabilistic prediction of conversion to Alzheimer's disease in patients with mild cognitive impairment. NeuroImage: Clin. **2**, 735–745 (2013)

31. Zhang, D., Wang, Y., Zhou, L., et al.: Multimodal classification of Alzheimer's disease and mild cognitive impairment. Neuroimage **55**(3), 856–867 (2011)

32. Zhang, D., Shen, D.: Multi-modal multi-task learning for joint prediction of multiple regression and classification variables in Alzheimer's disease. Neuroimage **59**(2), 60–67 (2012)

33. Zhang, X., Yang, Y., Wang, H., et al.: Analysis of senile dementia from the brain magnetic resonance imaging data with clustering. In: Proceedings of the 13th International FLINS Conference (FLINS 2018) and Intelligent Systems and Knowledge Engineering (ISKE 2018), pp. 1454–1461 (2018)

34. Zhu, X., Suk, H.I., Wang, L., et al.: A novel relational regularization feature selection method for joint regression and classification in AD diagnosis. Med. Image Anal. **38**, 205–214 (2017)

Knowledge Graph and Interpretable Data Mining

Granger Causality for Heterogeneous Processes

Sahar Behzadi[1(✉)], Kateřina Hlaváčková-Schindler[1], and Claudia Plant[1,2]

[1] Faculty of Computer Science, Data Mining, University of Vienna, Vienna, Austria
{sahar.behzadi,katerina.schindlerova,claudia.plant}@univie.ac.at
[2] ds:UniVie, University of Vienna, Vienna, Austria

Abstract. Discovery of temporal structures and finding causal interactions among time series have recently attracted attention of the data mining community. Among various causal notions graphical Granger causality is well-known due to its intuitive interpretation and computational simplicity. Most of the current graphical approaches are designed for homogeneous datasets i.e. the interacting processes are assumed to have the same data distribution. Since many applications generate heterogeneous time series, the question arises how to leverage graphical Granger models to detect temporal causal dependencies among them. Profiting from the generalized linear models, we propose an efficient **H**eterogeneous **G**raphical **G**ranger **M**odel (HGGM) for detecting causal relation among time series having a distribution from the exponential family which includes a wider common distributions e.g. Poisson, gamma. To guarantee the consistency of our algorithm we employ adaptive Lasso as a variable selection method. Extensive experiments on synthetic and real data confirm the effectiveness and efficiency of HGGM.

1 Introduction

Recently there is a significant interest in causal inference in various data mining tasks. Discovery of causal relations among different processes leads to characterize the evolution in time of regular instances. The regular pattern can be used to detect the deviated observations or outliers in anomaly detection [15]. A number of methods has been developed to infer causal relations from time series data by Granger causality [8] which is a popular method due to its computational simplicity. The presumption of this approach is that a cause helps to predict its effects in the future. Most of the existing methods in this area assume additive causal interactions among time series following a specific data type or a certain distribution. The well-know causality notion, Additive Noise Models (ANMs), have been proposed for either continuous [17] or discrete [14] time series. Moreover, most of the probabilistic approaches are designed for homogeneous datasets

Electronic supplementary material The online version of this chapter (https://doi.org/10.1007/978-3-030-16142-2_36) contains supplementary material, which is available to authorized users.

Q. Yang et al. (Eds.): PAKDD 2019, LNAI 11441, pp. 463–475, 2019.
https://doi.org/10.1007/978-3-030-16142-2_36

[4,5]. However, in reality the interacting processes do not have to be homogeneous (having the same distribution). Such situations can occur, for example, in climatology when various measurements are provided for different meteorological stations. Figure 1 shows 10 weather stations and three major weather systems in Austria. The monthly amount of precipitation as well as the number of sunny days have been measured for every station, each of which with a non-Gaussian distribution. One can be interested in investigating how the number of sunny days in a station, influenced by one of the weather systems, can impact the amount of precipitation in the other locations.

Fig. 1. Meteorological stations and three major weather systems influencing Austria.

Applying existing algorithms on such data sets can result an inaccurate Granger causal model since they have been designed for specific homogeneous data types. Moreover, the small set of algorithms, which are supposed to cope with the heterogeneity, mostly employ an exhaustive pairwise testing. This leads to inefficiency in a causal network discovery specially when the number of interacting processes is increasing. In between, graphical Granger models are popular due to their efficiency and effectiveness. They employ a penalized Vector Autoregression (VAR) to the Granger concept [1,3,7,18]. However, to the best of our knowledge, so far they have been designed only for homogeneous data sets. Thus, in this paper we propose a penalized VAR-based algorithm to detect the **H**eterogeneous **G**raphical **G**ranger **M**odel (HGGM) by employing generalized linear models (GLMs). Similar to the other graphical models, we assume that the interactions among the involved processes are additive. Moreover, to ensure the convergence of HGGM to the true causal graph (i.e. consistency) we employ the well-know penalization approach, adaptive Lasso, with oracle properties [20]. The paper brings the following contributions:

- **Heterogeneity:** Applying the GLM methodology, we propose a heterogeneous graphical Granger model to discover the causal interactions among a wide variety of heterogeneous time series from the exponential family;

- **Consistency:** Assessing the causal relations via adaptive Lasso ensures consistency of our method;
- **Scalability:** Unlike other existing algorithms, HGGM avoids an exhaustive pairwise causality testing by penalized estimation of VAR models. Due to the computational simplicity of HGGM, it is convenient to be used in practice. Moreover, its reasonable runtime complexity makes our algorithm scalable for the large data sets consisting of long time series;
- **Effectiveness:** Following the result of our extensive experiments on synthetic and real datasets, HGGM is an effective algorithm even by detecting sparse causal graphs.

In the following we specify the problem and the theoretical background and propose our HGGM model. Section 2 presents the related work. In Sect. 3, we introduce the problem and our proposed framework to deal with heterogeneous data. In Sect. 4 we introduce our integrative algorithm HGGM and the theoretical considerations of it. Extensive experiments on synthetic and real data are demonstrated in Sect. 5. Our conclusion is in Sect. 6.

2   Related Work

Among various approaches to infer causality, Granger causality [8] is well-known due to its simplicity and computational efficiency. It states that a cause efficiently improves the predictability of its effect. There are various approaches depending on how to assess the predictability. Probabilistic approaches interpret it as the improvement in the likelihood (i.e. probability). However, several methods in this group are distinguished based on the way how they employ probability. Information-theoretic methods detect the causal direction by introducing some indicators. Among them, compression-based algorithms apply the Kolmogorov complexity and define a causal indicator by mean of the Minimum Description Length (MDL) [4–6]. Essentially, these algorithms are designed to infer the pairwise causal relations. Therefore, employing them for discovery of causal networks leads to inefficiency, especially when the number of processes increases. Moreover, to the best of our knowledge, almost all the algorithms in this category deal with homogeneous data sets except *Crack* [10], the most recent compression-based algorithm to deal with multivariate and heterogeneous processes. Beside the pairwise testing and its drawbacks, this algorithms lacks the accurate causal relations since there is no lag parameter considered in this approach. Transfer entropy, shortly TEN, is another approach among information-theoretic methods which is based on Shannon's Entropy [16]. In this approach it is more likely that the causal direction with the lower entropy corresponds to the true causal relation. Given a lag variable, TEN can detect both linear and non-linear causal relations. However, due to pairwise testing and its dependency on the lag variable, the computational complexity of this algorithm is exponential in the lag parameter. Moreover, similar to compression-based methods, TEN is not designed to deal with bidirectional causalities. As another method in this category, the authors in [9] employ the log-likelihood ratio to detect any causal relations among processes. They propose a statistical framework (SFGC) for mixed type data and assessing the causal relations between multiple time series is accomplished by the false discovery rate (FDR). The statistical power of the FDR based methods rapidly decreases with increasing number of hypotheses and these methods are computationally intensive. As the consequence, the statistical efficiency of SFGC decreases for the increasing number of investigated time series.

Another approach to assess the predictability is the graphical Granger method where a penalized VAR model is supposed to be estimated [1,18]. Graphical Granger method is popular for its simplicity and efficiency since employing a penalized VAR model we avoid the pairwise testing. However most of the algorithms in this category are designed for Gaussian processes. Utilizing the advantages of this approach we introduced a graphical Granger algorithm for heterogeneous processes.

3   Theory

3.1   Granger Causality

Granger causality is a well-known notion of causality introduced by Granger in the area of econometrics [8]. Although the Granger causality is not meant to be equivalent to the true causality but it provides useful information capturing the temporal dependencies among time series. In a bivariate case let $x^{1:n} = \{x^t | t = 1, \ldots, n\}$ and $y^{1:n} = \{y^t | t = 1, \ldots, n\}$ denote two time series up to time n. Moreover, let the following two models represent two autoregressive models corresponding to time series y with and without taking past observations of x into consideration.

$$y^T = \alpha_1 y^1 + \cdots + \alpha_{T-1} y^{T-1} + \gamma_1 x^1 + \cdots + \gamma_{T-1} x^{T-1} + \varepsilon^T \tag{1}$$

$$y^T = \alpha_1 y^1 + \cdots + \alpha_{T-1} y^{T-1} + \varepsilon^T \tag{2}$$

Following the principle of Granger causality, x Granger-causes y if the Model 1 significantly improves the predictability of y comparing to the Model 2. The concept of Granger causality can be extended to more than two time series. Let $x_1^{1:n}, \ldots, x_p^{1:n}$ be p time series up to time n and X^T be the concatenated vector of all time series at time T, i.e. $X^T = (x_1^T, \ldots, x_p^T)$. The vector autoregressive (VAR) model is given by:

$$X^T = A_1 X^1 + \cdots + A_{T-1} X^{T-1} + \varepsilon^T \tag{3}$$

where A_t is a matrix of the regression coefficients at time $t = 1, \ldots, T-1$ and ε^t is a white noise. Thus, x_j Granger-causes x_i if at least one of the $(i,j)th$ elements in the coefficient matrices $A_1, \ldots, A_{T-1}$ is non-zero.

3.2 Causal Inference by Penalization

In order to detect the causal relations between several time series, one needs to estimate the coefficients of the VAR model introduced in the last section. Since this problem can be ill-posed, penalizing the VAR of order d (a time window) by means of a penalty function provides an efficient and sparse solution when the convergence to the true causal graph is ensured (e.g. [1,18]). The penalization approach is referred to as variable selection as well. Thus, given the window size d for any time series $x_i, i = 1, \ldots, p$, we consider the VAR model including all p time series. We slide the window over time series and get the corresponding VAR model. The fact is that the best regressors with the least squared error for any specific time series will have non-zero coefficients in the VAR model only for the dependent time series. More precisely, Let $X_{T,d}^{Lag} = \{x_i{}^{T-t} | i = 1, \ldots, p; t = 1, \ldots, d\}$ denote the concatenated vector of all the lagged variables up to time T for a given time window of length d. For simplicity we consider the same lag d for each time series. Applying the penalized optimization, the variable selection problem for the time series x_i is given by:

$$\hat{\beta}_i = \arg\min_{\beta_i} \sum_{T=d+1}^{n} \left(x_i^T - X_{T,d}^{Lag}\beta_i\right)^2 + \lambda R(\beta_i) \tag{4}$$

where $R(.)$ is the penalty function and λ is the regularization parameter. $\hat{\beta}_i = (\beta_1, \ldots, \beta_p)$ is a concatenated vector of the regression coefficients $\beta_1, \ldots, \beta_p$ corresponding to any time series $x_1, \ldots, x_p$. Back to the definition of Granger causality, x_j Granger-causes x_i if and only if at least one of the coefficients in β_j is non-zero.

3.3 Adaptive Lasso

One of the well-known variable selection methods is Lasso [19] where the penalty function considered in Eq. 4 is the L_1 norm of the coefficients, i.e. $R(\beta_i) = ||\beta_i||_1$. Despite the efficiency of Lasso, the consistency[1] of this approach is not ensured. Therefore, we employ adaptive Lasso [20], a modification of Lasso, as the variable selection method in our model due to its consistency as well as its oracle properties. In this approach we assign adaptive weights for penalizing the L_1 norm of different coefficients. The penalty function is given by:

$$R(\beta_i) := \sum_{j=1}^{p} w_j |\beta_j| \quad where \quad w_j = \frac{1}{|\hat{\beta}_j^{(mle)}|^\omega} \tag{5}$$

In fact, w_j is the weight vector for some $\omega > 0$ and $\hat{\beta}_j^{(mle)}$ is the maximum likelihood estimate of the parameters. The consistency of adaptive Lasso is guaranteed under some mild regularity conditions in the following theorem [20]:

---

[1] I.e. the resulting sequence of estimates does not have to converge in probability to the optimal solution for variable selection under certain conditions (Sect. 2 in [20]).

Theorem 1. *Let $\mathcal{A} = \{i : \hat{\beta}_i \neq 0\}$ be the set of all non-zero coefficient estimates. Suppose that $\lambda/\sqrt{n} \to 0$ and $\lambda n^{\frac{(\omega-1)}{2}} \to \infty$ then under some mild regularity conditions adaptive Lasso must be consistent for the variable selection.*

3.4 Heterogeneous Granger Causality

Most of the approaches to detect the Granger causality among time series have certain Gaussian assumptions for the interacting processes. However in many applications this assumption leads to an inaccurate causal model. Moreover, mostly the variable selection algorithms employed to penalize the VAR model are consistent under additional specific conditions on the Gaussian time series, see e.g. [1]. Profiting from the GLM framework, we propose a general integrative model to detect causal relations among a large number of heterogeneous time series. GLM, introduced by Nelder and Baker in [13], is a natural extension of linear regression to the cases when the regressed variables (time series) can have any distribution from the exponential family. In another word, the relation among the response variable and the covariates in a regression is not any more linear but defined by a link function g, a monotone twice differentiable function depending on concrete distribution functions from the exponential family.

In our model we assume the mean value of each time series at time T depends on its own history and the past values of the concurrent time series so that:

$$E(x_i^T) = g_i^{-1}(X_{T,d}^{Lag}.\beta_i). \tag{6}$$

Finally, our general objective function is defined as:

$$\hat{\beta}_i = \arg\min_{\beta_i} \sum_{T=d+1}^{n} \left[-x_i^T(X_{T,d}^{Lag}.\beta_i) + g_i^{-1}(X_{T,d}^{Lag}.\beta_i) \right] + \lambda.\sum_{j=1}^{p} w_j|\beta_j|. \tag{7}$$

The concrete form of our proposed objective function (7) concerning x_i to have binomial and Poisson distribution, respectively, is given by:

$$\hat{\beta}_i = \arg\min_{\beta_i} \sum_{T=d+1}^{n} \left[-x_i^T(X_{T,d}^{Lag}.\beta_i) + log(1 + e^{(X_{T,d}^{Lag}.\beta_i)}) \right] + \lambda.\sum_{j=1}^{p} w_j|\beta_j|, \tag{8}$$

$$\hat{\beta}_i = \arg\min_{\beta_i} \sum_{T=d+1}^{n} \left[-x_i^T(X_{T,d}^{Lag}.\beta_i) + \exp(X_{T,d}^{Lag}.\beta_i) \right] + \lambda.\sum_{j=1}^{p} w_j|\beta_j|. \tag{9}$$

4   HGGM Algorithm

Our method HGGM is summarized in Algorithm 1. At first it constructs the overall lagged matrix X^{Lag}, by sliding the window of size d over each time series. Then, HGGM solves the optimization problem (Eq. 7) for each time series by calling $GLM - penalize()$, [11]. This procedure applies Fisher scoring algorithm to estimate the coefficients. We set the maximum λ as an input of $GLM-penalize()$ and the procedure employs the cross-validation to find the best regularization parameter.

Essentially one needs to know the distribution of every time series in order to specify an appropriate link function g. We utilize a statistical fitting procedure to find the most accurate distribution for every time series. We assign to any time series the distribution from the exponential family with the least Akaike Information Criterion (AIC). Finally, based on the definition of Granger causality we get pairwise Granger-causal relations among p time series out of which we construct the adjacency matrix corresponding to the final causal graph.

Algorithm 1. Causal Detection by HGGM

$HGGM\ (x_i, g_i, i = 1, \ldots, p; d; \lambda_{max})$
$Adj :=$ adjacency matrix of the output graph
$X^{lag} :=$ lagged matrix of all temporal variables
// find Granger causalities for each feature
for all x_i **do**
    // solve the penalized optimization problem considering lagged variables
    $\beta_i = GLM - penalize(X^{Lag}, x_i, g_i, \lambda_{max}, d);$ // $\beta_i :=$ coefficients w.r.t x_i
    for all $\beta_i{}^j$ sub-vectors of β_i **do**
        $Adj(j, i) = 0$ //discover Granger-causalities
        if $(\exists t, 1 < t < d$ such that $\beta_i{}^j(t) > 0)$ **then**
            $Adj(j, i) = 1$
        end if
    end for
end for
return (Adj)

Consistency: The consistency of adaptive Lasso for the variable selection has been proven under some mild regularity conditions (Sect. 3). Thus, applying the adaptive Lasso for GLMs enables us to make the following statement about the consistency of HGGM.

Corollary 1. *Assume G be a true Granger causal graph corresponding to p time series, each of length n. Let the regularization parameter λ fulfils the conditions of Theorem 1. Then taking p time series as input, HGGM outputs a causal graph which converges to the true graph G with probability approaching 1 as $n \to \infty$.*

Proof. When $n \to \infty$ the conditions of Theorem 1 are fulfilled. Therefore it follows that the procedure GLM–$penalize(.)$ in Algorithm 1 converges to the true Granger causal graph. Thus, HGGM is consistent as well.

Computational Complexity: Based on the proposed objective function (7), we investigate causal relationships for any time series $x_i, i = 1, \ldots, p$ by fitting the most accurate VAR model. Therefore at any time we have p regression models each of which consists of d lagged variables corresponding to $x_1, \ldots, x_p$. Applying Fisher scoring to estimate the parameters of VAR models, the number of computations required to solve a VAR of order d is $\mathcal{O}(d^2)$. Thus, the computational complexity of HGGM is in order of $\mathcal{O}(np^2d^2)$.

5   Experimental Results

In this section the performance of HGGM in comparison to other algorithms will be assessed in terms of *F-measure* which takes both precision and recall into account. Although there are many approaches to detect the Granger causality, only few of them are designed for heterogeneous time series. Therefore, we compare our algorithm to three methods which are applicable to mixed time series, i.e. transfer entropy, shortly TEN [16], Crack [10] and SFGC [9]. To evaluate HGGM we investigate the effectiveness and efficiency of HGGM by extensive experiments on synthetic and real-world data sets. HGGM is implemented in MATLAB and for the other comparison methods we use their publicly available implementations and recommended parameter settings. The source code and data sets are publicly available at: https://bit.ly/2FkUB3Q.

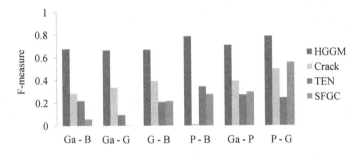

Fig. 2. Performance in various heterogeneous data sets. Ga: Gamma, G: Gaussian, B: Bernoulli, P: Poisson.

5.1   Synthetic Heterogeneous Data Sets

Firstly, we investigate the effectiveness of HGGM comparing to other algorithms in terms of *F-measure*. That is, we conduct various experiments each of which concerning a unique aspect. Then, we target the scalability of the algorithms varying the number of time series and the length of them. In any synthetic experiment, we report the average performance of 50 iterations performed on different data sets with the given characteristics. The length of generated time series n is always 1,000 except for the experiment on increasing the length. For any algorithm which requires to specify the lag variable we run the algorithm for various lags and take the average F-measure as the final result.

Effectiveness: HGGM is designed to deal with Gaussian as well as non- Gaussian time series having a distribution from the exponential family. In this experiment we generated time series with various combinations of Gaussian and non-Gaussian distributions in order to assess HGGM in various cases. Figure 2 shows that HGGM outperforms other algorithms in various combinations of Gaussian – non-Gaussian distributions and discrete - continuous time series. It confirms that our GLM-based objective function effectively copes with heterogeneity of time series comparing to the other methods. For the rest of the experiments we focus on Poisson - Gaussian combination as a representative for heterogeneous data sets.

Dependency: Figure 3a illustrates how various algorithms perform when the dependency among time series, i.e. the coefficients in VAR model, is increasing ranging from 0.1 to 1. As one can expect, HGGM and SFGC have an ascending trend. However, the effectiveness of Crack and TEN is surprisingly decreasing. Although the performance of HGGM is smaller than SFGC and TEN in a very early stage, it outperforms other algorithm for the dependencies higher than 0.3 with a high margin.

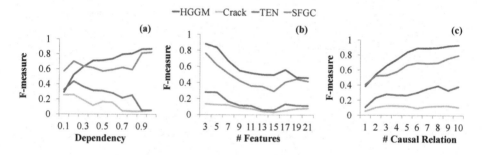

Fig. 3. Synthetic experiments.

Increasing the Number of Features: We increased the number of time series (features) iteratively in order to compare the performance of the algorithms when many time series are involved. Figure 3b shows that the F-measure of any algorithm is descending while HGGM is still more efficient than others in any case. There is a big gap among the performance of two algorithms, Crack and TEN, comparing to HGGM in this figure. One of the reasons for this is that they are not able to deal with the bidirectional causality and by increasing the number of time series it effects the performance more and more.

Causal Relations: How will the various algorithms behave when the true causal graph is sparse? In this experiment we vary the number of causal relations among 5 mixed time series from Poisson - Gaussian combination. As expected, the effectiveness of any algorithm is increasing when the density of the true causal graph is increasing too. However Fig. 3c shows the superiority of our algorithm comparing to others even for sparse graphs.

Scalability: The scalability is investigated in two experiments. First, we increase the number of time series iteratively where the length is set to 1,000 i.e. $n = 1,000$. Then we vary n while every time four time series are generated. By the first experiment the efficiency of HGGM is shown (Fig. 4a) when the number of features is bigger than 6 comparing to Crack and TEN and bigger than 9 comparing to SFGC. However, considering the next experiment (Fig. 4b) the efficiency of our algorithm is confirmed. HGGM is the fastest algorithm almost always for the time series longer than 2,000.

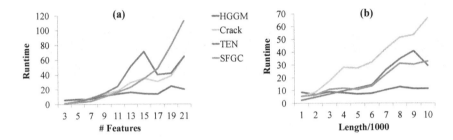

Fig. 4. Experiments on runtime in seconds

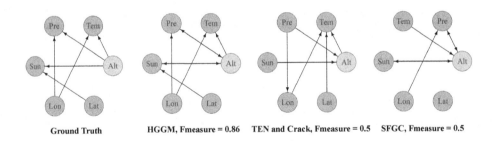

Fig. 5. Comparison on German weather data set.

5.2   Real-World Applications

We conducted the experiments on publicly available real data sets considering two cases, whether a ground truth is given or not. In order to be fair in the real experiments we set $d = 15$ for all the algorithms which require a lag variable.

Weather in Germany: The first data set DWD^2 is a climatological data consisting of 6 measurements, temperature, sunshine hours, altitude, precipitation, longitude and latitude for 394 weather stations all over Germany. The altitude measurement is already provided in a discrete time series while all other measurements are continuous. Applying the statistical fitting procedure (Sect. 4), we assign Gaussian distribution for all continuous time series and the Poisson distribution for the altitude. The ground truth is available in [12] which is provided by pairwise causal relations. In order to be fair by evaluating the results of the algorithms, we do not consider the causal interactions where no information is provided. Figure 5 shows the performance of HGGM comparing to other algorithms in terms of *F-measure*. HGGM ably finds all the existing causal relations. However, it detects causal relations where sunshine and temperature cause altitude.

Marks: The next two data sets together with the corresponding ground truth are publicly available[3]. *Marks* data set concerns the examination marks of 88 students on five different topics. The given true causal graph reveals any impacts the grades of a topic could have on the other topics. We assign Poisson distribution to any topic. In this experiment HGGM (*F-measure* = 0.74) was able to outperform TEN (0.55), Crack (0.6) and SFGC (0.71).

Gaussian: The Gaussian data set is a simulated data showing the causal interactions among 7 Gaussian time series. The time series are of the length 5,000. HGGM (*F-measure* = 0.4) performs more accurately comparing to other algorithms, TEN (0), Crack (0.14) and SFGC (0.14), although non of the algorithms was able to capture all the causal relations in the ground truth.

Austrian Climatological Data Set: As a real world application we investigate causal spatio-temporal interactions among climatological phenomena for 10 sites uniformly distributed in Austria (Fig. 1). For any site we used the monthly measurements of precipitation and of the number of sunny days for 26 months. Employing the statistical fitting, we consider a Gamma distribution for the precipitation and a Poisson distribution for the number of sunny days. Because of the space limit we randomly focus on one of the stations, *Feuerkogel*, and the complete experiment is provided in the supplementary material. Moreover, the real meteorological data set is publicly available[4]. Essentially, Austrian weather is influenced by three climatic systems while any system has its own characteristics. Concerning the interpretation of results for the selected station, we concentrate on the Atlantic maritime climate from the north-west which is characterized by low-pressure fronts, mild air from the Gulf Stream, and precipitation [2]. The northern slopes of the Alps, the Northern Alpine Foreland, and the Danube valley are influenced by the Atlantic weather system.

Figure 6 shows the causal graph discovered by HGGM, TEN and Crack. SFGC was not able to detect any causal relation therefore we exclude its result.

---

[2] http://www.dwd.de/DE/Home/home_node.html.
[3] http://www.bnlearn.com/documentation.
[4] https://www.zamg.ac.at.

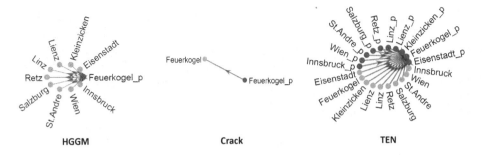

Fig. 6. Experiment on the Austrian climatological data. Blue circles: amount of precipitation and orange circles: number of sunny days. (Color figure online)

Considering the impact of the Atlantic weather system, one expects the influence on the neighbour sites of *Feuerkogel* and the sites in eastern Austria. The sites in southern slope cannot be influenced by this system since the Alps are located in between. Comparing HGGM to other algorithms, HGGM is successful to detect more influenced sites by finding the correct causal direction among *Linz, Salzburg, Retz, Wien* and *Eisenstadt*. However it detects an interaction between *Feuerkogel* and *Lienz* which is not likely due to the large mountain area between the sites. Regarding Crack, although the only causal relation discovered by this algorithm sounds reasonable, there are other stations, e.g. *Linz* and *Salzburg*, where it is plausible to consider a causal interaction among them. On the other hand, TEN discovers a dense causal graph among all 20 time series and *Feuerkogel* which is hard to interpret. Moreover considering the Atlantic weather system, there is no interpretation for the causal direction from *Retz* to *Feuerkogel* detected by TEN since its direction is exactly in the opposite.

6   Conclusions and Future Work

In this paper we proposed HGGM, a graphical Granger model for discovery of causal relations among a number of heterogeneous processes. Profiting of a GLM framework our approach is generalized for time series having distributions from exponential family. Moreover to ensure the consistency of HGGM we employ adaptive Lasso with a proven consistency. We investigated the performance of HGGM in terms of effectiveness and efficiency comparing to state-of-the-art methods. Extensive experiments on synthetic and real data sets demonstrates the advantages of HGGM. As already mentioned, one of the interesting applications of our algorithm can be utilizing HGGM to detect anomalies among heterogeneous time series. To the best of our knowledge none of the current algorithms deal with heterogeneous anomalies by means of graphical Granger causality.

References

1. Arnold, A., Liu, Y., Abe, N.: Temporal causal modelling with graphical Granger methods. In: KDD (2007)
2. Bacsó, N.: Das Klima des Donauraumes. Geoforum (1971)
3. Bahadori, M.T., Liu, Y.: Granger causality analysis in irregular time series. In: SDM (2012)
4. Budhathoki, K., Vreeken, J.: Causal inference by compression. In: ICDM (2016)
5. Budhathoki, K., Vreeken, J.: MDL for causal inference on discrete data. In: ICDM (2017)
6. Budhathoki, K., Vreeken, J.: Causal inference on event sequences. In: SDM (2018)
7. Cheng, D., Bahadori, M.T., Liu, Y.: FBLG: a simple and effective approach for temporal dependence discovery from time series data. In: KDD (2014)
8. Granger, C.W.: Investigating causal relations by econometric models and cross-spectral methods. Econometrica, 424–438 (1969)
9. Kim, S., Putrino, D., Ghosh, S., Brown, E.: A Granger causality measure for point process models of ensemble neural spiking activity. PLOS Comput. Biol. **7**, 1–13 (2011)
10. Marx, A., Vreeken, J.: Causal inference on multivariate and mixed-type data. In: Berlingerio, M., Bonchi, F., Gärtner, T., Hurley, N., Ifrim, G. (eds.) ECML PKDD 2018. LNCS, vol. 11052, pp. 655–671. Springer, Cham (2019). https://doi.org/10.1007/978-3-030-10928-8_39
11. McIlhagga, W.: penalized: a MATLAB toolbox for fitting generalized linear models with penalties. J. Stat. Softw. (2016). Articles
12. Mooij, J.M., Peters, J., Janzing, D., Zscheischler, J., Schölkopf, B.: Distinguishing cause from effect using observational data: methods and benchmarks. J. Mach. Learn. Res. **17**, 1103–1204 (2016)
13. Nelder, J.A., Baker, R.J.: Generalized linear models. In: Encyclopedia of Statistical Sciences (1972)
14. Peters, J., Janzing, D., Schölkopf, B.: Causal inference on discrete data using additive noise models. IEEE Trans. Pattern Anal. Mach. Intell. **33**, 2436–2450 (2011)
15. Qiu, H., Liu, Y., Subrahmanya, N.A., Li, W.: Granger causality for time-series anomaly detection. In: ICDM (2012)
16. Schreiber, T.: Measuring information transfer. Phys. Rev. Lett. **85**(2), 461 (2000)
17. Shimizu, S., Hoyer, P.O., Hyvärinen, A., Kerminen, A.: A linear non-Gaussian acyclic model for causal discovery. J. Mach. Learn. Res. **7**(Oct), 2003–2030 (2006)
18. Shojaie, A., Michailidis, G.: Discovering graphical Granger causality using the truncating lasso penalty. Bioinformatics **26**, i517–i523 (2010)
19. Tibshirani, R.: Regression shrinkage and selection via the Lasso. J. Roy. Stat. Soc. Ser. B (Methodol.) **58**, 267–288 (1996)
20. Zou, H.: The adaptive Lasso and its Oracle property. J. Am. Stat. Assoc. **101**, 1418–1429 (2008)

Knowledge Graph Embedding with Order Information of Triplets

Jun Yuan[1,2], Neng Gao[1], Ji Xiang[1(✉)], Chenyang Tu[1], and Jingquan Ge[1,2]

[1] Institute of Information Engineering, Chinese Academy of Sciences, Beijing, China
{yuanjun,gaoneng,xiangji,tuchenyang,gejingquan}@iie.ac.cn
[2] School of Cyber Security, University of Chinese Academy of Sciences,
Beijing, China

Abstract. Knowledge graphs (KGs) are large scale multi-relational directed graph, which comprise a large amount of triplets. Embedding knowledge graphs into continuous vector space is an essential problem in knowledge extraction. Many existing knowledge graph embedding methods focus on learning rich features from entities and relations with increasingly complex feature engineering. However, they pay little attention on the order information of triplets. As a result, current methods could not capture the inherent directional property of KGs fully. In this paper, we explore knowledge graphs embedding from an ingenious perspective, viewing a triplet as a fixed length sequence. Based on this idea, we propose a novel recurrent knowledge graph embedding method RKGE. It uses an order keeping concatenate operation and a shared sigmoid layer to capture order information and discriminate fine-grained relation-related information. We evaluate our method on knowledge graph completion on benchmark data sets. Extensive experiments show that our approach outperforms state-of-the-art baselines significantly with relatively much lower space complexity. Especially on sparse KGs, RKGE achieves a 86.5% improvement at *Hits@1* on FB15K-237. The outstanding results demonstrate that the order information of triplets is highly beneficial for knowledge graph embedding.

Keywords: Knowledge graph · Embedding · Order information · Recurrent model

1 Introduction

Knowledge graphs (KGs) such as WordNet [17] and Freebase [1] have been widely adopted in various applications such as web search, Q&A, etc. KG is a multi-relational directed graph and usually organized in the form of triplets, denoted by (h, r, t). h and t are head and tail entities, respectively, and r is the relation between h and t. For instance, (*William Shakespeare, Write, Hamlet*) denotes the fact that *William Shakespeare* wrote *Hamlet*.

However, current KGs are both extremely large and highly incomplete [6]. How can we predict missing entities based on the observed triplets in an incomplete graph presents a tough challenge for machine learning research. In addition,

© Springer Nature Switzerland AG 2019
Q. Yang et al. (Eds.): PAKDD 2019, LNAI 11441, pp. 476–488, 2019.
https://doi.org/10.1007/978-3-030-16142-2_37

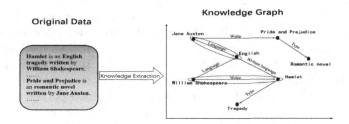

Fig. 1. A Simple Illustration of Knowledge Graph. Knowledge graph is directed graph and organized in the form of triplets. The meaning of a triplet should be determined by head entity, relation, tail entity and the order of three objects together, for example, (*Hamlet, Write, William Shakespeare*) is invalid.

due to the large amount data in real-world KGs, an efficient and scalable solution is crucial [16]. To address the challenge, knowledge graph embedding (KGE) has been widely adopted. The key idea of KGE is to embed entities and relation types of a KG into a continuous vector space. Therefore, we can do reasoning over KGs through algebraic computations.

Many KGE methods have been proposed to learn low-dimensional embeddings of entities and relations [2,19,22]. As shown in Fig. 1, the meaning of a triplet should be determined by head entity, relation, tail entity and the order of three objects together. Thus, the order information of triplets is significant for KGE. However, few researches pay attention on it. For example, DistMult [25] just ignores the directional property of KGs, so it is unable to model asymmetric relations. ComplEx [22] makes use of complex embeddings and Hermitian dot product to address the asymmetric problem of DistMult. Translation-based models can capture order information to some extent, since they treat heads and tails in different way. TransE [2] regards every relation as translation between the heads and tails. When (h, r, t) holds, the embedding $\mathbf{h}$ is close to the embedding $\mathbf{t}$ by adding the embedding $\mathbf{r}$, that is $\mathbf{h} + \mathbf{r} \approx \mathbf{t}$. In summary, current methods could not capture the inherent directional property of KGs fully.

On the other hand, indiscriminate methods, like TransE and ComplEX, process complex relations poorly (*i.e.*, one-to-many, many-to-one and many-to-many relations) [23]. Because the representation of an entity is the same when involved in any relations. In fact, an entity may have multiple aspects that related to different relations [14]. As a consequence, one key to improve embedding is developing a good mechanism to discriminate relation-related information from entities [11,12,14,23]. TransH [23] realizes discrimination by introducing relation-specific projection vectors and projecting entities to relation-specific hyperplanes. TransR [14] realizes discrimination by introducing relation-specific transformation matrix M_r and map entity vectors into different relation spaces. However, current methods are failed to create practical discriminate mechanisms, which are low efficient or low scalable [5,18].

In this paper, we explore knowledge graphs embedding from an ingenious perspective: since the meaning of a triplet is determined by head entity, relation, tail entity and the order of three objects, we believe a triplet in KG form a fixed length sequence virtually. Based on this idea, we develop a novel recurrent discriminate mechanism and propose a scalable and efficient method RKGE. It uses an order keeping concatenate operation and a shared sigmoid layer to capture order information and discriminate fine-grained relation-related information. Our method can avoid large number of discriminate parameters and has almost same space complexity as TransE. We evaluate our method on knowledge graph completion on benchmark data sets. Extensive experiments show that RKGE outperforms state-of-art baselines significantly.

The main contributions of this paper are summarized as follows:

1. We identify the significance of order information that is largely overlooked by existing literature. Furthermore, we explore KGE from an ingenious perspective, viewing triplets as fixed length sequences. As far as we know, we are the first research that models KG as sequences.
2. To appropriately leverage order information, we propose a scalable and efficient model RKGE by adopting a order keeping concatenate operation and developing a novel recurrent discriminate mechanism.
3. Our experimental results show that RKGE outperforms state-of-the-art baselines significantly. The outstanding results demonstrate that the order information in triplets is highly beneficial for knowledge graph embedding.

2   Related Work

Notation. We briefly introduce mathematical notations used in this paper here. A knowledge graph G is defined as a set of triplets of the form (h, r, t). Each triplet is composed of two entities $h, t \in E$ and relation $r \in R$, where E and R are the set of entities and relations, respectively. The embeddings are denoted with the same letters in boldface.

2.1   Researches on Order Information

Current KGE methods always pay little attention on order information and thus could not take full advantage of order information of triplets. DistMult [25] neglects the directional property of KGs and uses weighted element-wise dot product to define the score function $f_r(h, t) = \sum h_k r_k t_k$. DistMult is unable to model asymmetric relations, since (h, r, t) and (t, r, h) will get same score in DistMult, while (t, r, h) is usually invalid. ComplEx [22] makes use of complex valued embeddings and Hermitian dot product to address the antisymmetric problem in DistMult, but ComplEX is failed to model symmetry relations. TransE [2] regards each relation as translation between the heads and tails and wants $\mathbf{h} + \mathbf{r} \approx \mathbf{t}$ when (h, r, t) holds. The score function of TransE is $f_r(h, t) = ||\mathbf{h} + \mathbf{r} - \mathbf{t}||_{L_1/L_2}$. TransE treats heads and tails in different way, so it can capture order information to some extent.

Researchers have always attach great importance to the order information in sequence. The great ability of recurrent neural in sequential modeling has been shown in many researches [15]. In this paper, we view triplets as fixed length sequence and propose a recurrent model to utilize order information.

2.2 Discriminate Models

Indiscriminate models often process complex relations poorly [23]. Because the representation of an entity is the same when involved in various relations. Many discriminate models have been proposed recently.

TransH [23] extends TransE by mapping entity embedding into relation hyperplanes to discriminate relation-related information. The score function is defined as $f_r(h,t) = ||\mathbf{h} - \mathbf{w}_r^T\mathbf{h}\mathbf{w}_r + \mathbf{r} - (\mathbf{t} - \mathbf{w}_r^T\mathbf{t}\mathbf{w}_r)||_{L1/L2}$, where $\mathbf{w}_r$ is the normal vector of r's relation hyperplane. TransR [14] learns a mapping matrix $\mathbf{M}_r$ for each relation. It assumes entity and relation are in different spaces and maps entity embedding into relation space to realize the discrimination. The score function is $f_r(h,t) = ||\mathbf{M}_r\mathbf{h} + \mathbf{r} - \mathbf{M}_r\mathbf{t}||_{L_1/L_2}$. CTransR [14] is an extension of TransR. It clusters diverse head-tail entity pairs into groups and sets a relation vector for each group. TransD [11] constructs two mapping matrices dynamically for each triplet by setting projection vector for each entity and relation, that is $\mathbf{M}_{rh} = \mathbf{h}_p^\top\mathbf{r}_p + \mathbf{I}^{m\times n}$, $\mathbf{M}_{rt} = \mathbf{t}_p^\top\mathbf{r}_p + \mathbf{I}^{m\times n}$. The score function is $f_r(h,t) = ||\mathbf{M}_{rh}\mathbf{h} + \mathbf{r} - \mathbf{M}_{rt}\mathbf{t}||_{L1/L2}$. NTN [19] learns a 3-way tensor and two transfer matrices for each relation as discriminate parameter set. The score function defined as $f_r(h,r,t) = \mathbf{u}_r^T f(\mathbf{h}^\top\hat{\mathbf{M}}_r\mathbf{t} + \mathbf{M}_{r1}\mathbf{h} + \mathbf{M}_{r2}\mathbf{t} + \mathbf{b}_r)$, where $\hat{\mathbf{M}}_r \in \mathbb{R}^{m\times m\times s}$, and $f()$ is $tanh$ operation.

However, current methods are failed with creating practical discriminate mechanisms, which are low efficient or low scalable [5,18]. For example, it is shown by [18] that TransR learns thousands of mapping matrices, but they are all similar and approximate to the unit matrix.

2.3 Other Models

Some types of additional information have been used to improve embedding. The NLFeat model [21] is a log-linear model using simple node and link features. RUGE [10] utilizes the soft-rule between triplets to enhance ComplEX. TransAt [18] adopts K-means to collect category information in KGs to reduces candidate entities and improve prediction. Jointly [20] adopts three neural models to encode the text description of entity. But the extraction of additional information is always the bottleneck.

Many researches attempt to introduce some novel techniques of deep learning into knowledge graph embedding. KBGAN [3] introduces GAN to boost several embedding models. ConvE [5] and ConvKB [4] introduce convolution network in KGE and achieve the state-of-art results on knowledge graph completion. ConvKB concatenates head, tail and relation vectors into a $3 \times m$ matrix, then feeds it into a convolution network. The score function is defined as $f_r(h,t) = \text{concat}(g([\mathbf{h},\mathbf{r},\mathbf{t}] * \mathbf{\Omega}))\mathbf{w}$, where $*$ means convolution operation, $\mathbf{\Omega} \in \mathbb{R}^{\tau\times 1\times 3}$ is the set of filters, τ is the number of filters, and $\mathbf{w} \in \mathbb{R}^{\tau m\times 1}$. To learn expressive features, the magnitude of τ is same as m in above three convolution models.

RKGE is a shallow model without using any additional information. This make it scale to real-world KGs easier.

3   Our Model

3.1   Motivation

In essence, knowledge graph is a directed graph. As shown in Fig. 1, the meaning of a triplet should be determined by head entity, relation, tail entity and the order of three objects together. Therefore, order information of triplets is significant to KGE, but few researches pay attention on it. Based on the direction property of KGs, we believe a triplet in KGs form a fixed length sequence virtually. We believe only operating heads and tails in different ways, like translation-based methods, could not take full advantage of order information.

Additionally, as mentioned in "Introduction", one key to improve embedding is developing a good mechanism to discriminate relation-related information from entities. However, previous methods are failed with creating fine-grained discriminate mechanisms. Worse still, they potentially suffer from large parameters in real-world KGs. Because current models always introduce large discriminate parameters. In fact, the relation-related information should be determined by entity, relation and order between these two objects together. For example, in (*Jane Austen, Language, English*) the *Language*-related information of *Jane Austen* and *English* should be discriminated in different ways. Because *Language* is a antisymmetric relation and (*English, Language, Jane Austen*) is invalid.

Our objective in this paper is not only to learn appropriate embeddings of KGs, but also to propose a scalable and efficient KGE method. Based on the reasons in last paragraph, we explore to develop a novel recurrent discriminate mechanism to discriminate fine-grained relation-related information and avoid large number of discriminate parameters. More than this, we believe the directional property of KGs comes from original data, which is text mainly. In fact, recurrent neural networks (RNNs) have been extremely successfully in natural language processing [20].

3.2   KGE with Recurrent Discriminate Mechanism (RKGE)

The framework of RKGE is shown as Fig. 2 and the detailed descriptions are as follow:

Entities and relations are embedded into same continuous vector space $\mathbb{R}^m$. We input both entity embedding and relation embedding into sigmoid layer, while keeping the original order between them. Afterwards, output of sigmoid layer will be used to discriminate relation-related information. In this way, RKGE is able to capture the inherent directional property of KGs.

The sigmoid layer is fully connected layer, and its parameters denoted as $\mathbf{W} \in \mathbb{R}^{m \times 2m}$ and $\mathbf{b} \in \mathbb{R}^m$. $\mathbf{W}$ and $\mathbf{b}$ are shared parameters and independent of embeddings. Then we define the discriminated vectors of entities as

$$\mathbf{h}_r = \mathbf{h} \odot \sigma(\mathbf{W}[\mathbf{h}, \mathbf{r}]),\tag{1}$$

$$\mathbf{t}_r = \mathbf{t} \odot \sigma(\mathbf{W}[\mathbf{r}, \mathbf{t}]). \tag{2}$$

The sigmoid function $\sigma(\cdot)$ is applied element-wise and $\sigma(x) = \frac{1}{1+\exp(-x)}$. $[,]$ means concatenate operation. $\odot$ means the element-wise product.

As we view triplets as fixed length sequences, we just view relation as the predecessor of tail and successor of head. Note that the concatenate operation will keep the order of head, relation and tail in triplet. Then we share the sigmoid layer between $[\mathbf{h}, \mathbf{r}]$ and $[\mathbf{r}, \mathbf{t}]$, forming a recurrent layer. The sigmoid operation sets its output between 0 and 1, describing how much information should be let through. Thus, we use element-wise product to discriminate relation-related information.

Given a triplet (h,r,t), the log-odd of the probability that G holds the triplet is true is:

$$P((h,r,t) \in G) = \sigma(f_r(h,r)). \tag{3}$$

Thus, the f_r is the score function and score is expected to be higher for valid triplets and lower for invalid ones. Formally, we define the RKGE score function f_r as follows:

$$f_r(h,r) = <\mathbf{h}_r, \mathbf{r}, \mathbf{t}_r> = \sum_{k=1}^{m} h_{rk} r_k t_{rk}. \tag{4}$$

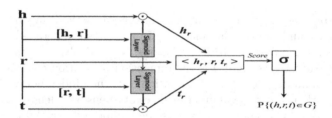

Fig. 2. RKGE Architecture. We unfold recurrent sigmoid layer and dashed arrow means no information transmission. The concatenate operation keeps the order in triplets.

As described above, RKGE can avoid huge discriminate parameters and achieve recurrent relation-related discrimination by sharing discriminate parameters across different triplets.

3.3   Training

To learn above embeddings, we use the Adam optimizer to minimizing following common loss function [22] with L_2 regularization on weight matrix $\mathbf{W}$ of sigmoid layer.

$$\mathcal{L} = \sum_{(h,r,t) \in \{G \cup G'\}} \log(1 + \exp(-Y_{hrt} f_r(h,t))) + \frac{\lambda}{2} ||\mathbf{W}||_2^2, \tag{5}$$

where $Y_{hrt} = 1$ if $(h,r,t) \in G$, and $Y_{hrt} = -1$ otherwise.

G' is a collection of invalid triplets generated by replacing entities or relations in training triplets randomly. That is

$$G' = \{(h',r,t)|h' \in E\} \cup \{(h,r,t')|t' \in E\} \cup \{(h,r',t)|r' \in R\} \qquad (6)$$

It should be noted that $|G'|$ can be different from $|G|$. We use η to represent negative samples per training sample, that is $\eta = \frac{|G'|}{|G|}$. Our experiments have shown that η is an important hyperparameter for RKGE. Please refer to Fig. 3 and "Experiment" section for details.

In practice, we enforce constraint on the norm of the relation embeddings, *i.e.* $\forall r \in R$, $||\mathbf{r}||_2 = 1$. Because the element-wise product will reduce the element values of entity embeddings. We make no constraint on entity embeddings and bias of sigmoid layer.

The embeddings and weight matrix are initialized by sampling from a truncated standard normal distribution, rather than using Xavier initialization [9]. Since it assumes the network consists only of a chain of matrix multiplications. Experiments also show a higher rate of convergence. In addition, the bias is initialized as a vector that all elements are 1. The training is stopped based on the performance on validation set.

3.4   Complexity Analysis

We compare our method with several state-of-the-art methods in parameter size, time complexity and pre-training method. Table 1 lists the detailed results. N_e, N_r represent the number of entities, relations, respectively. m is the dimension of entity embedding space and n is the dimension of relation embedding space. d denotes the number of clusters of a relation. k is the hidden nodes' number of a neural network and s is the number of slices of a tensor. $\hat{\theta}$ denotes the average sparse degree of all transfer matrices. τ is the number of convolutional kernels.

Table 1. Complexities (the parameter size, time complexity in an epoch and pre-training method) of several embedding models.

Model	Parameter size	Time complexity	Pre-training method
NTN [19]	$O(N_e m + N_r(n^2 s + 2nk + 2k))(m = n)$	$O(m^2 s + mk)$	Word embedding
TransE [2]	$O(N_e m + N_r n)(m = n)$	$O(m)$	None
TransH [23]	$O(N_e m + 2N_r n)(m = n)$	$O(m)$	TransE
DistMult [25]	$O(N_e m + N_r n)(m = n)$	$O(m)$	None
TransR [14]	$O(N_e m + N_r(m + 1)n)$	$O(mn)$	TransE
CTransR [14]	$O(N_e m + N_r(m + d)n)$	$O(mn)$	TransR
TransD [11]	$O(2N_e m + 2N_r n)$	$O(m)$	TransE
ComplEx [22]	$O(2N_e m + 2N_r n)(m = n)$	$O(m)$	None
ConvKB [4]	$O(N_e m + N_r n + (\tau + 3)m)$(m=n)	$O(\tau m)$	None
TransAt [18]	$O(N_e m + 4N_r n)(m = n)$	$O(m)$	K-means
RKGE	$O(N_e m + N_r n + 2m^2))(m = n)$	$O(m^2)$	None

It can be seen that our method does not significantly increase the space or time complexity. Since $m \ll N_r \ll N_e$ among existing KGs, parameter size of RKGE is almost same as TransE, which has been applied on real-word KGs [24]. The time complexity of RKGE is similar to TransR in an epoch. Note that our method is self-contained method, $i.e.$, RKGE does not require pre-trained embeddings from prerequisite models. In contrast, TransR needs pre-trained embeddings from TransE, which further increases the cost of training. In practice, our model can be applied on real-world KGs with less computing resources and running time than TransR.

In summary, RKGE has relatively much lower time complexity and tolerable space complexity, making it a scalable and efficient KGE method.

4   Experiments

4.1   Data Sets

We evaluate our model on knowledge graph completion using two commonly used large-scale knowledge graph, namely Freebase and WordNet. Table 2 lists statistics of the data sets used in this paper.

Table 2. Statistics of data sets.

Dataset	#Rel	#Ent	#Train	#Valid	#Test
WN18RR	11	40,943	86,835	3,034	3,134
FB15K-237	237	14,541	272,115	17,535	20,466

FB15K-237. FB15K [2] is a subset of Freebase which contains about 14,951 entities, 1,345 different relations and 592,213 triplets. It is firstly discussed by [21] that FB15K suffers from test leakage through inverse relations, $i.e.$ many test triplets can be obtained simply by inverting triplets in the training set. To address this issue, Toutanova $et\ al.$ [21] generated FB15K-237 by removing redundant relations in FB15K and greatly reducing the number of relations.

WN18RR. WordNet provides semantic knowledge of words. WN18 [2] is a subset of WordNet and contains 40,943 entities, 18 relation types and 151,442 triplets. Like FB15K, WN18 suffers from test leakage through inverse relations too. Therefore, Dettmers $et\ al.$ [5] removed reversing relations in WN18 and generated WN18RR. As a consequence, the difficulty of reasoning on the data set is increased dramatically.

4.2   Knowledge Graph Completion

Knowledge graph completion aims to predict the missing h or t for a triplet (h, r, t). In this task, the model is asked to rank a set of candidate entities from

Table 3. Experimental results of knowledge graph completion.

Model	WN18RR			FB15K-237		
	MRR	Hits@10(%)	Hits@1(%)	MRR	Hits@10(%)	Hits@1(%)
TransE [2]	0.226	50.1	39.1	0.294	46.5	14.7
DistMult [25]	0.43	49.0	39.0	0.241	41.9	15.5
TransD [11]	-	42.8	-	-	45.3	-
ComplEX [22]	0.44	51.0	41.0	0.247	42.8	15.8
KB-LRN [8]	-	-	-	0.309	49.3	21.9
NLFeat [21]	-	-	-	0.249	41.7	-
KBGAN [3]	0.213	48.1	-	0.278	45.8	-
ConvE [5]	0.43	52.0	40.0	0.325	50.1	23.7
ConvKB [4]	0.248	52.5	-	0.396	51.7	-
RKGE	**0.44**	**53.0**	**41.9**	**0.477**	**55.4**	**44.2**

the KG, instead of giving one best result. We report two common measures as our evaluation metrics: the mean reciprocal rank of all correct entities (*MRR*), and the proportion of correct entities ranked in top K (*Hits@K*). A good KGE method should achieve high results on both *MRR* and *Hits@K*.

For each test triplet (h, r, t), we replace the head/tail entity by all possible candidates in the KG, and rank these entities in descending order of scores calculated by score function f_r. We follow the evaluation protocol in [2] to report filtered results. Because a corrupted triplet, generated in the aforementioned process of removal and replacement, may also exist in KG, and it should be considered as correct. In other word, we filtered out the valid triplets from corrupted triplets which exist in the training set.

Hyperparameters of RKGE are selected via grid search according to the *MRR* on the validation set. For two data sets, we traverse all the training triplets for at most 1000 epochs. We search the initial learning rate α for Adam among $\{0.01, 0.1, 0.5\}$, the L_2 regularization parameter λ among $\{0, 0.01, 0.03, 0.1, 0.3, 0.5, 1\}$, the embedding dimension m among $\{32, 50, 100, 200\}$, the batch size B among $\{120, 480, 1440\}$, and the number of negative samples generated per training sample η among $\{1, 2, 5, 10, 20\}$.

The best configurations are as follow: on WN18RR, $\lambda = 0.01$, $\alpha = 0.01$, $m = 200$, $B = 120$ and $\eta = 20$; on FB15K-237, $\lambda = 0.1$, $\alpha = 0.1$, $m = 200$, $B = 480$ and $\eta = 20$. Table 3 shows the evaluation results on knowledge graph completion. Since the data sets are same, we directly copy experimental results of several baselines from [3–5]. It can be observed from Table 3 that:

1. RKGE outperforms all baselines on WN18RR and FB15K-237 at every metric. The results indicate that RKGE learns better embeddings and achieves more fine-grained model relation-related discrimination.

2. Our method outperforms baselines significantly on sparse data set, *i.e.*, FB15K-237, 20.5% higher at *MRR* and 86.5% higher at *Hits@1* than previous best result.

3. RKGE is able to better handle both sparse graph FB15K-237 and dense graph WN18RR. This indicates the good generalization of our methods.

4. RKGE outperforms TransE on both data sets. Especially on FB15K-237. RKGE is 62.2% higher at *MRR* and 200% higher at *Hits@1* than TransE. These results indicate translation-based method could not capture inherent directional property of KGs fully.

5. The outstanding results demonstrate that the order information in triplets is highly beneficial for KGE. These results also indicate that recurrent discriminate mechanism may be a good direction for improving KGE.

4.3   Influence of Negative Samples

We further investigate the influence of the number of negative samples generated per training sample. Due to computational limitations, η was validated among $\{1,2,5,10,20\}$ in the previous experiments. We want to explore the influence here from two aspects: the performance and the training time of our method. To do so, experiments focus on FB15K-237, with the best configurations obtained from the previous experiment. Then we let η vary in $\{1,2,5,10,20,50,100,150,200\}$. Training was stopped using early stopping based on filtered *MRR* on the validation set, computed every 50 epochs with a maximum of 1000 epochs. Note that our

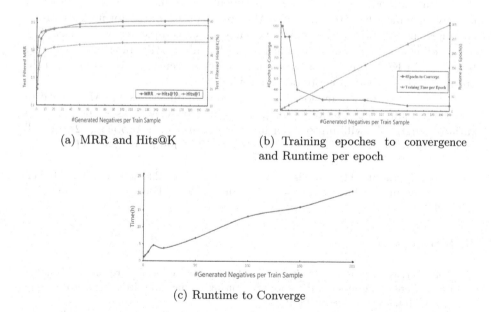

(a) MRR and Hits@K

(b) Training epoches to convergence and Runtime per epoch

(c) Runtime to Converge

Fig. 3. Influence of the number of negative triples generated per training sample

method is implemented in TensorFlow, and all models were run on a standard hardware of Inter(R) Core(TM) i7 2.6 GHz + GeForce GTX 960M, with 480 at mini-batch size. The evaluation results are shown at Fig. 3.

It can be observed from Fig. 3(a) that generating more negatives clearly improves the results. RKGE achieve the best results when η is 200: 0.511 at filtered *MRR*, 57.5% at *Hits@10* and 47.7% at *Hits@1*. It should be noted that when η is 5, RKGE has outperformed all state-of-art baselines with 0.442 at filtered *MRR*, 54.2% at *Hits@10* and 38.8% at *Hits@1*. However, results improve slowly as the number of negatives increases when η is larger than 10. As the training time, Fig. 3(b) shows that our method basically converges with fewer epochs when η is larger. RKGE reduces training epochs sharply when η is 20 and reduces epochs relatively slow in other cases. Meanwhile, the training time per epoch grows linearly as the η increases. As a result, runtime to converge increases as the growth of η, except η is 20. In conclusion, $\eta = 5$ makes a good trade-off between accuracy and training time.

5   Conclusion and Future Work

In this paper, we identify the significance of order information that is largely overlooked by existing literature. To appropriately leverage the information, we ingeniously view a triplet as a fixed length sequence. We develop a novel recurrent discriminate mechanism and propose a scalable and efficient method RKGE. As far as we know, we are the first research tring model KGs as sequences.

Our method is evaluated on knowledge graph completion. To deeper understand our method, we also make experiments in influence of the number of negative samples. Experiments show that our method outperforms state-of-art baselines significantly with almost the same space complexity as TransE. The outstanding results demonstrate that the order information in triplets is highly beneficial for KGE. These results also indicate that recurrent discriminate mechanism may be a good direction for improving KGE.

In the future, we will explore following research directions: (1) Many entities may have multi-step relations [7,13], which can be viewed as sequences with variable length. We will improve our method to handle this problem; (2) We will explore sophisticated RNNs to propose better KGE methods, like LSTM.

Acknowledgement. This work is supported by the National Key Research and Development Program of China, and National Natural Science Foundation of China (No. U163620068).

References

1. Bollacker, K., Evans, C., Paritosh, P., Sturge, T., Taylor, J.: Freebase: a collaboratively created graph database for structuring human knowledge. In: Proceedings of SIGMOD, pp. 1247–1250 (2008)
2. Bordes, A., Usunier, N., Weston, J., Yakhnenko, O.: Translating embeddings for modeling multi-relational data. In: Proceedings of NIPS, pp. 2787–2795 (2013)

3. Cai, L., Wang, W.Y.: KBGAN: adversarial learning for knowledge graph embeddings. In: Proceedings of NAACL (2018)
4. Dai, Q.N., Tu, D.N., Nguyen, D.Q., Phung, D.: A novel embedding model for knowledge base completion based on convolutional neural network. In: Proceedings of NAACL (2018)
5. Dettmers, T., Pasquale, M., Pontus, S., Riedel, S.: Convolutional 2D knowledge graph embeddings. In: Proceedings of AAAI, pp. 1811–1818 (2018)
6. Dong, X., et al.: Knowledge vault: a web-scale approach to probabilistic knowledge fusion. In: Proceedings of SIGKDD, pp. 601–610 (2014)
7. Garciaduran, A., Bordes, A., Usunier, N.: Composing relationships with translations. In: Proceedings of EMNLP, pp. 286–290 (2015)
8. Garciaduran, A., Niepert, M.: KBLRN: end-to-end learning of knowledge base representations with latent, relational, and numerical features. In: Proceedings of UAI (2017)
9. Glorot, X., Bengio, Y.: Understanding the difficulty of training deep feedforward neural networks. In: Proceedings of the Thirteenth International Conference on Artificial Intelligence and Statistics, pp. 249–256 (2010)
10. Guo, S., Wang, Q., Wang, L., Wang, B., Guo, L.: Knowledge graph embedding with iterative guidance from soft rules. In: Proceedings of AAAI (2018)
11. Ji, G., He, S., Xu, L., Liu, K., Zhao, J.: Knowledge graph embedding via dynamic mapping matrix. In: Proceedings of ACL, pp. 687–696 (2015)
12. Ji, G., Liu, K., He, S., Zhao, J.: Knowledge graph completion with adaptive sparse transfer matrix. In: Proceedings of AAAI, pp. 985–991 (2016)
13. Lin, Y., Liu, Z., Luan, H., Sun, M., Rao, S., Liu, S.: Modeling relation paths for representation learning of knowledge bases. In: Proceedings of EMNLP, pp. 705–714 (2015)
14. Lin, Y., Liu, Z., Zhu, X., Zhu, X., Zhu, X.: Learning entity and relation embeddings for knowledge graph completion. In: Proceedings of AAAI, pp. 2181–2187 (2015)
15. Lipton, Z.C.: A critical review of recurrent neural networks for sequence learning. CoRR abs/1506.00019 (2015). http://arxiv.org/abs/1506.00019
16. Liu, H., Wu, Y., Yang, Y.: Analogical inference for multi-relational embeddings. In: Proceedings of ICML (2017)
17. Miller, G.A.: WordNet: a lexical database for English. Commun. ACM **38**(11), 39–41 (1995)
18. Qian, W., Fu, C., Zhu, Y., Cai, D., He, X.: Translating embeddings for knowledge graph completion with relation attention mechanism. In: Proceedings of IJCAI, pp. 4286–4292 (2018)
19. Socher, R., Chen, D., Manning, C.D., Ng, A.Y.: Reasoning with neural tensor networks for knowledge base completion. In: Proceedings of NIPS, pp. 926–934 (2013)
20. Socher, R., et al.: Recursive deep models for semantic compositionality over a sentiment treebank. In: Proceedings of EMNLP, pp. 1631–1642 (2013)
21. Toutanova, K., Chen, D.: Observed versus latent features for knowledge base and text inference. In: Proceedings of CVCS (2015)
22. Trouillon, T., Welbl, J., Riedel, S., Gaussier, E., Bouchard, G.: Complex embeddings for simple link prediction. In: Proceedings of ICML, vol. 48, pp. 2071–2080 (2016)
23. Wang, Z., Zhang, J., Feng, J., Chen, Z.: Knowledge graph embedding by translating on hyperplanes. In: Proceedings of AAAI, pp. 1112–1119 (2014)

24. Xu, H., Yankai, L., Ruobing, X., Zhiyuan, L., Maosong, S.: OpenKE: an open-source framework for knowledge embedding (2017). http://openke.thunlp.org/home
25. Yang, B., Yih, W., He, X., Gao, J., Deng, L.: Embedding entities and relations for learning and inference in knowledge bases. In: Proceedings of ICLR (2015)

Knowledge Graph Rule Mining
via Transfer Learning

Pouya Ghiasnezhad Omran[1(✉)], Zhe Wang[1,2], and Kewen Wang[1]

[1] Griffith University, Brisbane, QLD, Australia
`pouya.ghiasnezhadomran@griffithuni.edu.au,`
`{zhe.wang,k.wang}@griffith.edu.au`
[2] State Key Laboratory of Computer Science, Institute of Software,
Chinese Academy of Sciences, Beijing, China

Abstract. Mining logical rules from knowledge graphs (KGs) is an important yet challenging task, especially when the relevant data is sparse. Transfer learning is an actively researched area to address the data sparsity issue, where a predictive model is learned for the target domain from that of a similar source domain. In this paper, we propose a novel method for rule learning by employing transfer learning to address the data sparsity issue, in which most relevant source KGs and candidate rules can be automatically selected for transfer. This is achieved by introducing a similarity in terms of embedding representations of entities, relations and rules. Experiments are conducted on some standard KGs. The results show that proposed method is able to learn quality rules even with extremely sparse data and its predictive accuracy outperformed state-of-the-art rule learners (AMIE+ and RLvLR), and link prediction systems (TransE and HOLE).

Keywords: Knowledge graph · Transfer learning ·
Representation learning

1 Introduction

Following Google's success in semantic search empowered by its Knowledge Graph, more and more companies have started to develop their own knowledge graphs (KGs) [1]. Well-known examples of large-scale and comprehensive KGs include YAGO [2], DBpedia [3], and Freebase [4]. In a KG, entities such as persons and places are nodes linked through predicates representing relationships among the entities; a pair of entities and a predicate linking them form a fact about the world.

Despite of the large sizes of these comprehensive KGs, the issue of data sparsity exists, that is, when only a small number of *relevant* facts are available. For example, while large KGs contain many entities, most entities are associated with a small number of facts [5], and for a large portion of predicates, only a few facts are available for each of them. For instance, only few facts are available for

© Springer Nature Switzerland AG 2019
Q. Yang et al. (Eds.): PAKDD 2019, LNAI 11441, pp. 489–500, 2019.
https://doi.org/10.1007/978-3-030-16142-2_38

the predicates organismsOfThisType and exhibitionSponsor in Freebase (FB15K). Also, when companies develop their business related KGs, these KGs often need to be developed from scratch and from small sizes, where both entities and available facts are few.

The importance of coupling KGs with logical rules has been highlighted in the literature [1]. A rule can present an abstract pattern mined from the data. For instance, if a person was born in New York and we know that New York is a city of the United States, then the person's nationality is the United States with a high possibility. This pattern can be expressed as the following first-order rule with a confidence degree such as 0.8:

$$\mathsf{bornInCity}(x, y) \wedge \mathsf{cityOfCountry}(y, z) \rightarrow \mathsf{nationality}(x, z).$$

While it is well known that rules are important for reasoning, they are also useful for KG completion [6], which is the process to automatically extract new facts from existing ones (e.g., link prediction). Recently, several approaches to rule mining in KGs have been proposed, such as SWARM [7], RDF2rules [8], ScaleKB [9], AMIE+ [10] and RLvLR [11]. These methods are capable to mine firs-order rules augmented with confidence degrees.

However, these approaches to KG rule mining all suffer from the so-called data starvation problem [12]. That is, they typically require very large datasets for rule mining and struggle to mine over sparse data.

Transfer learning is a promising paradigm for overcoming the data starvation problem. The goal of transfer learning is to reuse knowledge learned from one task (source task) and apply it in a different and unlearned task (target task). This paradigm of learning is mostly pursued in feature vector machine learning, but some attempts have been made to learn relational models. These methods usually assume that the source and target domains are highly similar in terms of models of interest. Once such strict conditions are violated, unfavourable effects can be caused. This is usually referred to as negative transfer [12,13].

In this paper, we tackle the problem of rule mining with sparse data by combining transfer learning with the technique of embedding in representation learning. Given a KG, a *learning task* (or just a *task*) is that of learning rules whose head is about a specific predicate. Given a set of source tasks and a target task, we assume that each source task is learned, that is, a set of rules have been learned for the source task, and we aim to learn a set of rules for the target task based on the source tasks. To facilitate knowledge transfer and reduce the effect of negative transfer, we make use of embedding techniques.

The basic idea of embedding techniques is to encode relational information as low-dimensional representations (embeddings) of entities and predicates [6]. Such representation learning techniques have been applied in rule mining over KGs [14,15]. Instead of directly mining rules from data and embeddings, we utilize the pre-trained embeddings to measure structural similarity between source and target tasks and relevant predicates, in order to transfer logical rules from source tasks to the target task.

The novelty of our approach lies in that the paradigm of transfer learning and embedding techniques are combined for rule mining which is not trivial. Transfer learning can help address the data starvation problem while embedding techniques can provide useful similarity measures that reduce negative transfer. As a result, this combination makes it possible to develop effective systems for rule mining over sparse data in KGs. In fact, our experiments demonstrate that on sparse data, our system outperformed state-of-the-art rule learner AMIE+ [10] and RLvLR [11] in terms of the number of quality rules and the accuracy in link prediction. We also compared our system with embedding-based link prediction systems, our system outperformed TransE [16] and HOLE [17] in link prediction on sparse predicates regarding accuracy and runtime.

The paper is organised as follows. We introduce some basics of KGs, rules and embeddings in Sect. 2. Then we provide an overview of our proposed method in Sect. 3. We define measures for similarity using embeddings in Sect. 4. The experimental evaluation is reported in Sect. 5. Finally, we discuss some related works and conclude our work in Sect. 6.

2   Preliminaries

In this section, we briefly recall some basics of knowledge graphs and embeddings as well as fixing some notations to be used later.

2.1   Knowledge Graphs and Rules

An entity e is an object such as a place, a person, etc., and a fact is an RDF triple (e, P, e'), which means that the entity e is related to another entity e' via the binary predicate P. Following the convention in knowledge representation, we denote such a fact as $P(e, e')$. A *knowledge graph* $(\mathcal{K})$ is a set of facts.

A *Horn rule* or just a *rule* r is of the form $a_1 \wedge \ldots \wedge a_n \to a$ where $a, a_1, \ldots, a_n$ are atoms of the form $P(t_1, t_2)$ with each of the t_1 and t_2 being an entity or a variable. Intuitively, the rule r reads that if $a_1, \ldots, a_n$ hold, then a holds too. Atom a is referred to as the head of r and atoms $a_1, \ldots, a_n$ as the body of r.

To assess the quality of mined rules, standard confidence (SC) and head coverage (HC) are used in some major approaches of rule learning [9,10]. Formally, assume the head of r is of the form $P(x, y)$, a pair of entities (e, e') satisfies the body of r in KG $\mathcal{K}$, denoted $body(r, e, e', \mathcal{K})$, if there is a way of substituting variables in the body of r with entities in $\mathcal{K}$ such that (i) (e, e') substitutes (x, y) and (ii) all atoms in the body of r (after substitution) are facts in $\mathcal{K}$. (e, e') satisfying the head of r in KG $\mathcal{K}$ is defined in the same way and denoted as $head(r, e, e', \mathcal{K})$. Then the support degree of r is defined as

$$supp(r, \mathcal{K}) = \#(e, e') : body(r, e, e', \mathcal{K}) \wedge head(r, e, e', \mathcal{K})$$

To normalize this degree, the degrees of standard confidence (SC) and head coverage (HC) are defined as follows:

$$SC(r, \mathcal{K}) = \frac{supp(r, \mathcal{K})}{\#(e, e') : body(r, e, e', \mathcal{K})} \qquad HC(r, \mathcal{K}) = \frac{supp(r, \mathcal{K})}{\#(e, e') : head(r, e, e', \mathcal{K})}$$

2.2 Embeddings

Various approaches have been proposed to construct embeddings, which include translation based embeddings [16] and matrix factorization based embeddings [17,18]. The translation based embeddings use additive calculus and uses vectors to represent the embeddings of predicates, whereas the matrix factorization based embeddings use dot calculus and uses matrices to represent the embeddings of predicates. Since our rule mining approach requires a relatively expressive form of embeddings, we adopt matrix factorization based embeddings. In particular, we employ the state-of-the-art RESCAL system [17,18] to construct such embeddings.

RESCAL embeds each entity e to a vector $\mathbf{e}$ and each predicate P to a matrix $\mathbf{P}$. In [11], a notion of argument embeddings is introduced by aggregating the entities embeddings. Formally, the embeddings of the subject and object argument of a predicate P are defined as:

$$\mathbf{p}^{(1)} = \frac{1}{n} \sum_{e \in S_P} s_e . \mathbf{e} \quad \text{and} \quad \mathbf{p}^{(2)} = \frac{1}{n} \sum_{e \in O_P} o_e . \mathbf{e}$$

where n is the number of facts in the KG $\mathcal{K}$, S_P and O_P are the sets of entities occurring as respectively subjects and objects of P in $\mathcal{K}$ (more precisely, $S_P = \{e \mid \exists e' \text{ s.t. } P(e, e') \in \mathcal{K}\}$ and $O_P = \{e' \mid \exists e \text{ s.t. } P(e, e') \in \mathcal{K}\}$), and s_e and o_e are the numbers of occasions for entity e to occur as respectively a subject and a object in $\mathcal{K}$ (more precisely, $s_e = \#\{P(e, e') \in \mathcal{K}\}$ and $o_e = \#\{P(e', e) \in \mathcal{K}\}$).

3 An Overview of Our Approach

Unlike a standard transfer learning setting, we do not assume the source domain for transfer is pre-known; instead, we generate a pool of potential sources from existing KGs and select the sources that are most similar to the target.

We consider a *learning task* (or simply a *task*) T to be a triple $T = (\mathcal{K}, P, \mathcal{R})$ where $\mathcal{K}$ is a KG, P is a predicate in $\mathcal{K}$ and $\mathcal{R}$ is a set of rules about P (i.e., having P in its head). Intuitively, $T = (\mathcal{K}, P, \mathcal{R})$ represents the task of learning rules about predicate P over KG $\mathcal{K}$, and $\mathcal{R}$ consists of the rules learnt so far. For transfer learning, the input consists of a target task $T_t = (\mathcal{K}_t, P_t, \mathcal{R}_t)$ and a set of n (potential source) tasks $T_i = (\mathcal{K}_i, P_i, \mathcal{R}_i)$ with $1 \leq i \leq n$. For convenience, we assume $\mathcal{R}_t = \emptyset$ initially, and P_t and each P_i are pairwise distinct. We call P_t and each P_i the *goal predicate* respectively for task T_t and each task T_i. The output of learning is a set of rules $\mathcal{R}_t$ about P_t over $\mathcal{K}_t$.

In Algorithm 1, we illustrate the data flow and major components of our algorithm.

In line 1, the Embedding method computes the embeddings for all (source and target) KGs. It employs a sampling method [11] to filter out irrelevant facts in $\mathcal{K}_t$ and each $\mathcal{K}_i$, and then computes predicate embeddings as well as subject and object argument embeddings as in [11]. After sampling, the (source and target)

Algorithm 1. Transfer rule learning for KGs

Input: a target task $T_t = (\mathcal{K}_t, P_t, \emptyset)$ and a set of learnt tasks $\mathcal{T} = \{T_1, \ldots, T_n\}$ where
$\quad T_i = (\mathcal{K}_i, P_i, \mathcal{R}_i)$ for $1 \leq i \leq n$
Output: a set of rules $\mathcal{R}_t$ about P_t over $\mathcal{K}_t$
1: $(\mathcal{E}_t, \mathcal{E}_1, \ldots, \mathcal{E}_n) := \mathsf{Embedding}(\mathcal{K}_t, \mathcal{K}_1, \ldots, \mathcal{K}_n)$ /* compute embeddings */
2: $\mathcal{T}_s := \mathsf{Retrieve}(\mathcal{E}_t, \mathcal{E}_1, \ldots, \mathcal{E}_n)$ /* retrieve source tasks */
3: $\mathcal{R}_t := \emptyset$ /* initialise target rule set */
4: **for** each $T_i \in \mathcal{T}_s$ **do**
5:     $\mathcal{M} := \mathsf{Map}(\mathcal{E}_i, \mathcal{E}_t)$ /* build a mapping between source and target */
6:     $\mathcal{R} := \mathsf{Transfer}(\mathcal{R}_i, \mathcal{M})$ /* transfer rules from source to target */
7:     $\mathcal{R}_t := \mathcal{R}_t \cup \mathsf{Validate}(\mathcal{R}, \mathcal{K}_t)$ /* validate the transferred rules */
8: **end for**
9: **return** $\mathcal{R}_t$

KGs contain only predicates related to the corresponding goal predicates, and we measure similarity (for transfer) only among these predicates.

Then, in line 2, the $\mathsf{Retrieve}$ method selects a set of tasks $\mathcal{T}_s$ from the pool of tasks $\mathcal{T}$ that are considered most similar to the target task T_t. Tasks in $\mathcal{T}_s$ are called *source tasks*. Since our goal is to learn rules about P_t, we consider a task T_i similar to T_t if the goal predicate P_i (of T_i) is similar to P_t, and such similarity is measured by their embeddings (ref. Sect. 4.1).

After that, in lines 4 and 5, for each source task T_i, the Map method establishes a mapping $\mathcal{M}$ between the predicates in respectively the source KG $\mathcal{K}_i$ and the target KG $\mathcal{K}_t$. We developed two effective similarity measures based on embeddings ($\mathcal{E}_i$ and $\mathcal{E}_t$) for constructing such a mappings and both similarity measure ensure the goal predicates (P_i and P_t) are mapped to each other (ref. Sect. 4.2). Then in line 6, the $\mathsf{Transfer}$ method transfers the source rules in $\mathcal{R}_i$ to generate candidate rules through the mapping $\mathcal{M}$, by substituting the source predicates of a rule with the corresponding target predicates.

Finally, in line 7, the $\mathsf{Validate}$ method validates the transferred rules and adds validated rules to $\mathcal{R}_t$. Standard validation is ineffective over sparse data, where both SC and HC measures have limited significance. Hence, instead of validating single candidate rules, we measure for each source task T_i the percentage of transferred rules that have sufficient support in the target KG $\mathcal{K}_t$. We call this measure *Transfer Confidence* (TC), and it essentially validates the selected source task and the established mapping. When the TC is above a certain threshold, all the rules transferred from T_i are considered valid. We define TC as follows:

$$TC(\mathcal{R}, \mathcal{K}_t) = \frac{\#r \in \mathcal{R} : supp(r, \mathcal{K}_t) \geq \alpha}{|\mathcal{R}|}$$

where α is a learning parameter.

4   Similarity Measures via Embeddings

In this section, we present our approach for measuring the similarity between two predicates (in possibly two different KGs), which is central to both Retrieve and Map methods in Algorithm 1.

4.1   Similarity Between Goal Predicates

For a source tasks $T_i = (\mathcal{K}_s, P_s, \mathcal{R}_s)$ to be considered similar to the target task $T_t = (\mathcal{K}_t, P_t, \mathcal{R}_t)$, intuitively it requires that the goal predicate P_s associates entities in $\mathcal{K}_s$ in a similar way as the goal predicate P_t does in $\mathcal{K}_t$. Inspired by entity alignment measures [19], where similarity between entities are measure through their embeddings, we show that similarity on how predicates associate entities can be captured by predicate embeddings.

For two matrices $\mathbf{M}_1$ and $\mathbf{M}_2$, the closeness between them can be defined in a standard way using the Frobenius norm:

$$close(\mathbf{M}_1, \mathbf{M}_2) = -\|\mathbf{M}_1 - \mathbf{M}_2\|_F.$$

We also measure the (absolute value of the) difference between the numbers of predicates in two KGs $\mathcal{K}_1$ and $\mathcal{K}_2$, denoted as $diff(\mathcal{K}_1, \mathcal{K}_2)$. Note that after sampling, this measure shows the difference between two goal predicates regarding how many other predicates each of them are related to in their corresponding KGs. Then, the similarity degree between two tasks $T_1 = (\mathcal{K}_1, P_1, \mathcal{R}_1)$ and $T_2 = (\mathcal{K}_2, P_2, \mathcal{R}_2)$ are defined as

$$sim(T_1, T_2) = exp(close(\mathbf{P}_1, \mathbf{P}_2)) - \beta \cdot diff(\mathcal{K}_1, \mathcal{K}_2)$$

where $\mathbf{P}_1$ and $\mathbf{P}_2$ are predicate embeddings of P_1 and P_2 respectively, and β is a small coefficient which adjusts the effect of the second measure.

4.2   Similarity Between Other Predicates

To build a mapping between other predicates in the source and target tasks, we measure the similarity between each pair of predicates in respectively source and target KGs ($\mathcal{K}_s$ and $\mathcal{K}_t$). The mapping will be used to transfer rules about the source goal predicate (P_s) to construct rules about the target goal predicate (P_t). Hence, the source and target goal predicates should be mapped to each other under our similarity measure. Also, the similarity between other predicates should be measured by how they relate to the goal predicates; that is, a predicate P_1 in $\mathcal{K}_s$ is considered similar to a predicate P_2 in $\mathcal{K}_t$ if P_1 relates to P_s (through joint entities) in a similar way as how P_2 relates to P_t.

We propose two similarity measures: linear transformation and translation with argument embeddings, one based on the intuition that predicate embedding transformations [11,20] and the other is based on the intuition of embedding translations [16].

Linear Transformation. It uses a single matrix, $\mathbf{T}$, to represent the transformation from the source goal predicate P_s to the target goal predicate P_t, and then apply the same transformation to each source predicate P_1 to search for the corresponding target predicate P_2. Formally, a transformer $\mathbf{T}$ from P_s to P_t satisfies the following condition:

$$\mathbf{P}_s \cdot \mathbf{T} = \mathbf{P}_t$$

That is, $\mathbf{T}$ can be computed through $\mathbf{T} = \mathbf{P}_s^{-1} \cdot \mathbf{P}_t$ whenever the matrix $\mathbf{P}_s$ is invertible (which is the case for all matrices we had in our experiments). If $\mathbf{P}_s$ is not invertible and a $\mathbf{T}$ cannot be computed, we simply skip this source task.

Then, for a pair of source and target predicates P_1 and P_2, their similarity degree is defined as follows:

$$sim_{LT}(P_1, P_2) = close(\mathbf{P}_1 \cdot \mathbf{T}, \mathbf{P}_2).$$

Translation with Argument Embeddings. It uses predicate embeddings as well as argument embeddings, i.e., a matrix $\mathbf{T}$ and two vectors $\mathbf{t}^{(1)}$ and $\mathbf{t}^{(2)}$, to represent the translations of respectively the predicate embedding, the subject argument embedding and the object argument embedding of the source goal predicate P_s to those of the target goal predicate P_t. For the translation, we use the additive calculus instead of the dot calculus. In particular, the three translations are as following:

$$\mathbf{P}_s + \mathbf{T} = \mathbf{P}_t \qquad \mathbf{p}_s^{(1)} + \mathbf{t}^{(1)} = \mathbf{p}_t^{(1)} \qquad \mathbf{p}_s^{(2)} + \mathbf{t}^{(2)} = \mathbf{p}_t^{(2)}$$

Here, $\mathbf{p}_s^{(1)}$ and $\mathbf{p}_s^{(2)}$ are the subject and object argument embeddings of P_s, and $\mathbf{p}_t^{(1)}$ and $\mathbf{p}_t^{(2)}$ are the subject and object argument embeddings of P_t.

Then, for a pair of source and target predicates P_1 and P_2, their similarity degree is defined as follows:

$$sim_{TAE}(P_1, P_2) = close(\mathbf{P}_1 + \mathbf{T}, \mathbf{P}_2) + close(\mathbf{p}_1^{(1)} + \mathbf{t}^{(1)}, \mathbf{p}_2^{(1)}) + close(\mathbf{p}_1^{(2)} + \mathbf{t}^{(2)}, \mathbf{p}_2^{(2)}).$$

5   Experiments

We have implemented a Transfer Rule Learner (TRL) based on the described algorithms and conducted two sets of experiments on inter- and intra-KG transfer rule learning. The datasets and detailed results can be found at https://www.ict.griffith.edu.au/aist/TRL/.

The adopted benchmark datasets for our experiments include modified versions of Freebase and YAGO. The two benchmark datasets are commonly used for rule mining and link prediction, whose specifics are in Table 1. FB15K is the same as used in [16, 20], and YAGO2s is the same as in [10].

Since existing transfer rule learners cannot handle the adopted datasets due to their lack of scalability (ref. Sect. 6 for detailed explains), we compared TRL with state-of-the-art non-transfer rule learners AMIE+ [10] and RLvLR [11], as well as statistical link predictors TransE [16] and HOLE [17]. Our experiments were designed to validate the following statements:

Table 1. Benchmark specifications

KG	# Facts	# Entities	# Predicates
FB15K	541K	15K	1345
YAGO2s	4.12M	1.65M	37

1. For small KGs (i.e., with small numbers of facts), TRL is able to learn, through inter-KG transfer, more quality rules than AMIE+ and RLvLR.
2. For sparse predicates (i.e., associated with limited numbers of facts) in large KGs, TRL can also provide, through intra-KG transfer, better accuracy in link prediction than RLvLR, TransE and HOLE.

All experiments were conducted on a desktop with Intel Core i5-4590 CPU at 3.3 GHz (one thread) and with 8 GB of RAM, running Ubuntu 14.04.

5.1   From FB15K to YAGO2s

For the first set of experiments, we evaluated inter-KG transfer rule learning from large source KGs to small target KGs. To this end, we pruned YAGO2s by eliminating entities and predicates with fewer than respectively 10 and 100 occurrences, and the pruned version contains 24 predicates. For each of these 24 predicates P_t, we constructed a target task with P_t as the goal predicate as follows: we first set apart 30% of the facts about P_t as the test data and the remaining facts as evaluation data. From the evaluation data, we extracted three subsets of YAGO2s with decreasing sizes, i.e, with 500, 200 and 100 entities, as the target KGs, through a sampling procedure similar to that in [11].

To obtain a pool of potential source tasks for transfer, we coupled FB15K with rules learnt by RLvLR. For generating source rules, other rule learner like AMIE+ can be also used, and we deployed only RLvLR for implementation convenience. For each target task, top 100 most similar source tasks were selected from the pool for transfer.

We compared TRL with AMIE+ and RLvLR, where those two learned rules directly from the target KGs. To evaluate the quality of learnt rules, we first evaluated their SC and HC over the evaluation data. Note that the evaluation data are not sparse, and thus the SC and HC scores indeed reflect the quality of the learnt rules. We recorded the numbers of rules with $SC \geq 0.1$ and $HC \geq 0.01$.

We also evaluated the quality of rules through link prediction, through two queries $P_t(e, ?)$ and $P_t(?, e')$ for each goal predicate P_t and each fact $P_t(e, e')$ from the test data. We applied learnt rules on the evaluation data to predict the missing entities (instead of the sparse data in the target KGs for all three learners) for better evaluation of the predictive power of the learnt rules. The predictions were measured using Mean Reciprocal Rank (MRR) and Hits@10, by ranking each inferred fact based on the numbers of rules inferring it [20]. MRR indicates the average of the reciprocal ranks of the missing entities and Hits@10 expresses the percentage of missing entities being ranked among the top 10.

Table 2 summarizes the results, where #E, #P and #F are respectively the numbers of entities, predicates and facts in the target KGs, #QR is the number of quality rules, and H@10 is the Hits@10 score in percentage. All the numbers are averaged over the 24 target tasks.

Table 2. Transfer rule learning from large KGs to small KGs.

#E	#P	#F	TRL			AMIE+			RLvLR		
			#QR	MRR	H@10	#QR	MRR	H@10	#QR	MRR	H@10
500	17.2	44K	**16.3**	0.18	30	0.7	0.12	19	8.7	**0.19**	**31**
200	12.6	17K	**13.6**	**0.17**	**28**	0.7	0.12	19	5.3	0.15	25
100	11.3	6K	**10.0**	**0.17**	**29**	0.2	0.08	12	1.4	0.06	12

While AMIE+ struggled on such sparse data (learning averagely fewer than a single rule), TRL managed to learn more than 10 rules on average. On relatively larger KGs with 500 entities, the predictive power of rules produced by RLvLR was comparable to those produced by TRL. Yet as the sizes of target KGs decrease, the performance of both AMIE+ and RLvLR dropped significantly, whereas that of TRL is relatively stable due to the nature of transfer learning.

Among the rules learnt by TRL, we discovered some interesting rule patterns (i.e., transferable knowledge). For instance, a large number of rules are like the following rules (with distinct predicates):

$$\text{wasBornIn}(x, z) \wedge \text{wasBornIn}(t, z) \wedge \text{livesIn}(t, y) \rightarrow \text{livesIn}(x, y),$$

which states if two persons x and t were born in the same city z then it is likely that they both live in the same city y. There are also quite a few rules stating symmetry property of predicates such as

$$\text{influences}(y, x) \rightarrow \text{influences}(x, y),$$

and those stating association between predicates such as

$$\text{isAffiliatedTo}(x, y) \rightarrow \text{playsFor}(x, y).$$

5.2   From FB15K to FB15K

For the second set of experiments, we evaluated TRL on intra-KG transfer learning, i.e., to transfer rules from rich data to sparse data within the same KG. FB15K includes a large number of predicates that are each associated with only a small number of facts, which we call sparse predicates. We consider the top 5%, 10%, 15%, and 20% most sparse predicates in FB15K from all of its 1345 predicates, and use them as goal predicates P_t. FB15K comes with its own separate training and test data [16], and we used its training data for both the target

KG and the source KG. We eliminated from the pool of source tasks those with same goal predicates as P_t to avoid trivial transfer.

Since RLvLR outperforms AMIE+ on large KGs [11], we compared TRL with RLvLR and focused on link prediction. We added two statistical link predictors (not rule learners) TransE and HOLE in the comparison, and measure the prediction only against goal predicates P_t (by removing irrelevant facts in test data). For fine-grained comparisons of prediction accuracy, we also measured the Hits@1 and Hits@3 scores, which are the proportions of correctly predicted entities that are ranked respectively, top one and top three.

Table 3 summaries the results, where %P and #P are respectively the percentage and the number of most sparse predicates as goal predicates, Time is the system time for model learning and link prediction (in hours, disregarding the times for computing source embeddings which could be done offline), and H@1, H@3, and H@10 are the Hits@1, Hits@3, and Hits@10 scores in percentage.

Table 3. Transfer rule learning from dense predicates to sparse predicates.

%P	#P	TRL					RLvLR				
		Time	MRR	H@1	H@3	H@10	Time	MRR	H@1	H@3	H@10
20%	272	**3.5**	0.18	13	23	24	6.8	0.14	9	16	21
15%	204	**2.2**	**0.17**	**13**	**20**	22	3.9	0.11	7	12	17
10%	136	**1.6**	**0.11**	9	**14**	14	2.6	0.05	2	7	9
5%	68	**1.0**	**0.11**	8	**14**	14	1.9	0.03	0	6	6
%P	#P	TransE					HOLE				
		Time	MRR	H@1	H@3	H@10	Time	MRR	H@1	H@3	H@10
20%	272	5.4	0.30	25	33	33	15.6	**0.48**	42	42	**75**
15%	204	5.3	0.02	0	0	0	15.5	0.08	0	0	**50**
10%	136	5.3	0.02	0	0	0	15.1	0.08	0	0	**50**
5%	68	5.5	0.02	0	0	0	15.3	0.08	0	0	**50**

On such sparse predicates, TRL again outperformed RLvLR regarding both prediction accuracy and time efficiency. While HOLE and TransE showed better accuracy on 20% most sparse predicates, as the sparsity increases, TRL outperforms them on Hits@3 and Hits@1 accuracy, which means top ranked predictions from TRL are more likely to be correct. Note that for top 5% sparse predicates, TRL could hit the target with its top one prediction (i.e., Hits@1) for 8% of the cases, whereas the other systems failed in all cases.

To analyse the contributions of the two similarity measures presented in Sect. 4.2, we tested three configurations of TRL on 10% most sparse predicates: with linear transformation (LT) only, with translation with argument embeddings (TAE) only, and with both. Other settings are as before, and Table 4 reports the prediction accuracy of learnt rules.

Table 4. Comparison of different similarity measures.

Similarity measures	MRR	H@1	H@3	H@10
LT	0.08	5	7	14
TAE	0.10	8	11	14
LT, TAE	**0.11**	9	14	14

Using both LT and TAE showed the best prediction accuracy, which justifies our default configuration of combining them. The performance is suboptimal when using either LT or TAE alone, while the performance of TAE alone is slightly better than that of LT alone. This demonstrates the usefulness of argument embeddings in measuring similarity.

6 Conclusion and Discussion

In this paper, based on embedding in representation and transfer learning, we have proposed a method TRL (Transfer Rule Learner) for mining first order rules in KGs with sparse data. In our approach, most similar source tasks to the target task can be retrieved and high quality mappings between predicates in source tasks and the target task can be built to reduce negative transfer. We evaluated our method TRL for the tasks of rule mining and link prediction in widely used large KGs. Our experimental results demonstrate the superior performance of TRL on sparse data over AMIE+ and RLvLR, two state-of-the-art rule learners, and over link prediction systems TransE and HOLE (although our system is not specifically design for link prediction).

We are unaware of any approach on embedding-based transfer rule learning for KGs. Transfer learning on relational data has been intensively studied in the literature, examples include [21,22]. These approaches transfer logical probabilistic models such as Markov Logic Network (MLN), which can be expressed as a form of rule languages. However, these transfer learners assume a known source domain that is similar enough to the target domain for transfer, which may work for specific domains but not for KGs with general knowledge. Our system, on the other hand, can explore the existing KGs and retrieve source tasks.

For future work, we plan to extend our system to parallel processing. We are also looking into iterated learning, where rules learned via transfer can be applied to learn new rules in the next iteration.

Acknowledgments. We would like to thank the anonymous referees for their helpful comments. This work was supported by the Australian Research Council (ARC) under DP130102302.

References

1. Bellomarini, L., Gottlob, G., Pieris, A., Sallinger, E.: Swift logic for big data and knowledge graphs. In: Proceedings of IJCAI, pp. 2–10 (2017)
2. Suchanek, F.M., Kasneci, G., Weikum, G.: YAGO: a core of semantic knowledge. In: Proceedings of WWW, pp. 697–706 (2007)
3. Auer, S., Bizer, C., Kobilarov, G., Lehmann, J., Cyganiak, R., Ives, Z.: DBpedia: a nucleus for a web of open data. In: Proceedings of ISWC, pp. 722–735 (2007)
4. Bollacker, K., Evans, C., Paritosh, P., Sturge, T., Taylor, J.: Freebase: a collaboratively created graph database for structuring human knowledge. In: Proceedings of SIGMOD, pp. 1247–1250 (2008)
5. Pujara, J., Augustine, E., Getoor, L.: Sparsity and noise: where knowledge graph embeddings fall short. In: Proceedings of EMNLP, pp. 1751–1756 (2016)
6. Nickel, M., Murphy, K., Tresp, V., Gabrilovich, E.: A review of relational machine learning for knowledge graphs. Proc. IEEE **104**(1), 11–33 (2016)
7. Barati, M., Bai, Q., Liu, Q.: SWARM: an approach for mining semantic association rules from semantic web data. In: Booth, R., Zhang, M.-L. (eds.) PRICAI 2016. LNCS (LNAI), vol. 9810, pp. 30–43. Springer, Cham (2016). https://doi.org/10.1007/978-3-319-42911-3_3
8. Wang, Z., Li, J.-Z.: RDF2Rules: learning rules from RDF knowledge bases by mining frequent predicate cycles. CoRR abs/1512.07734 (2015)
9. Chen, Y., Wang, D.Z., Goldberg, S.: ScaLeKB: scalable learning and inference over large knowledge bases. VLDB J. **25**, 893–918 (2016)
10. Galárraga, L., Teflioudi, C., Hose, K., Suchanek, F.M.: Fast rule mining in ontological knowledge bases with AMIE+. VLDB J. **24**, 707–730 (2015)
11. Omran, P.G., Wang, K., Wang, Z.: Scalable rule learning via learning representation. In: Proceedings of IJCAI, pp. 2149–2155 (2018)
12. Weiss, K., Khoshgoftaar, T.M., Wang, D.D.: A survey of transfer learning. J. Big Data **3**, 9 (2016)
13. Pan, S.J., Yang, Q.: A survey on transfer learning. TKDE **22**, 1345–1359 (2010)
14. Yang, B., Yih, W., He, X., Gao, J., Deng, L.: Embedding entities and relations for learning and inference in knowledge bases. CoRR abs/1412.6575 (2015)
15. Neelakantan, A., Roth, B., McCallum, A.: Compositional vector space models for knowledge base inference. In: Proceedings of AAAI Spring Symposia (2015)
16. Bordes, A., Usunier, N., Garcia-Duran, A., Weston, J., Yakhnenko, O.: Translating embeddings for modeling multi-relational data. In: Proceedings of NIPS, pp. 2787–2795 (2013)
17. Nickel, M., Rosasco, L., Poggio, T.: Holographic embeddings of knowledge graphs. In: Proceedings of AAAI, pp. 1955–1961 (2016)
18. Nickel, M., Tresp, V., Kriegel, H.-P.: A three-way model for collective learning on multi-relational data. In: Proceedings of ICML, pp. 809–816 (2011)
19. Zhu, H., Xie, R., Liu, Z., Sun, M.: Iterative entity alignment via knowledge embeddings. In: Proceedings of IJCAI, pp. 4258–4264 (2017)
20. Yang, F., Yang, Z., Cohen, W.W.: Differentiable learning of logical rules for knowledge base reasoning. In: Proceedings of NIPS, pp. 2316–2325 (2017)
21. Van Haaren, J., Kolobov, A., Davis, J.: TODTLER: two-order-deep transfer learning. In: Proceedings of AAAI, pp. 3007–3015 (2015)
22. Omran, P.G., Wang, K., Wang, Z.: Transfer learning in probabilistic logic models. In: Kang, B.H., Bai, Q. (eds.) AI 2016. LNCS (LNAI), vol. 9992, pp. 378–389. Springer, Cham (2016). https://doi.org/10.1007/978-3-319-50127-7_33

Knowledge Base Completion by Inference from Both Relational and Literal Facts

Zhichun Wang[1]([✉]) and Yong Huang[2]

[1] Beijing Normal University, Beijing, China
zcwang@bnu.edu.cn
[2] China UnionPay Co., Ltd., Shanghai, China
huangyong@unionpay.com

Abstract. Knowledge base (KB) completion predicts new facts in a KB by performing inference from the existing facts, which is very important for expanding KBs. Most previous KB completion approaches infer new facts only from the relational facts (facts containing object properties) in KBs. Actually, there are large number of literal facts (facts containing datatype properties) besides the relational ones in most KBs; these literal facts are ignored in the previous approaches. This paper studies how to take the literal facts into account when making inference, aiming to further improve the performance of KB completion. We propose a new approach that consumes both relational and literal facts to predict new facts. Our approach extracts literal features from literal facts, and incorporates them with path-based features extracted from relational facts; a predictive model is then trained on all the features to infer new facts. Experiments on YAGO KB show that our approach outperforms the compared approaches that only take relational facts as input.

Keywords: Knowledge base completion · Path ranking · Relational facts · Literal facts

1 Introduction

Recently, a number of large-scale knowledge bases (KBs) have been created, such as DBpedia [1], YAGO [16], and Freebase [2]. These KBs contain large amounts of facts regarding various entities, and they are becoming useful resources for many applications, such as question answering, semantic relatedness computations, and entity linking. Large-scale KBs are usually constructed automatically based on information extraction techniques. Although KBs may contain huge amounts of facts, most of them are still incomplete, missing many important facts. To add more facts to KBs, much work has been undertaken regarding KB completion, aiming to automatically infer new facts from the existing ones in KBs.

In general, existing KB completion approaches fall into two major groups: symbolic approaches and embedding approaches. Symbolic approaches use symbolic rules or relation paths to infer new facts in KBs. For example, Galárraga et

© Springer Nature Switzerland AG 2019
Q. Yang et al. (Eds.): PAKDD 2019, LNAI 11441, pp. 501–513, 2019.
https://doi.org/10.1007/978-3-030-16142-2_39

al. [5] proposed a rule mining system, AMIE, which extracts logical rules based on their support in a KB. The learned rules are then used to infer new facts in a KB. Lao et al. [11] introduced the path ranking algorithm (PRA), which uses random walks to search relation paths connecting entity pairs. These paths are then used as features in a classifier to predict new facts. Embedding approaches learn low-dimensional representations of entities and relations in KBs, which can be used to infer new facts. TransE [3] is a representative embedding approach, which learns to represent both entities and relations as vectors in $\mathbb{R}^k$. If a triple (h, r, t) holds, then TransE wants that $h + r \approx t$. After the embedding representations are learned, new facts can be predicted based on computations over embeddings. Most recently, some methods have attempted to combine symbolic and embedding techniques, including path-based TransE [12] and recurrent neural network (RNN)-based relation path composition [6].

Existing KB completion approaches of both the symbolic and embedding variety only consider relational facts in KBs when inferring new facts. Here, relational facts refer to those facts using object properties to describe relations between entities. However, in most KBs, there are literal facts as well, which describe certain datatype properties of entities, such as ages of people or areas of a city. Table 1 shows the numbers of relational facts and literal facts in three well-known KBs. This shows that the number of literal facts is close to or even bigger than the number of relational facts. We believe that such a huge number of literal facts in KBs must be useful for inferring new facts as well.

Table 1. Numbers of relational and literal facts in KBs

Knowledge base	#Relational facts	#Literal facts
YAGO	4.48M	3.35M
DBpedia	14.8M	17.3M
Freebase	1.3B	1.8B

Based on the above observations, we propose a new KB completion approach named IRL (inference from relational and literal facts). The most significant feature of our approach is its ability to extract useful information from literal facts in order to improve the KB completion performance. Our approach first finds a set of path types from the relational facts in KBs, following the same method in PRA. Then, it extracts useful features from the literal facts. Path types and literal features are combined as the input of a prediction model, which is trained to predict new facts in KBs. Experiments on Freebase and YAGO show that IRL outperforms comparable approaches that only use relational facts.

The remainder of this paper is organized as follows. Section 2 introduces the path ranking algorithm, which generates path types in our approach. Section 3 describes the proposed approach. Section 4 presents the experiment results. Section 5 discusses some related work, and finally Sect. 6 concludes this paper.

2   Background Knowledge

2.1   RDF and RDF KB

The resource description framework (RDF) is a framework for the conceptual description or modeling of information in web resources. RDF expresses information by making statements about resources in the form of

$$\langle subject \rangle \langle predicate \rangle \langle object \rangle.$$

The *subject* and *object* represent two resources, and the *predicate* represents the relationship (directional) between the *subject* and *object*. RDF statements are called triples, because they consist of three elements. RDF is a graph-based data model. A set of RDF triples constitutes an RDF graph, where nodes represent resources and directed vertices represent predicates. There can be three types of nodes (resources) in an RDF graph: IRIs, literals, and blank nodes. An IRI is a global identifier for a resource, such as people, organizations, and places. Literals are basic values, including strings, dates, and numbers. Blank nodes in RDF represent recourses without global identifiers. Predicates in RDF are also represented by IRIs, because they can be considered as resources specifying binary relations.

An RDF KB is a well-defined RDF dataset that consists of RDF statements (triples). The statements in an RDF KB are usually divided into two groups: T-box statements, which define a set of domain specific concepts and predicates; and A-box statements, which describe facts about instances of the concepts. A-box triples excluding triples with literals are employed by our approach to learn inference rules. Unlike AMIE, our approach also takes triples having *rdf:type* predicate as input. *rdf:type* is a special predicate, which is used to state that a resource is an instance of a concept. The entity-type information specified by *rdf:type* predicate is very useful and important for rule learning from RDF KBs, which is verified by our experiments.

2.2   Path Ranking Algorithm

PRA is a state-of-the-art KB completion approach, which infers new relational facts from the existing ones. In this work, we propose to extract literal features from KBs and then combine them with path types generated by PRA for KB completion. Therefore, we first briefly introduce PRA in this section.

PRA was first proposed by Lao et al. [11]. PRA predicates new relations between two entities based on a set of relation paths that connect the entity pair. A relation path is a sequence of relations $\langle r_1, r_2, ..., r_k \rangle$. Such a path can be an indicator of a new relation between entities that are linked by that path. For example, $\langle bornInCity, CityInCountry \rangle$ is a path that may indicate the *nationality* relation between the entities it connects. PRA first finds a set of potentially useful relation paths that connect the entity pairs, and then uses the discovered paths as features in a classification model to predict whether or not a specific relation holds. Formally, there are three basic steps in PRA: (1) path feature selection, (2) path feature computation, and (3) classification model training.

Path Feature Selection. Given a target relation r and a set of its instances $I_r = \{(s_j, t_j)|\langle s_j, r, t_j \rangle \in KB\}$, the total number of path types that connect the entity pairs can be considerably large. Therefore, the first step of PRA is to select a set of path types as features in the prediction model. PRA selects useful path types by performing random walks on the graph, starting at the source and target entities. If the walks from the source entities and target entities reach the same intermediate entities, then the corresponding path types are recorded, and measures of precision and recall will be computed for each path type. Path types whose precisions and recalls are not lower than predefined thresholds will be selected as path features in the prediction model.

Path Feature Computation. After a set of path features are selected, PRA computes the feature values of each entity pair in I_r. In this step, PRA will generate a feature vector for each entity pair, where each feature in the generated feature vector corresponds to a path type selected in the first step. Specifically, for an entity pair (s_j, t_j), the value of a path type π is computed as the probability of a random walk starting from s_j and arriving at t_j following the path type π, which is denoted as $p(t_j|s_j, \pi)$. In a recent extension of PRA [7], it is shown that using binary features instead of random walk probabilities improves the efficiency of PRA, and there is no statistically significant difference in the performance for KB completion.

Classifier Training. The final step of PRA is to train a classification model on the feature vectors of entity pairs in I_r. Technically, any classifier can be used in this step, but PRA simply uses a logistic regression model.

3   Proposed Approach

This section presents our proposed approach, IRL. This infers new facts using both relational and literal facts in KBs. Figure 1 illustrates the framework of our approach. To predict a relation r between the entities h and t, our approach first extracts relational features by finding path types that connect h and t, following the same method in PRA. Then, it extracts literal features from related literal facts in the KB. After the two types of features are extracted, they are merged to form the combined feature vector of the entity pair (h, t), which is then fed into a classification model for the relation r. The classification model decides whether or not the fact $\langle h, r, t \rangle$ holds. The remainder of this section introduces our proposed approach in further detail.

3.1   Extract Relational Features

Previous KB completion approaches have already investigated how to extract useful features from relational facts to infer new facts. In our approach, we

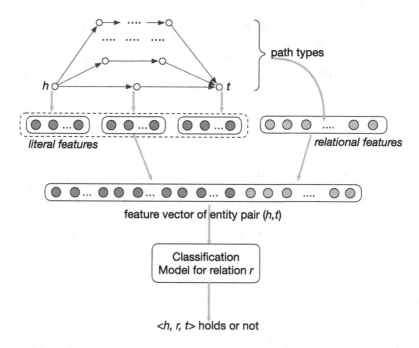

Fig. 1. Framework of our IRL approach

follow the method in PRA to selects a set of path types as features for fact prediction. Given two entities h and t, the path types that connect them are located in the KB. Instead of computing the random walk probabilities of path types as the feature values, we compute binary features for the path types, as introduced in [7]. It was reported that computing binary features costs less time, and will not decrease the performance. In the work of [7], the authors proposed a subgraph feature extraction (SFE) method, which leads to better results than PRA. Thus, we also test subgraph features as relational features in our approach. Here, we use $v_r(h, t)$ to denote the extracted relational features of an entity pair (h, t).

3.2   Extract Literal Features

To utilize literal facts when predicting relations between (h, t), our approach extracts literal features from the related literal facts. The process of extracting literal features includes the following steps:

(1) Literal Value Preprocessing

Most literal facts contain numeric values, or contain literals that can be transformed into numeric values. Before extracting features from the literal facts, our approach first preprocess the literal values contained in them. Specifically, numeralization and normalization are performed on the literal values. Numeralization transforms non-numeric literal values into numeric values.

After transforming non-numeric literal values into numeric ones, all the numeric values are then normalized to the same scale, i.e., $[0.1, 1]$. Normalization is performed for values of each datatype property p separately. Let X^p_{min} and X^p_{max} denote the minimum and maximum property values of p. Then, all the property values of p are normalized by

$$X' = 0.1 + 0.9 \frac{X - X^p_{min}}{X^p_{max} - X^p_{min}} \tag{1}$$

(2) Literal Feature Selection

To extract literal features from literal facts, we first have to determine which literal facts are useful in predicting new instances of the target relation. Intuitively, if a fact $\langle h, r, t \rangle$ is to be predicted, then literal facts for the entities h and t should be considered in the prediction model. For example, if the target relation is $hasCapital$, then the populations and areas of the entities h and t provide important clues for predicting the $hasCapital$ relation between them. In addition, if the entities h and t are connected by multiple paths, then the literal facts for the intermediate entities in these paths might also be useful. Therefore, our approach selects literal features from the literal facts of subject entities, object entities, and intermediate entities in paths that connect the subject and object entities. We use E_{sub}, E_{obj}, and E_{inter}, respectively, to denote the above three groups of entities.

Entities in a KB are described by different datatype properties, even for entities of the same type. In order to select literal features of a fixed length, we have to select commonly used datatype properties. This is done by finding frequent datatype properties for entities in E_{sub}, E_{obj}, and E_{inter}. More specifically, our approach will enumerate all the entities in E_{sub}, E_{obj}, and E_{inter}, and takes all the literal facts these entities appear in from the KB. Then, datatype properties appearing in these literal facts and their frequencies will be recorded. By setting a frequency threshold δ, our approach selects frequent datatype properties whose frequencies are no lower than the threshold as the literal features. Literal features are selected for entities in E_{sub}, E_{obj}, and E_{inter} separately. Correspondingly, the selected sets of features are denoted as L_{sub}, L_{obj}, and L_{inter}.

(3) Literal Feature Computation

Once the literal features have been selected for the target relation r, the next step is to compute their values. Given an entity pair (h, t), the values of the literal features in L_{sub} or L_{obj} are computed by directly taking the values of the datatype properties of h and t in the KB. More specifically, for a datatype property $l \in L_{sub}$, if there is a literal fact $\langle h, l, x \rangle \in KB$, then x is taken as the feature value of l. For datatype properties L_{obj}, literal facts of t are used to obtain their values. As introduced above, numeralization and normalization are performed on literal values. The values of datatype properties used here have already been preprocessed. If the value of a selected datatype property does not exist in the KB (i.e., $\langle h, l, x \rangle \notin KB$ or $\langle t, l, x \rangle \notin KB$), then the value of the literal feature l is set to 0. If there are multiple literal facts for the same datatype property of an entity, then the feature value is computed as the average of the literal values in these facts.

To compute the feature values of intermediate entities of (h, t), our approach first obtains all the intermediate entities that appear in the paths connecting h and t. These paths are discovered in the step of extracting relational features, so the intermediate entities for each entity pair will be recorded in that step, denoted by $I_{(h,t)}$. For each entity $e \in I_{(h,t)}$, a feature vector $v_{L_{inter}}(e)$ is generated by using the same method as for computing literal features of subject and object entities. After obtaining feature vectors for all the intermediate entities, our approach computes the mean vector of these vectors. The mean vector is computed by averaging over all the non-zero elements along each dimension. If the elements along one dimension in all the feature vectors are zeros, this means that none of the entities have the corresponding datatype property in the KB, and the feature value in the mean vector will be set to 0. The mean vector for all the feature vectors is taken to represent the feature values of the intermediate entities of (h, t). In this manner, feature vectors of fixed length can be computed for different entity pairs, although they may have different numbers of intermediate entities.

3.3   Prediction Model Training

Predicting new facts in a KB is considered as a classification problem in our approach. Before predicting new facts, the classification model has to be trained. Our approach trains separate classifiers for different relations. For a target relation r, let the set of training examples for this be $T_r = \{(h_i, t_i), y_i\}_{i=1}^{N}$, where (h_i, t_i) denotes an entity pair, and $y_i \in \{0, 1\}$ is the class label (0 indicates a negative example and 1 indicates a positive example). After computing the relational and literal features for every entity pair in T_r, we obtain the feature matrix $M_r = [M_{rel}\ M_{lit}]$ for the training examples, where M_{rel} is the sub-matrix of relational features and M_{lit} is the sub-matrix of literal features. The computed feature matrix will be used to train the classification model to predict new facts in the KB.

In this work, we test both logistic regression and random forest models as the classification model in our approach. Logistic regression is linear classification model, which was adopted in PRA and SFE to predict new facts. Random forest [4] is an ensemble learning method for classification and regression. We use random forest to handle the possible complicated dependencies among relational and literal features.

4   Experiments

4.1   Experimental Settings

Datasets. To evaluate our approach, we used the data from YAGO knowledge base. YAGO is built automatically from Wikipedia, GeoNames, and WordNet.

Currently, YAGO contains more than 10 million entities and more than 120 million facts about these entities. In our experiment, we generate evaluation datasets from two YAGO's datasets named *yagoFacts* and *yagoLiteralFacts*, which can be downloaded from the website of YAGO[1]. *yagoFacts* contains all the relational facts, and *yagoLiteralFacts* contains all the literal facts (except entities' labels). There are 4,484,914 relational facts describing 38 kinds of relations about entities, and there are 3,353,659 literal facts describing 35 datatype properties of entities.

To generate the evaluation datasets, we first extract positive examples for each relation from the relational facts. Given a relation r and a fact $\langle h, r, t \rangle$, the entity pair (h, t) is then taken as a positive example of the relation r. For each positive example, we generate 10 negative examples; 5 of them are generated by replacing the subject entity h with random entities, and the rest are generated by replacing the object entity t with random entities. 80% of positive examples are used for training and 20% of them are used for testing. Negative examples are split into training and testing sets according to their corresponding positive examples.

Using the above method, we generated two datasets for evaluation. The first one is generated by using all the available entity pairs of each relation as the positive examples, which is denoted as YAGO$_{all}$ here. Another one is generated by removing from YAGO$_{all}$ the entity pairs having not enough literal facts in the KB; the second dataset is denoted YAGO$_{lit}$. The reason of building YAGO$_{lit}$ is that our approach has to extract literal features from literal facts, if entities don't have enough literal facts, our approach will not perform well. So we think it's more fare to use YAGO$_{lit}$ for evaluating our approach. Every entity in YAGO$_{lit}$ has no less than 3 literal facts. In both two datasets, 32 out of 38 relations are selected as the tested relation; 6 relations are not chosen because they have few or too many positive examples[2]. Table 2 outlines some details of YAGO$_{all}$ and YAGO$_{lit}$.

Table 2. Details of evaluation datasets

	YAGO$_{all}$	YAGO$_{lit}$
# Tested relations	32	32
# Avg. train examples/relation	106, 711	63, 479
# Avg. test examples/relation	26, 557	15, 706

Evaluation Metric. We use Mean Average Precision (MAP) as the evaluation metric, which is widely used in recent work on knowledge base completion. MAP

---

[1] http://www.mpi-inf.mpg.de/departments/databases-and-information-systems/ research/yago-naga/yago/downloads/.

[2] Not tested relations: Earth, hasGender, wasBornIn, isLeaderOf, participatedIn, isAffiliatedTo.

is computed based on the ranks of positive examples according to the predictions, and it reflects both precision and recall [17].

Compared Methods. We compare our approach IRL with PRA and SFE. These two approaches achieve the state-of-art performance in KB completion; and another important reason for comparing with these two approaches is that the relational features used in our approach are computed in the same way as they do. PRA and SFE are both implemented using the code provided by Matt Gardner[3]. Our approach has two variants in the experiments, IRL_{pra} using relational features of PRA and IRL_{sfe} using relational features of SFE.

4.2 Experiment Results

Table 3 shows the results of different methods on $YAGO_{all}$ and $YAGO_{lit}$. MAP of different methods on the 32 tested relations are outlined in the table. Methods are compared in two groups, $\{PRA, IRL_{pra}\}$ and $\{SFE, IRL_{sfe}\}$; because methods in the same group use the same kind of relational features. In Table 3, numbers in bold are the single best results for relations among two compared methods. Average results of all the methods are outlined at the bottom of the table, the best average results are also in bold. Paired t-test with significance level $p < 0.05$ is performed to find whether the overall improvements of the winner methods are statistically significant. If a method with the best result significantly outperform the corresponding baseline method (PRA or SFE), a symbol "*" is marked on its average result.

The results show that the performance of KB completion is improved by taking literal features into account. On $YAGO_{all}$ dataset, IRL_{pra} performs the best among four compared methods, and its MAP is significantly better than PRA; when using the relational features from SFE, IRL_{sfe} gets higher average MAP than SFE and IRL_{sfe}, but the improvements are not significant according to the paired t-test. On $YAGO_{lit}$ dataset, IRL_{pra} and IRL_{pra} both get the highest average MAP among the compared methods; and the differences between them and their baseline methods (PRA and SFE) are both significant. IRL_{pra} gets a 5% improvement of MAP over PRA, and IRL_{sfe} gets a 6% improvement of MAP.

Based on the experiment results, we get the following observations. (1) The predicted relations between entities can be more accurate if entities' literal features are also taken as the input of the prediction model. Since the improvements of IRL-series methods on $YAGO_{lit}$ are bigger than on $YAGO_{all}$, it is obvious that more literal facts are helpful for improving the KB completion performance. (2) Using relational features from SFE leads to better results. SFE uses more expressive features than PRA does, and SFE performs better than PRA according to previous published work. So combining literal features with more expressive relational features is helpful for getting better results.

---

[3] https://github.com/matt-gardner/pra.

Table 3. KB completion results on YAGO$_{all}$ and YAGO$_{lit}$

Relation	YAGO$_{all}$				YAGO$_{lit}$			
	PRA	IRL$_{pra}$	SFE	IRL$_{sfe}$	PRA	IRL$_{pra}$	SFE	IRL$_{sfe}$
actedIn	0.2924	**0.3243**	0.3179	**0.3337**	0.2712	**0.3659**	0.3080	**0.4281**
created	0.3312	**0.3366**	0.3432	0.3397	0.7743	**0.8099**	0.8094	**0.8225**
dealsWith	0.0617	**0.0669**	0.0562	**0.0680**	0.0691	0.0527	**0.0581**	0.0524
diedIn	0.5312	**0.5369**	**0.5304**	0.5168	0.5367	**0.5520**	0.4949	**0.5270**
directed	0.7443	**0.7485**	0.8020	**0.8036**	0.8134	**0.8236**	0.8675	**0.9131**
edited	0.7627	**0.7646**	0.7712	**0.7731**	0.7089	**0.8410**	0.7091	**0.8430**
exports	0.0485	**0.1078**	0.0740	**0.1003**	0.0344	**0.0751**	0.0344	**0.0927**
graduatedFrom	0.5849	**0.6407**	0.6138	**0.6223**	0.5421	**0.6504**	0.5426	**0.6483**
happenedIn	0.6481	**0.6600**	0.6545	**0.6618**	**0.7812**	0.7808	0.7870	**0.7909**
hasAcademicAdvisor	0.8826	**0.9381**	0.9039	**0.9362**	0.8396	**0.9051**	0.9232	**0.9510**
hasCapital	0.6242	**0.6356**	0.6364	**0.6408**	**0.9189**	0.9154	0.9308	**0.9333**
hasChild	0.8187	0.8187	0.8170	**0.8205**	0.8557	**0.8591**	0.8580	**0.8666**
hasCurrency	0.3519	**0.4004**	0.3742	**0.4108**	0.6976	**0.7505**	0.7337	**0.8319**
hasMusicalRole	**0.0030**	0.0001	0.0006	**0.0012**	**0.9975**	0.9950	0.9765	**0.9975**
hasOfficialLanguage	0.3709	**0.4393**	0.4331	**0.4570**	0.5829	**0.5905**	0.5874	**0.6566**
hasWebsite	**0.0331**	0.0304	0.0301	**0.0306**	0.8500	**0.9500**	0.7500	**0.9500**
hasWonPrize	**0.1286**	0.1217	**0.1417**	0.1157	0.4783	**0.8913**	0.5000	**0.9783**
holdsPoliticalPosition	0.5247	**0.5341**	0.5381	**0.5446**	0.4016	**0.5543**	0.4815	**0.5509**
imports	0.0203	**0.0353**	**0.0415**	0.0188	0.0208	**0.0441**	**0.0271**	0.0143
influences	0.3768	**0.4215**	0.4111	**0.4181**	0.3173	**0.3770**	0.3299	**0.3878**
isCitizenOf	0.4626	**0.7720**	0.4776	**0.4844**	**0.8598**	0.8582	0.5206	**0.5376**
isConnectedTo	0.4505	**0.4599**	0.4577	**0.4660**	0.6944	**0.6964**	0.7039	**0.7082**
isInterestedIn	0.1049	**0.1705**	0.1352	**0.1405**	0.2656	**0.3750**	0.3594	**0.4063**
isKnownFor	0.2204	**0.2644**	0.2429	**0.3259**	0.4615	**0.5128**	0.6154	**0.7179**
isLocatedIn	0.5322	**0.5368**	0.5353	**0.5415**	**0.8936**	0.8920	0.9024	**0.9066**
isMarriedTo	0.3493	**0.3621**	0.3804	**0.4031**	**0.5862**	0.5065	**0.6272**	0.5522
isPoliticianOf	0.5206	**0.7285**	0.7272	**0.7337**	0.5395	**0.7474**	0.5378	**0.7621**
livesIn	0.6751	**0.6768**	0.6908	**0.6983**	0.6938	**0.6965**	0.7098	**0.7305**
owns	0.4020	**0.4079**	0.4098	**0.4109**	**0.8211**	0.8198	0.8294	**0.8312**
worksAt	0.5409	**0.5572**	0.5407	**0.5472**	0.5281	**0.5746**	0.5151	**0.5787**
wroteMusicFor	**0.7243**	0.7170	**0.7648**	0.7632	0.7730	**0.7926**	0.8445	**0.8632**
playsFor	1.0000	1.0000	1.0000	1.0000	1.0000	1.0000	1.0000	1.0000
Avg	0.4413	**0.4755***	0.4642	**0.4728**	0.6128	**0.6642***	0.6211	**0.6822***

5   Related Work

As mentioned in Sect. 1, KB completion approaches can be generally divided into two groups, symbolic approaches and embedding approaches. Symbolic approaches use symbolic rules or relation paths to infer new facts in KBs. For example, AMIE learns logic rules to infer new facts; PRA infers new facts by training classification model based on relation paths. Embedding approaches learn embeddings of entities and relations, and then predict new facts by computations over embeddings. TransE [3] is a representative embedding model, which is simple but powerful. Recently, several extensions of TransE have been proposed, including TransR [14], TransH [19], etc. Most recently, some approaches have been proposed to combine symbolic and embedding technique, aiming to get better performance in KB completion. Nickel et al. gave a comprehensive review of different kinds of KB completion approaches [15].

Most existing approaches only use the relational facts in KBs to infer new facts. Recently, there have been several approaches that utilize information other than the relational facts, but the additional information usually comes from resource outside KBs. For example, Gardner et al. proposed approaches that incorporate latent features mined from large corpus in PRA to improve the performance [8] and use vector space similarity in the random walk inference in PRA [9]; these improved approaches of PRA use extra information in texts, literal facts in KBs are not used. There are also several embedding approaches that use information from texts to improve the performance. Wang et al. proposed a method of jointly embedding entities and words into the same continuous vector space [18]. Their approach attempts to learn embeddings preserving the relations between entities in the knowledge graph and the concurrences of words in the text corpus. Approaches proposed in [10,21] also used information in text when learn the embedded representations of entities and relations.

Before our work, there have been approaches also use literal facts in KBs. In the work of SFE, Gardner et al. tested one-side feature comparisons as a kind of new features, which involves comparing datatype properties of subject and object entities. But in their work, they just computed the differences between the origin values of datatype properties shared by two entities; and the experimental results show that the performance actually drops after adding one-side feature comparisons in the model. Compared with SFE, our approach provides a more general and effective way to use literal facts in KB completion; our approach uses not only the datatype properties of subject and object entities, but also the datatype properties of intermediate entities in the paths connecting subject and object entities. Xie et al. [20] proposed an embedding approach that uses the text descriptions of entities in KBs when learns the representations of KBs; the descriptions of entities are from literal facts in KBs, but they are the only kind of literal facts used in their approach. Our work provides method to incorporate literal different kinds of facts into the relation prediction model, and to further improve the precision and recall of new predicted facts. In the work of Lin et al. [13], object properties are divided into two groups, relations and attributes. Attributes in their work are not datatype properties in the KB.

6   Conclusion

In this paper, we studied how to perform inference from both relational and literal facts for knowledge base completion. We propose a new approach IRL, which extracts relational and literal features from two kinds of facts in KBs for predicting new facts. By taking literal facts into account, IRL effectively improves the results of KB completion. Experiments on YAGO shows that our approach outperforms the compared state-of-art approaches.

Acknowledgments. The work is supported by the National Key Research and Development Program of China (No. 2017YFC0804000) and the National Natural Science Foundation of China (No. 61772079).

References

1. Bizer, C., et al.: Dbpedia-a crystallization point for the web of data. Web Semant.: Sci. Serv. Agents World Wide Web **7**(3), 154–165 (2009)
2. Bollacker, K., Evans, C., Paritosh, P., Sturge, T., Taylor, J.: Freebase: a collaboratively created graph database for structuring human knowledge. In: Proceedings of the 2008 ACM SIGMOD International Conference on Management of Data, pp. 1247–1250. ACM (2008)
3. Bordes, A., Usunier, N., Garcia-Duran, A., Weston, J., Yakhnenko, O.: Translating embeddings for modeling multi-relational data. In: Advances in Neural Information Processing Systems, pp. 2787–2795 (2013)
4. Breiman, L.: Random forests. Mach. Learn. **45**, 5–32 (2001)
5. Galárraga, L.A., Teflioudi, C., Hose, K., Suchanek, F.: AMIE: association rule mining under incomplete evidence in ontological knowledge bases. In: Proceedings of the 22nd International Conference on World Wide Web, pp. 413–422. International World Wide Web Conferences Steering Committee (2013)
6. Garcıa-Durán, A., Bordes, A., Usunier, N.: Composing relationships with translations. In: Proceedings of the Conference on Empirical Methods in Natural Language Processing (EMNLP 2015) (2015)
7. Gardner, M., Mitchell, T.: Efficient and expressive knowledge base completion using subgraph feature extraction. In: Proceedings of the 2015 Conference on Empirical Methods in Natural Language Processing, pp. 1488–1498 (2015)
8. Gardner, M., Talukdar, P.P., Kisiel, B., Mitchell, T.: Improving learning and inference in a large knowledge-base using latent syntactic cues. In: Proceedings of the 2013 Conference on Empirical Methods in Natural Language Processing, pp. 833–838 (2013)
9. Gardner, M., Talukdar, P.P., Krishnamurthy, J., Mitchell, T.: Incorporating vector space similarity in random walk inference over knowledge bases. In: Proceedings of the 2014 Conference on Empirical Methods in Natural Language Processing (2014)
10. Han, X., Liu, Z., Sun, M.: Joint representation learning of text and knowledge for knowledge graph completion. arXiv preprint arXiv:1611.04125 (2016)
11. Lao, N., Mitchell, T., Cohen, W.W.: Random walk inference and learning in a large scale knowledge base. In: Proceedings of the Conference on Empirical Methods in Natural Language Processing, EMNLP 2011, pp. 529–539. Association for Computational Linguistics, Stroudsburg, PA, USA (2011). http://dl.acm.org/citation.cfm?id=2145432.2145494
12. Lin, Y., Liu, Z., Luan, H.B., Sun, M., Rao, S., Liu, S.: Modeling relation paths for representation learning of knowledge bases. In: Proceedings of the Conference on Empirical Methods in Natural Language Processing (EMNLP 2015) (2015)
13. Lin, Y., Liu, Z., Sun, M.: Knowledge representation learning with entities, attributes and relations. In: Proceedings of the 25th International Conference on Artificial Intelligence (2016)
14. Lin, Y., Liu, Z., Sun, M., Liu, Y., Zhu, X.: Learning entity and relation embeddings for knowledge graph completion. In: Proceedings of the 29th AAAI Conference on Artificial Intelligence (AAAI 2015) (2015)

15. Nickel, M., Murphy, K., Tresp, V., Gabrilovich, E.: A review of relational machine learning for knowledge graphs. Proc. IEEE **104**(1), 11–33 (2016). https://doi.org/10.1109/JPROC.2015.2483592

16. Suchanek, F.M., Kasneci, G., Weikum, G.: YAGO: a large ontology from wikipedia and wordnet. Web Seman.: Sci. Serv. Agents World Wide Web **6**(3), 203–217 (2008)

17. Turpin, A., Scholer, F.: User performance versus precision measures for simple search tasks. In: Proceedings of the 29th Annual International ACM SIGIR Conference on Research and Development in Information Retrieval, SIGIR 2006, pp. 11–18. ACM, New York (2006). https://doi.org/10.1145/1148170.1148176

18. Wang, Z., Zhang, J., Feng, J., Chen, Z.: Knowledge graph and text jointly embedding. In: Proceedings of the 2014 Conference on Empirical Methods in Natural Language Processing, pp. 1591–1601 (2014)

19. Wang, Z., Zhang, J., Feng, J., Chen, Z.: Knowledge graph embedding by translating on hyperplanes. In: Proceedings of the 28th AAAI Conference on Artificial Intelligence (AAAI 2014), pp. 1112–1119 (2014)

20. Xie, R., Liu, Z., Jia, J., Luan, H., Sun, M.: Representation learning of knowledge graphs with entity descriptions. In: Proceedings of the 26th AAAI Conference on Artificial Intelligence (AAAI 2016) (2016)

21. Xu, J., Chen, K., Qiu, X., Huang, X.: Knowledge graph representation with jointly structural and textual encoding. arXiv preprint arXiv:1611.08661 (2016)

EMT: A Tail-Oriented Method for Specific Domain Knowledge Graph Completion

Yi Zhang, Zhijuan Du, and Xiaofeng Meng[✉]

Renmin University of China, Haidian District, Beijing 100872, China
{yizhang1208,xfmeng}@ruc.edu.cn, nmg-duzhijuan@163.com

Abstract. The basic unit of knowledge graph is triplet, including head entity, relation and tail entity. Centering on knowledge graph, knowledge graph completion has attracted more and more attention and made great progress. However, these models are all verified by open domain data sets. When applied in specific domain case, they will be challenged by practical data distributions. For example, due to poor presentation of tail entities caused by their relation-oriented feature, they can not deal with the completion of enzyme knowledge graph. Inspired by question answering and rectilinear propagation of lights, this paper puts forward a tail-oriented method - Embedding for Multi-Tails knowledge graph (EMT). Specifically, it first represents head and relation in question space; then, finishes projection to answer one by tail-related matrix; finally, gets tail entity via translating operation in answer space. To overcome time-space complexity of EMT, this paper includes two improved models: EMT^v and EMT^s. Taking some optimal translation and composition models as baselines, link prediction and triplets classification on an enzyme knowledge graph sample and Kinship proved our performance improvements, especially in tails prediction.

Keywords: Knowledge graph · Knowledge graph completion · Specific domain knowledge graph · Embedding · Tail-oriented

1 Introduction

With constantly booming, Knowledge Graph (KG) has been widely used in search engine [7], question answering and others. Before made full use of, they should be built firstly. As a necessary building part, KG completion has become

This research was partially supported by the grants from the Natural Science Foundation of China (No. 91846204, 61532010, 61532016, 61379050, 91646203, 61762082); the National Key Research and Development Program of China (No. 2016YFB1000602, 2016YFB1000603); the Fundamental Research Funds for the Central Universities, the Research Funds of Renmin University (No. 11XNL010); and the Science and Technology Opening Up Cooperation Project of Henan Province (172106000077).

© Springer Nature Switzerland AG 2019
Q. Yang et al. (Eds.): PAKDD 2019, LNAI 11441, pp. 514–527, 2019.
https://doi.org/10.1007/978-3-030-16142-2_40

a hot spot and gotten great theoretical developments in recent years. And almost all existing completing models are verified on open domain (for example, Freebase [2], YAGO [21], Nell [6] and Probase [24]) data samples, like WN18 [4] and FB15k [4], whose data distributions are different from specific domain ones, such as bioinformatics [1], life science [15] and biomedicine [19]. What if existing models are applied on specific domain?

As Table 1 shows, heads and tails number in open domain KGs, like WN18 [4] and FB15k [4], are nearly equal. But it is not true for specific domain. For example, the Nations [18] describes social changes from 1950 to 1965. To exclude time influence, we sampled data in 1965, getting the Nations65, showing one head has about 5 tails on average. In microbiology Enzyme KG (EnzymeKG), it is about 1,002. For typicality, we will take EnzymeKG as a case to analyze.

Table 1. Open v.s. specific domain KGs.

Data sets	#Ent.	#Head	#Tail	Ratio
WN18	40,943	40,940	40,939	1
FB15k	14,951	14,866	14,913	1
Nations65	1,106	182	924	5
EnzymeKG	6,482,370	6,463	6,475,907	1,002

Ratio means #Tail to #Head.

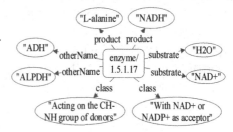

Fig. 1. An EnzymeKG illustration.

Taking head *enzyme/1.5.1.17* as an example, Fig. 1 shows that same head can be linked to various tails by same relation. Furthermore, based on EnzymeKG, Table 2 statistics the proportion of each relation; the largest, smallest and average #Head (#Tail) when sharing same tail (head) and relation. Largest #Head of *type* and *class* are 6,436 and 1,840, much larger than others. Things are totally different for tails: *x-gene*, *ncbiGene* and *keggGene* are respectively 31,809, 39,051 and 40,635. And proportions of these three relations can make about 98.58%. So, it is reasonable to say they can represent most triplets in EnzymeKG.

As the most widely recognized completing genre, embedding representation includes translation, composition and neural network. From Fig. 1, we can see heads and tails are not invertible in EnzymeKG, conflicting with similar operations on heads and tails in composition models. Neural network models have many parameters to learn, bringing high cost. So we focus on translation ones.

TransE [4] is a classical translation model. Taking it for an example, we did entities prediction on an EnzymeKG sample, *filt Hit@10* was 57.2%. However, TransE on WN18 can make 89.2% [4]. Combining with the above data distribution features, we found following reasons. By training embeddings **h**, **r** and **t** for head, relation and tail, TransE fits triplet $<h, r, t>$ by $\mathbf{h} + \mathbf{r} \approx \mathbf{t}$. When one head connects only one tail by one relation, it performs well. However, as Fig. 2 shows, it is not good at handling the situation that one head links 3 tails

Table 2. Data distributions of EnzymeKG.

Relations	Proportion	#Head	#Tail
		Largest/Smallest/Avg. (N_h)	Largest/Smallest/Avg. (N_t)
sysname	0.07%	11/1/1.0	3/1/1.0
description	0.08%	37/1/1.0	1/1/1.0
name	0.09%	298/1/1.1	1/1/1.0
history	0.09%	1/1/1.0	1/1/1.0
type	0.09%	6,436/6,436/6,436.0	1/1/1.0
x-pathway	0.10%	1,497/1/46.1	21/1/2.4
substrate	0.16%	1,129/1/3.0	8/1/2.3
product	0.17%	702/1/2.9	9/1/2.4
class	0.28%	1,840/1/271.0	3/2/3.0
otherName	0.28%	5/1/1.0	206/1/4.1
x-gene	12.88%	5/1/1.1	31,809/1/275.5
ncbiGene	39.36%	5/1/1.1	39,051/1/821.8
keggGene	46.34%	5/1/1.1	40,635/1/966.3

by same relation. Because these 3 tails can be very close and even overlap each other to meet $\mathbf{h} + \mathbf{r} \approx \mathbf{t}$, influencing semantic expression.

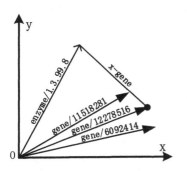

Fig. 2. Basic idea of TransE.

Table 3. Mathematical notations.

Notations	Descriptions
$h/r/t$	Head entity/relation/tail entity
$\langle h, r, t \rangle$	A triplet with h, r and t
$\mathbf{h}/\mathbf{r}/\mathbf{t}$	Embeddings of $h/r/t$
$f_{r/t}(\mathbf{h}, \mathbf{t}/\mathbf{r})$	Score function
$\| \|_{L_1/L_2}^2$	First or second order distance

So, it is necessary to deal with specific domain KG completion by new models. Based on the above analysis, its key is to distinguish tails for same head-relation. Different from traditional relation-oriented models, inspired by question answering and rectilinear propagation of lights, we put forward a tail-oriented model, i.e. first embedding head and relation in question space; then finishing projection to answer one by tail-related matrix; and finally getting corresponding tail by translation.

Contributions of this paper are listed as follows:

- A tail-oriented completing method: EMT. It distinguishes multiple tails for same head-relation with various projecting matrices. Experiments proved its superiority over translation and composition models.
- Improved EMT: EMT^v and EMT^s. Instead of dense matrix $\mathbf{M}_t$ and matrix-vector multiplication in EMT with vectors multiplication and sparse matrix respectively, EMT^v and EMT^s achieve a performance-complexity balance.

The rest is structured as follows. Related work is discussed in Sect. 2. After fundamental definitions, Sect. 3 elaborates EMT method. Then Sect. 4 demonstrates experimental results. Finally, Sect. 5 concludes this paper.

2   Related Work

To be clear, we define some general notations in Table 3. Others will be defined near their first use. Generally, KG completing models contain translation, composition and neural network.

TransE [4], a typical translation model, believes $\mathbf{h} + \mathbf{r} \approx \mathbf{t}$. Considering relation properties, TransH [22] distinguishes reflexive, 1-N, N-1 and N-N relations [4] by a hyperplane. But an entity may have many aspects. And relations may focus on any one of them. TransR [12] gets projected entities in relation space by $\mathbf{M}_r\mathbf{h}$ and $\mathbf{M}_r\mathbf{t}$. TransD [9] and Sparse [10] are TransR variants.

As a composition model, RESCAL [17] denotes triplets as a three-way tensor χ, where an entry χ_{ijk} represents existing triplets as 1; otherwise, 0. Based on relation, χ is sliced. For k-th relation, slice $\chi_k \approx \mathbf{A}\mathbf{R}_k\mathbf{A}^T$. Specifically, $\mathbf{A}$ is a $n \times r$ matrix containing entity embeddings, $\mathbf{R}_k$ is an asymmetric $r \times r$ matrix modeling the interactions of entities linked by k-th relation. By modeling relation as bilinear matrix $\mathbf{M}_r$, score function of LFM [8] is $f_r(\mathbf{h}, \mathbf{t}) = \mathbf{h}^T\mathbf{M}_r\mathbf{t}$. Making $\mathbf{M}_r$ be a diagonal matrix, DistMult [25] reduces complexity and improves performance. HolE [16] captures rich heads and tails interactions by circular correlation, while Complex [23] replaces real embeddings with complex ones. By considering analogical properties of entities and relations, ANALOGY [13] is an integration of DistMult, HolE and Complex.

Compared with SE [5], SLM [20] describes entity-relation semantic links by nonlinear operations. But it costs more by score function $f_r(\mathbf{h}, \mathbf{t}) = \mathbf{u}_r^T g(\mathbf{M}_{r,1}\mathbf{l}_h + \mathbf{M}_{r,2}\mathbf{l}_t)$. With higher cost, NTN [20] connects entity vectors across multiple dimensions by bilinear tensor layer, getting corrected probabilities of triplets, while SME [3] links entity and relation by multiple projecting matrices. Differently, including DNN and RMNN, NAM [14] models entities connection as a conditional probability via multilayer nonlinear activations in deep neural nets.

Therefore, translation models finish translating operations from head to tail via relation. Composition ones describe heads and tails from the view of relations. As for neural network, it focuses on entity-relation interactions. Here, we call them **relation-oriented** models that are indeed powerful in relations prediction, but not necessarily for heads and tails one. Specifically, as mentioned in Sect. 1, they can not distinguish multiple tails for same head-relation. Besides, composition and neural network models, especially the later, often bring high computing cost.

3   EMT Method

Before modeling, we first abstract the phenomenon that one head links multiple tails via same relation. Let N_h and N_t be the average #Head and #Tail in Table 2. Then, taking δ as boundary, [4] divides triplets into *1-1*, *1-N*, *N-1* and *N-N* by fixing only relation, especially *N-N*. However, KG completion aims to identify the third one of a triplet via referring to the other two. So, fixing only one element can lead to distraction. Moreover, Sect. 1 shows tails share same head-relation in Nations65 and EnzymeKG. Therefore, it is necessary to make new categorizations by fixing heads (tails) and relations.

Considering N_h, N_t and δ, we can get new categorizations by

$$N_h < \delta \leftrightarrow \textit{1-}\underline{\textit{1-1}} \quad N_h \geq \delta \leftrightarrow \textit{N-}\underline{\textit{1-1}} \quad N_t < \delta \leftrightarrow \underline{\textit{1-1}}\textit{-1} \quad N_t \geq \delta \leftrightarrow \underline{\textit{1-1}}\textit{-N}, \quad (1)$$

where underline parts are the two given elements. Here makes $\delta = 1.5$. Mapping of [4] and ours is shown in Fig. 3, including 3 complete sets: {*1-1*, *1-N*, *N-1*, *N-N*}, {*1-*$\underline{\textit{1-1}}$, *N-*$\underline{\textit{1-1}}$} and {$\underline{\textit{1-1}}$*-1*, $\underline{\textit{1-1}}$*-N*}.

With more *1-*$\underline{\textit{1-N}}$ triplets than *N-*$\underline{\textit{1-1}}$ ones, KG like EnzymeKG (6,986,933 v.s. 53,944) is **Multi-Tails KG (MTKG)**. Otherwise, Multi-Heads KG (MHKG). Since our data sets are typical MTKGs, this paper will focus on **MTKG completion**, i.e. representation learning of $\underline{\textit{1-1}}$*-1* and $\underline{\textit{1-1}}$*-N* triplets. According to Sect. 1, existing models can deal with $\underline{\textit{1-1}}$*-1* well. So our problem can be narrowed to **the representation learning of $\underline{\textit{1-1}}$-N**.

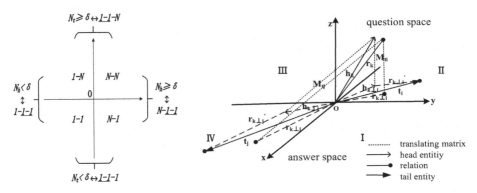

Fig. 3. Triplets categorizations.

Fig. 4. A simple EMT illustration.

3.1   Embedding for MTKG

As mentioned above, the crucial point of representing *1-1-N* is to find the most suitable tail for given head-relation. It is similar to the problem of picking out the wanted one among many close answers for a fixed question. So, by regarding head-relation as question, while the wanted tail as the answer, we construct a model for *1-1-N* triplets, called **Embedding for MTKG (EMT)**.

Figure 4 details EMT. Divided into four quadrants by x and y axes, plane xOy is answer space, above which is question one. There are two triplets $<h_k, r_k, t_i>$ and $<h_k, r_k, t_j>$ in this figure. Located at question space, head and relation embeddings $\mathbf{h}_k$, $\mathbf{r}_k \in \mathbb{R}^n$ start from the origin point. According to geometrical optics, light propagates along straight lines in homogeneous medium. So things get different shadows when irradiated by parallel rays from various directions. Projecting matrices $\mathbf{M}_{ti}$ and $\mathbf{M}_{tj} \in \mathbb{R}^{m \times n}$ in Fig. 4 play the role of parallel rays. Here considers $<h_k, r_k, t_i>$. Based on

$$\mathbf{h}_\perp = \mathbf{M}_t \mathbf{h} \quad \mathbf{r}_\perp = \mathbf{M}_t \mathbf{r}, \tag{2}$$

model gets projected head and relation $\mathbf{h}_{k \perp i}$ and $\mathbf{r}_{k \perp i}$ to predict tail t_i in quadrant II of answer space. Then, $\mathbf{t}_i$ can be obtained by $\mathbf{h}_{k \perp i} + \mathbf{r}_{k \perp i}$, during which $\mathbf{r}_{k \perp i}'$ is produced by moving $\mathbf{r}_{k \perp i}$ parallelly. Similarly, $\mathbf{t}_j$ is born in quadrant IV.

Corresponding score function is

$$f_t(\mathbf{h}, \mathbf{r}) = ||\mathbf{h}_\perp + \mathbf{r}_\perp - \mathbf{t}||^2_{L_1/L_2}. \tag{3}$$

Constraints added in experiments are $||\mathbf{h}||^2_{L_1/L_2}$, $||\mathbf{r}||^2_{L_1/L_2}$, $||\mathbf{t}||^2_{L_1/L_2}$, $||\mathbf{h}_\perp||^2_{L_1/L_2}$ and $||\mathbf{r}_\perp||^2_{L_1/L_2} \leq 1$.

Figure 4 shows that though starting from the origin point like $\mathbf{h}_k$ and $\mathbf{r}_k$, various tail embeddings, i.e. $\mathbf{t}_i$ and $\mathbf{t}_j$, have different directions and lengths. Since projecting heads and relations to tail space, EMT is a **tail-oriented** model.

3.2   EMT by Vectors Multiplication

According to Eq. (2), EMT includes a dense matrix and two matrix-vector multiplications, leading to a bottleneck when completing large-scale KGs. By replacing the dense matrix with vectors multiplication, we put forward **EMT by Vectors Multiplication ($\mathbf{EMT}^v$)**.

It respectively represents head, relation and tail with 2 vectors. One is for semantic ($\mathbf{h}$, $\mathbf{r}$ and $\mathbf{t}$). The other one is for projecting ($\mathbf{h}_p$, $\mathbf{r}_p$ and $\mathbf{t}_p$). To distinguish heads and relations, their projecting matrices are constructed separately by

$$\mathbf{M}_t^h = \mathbf{t}_p \mathbf{h}_p^T + \mathbf{I}^{m \times n} \quad \mathbf{M}_t^r = \mathbf{t}_p \mathbf{r}_p^T + \mathbf{I}^{m \times n}, \tag{4}$$

where $\mathbf{M}_t^h$, $\mathbf{M}_t^r \in \mathbb{R}^{m \times n}$. Like Eq. (2), projected $\mathbf{h}$ and $\mathbf{r}$ are

$$\mathbf{h}_\perp = \mathbf{M}_t^h \mathbf{h} \quad \mathbf{r}_\perp = \mathbf{M}_t^r \mathbf{r}. \tag{5}$$

After replacing matrices in Eq. (5) with Eq. (4), we can get

$$\begin{aligned}
\mathbf{h}_\perp &= t_p \mathbf{h}_p^T \mathbf{h} + \mathbf{I}^{m \times n} \times \mathbf{h} \\
\mathbf{r}_\perp &= t_p \mathbf{r}_p^T \mathbf{r} + \mathbf{I}^{m \times n} \times \mathbf{r},
\end{aligned} \Rightarrow \begin{aligned}
\mathbf{h}_\perp &= t_p(\mathbf{h}_p^T \mathbf{h}) + \mathbf{I}^{m \times n} \times \mathbf{h} \\
\mathbf{r}_\perp &= t_p(\mathbf{r}_p^T \mathbf{r}) + \mathbf{I}^{m \times n} \times \mathbf{r},
\end{aligned} \tag{6}$$

where $\mathbf{h}_p^T \mathbf{h}$ and $\mathbf{r}_p^T \mathbf{r}$ are dot products, getting scalars. Then $t_p(\mathbf{h}_p^T \mathbf{h})$ and $t_p(\mathbf{r}_p^T \mathbf{r})$ are scalar products. In other words, EMT^v replaces original matrix-vector multiplication with dot and scalar products, reducing time-space complexity. With same score function as EMT, its constraints added in experiments are $||\mathbf{h}||_{L_1/L_2}^2$, $||\mathbf{r}||_{L_1/L_2}^2$, $||\mathbf{t}||_{L_1/L_2}^2$, $||\mathbf{h}_p||_{L_1/L_2}^2$, $||\mathbf{r}_p||_{L_1/L_2}^2$, $||\mathbf{t}_p||_{L_1/L_2}^2$, $||\mathbf{h}_\perp||_{L_1/L_2}^2$ and $||\mathbf{r}_\perp||_{L_1/L_2}^2 \leq 1$.

3.3   EMT by Sparse Matrix

EMT^v gets projecting matrices by vectors multiplication. According to Sylvester's Inequality, if $\mathbf{A}$ is $m \times n$ and $\mathbf{B}$ is $n \times r$, then $rank(\mathbf{AB}) \leq min\{rank(\mathbf{A}), rank(\mathbf{B})\}$. So projecting matrices in EMT^v are low-rank ones with $rank \leq 1$, reducing expressiveness. As for sparse matrix, some of its elements are made zeros and others not. Zero elements are kept and non-zero ones are updated during training. Therefore, its values are more free, corresponding model is more expressive than EMT^v. Next will focus on **EMT by Sparse Matrix (EMT^s)**.

More complex a tail is, more parameters it should have. We think that semantic complexity of a tail is related to corresponding numbers of head-relation (h-r) pairs and relations. So EMT^s constructs two corresponding sparse matrices $\mathbf{M}_t^{hr}(\theta_t^{hr})$ and $\mathbf{M}_t^r(\theta_t^r)$ for each tail, where θ_t^{hr} and θ_t^r are sparse degrees, i.e. proportions of zero elements. For a given tail t, N_t^{hr} is corresponding h-r pairs number. $N_{t^*}^{(hr)^*}$ is the max one of N_t^{hr}, i.e. tail t^* links most h-r pairs at $(hr)^*$. So do N_t^r and $N_{t^*}^{r^*}$. Given min sparse degrees of projecting matrices θ_{min}^{hr}, $\theta_{min}^r \in (0,1]$, then final sparse degrees are

$$\theta_t^{hr} = \theta_{min}^{hr} N_t^{hr}/N_{t^*}^{(hr)^*} \quad \theta_t^r = \theta_{min}^r N_t^r/N_{t^*}^{r^*}. \tag{7}$$

Projected heads and relations can be computed by

$$\mathbf{h}_\perp = \mathbf{M}_t^{hr}(\theta_t^{hr})\mathbf{h} \quad \mathbf{r}_\perp = \mathbf{M}_t^r(\theta_t^r)\mathbf{r}. \tag{8}$$

With same score function as EMT, its constraints added in experiments are $||\mathbf{h}||_{L_1/L_2}^2$, $||\mathbf{r}||_{L_1/L_2}^2$, $||\mathbf{t}||_{L_1/L_2}^2$, $||\mathbf{h}_\perp||_{L_1/L_2}^2$ and $||\mathbf{r}_\perp||_{L_1/L_2}^2 \leq 1$.

3.4   Time-Space Complexity Analysis

Up to now, EMT method has been introduced. As you can see, there are dense matrices and matrix-vector multiplications in EMT. Based on this fact, EMT^v and EMT^s are put forward to deal with its time-space complexity problem. So this subsection will analyze time-space complexity of these three models.

This paper defines time complexity by multiplication operating times in an epoch and space one by parameters number. Specifically, N, N_r, N_t and N_e are respectively numbers of triplets, relations, tail entities and entities. Besides, m is the dimension of head and relation vectors and n is for tail. In EMT^s, θ_{avg} represents the average sparse degree of all projecting matrices.

Table 4. Time-space complexity analysis on EMT method

Model	Time complexity	Space complexity
EMT	$O(2mnN)$	$O((N_e + N_r)m + N_t mn)$
EMT^v	$O(2nN)$	$O(2(N_e + N_r)m + 2N_t n)$
$EMT^s (0 \ll \theta_{avg} \leq 1)$	$O(2(1 - \theta_{avg})mnN)$	$O((N_e + N_r)m + 2N_t(1 - \theta_{avg})mn)$

According to Table 4, we can see that EMT^v has lowest time and space cost. Besides, compared with EMT, EMT^v and EMT^s have some improvements. As explained before, our models are tail-oriented. That is to say, all of them project heads and relations to tail space. We do cost more than relation-oriented model. Because, generally, tail entities are much more than relations in KGs. However, in reality, performance can be as important as cost. Exactly, our models can improve performance a lot, which will be introduced in Sect. 4.

3.5 Training

Assuming that a train set includes s triplets, where the i-th one is $<h_i, r_i, t_i>$ $(i = 1, \ldots, s)$. Each of them has a y_i to denote it is positive ($y_i = 1$) or not ($y_i = 0$). Positive and negative examples are represented as Δ and Δ'. But our data only contain positive samples. Based on *bern* [22], negative ones were produced by replacing head or tail with different probabilities.

According to margin-based ranking loss, shared loss function is defined as

$$L = \sum_{<h,r,t> \in \Delta} \sum_{<h',r,t'> \in \Delta'} [f_t(\mathbf{h}, \mathbf{r}) + \gamma - f_{t'}(\mathbf{h'}, \mathbf{r})]_+, \tag{9}$$

where $[x]_+ \overset{\Delta}{=} max(0, x)$, γ is the separating margin of positive and negative examples. Objective function is optimized by mini-batch SGD. To exclude initialization influence, we initiate all vectors randomly, matrices by identity ones.

With more complicated training process, EMT^s will be detailed. When constructing sparse matrices, non-zero elements distribute symmetrically along the diagonal direction, and corresponding locations are stored. Based on mini-batch SGD, model updates these values by indexing locations during training.

4 Experiments

Like NTN [20], almost all neural network models have many parameters to learn, leading to high cost. So five translation models mentioned in Sect. 2

and four better composition ones (DistMult [25], HolE [16], Complex [23] and ANALOGY [13]) will be compared with ours via link prediction [5] and triplets classification [20].

4.1 Data Sets

Taking EnzymeKG as a case, we model EMT to deal with MTKG completion. For simplicity, here considered an EnzymeKG Sample (ES). Firstly, *description*, *history* and *type* were deleted from Table 2. Because tails linked by *description* and *history* are almost all long text. It is more reasonable to use them as auxiliary information. As for *type*, its heads are all enzymes, and corresponding tails are *EnzymeNode*. So it is not necessary to consider *type*. Then, to guarantee entities in valid and test data have been trained, we removed entities appearing only once, randomly sampled twice ones and kept those appearing more than twice.

Kinship [11] is another specific domain experimental data set, but it is a common one. Due to limiting space, we only demonstrate results of these twos. Their final statistics are shown in Table 5 where #*1-1-1* and #*1-1-N* are corresponding triplets number in test set.

Table 5. Statistics of experimental data sets.

Data sets	#Rel	#Ent	#Train	#Valid	#Test	#*1-1-1*	#*1-1-N*
ES	10	57,066	155,417	5,000	5,000	227	4,773
Kinship	25	104	6,411	2,137	2,138	14	2,124

4.2 Link Prediction

Usually, link prediction identifies missing heads and tails. Here added relations prediction. As previous works [4,9,10,12,22] did, this task had two evaluating protocols, i.e. *Mean Rank* (*MR*) and *Hit@k* (%). Lower *MR* and higher *Hit@k* mean better performance. Dealing with ES, we made $k = 10$ for heads and tails prediction. With only 10 relations totally, $k = 1$ was taken for relations one. As for Kinship, $k = 10$ for all. Besides, like [4], we reported both *raw* and *filt* results.

To be fair, we took all embedding dimensions as 20, and adopted mini-batch 100, max training epochs 1,000. Based on related papers of baselines, we adjusted other optimal parameters (like learning rate) of open domain KGs for experimental ones. As for ours, we selected learning rate λ in $\{0.1, 0.01, 0.001, 0.0001\}$, margin γ in $\{1, 1.5, 2, 2.5, 3, 3.5, 4, 4.5, 5, 5.5, 6\}$ and $\theta_{\min}^{hr} = \theta_{\min}^{r}$ in $\{0.1, 0.3, 0.5, 0.7, 0.9\}$ via grid searching. Then, we found best configurations for ES: $\lambda = 0.01$, $\gamma = 2$ for EMT; $\lambda = 0.001$, $\gamma = 2$ for EMT^v; $\theta_{\min}^{hr} = \theta_{\min}^{r} = 0.3$, $\lambda = 0.001$, $\gamma = 1$ for EMT^s. Similarly, best configurations for Kinship were $\lambda = 0.001$, $\gamma = 3.5$ for EMT; $\lambda = 0.001$, $\gamma = 1$ for EMT^v; $\theta_{\min}^{hr} = \theta_{\min}^{r} = 0.7$, $\lambda = 0.01$, $\gamma = 2$ for EMT^s.

Table 6. Link prediction results by heads, tails and relations.

Data sets	Models	Heads prediction		Tails prediction		Relations prediction	
		MR	Hit@10(%)	MR	Hit@10(%)	MR	Hit@1/10(%)
		raw/filt	raw/filt	raw/filt	raw/filt	raw/filt	raw/filt
ES	TransE	21/18	84.7/85.9	2,531/1,929	5.7/28.5	1/1	92.1/92.1
	TransH	25/22	86.6/87.4	1,726/1,114	5.4/49.6	1/1	93.5/93.5
	TransR	26/24	89.0/90.0	820/202	5.2/36.8	1/1	98.3/98.3
	TransD	38/35	96.2/96.5	728/137	7.3/76.1	1/1	92.4/93.7
	TranSparse	14/11	95.2/95.7	731/117	7.0/50.6	1/1	94.6/94.6
	DistMult	52/49	93.9/94.2	758/130	7.8/82.3	1/1	98.4/98.7
	HolE	17/11	94.2/96.4	732/124	9.0/83.0	1/1	99.5/99.8
	Complex	29/26	94.0/94.1	741/128	8.6/83.9	1/1	99.6/99.7
	ANALOGY	16/13	94.5/94.7	735/126	8.9/84.8	1/1	**99.9/99.9**
	EMT	**5/2**	**97.4/98.6**	621/11	**9.7**/86.7	1/1	99.7/99.7
	EMT^v	12/9	96.7/97.1	628/16	9.1/83.9	1/1	98.0/98.0
	EMT^s	10/7	96.7/97.7	**620**/21	8.8/**86.9**	1/1	99.3/99.3
Kinship	TransE	23/19	40.2/51.3	27/21	33.6/46.0	5/5	84.8/84.8
	TransH	20/16	42.9/56.1	22/17	35.5/53.3	4/4	91.4/91.4
	TransR	14/9	52.0/76.1	16/10	44.7/72.1	3/3	94.3/94.3
	TransD	15/8	51.5/65.2	13/12	45.9/64.4	3/3	92.7/92.7
	TranSparse	**9/5**	69.4/**90.2**	**11/5**	**57.7**/88.5	2/2	98.4/98.4
	DistMult	16/8	65.6/88.7	12/6	51.6/84.6	**2/2**	**98.6/98.8**
	HolE	14/7	66.6/87.3	15/7	56.3/88.6	3/3	97.6/97.6
	Complex	11/6	71.1/88.8	13/6	56.3/86.3	2/2	98.4/98.4
	ANALOGY	10/4	**74.1**/88.9	12/5	57.4/87.4	2/2	98.3/98.3
	EMT	11/5	62.6/87.4	12/**5**	53.7/**89.9**	2/2	97.9/97.9
	EMT^v	13/8	55.6/76.8	15/9	45.9/74.1	3/3	95.3/95.3
	EMT^s	12/7	56.3/82.9	13/6	51.9/85.1	2/2	97.8/97.8

Link prediction results are shown in Tables 6 and 7, where bold ones are the best results under given conditions. Overall, our models had obvious advantages over translation ones, and slight advantages over composition ones, especially in tails prediction of ES. It is because that ours fit MTKG very well. Distinctions between translation and composition ones are caused by richer interaction description among entities and relations, like circular correlation, in composition models. Among ours, EMT performed best, EMT^s took the second place, according with theoretical analysis on low-rank and sparse matrices in Sect. 3.3.

Table 6 shows that, in ES, *raw* results were close to *filt* ones in heads and relations prediction. However, when predicting tails, *raw MR* were over 600, but *filt* ones even can be 11. In terms of *Hit@10*, *raw* values were lower than 10%, while *filt* ones even near to 90%. According to [4], the only difference between *raw* and *filt* is whether deleting corrupted triplets in train, valid and test or not. Like heads and relations prediction, tails one replaced the linked tail by any other entities. But tails dominated entities. Moreover, same head and relation can have various tails. So when predicting tails, probability of deleting corrupted triplets was higher, leading to more distinctions between *raw* and *filt* values.

According to Table 6, ours outperformed translation models, especially *MR* of heads prediction and both metrics of tails one, on ES. Among translation models, overall, Trans R, D and Sparse were better than Trans E and H. Taking matrix to translate heads and tails, TransR is more expressive. As for TransD, its translating matrices are determined by both entities and relations. TranSparse believes that the semantic complexity of a relation is related to corresponding head and tail numbers. So these three models had better performance. As for composition ones, various models had similar performance.

In Table 6, dealing with Kinship, TranSparse was the best one among all models in some metrics, like *raw MR* and *filt Hit@10* of heads prediction. Compared with TranSparse, training a matrix for each tail, EMT had more parameters. However, according to Table 5, Kinship only has 6,411 training triplets, is about 23 times less than ES. With too many parameters to learn enough, EMT did not get best performance in heads prediction. Even though, it still had best tails prediction in *filt MR* and *Hit@10*, proving strengths of tail-oriented models. Besides, among our models, EMT^v did not perform well as the other two. Getting corresponding projecting matrices by vectors multiplication, EMT^v uses low-rank matrices to finish projecting process, limiting its expressiveness.

Table 7. Link prediction results by relations.

Data sets	ES				Kinship			
filt Hit@10 (%)	Heads prediction		Tails prediction		Heads prediction		Tails prediction	
Relations	*1-1-1*	*1-1-N*	*1-1-1*	*1-1-N*	*1-1-1*	*1-1-N*	*1-1-1*	*1-1-N*
TransE	22.9	88.9	79.3	26.0	57.1	51.3	50.0	46.0
TransH	18.9	90.7	75.3	48.4	57.1	56.1	71.4	53.2
TransR	19.4	93.4	71.8	35.2	50.0	76.3	57.1	72.3
TransD	31.3	99.6	61.2	76.8	50.0	65.3	42.9	64.5
TranSparse	30.0	98.8	85.9	48.9	78.6	**90.3**	78.6	88.6
DistMult	58.1	95.9	85.5	82.1	78.6	88.8	71.4	84.7
HolE	60.8	98.1	88.5	82.7	78.6	88.7	**85.7**	88.7
Complex	64.8	95.5	86.8	83.8	**85.7**	88.8	42.9	86.5
ANALOGY	**88.1**	95.0	87.2	84.7	**85.7**	88.9	50.0	87.6
EMT	69.6	**100.0**	**100.0**	86.1	71.4	87.5	71.4	**90.0**
EMT^v	35.7	**100.0**	90.7	83.6	64.3	76.9	50.0	74.2
EMT^s	54.6	99.7	98.2	**86.3**	82.6	82.9	84.6	85.2

Focusing on MTKG completion, furthermore, representation learning of *1-1-N*, different from [4], referring to *filt Hit@10*, Table 7 groups link predication into *1-1-1* and *1-1-N*. It shows that our models outperformed others on ES in *1-1-N*. Specifically, EMT^s got the best tails prediction of *1-1-N*, proving its superiority. As for the Kinship, it is similar to Table 6. With too many parameters to learn enough, EMT only got best performance in tails predication for *1-1-N* triplets.

By considering KG completion as a question answering problem, with tail-related projecting matrices, we project heads and relations from question space to answer one where tails are. Compared with baselines, ours had better performance in nearly all metrics on ES and tails prediction on Kinship. Therefore, tail-oriented models are good at handling MTKG.

4.3 Triplets Classification

According to [20], triplets classification is a binary classification problem, judging triplets are positive or not. Experimental details, including the metric (accuracy), in this paper were as same as those in [20].

With same baselines and similar parameters optimizing process, triplets classification was carried out like link prediction. Specifically, to be fair, we took all embedding dimensions as 20, adopted mini-batch 100 and max training epochs 1,000. Also, the gird searching were same: λ in $\{0.1, 0.01, 0.001, 0.0001\}$, γ in $\{1, 1.5, 2, 2.5, 3, 3.5, 4, 4.5, 5, 5.5, 6\}$ and $\theta^{hr}_{\min} = \theta^r_{\min}$ in $\{0.1, 0.3, 0.5, 0.7, 0.9\}$.

After optimized, we found best configurations for ES: $\lambda = 0.01$, $\gamma = 2.5$ for EMT; $\lambda = 0.01$, $\gamma = 1.5$ for EMT^v; $\theta^{hr}_{\min} = \theta^r_{\min} = 0.7$, $\lambda = 0.01$, $\gamma = 2$ for EMT^s. As for Kinship, best configurations were: $\lambda = 0.001$, $\gamma = 2$ for EMT; $\lambda = 0.01$, $\gamma = 2$ for EMT^v; $\theta^{hr}_{\min} = \theta^r_{\min} = 0.9$, $\lambda = 0.0001$, $\gamma = 2$ for EMT^s.

Table 8. Triplets classification results (%).

Models	TransE/H/R/D/Sparse	DistMult/HolE/Complex/ANALOGY	EMT/EMT^v/EMT^s
ES	87.34/89.98/75.89/96.29/94.64	93.31/93.8/93.36/93.58	94.03/96.32/**97.20**
Kinship	66.77/70.95/66.46/60.24/**72.82**	67.50/70.12/69.97/68.11	69.76/71.66/64.5

Triplets classification results are shown in Table 8 where bold results are the best one under given conditions. With 97.20%, EMT^s got highest accuracy on ES. As for Kinship, the best score 72.82% was from TranSparse. Our EMT^v was the runner up whose accuracy was 71.66%. It is related to non-enough training of our models (see the similar detailed analysis in Paragraph 6 of Sect. 4.2).

5 Conclusion

Focusing on specific domain KG, this paper practically pays attention to MTKG completion and puts forward a tail-oriented method EMT. Considering its time-space complexity, EMT^v and EMT^s were born. In link prediction, our models, especially EMT, can get best performance on ES in nearly all metrics. When handling Kinship, with limiting training data, our model can still get best performance in *filt Hit@10* and *MR* of tails predication. As for triplets classification, EMT^s performed best on ES; EMT^v was the runner up for Kinship, 1.16% lower than TranSparse. Although explained via EnzymeKG examples, EMT can be applied on any MTKG.

Based on our research, EMT can be strengthened by referring to local topological structure. In ES of our experiments, tails of *description* and *history* are all long text. So these two relations were removed. But if these text can be considered as auxiliary information, corresponding performance will be improved.

We define MHKG and MTKG, furthermore, put forward a tail-oriented method EMT for the later. What about a head-oriented one EMH? If numbers of *N-1-1* and *1-1-N* are very close, can we combine EMH and EMT together to form new strong adaptable models? If yes, how?

References

1. Belleau, F., Nolin, M.A., Tourigny, N., Rigault, P., Morissette, J.: Bio2RDF: towards a mashup to build bioinformatics knowledge systems. J. Biomed. Inform. **41**(5), 706–716 (2008)
2. Bollacker, K., Evans, C., Paritosh, P., Sturge, T., Taylor, J.: Freebase: a collaboratively created graph database for structuring human knowledge. In: SIGMOD, pp. 1247–1250. ACM (2008)
3. Bordes, A., Glorot, X., Weston, J., Bengio, Y.: A semantic matching energy function for learning with multi-relational data. Mach. Learn. **94**(2), 233–259 (2014)
4. Bordes, A., Usunier, N., Garcia-Duran, A., Weston, J., Yakhnenko, O.: Translating embeddings for modeling multi-relational data. In: NIPS, pp. 2787–2795 (2013)
5. Bordes, A., Weston, J., Collobert, R., Bengio, Y.: Learning structured embeddings of knowledge bases. In: AAAI, pp. 301–306 (2011)
6. Carlson, A., Betteridge, J., Kisiel, B., Settles, B.: Toward an architecture for never-ending language learning. In: AAAI (2010)
7. Dong, X., et al.: Knowledge vault: a web-scale approach to probabilistic knowledge fusion. In: KDD, pp. 601–610. ACM (2014)
8. Jenatton, R., Roux, N.L., Bordes, A., Obozinski, G.R.: A latent factor model for highly multi-relational data. In: NIPS, pp. 3167–3175 (2012)
9. Ji, G., He, S., Xu, L., Liu, K., Zhao, J.: Knowledge graph embedding via dynamic mapping matrix. In: ACL and IJCNLP, pp. 687–696 (2015)
10. Ji, G., Liu, K., He, S., Zhao, J.: Knowledge graph completion with adaptive sparse transfer matrix. In: AAAI, pp. 985–991 (2016)
11. Kemp, C., Tenenbaum, J.B., Griffiths, T.L., Yamada, T., Ueda, N.: Learning systems of concepts with an infinite relational model. In: AAAI, pp. 381–388. AAAI Press (2006). http://dl.acm.org/citation.cfm?id=1597538.1597600
12. Lin, Y., Liu, Z., Sun, M., Liu, Y., Zhu, X.: Learning entity and relation embeddings for knowledge graph completion. In: AAAI, pp. 2181–2187 (2015)
13. Liu, H., Wu, Y., Yang, Y.: Analogical inference for multi-relational embeddings. CoRR abs/1705.02426 (2017). http://arxiv.org/abs/1705.02426
14. Liu, Q., Jiang, H., Ling, Z., Wei, S., Hu, Y.: Probabilistic reasoning via deep learning: neural association models. CoRR abs/1603.07704 (2016). http://arxiv.org/abs/1603.07704
15. Momtchev, V., Peychev, D., Primov, T., Georgiev, G.: Expanding the pathway and interaction knowledge in linked life data. In: ISWC (2009)
16. Nickel, M., Rosasco, L., Poggio, T.: Holographic embeddings of knowledge graphs. In: AAAI, pp. 1955–1961 (2016)
17. Nickel, M., Tresp, V., Kriegel, H.P.: A three-way model for collective learning on multi-relational data. In: ICML, pp. 809–816 (2011)
18. Rummel, R.J.: Dimensionality of nations project: attributes of nations and behavior of nation dyads, 1950–1965 (1992). https://doi.org/10.3886/ICPSR05409.v1

19. Ruttenberg, A., Rees, J.A., Samwald, M., Marshall, M.S.: Life sciences on the Semantic Web: the Neurocommons and beyond. Brief. Bioinform. **10**(2), 193–204 (2009)

20. Socher, R., Chen, D., Manning, C.D., Ng, A.Y.: Reasoning with neural tensor networks for knowledge base completion. In: NIPS, pp. 926–934 (2013)

21. Suchanek, F.M., Kasneci, G., Weikum, G.: YAGO: a large ontology from Wikipedia and WordNet. Web Semant.: Sci. Serv. Agents World Wide Web **6**(3), 203–217 (2008)

22. Wang, Z., Zhang, J., Feng, J., Chen, Z.: Knowledge graph embedding by translating on hyperplanes. In: AAAI, pp. 1112–1119. AAAI Press (2014)

23. Welbl, J., Riedel, S., Bouchard, G.: Complex embeddings for simple link prediction. In: ICML, pp. 2071–2080 (2016)

24. Wu, W., Li, H., Wang, H., Zhu, K.Q.: Probase: a probabilistic taxonomy for text understanding. In: SIGMOD, pp. 481–492. ACM (2012)

25. Yang, B., Yih, W., He, X., Gao, J., Deng, L.: Embedding entities and relations for learning and inference in knowledge bases. CoRR abs/1412.6575 (2014). http://arxiv.org/abs/1412.6575

An Interpretable Neural Model
with Interactive Stepwise Influence

Yin Zhang$^{(\boxtimes)}$, Ninghao Liu, Shuiwang Ji, James Caverlee, and Xia Hu

Texas A&M University, College Station, TX, USA
{zhan13679,nhliu43,sji,caverlee,xiahu}@tamu.edu

Abstract. Deep neural networks have achieved promising prediction performance, but are often criticized for the lack of interpretability, which is essential in many real-world applications such as health informatics and political science. Meanwhile, it has been observed that many shallow models, such as linear models or tree-based models, are fairly interpretable though not accurate enough. Motivated by these observations, in this paper, we investigate how to fully take advantage of the interpretability of shallow models in neural networks. To this end, we propose a novel interpretable neural model with Interactive Stepwise Influence (ISI) framework. Specifically, in each iteration of the learning process, ISI interactively trains a shallow model with soft labels computed from a neural network, and the learned shallow model is then used to influence the neural network to gain interpretability. Thus ISI could achieve interpretability in three aspects: importance of features, impact of feature value changes, and adaptability of feature weights in the neural network learning process. Experiments on both synthetic and two real-world datasets demonstrate that ISI could generate reliable interpretation with respect to the three aspects, as well as preserve prediction accuracy by comparing with other state-of-the-art methods.

Keywords: Neural network · Interpretation · Stepwise Influence

1 Introduction

Neural networks (NNs) have achieved extraordinary predictive performance in many real-world applications [19]. Despite the superior performance, NNs are often regarded as black-boxes and difficult to interpret, due to their complex network structures and multiple nested layers of non-linear transformations. This "interpretability gap" poses key roadblocks in many domains – such as health informatics, political science, and marketing – where domain experts prefer to have a clear understanding of both the underlying prediction models as well as the end results [5]. In contrast, many "shallow" models, such as linear regression or tree-based models, do provide easier interpretability [3] (e.g., through inspection of the intermediate decision nodes) but may not achieve accuracy on par with deep models. To bridge this gap, we investigate how to take advantage

© Springer Nature Switzerland AG 2019
Q. Yang et al. (Eds.): PAKDD 2019, LNAI 11441, pp. 528–540, 2019.
https://doi.org/10.1007/978-3-030-16142-2_41

of the interpretability of shallow models in developing interpretable deep neural networks.

Recently, several efforts have been devoted to enable interpretability of deep models, including visualization for feature selection in computer vision area [2, 24], prediction-level interpretation [18] and attention models [7] in medical and other areas. These and related methods typically focus on *results interpretability* which explains results of each individual sample separately [18]. In contrast, we focus on *model interpretability* which can show the features influences to response variables regardless of individual samples; that is, we aim to identify for each feature its importance (the contribution to the result) and its influence (the impact of changes in the feature on changes in the result) [5]. Additionally, we aim to uncover aspects of the internal mechanism of the NN "black box" by capturing how each feature adapts over training iterations. Recently, a widely-used way to build such an interpretable neural network is to firstly train a complex but accurate deep NN, and then transfer its knowledge to a much smaller but interpretable model [6]. However, this approach has several limitations. First, it makes use of the soft labels computed from the deep model to train another shallow model, which ignores the fact that the "dark" knowledge [1] learned at the end may or may not be the best to train an effective shallow model. Second, parameters in NN are usually learned by complex process, which makes NN hard to be understood while the method does not consider that. So if we could show how each features is learned in NN, it can help interpret NN.

Motivated by these observations, we propose a novel framework ISI – an Interactive Stepwise Influence model, that can interactively learn the NN and shallow models simultaneously to realize both interpretability and accuracy. Specifically, ISI first uses a shallow model to approximate the neural network's predictions in a forward propagation. Then, ISI uses fitted values of the shallow model as prior knowledge to train the next learning step. In sum, the two parts in ISI – shallow models and the NN, interactively influence each other in each training iteration.

During the process, ISI can be interpreted in three aspects: (i) *Importance:* ISI calculates the contribution of each input feature; (ii) *Impact:* ISI gives the value changes of predicted variable based on different feature value changes by a relatively simple relationship; and (iii) *Adaptability:* ISI shows variations of feature weights changes in learning process of NN. In experiment, we evaluate ISI on both synthetic and two real-world datasets for classification problems. Specifically, we first evaluate the reliability of ISI interpretability based on the correctness of feature importance and feature influence. We also show the variations of feature weights changes in ISI updating process. At last, we compare the prediction accuracy of ISI with traditional machine learning methods and state-of-the-art methods such as CNN and MIMIC learning [6]. Our results show that ISI can give utility interpretations from the three aspects and outperforms all the other interpretable state-of-the-art methods in AUPRC and AUROC.

2   Related Work

NNs are widely used because of their extraordinary performance in fitting non-linear relationships and extracting useful patterns [12]. However, in some real world applications, such as health care, marketing, political science and education, interpretability provides significant insights behind the predictions. In such situations, interpretation can be more important than prediction accuracy. NNs are limited used [4,6,7,9] in those areas because they are hard to interpret.

Some researchers have been working on the interpretability of models [8]. There is an overview about making traditional classification models more comprehensible [10]. Specifically, Wang *et al.* built an oblique treed sparse addictive model to make the interpretable model more accurate [22]. [3] analyzed tree-based models by using a training selected set to make the original model interpretable. [7] proposed an end-to-end interpretable model RETAIN by using reverse time attention mechanism. Some methods use visualization to find the good qualitative interpretations of intermediate features [15]. [18] proposed LIME to learn an interpretable model locally around each prediction. [9] investigated a guided feature inversion framework which could show the NN decision-making process for interpretation. Another approach for the interpretation methods are based on calculating the sensitivity of the output in terms of the input. For example, if an input feature change can bring a significant prediction difference, it means the feature is important to the prediction, such as [20]. Among those methods, "distilled" methods [1,11,13] become popular because of their extraordinary performance and strong interpretability. [1] "distilled" a Monte Carlo approximation in Bayesian parameter estimation to consider the dark knowledge inside the deep NN. Meanwhile, recent work showed that by distilling the knowledge, models not only gained a good accuracy, but also maintained interpretability in the shallow models [6].

A popular interpretable "distilled" method [6] uses a shallow model as the mimic model to interpret the final neural network results. However, since only final results are learned, there could be a large gap between soft prediction score of NN and results of the mimic shallow model, which may have an influence on the interpretation. Secondly, parameters in NN are calculated by complicated training process (propagation) which makes it harder to understand while traditional methods could not interpret that. If we can show how the influence changes of input features in the training process, it can help users better understand the neural network.

3   Preliminaries

Before we introduce the interpretable framework ISI, it is important to clarify the kind of interpretability that we aim to achieve. Specifically, following previous work [6], we focus on three aspects of interpretability which is the input feature importance, their impacts and the adaptability for neural networks.

Formally, given a supervised neural network $f : \mathcal{X} \rightarrow \mathcal{Y}$. We assume all input features in $\mathcal{X}$ are explainable. x_i represents i^{th} input feature variable and $i \in$

$\{1, 2, \ldots q\}$, where q is the number of input features. Let $\mathbf{x} = [x_1, x_2, \ldots x_q] \in \mathcal{X}$ represents corresponding feature vector, and $\mathbf{y} = f(\mathbf{x})$. Our proposed NN targets the following three aspects of interpretation:

- *Importance:* For each feature X_i, f can provide the corresponding contribution $\beta_i \in \mathcal{R}$ to y;
- *Impact:* If feature X_i changes $\triangle X_i$, f can provide the change $\triangle \mathbf{y}$ of $\mathbf{y}$ in a linear/tree based relationship;
- *Adaptability:* Since f is a NN, f has a learning process to update its parameters. f can provide how each β_i changes in each iteration.

Here we target to perform "model interpretability" rather than "results/local interpretability" since latter explains results of each example separately [18] while the former shows the impact of features to response variable and the interpretation is not constrained by a single sample. For example, the interpretable linear models [21] can be

Table 1. Notations.

Notations	Definitions
$\mathbf{X} \in R^{n \times k}$	Input matrix for sample $\mathbf{x}_1, \ldots \mathbf{x}_n$
$\mathbf{y} \in R^n$	Output vector for sample $\mathbf{x}_1, \ldots \mathbf{x}_n$
$g(\cdot)$	Ground true relationship from $\mathcal{X}$ to $\mathcal{Y}$
$\theta_N = \{\mathbf{w}_N, \ldots\}$	Parameters set of neural network
$f(\mathbf{X}; \theta_N^{(i)})$	Learned neural network in i^{th} iteration
$\mathbf{w}_N^i \in R^i$	Weight in layer i of $f(\mathbf{X}; \theta_N), i \in 1, \ldots h$
$\hat{y}^{(i)}$	Output of $f(\mathbf{X}; \theta_N^{(i)})$
$\pi_S = \{\mathbf{w}_S^i, \ldots\}$	Parameters of mimic shallow model
$\xi(\mathbf{X}; \pi_S^{(i)})$	Mimic shallow model of $f(\mathbf{X}; \theta_N^{(i)}, \mathbf{y})$
$\tilde{y}^{(i)}$	Output of $\xi(\mathbf{X}; \pi_S^{(i)})$

used to explain the relationship between diabetes and lab test variables. Furthermore, humans are limited to understand complex associations between variables [14]. Shallow models are considered as more interpretable since they have simple structures explicitly expressing how features influence the prediction [6]. So for the second aspect, we are tying to find similar variable associations as shallow models to explain the feature impact. By combining the first two aspects of interpretation, f can identify features that are highly related to response variable. For the third aspect, we target to learn the changes of each input feature influence during the NN parameter updating process (Table 1).

4 Interpretable Neural Networks with Interactive Stepwise Influence

The key idea of the proposed framework ISI is to use an interpretable model to approximate the NN outputs in the forward propagation, and then, update NN parameters according to the output of the interpretable model. So in ISI, a NN f for gaining prediction accuracy, and the shallow but interpretable model ξ for tuning f parameters to make it interpretable. In this section, we first introduce our proposed framework ISI and show how to utilize the ISI framework to gain the three interpretation aspects. Then we provide details of ISI optimization.

4.1 The Proposed ISI Framework

In this section, we first introduce our novel ISI framework (shown in Fig. 1) in details. Suppose $g : \mathcal{X} \rightarrow \mathcal{Y}$ denote the prediction function, where $\mathcal{X}, \mathcal{Y}$ are

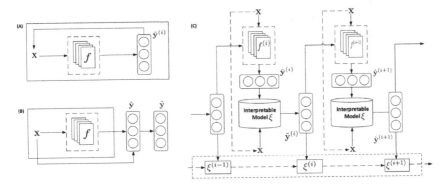

Fig. 1. (A) The standard NN learning architecture: update parameters through back-propagation from a NN output in each iteration. (B) MIMIC learning: first train a NN, and then train an interpretable model using the output of the NN as soft labels. (C) ISI architecture: the first module is a NN f used to gain accuracy. The second module is interpretable models $\xi^{(i)}$ embedded in f. Instead of using the difference between forward propagation and ground truths for backpropagation, we use forward propagation output as soft labels to train ξ, and then use the fitted output of ξ to replace forward propagation output in backpropagation. ξ can be used to adjust f to provide interpretations for f.

its domain and codomain, respectively. Samples $(\mathbf{x}_1, y_1), (\mathbf{x}_2, y_2), \ldots (\mathbf{x}_n, y_n) \in (\mathcal{X}, \mathcal{Y})$ constitute the dataset $(\mathbf{X}, \mathbf{y})$. The goal is to train a traditional NN $f(\mathbf{X}; \theta_N)$, which is parameterized by $\theta_N = \{\mathbf{w}_N^1, \mathbf{w}_N^2, \ldots \mathbf{w}_N^h\}$, $\mathbf{w}_N^j$ is the jth hidden layer parameter for f. Parameters in θ_N are get by minimizing the loss function $L_P(f(\mathbf{X}; \theta_N), \mathbf{y})$. For example, it can be the cross entropy loss function $L_P(f(\mathbf{X}; \theta_N), \mathbf{y}) = -\sum_i y_i log \hat{y}_i$, and we minimize it to get the optimal solution θ_N'.

Based on the interpretation that we target, we dig into the neural network learning process (backpropagation for parameters updating). For traditional neural network, the optimized parameters θ_N' is calculated from:

$$\theta_N' = \underset{\theta_N^{(i)}}{\operatorname{argmin}} L_P(f(\mathbf{X}; \theta_N), \mathbf{y}), \tag{1}$$

by backpropagations of iterations until it converges. Specifically, for each iteration $i > 1$ of backpropagation, it includes two parts:

(1) A forward pass to use learned $\theta_N^{(i)}$ in i^{th} iteration and generate the current prediction output: $\hat{\mathbf{y}}^{(i)} = f(\mathbf{X}; \theta_N^{(i)})$;
(2) Then a backward pass to update $\theta_N^{(i)}$ in f by minimizing the current loss function value: $\theta_N^{(i+1)} = \theta_N^{(i)} - \eta \nabla_{\theta_N} L(\hat{\mathbf{y}}^{(i)}, \mathbf{y})$, where γ is the learning rate;

Repeat (1)(2), we can get a sequence of $\theta_N^{(1)} \rightarrow \theta_N^{(2)} \rightarrow \ldots \rightarrow \theta_N^{(k)}, \ldots$ until to a stable state that $|L_P(\hat{\mathbf{y}}^{(i+1)}, \mathbf{y}) - L_P(\hat{\mathbf{y}}^{(i)}, \mathbf{y})| < \epsilon$, where $\epsilon \in R$ is the threshold.

As shown above, the training process is complicated and it is hard to find how each input feature in $\mathbf{X}$ influences θ_N during the two parts of backpropagation, which also makes the final neural network model hard to be interpreted. In our proposed ISI (shown in Fig. 1(c)), a mimic shallow model $\xi(\mathbf{X}; \pi_S)$ is embedded in f training process to adjust parameter updates in each iteration of f, where π_S denote the parameters of the shallow model $\xi(\mathbf{X}; \pi_S)$, respectively. Based on that, we propose a new loss function that can jointly train the shallow model $\xi(\mathbf{X}; \pi_S)$ and neural network $f(\mathbf{X}; \theta_N^{(i)})$ to gain the interpretation:

$$\theta_N^*, \pi_S^* = \underset{\theta_N, \pi_S}{\arg\min} L_P(\xi(\mathbf{X}; \pi_S, f(\mathbf{X}; \theta_N, \mathbf{y})), \mathbf{y}), \qquad (2)$$

where $L_P(\cdot)$ is the total loss function. Specifically, Eq. 2 includes three parts: neural network $f(\mathbf{X}; \theta_N)$ is trained based on ground truth $\mathbf{y}$ to ensure the accuracy of ISI. Then different from mimic learning where shallow model $\xi(\mathbf{X}; \pi_S)$ is fitted by the final results of f and is used to interpret f, we jointly train $\xi(\mathbf{X}; \pi_S)$ in the training process of f, to ensure the close connection between mimic model and neural network, since it decreases the differences between fitted $\hat{\xi}$ and f. Therefore, the shallow model can better approximate and interpret the NN than mimic learning model. Finally, we trained our joint model ISI by $L(\cdot, \mathbf{y})$ to gain interpretation. Details of ISI training process is explained in Sect. 4.2.

In sum, compared with the other interpretable methods, there are two major benefits of ISI: (1) The mimic shallow model $\xi(\mathbf{X}; \pi_S)$ is jointly trained with neural network f to ensure the close connection between them, rather than use the final results of f and directly fitted $\xi(\mathbf{X}; \pi_S)$ in traditional mimic learning process. Then $\xi(\mathbf{X}; \pi_S)$ can better be used for interpretation of f; (2) We can use the trained $\xi(\mathbf{X}; \pi_S^{(i)})$ in each parameter updating process to explain the feature influence in each iteration since they are jointly trained. Specifically, we can record ξ in each iteration: instead of representing the learning process as complex $f^{(1)} \rightarrow f^{(2)} \rightarrow \ldots \rightarrow f^{(k)} \ldots$, it can be represented by shallow models as $\xi^{(1)} \rightarrow \xi^{(2)} \rightarrow \ldots \rightarrow \xi^{(k)} \ldots$ which is easier to show the feature influence in each iteration. For example, if the shallow models are linear models, their corresponding parameters $\pi_S^{(1)} \rightarrow \pi_S^{(2)} \rightarrow \ldots \rightarrow \pi_S^{(k)} \ldots$ represent variations of feature contributions of each input feature; if the shallow models are tree-based models, we can use Gini importance to calculate the variations.

4.2   Optimization of ISI

Directly optimizing Eq. 2 is hard and time-consuming. In this section, we discuss how to optimize it. Specifically, we divide each learning iteration in three parts for Eq. 2 and formulate them as below:

1. **Train the shallow model with soft labels:** At the ith iteration, we utilize a loss function $\pi_S^{(i)} = \underset{\pi_S}{\arg\min} L_I(\xi(\mathbf{X}; \pi_S), \hat{\mathbf{y}}^{(i)})$ to train the shallow model part $\xi(\mathbf{X}; \pi_S)$. Here $\hat{\mathbf{y}}^{(i)}$ is the ith iteration output of f, so it contains the knowledge acquired by f.

Algorithm 1. Interactive Stepwise Influence (ISI) Model

Input : Data $\mathbf{X} = [\mathbf{x}_1^T, \mathbf{x}_2^T, ...\mathbf{x}_n^T]$, $\mathbf{y}$ is the true label, C is the number of class $\mathbf{y}$ has, η is the stepsize, $\gamma \in (0, 1]$ is the fitting parameter, T is the maximum number of iterations, h is the number of hidden layer

Output: $\mathbf{y}^{(f,\xi)}$ is the output

1　Initialized $\mathbf{W}^{total} = \{\mathbf{W}_1, \mathbf{W}_2, ...\mathbf{W}_h\}$, $\mathbf{b}^{total} = [\mathbf{b_1}, \mathbf{b_2}...\mathbf{b_h}]$;

2　Pick explainable model ξ;

3　**for** i *from* 1 *to* T **do**

4　　Assign $\hat{\mathbf{y}}^{(i)}$ by using forward-propagate of the inputs over the whole unfolded network;

5　　**for** $c \in Class$ **do**

6　　　Optimize objective function of $L_I(\mathbf{X}; \xi(\pi_S), \hat{\mathbf{y}}^{(i)})$ based on ξ to get $\hat{\xi}_c$

7　　　Calculate the fitted value $\tilde{\mathbf{y}} \leftarrow \hat{\xi}_c(\mathbf{X}, \hat{\mathbf{y}}^{(i)})$

8　　Calculate gradient $\frac{dL_P(\mathbf{y}, \hat{\mathbf{y}}^{(i)})}{d\mathbf{W}^{total}}, \frac{dL_P(\mathbf{y}, \hat{\mathbf{y}}^{(i)})}{d\mathbf{b}^{total}}$;

9　　Update $\mathbf{W}^{total} \leftarrow \mathbf{W}^{total} - \eta \tilde{d}L_P(\mathbf{y}, \tilde{\mathbf{y}}^{(i)})/\tilde{d}\mathbf{W}^{total}$;

10　　$\mathbf{b}^{total} \leftarrow \mathbf{b}^{total} - \eta \tilde{d}L_P(\mathbf{y}, \tilde{\mathbf{y}}^{(i)})/\tilde{d}\mathbf{b}^{total}$ based on previous step;

11　　Assign $\hat{\mathbf{y}}^{(i+1)}$ by using forward-propagate using updated parameter $\mathbf{W}^{total}, \mathbf{b}^{total}$;

12　　Calculate loss function $L_P(\mathbf{y}, \hat{\mathbf{y}}^{(i+1)})$;

13　　**if** $L_P(\mathbf{y}, \hat{\mathbf{y}}^{(i+1)})$ *increase* **then**

14　　　Update $\eta \leftarrow \gamma\eta$;

15　Use updated $\mathbf{W}^{total}, \mathbf{b}^{total}$ or π_S to calculate $\mathbf{y}^{(f,\xi)}$ based on performance.

---

2. **Obtain predictions from the shallow model:** The fitted output of the shallow model is obtained by computing $\tilde{\mathbf{y}}^{(i)} = \xi(\pi_S, \mathbf{X})$ with optimized π_S. The interpretable patterns are contained in ξ, and it can also be used as a snapshot of the learning process.

3. **Update parameters of the neural network:** We use the outputs $\tilde{\mathbf{y}}^{(i)}$ from the shallow model, instead of $\hat{\mathbf{y}}^{(i)}$ from the neural network, as an approximation of NN forward prediction to compute errors and update NN parameters: $\theta_N = \arg\min_{\theta_N} L_P(\tilde{\mathbf{y}}^{(i)}, \mathbf{y})$. Due to the relatively simple structure of $\tilde{\mathbf{y}}^{(i)}$, $\tilde{\mathbf{y}}^{(i)}$ makes NN easier to be interpreted [6].

The procedure above is formally presented in Algorithm 15. We first initialize parameters $\mathbf{w}_k, \mathbf{b}_k$ in each hidden layer k, then select a shallow model to be trained in line 2 and 3. From line 4 to 7, we optimize parameters in the shallow model ξ based on loss function $L_I(\xi(\mathbf{X}; \pi_S), \hat{\mathbf{y}}^{(i)})$. From line 8 to 14, we update the parameters in f using backward propagation. We use gradient descent as an example. Note here, if traditional gradient descent is used in $L_P(\tilde{\mathbf{y}}_{(S)}^{(i)}, \mathbf{y})$ to update parameters $\mathbf{w}_N$ in f, we should calculate the derivative of ξ trained by $L_I(\xi(\mathbf{X}; \pi_S), \hat{\mathbf{y}}^{(i)})$. Even if ξ is differentiable, calculating its gradient is time consuming. So instead of letting $\theta_N \leftarrow \theta_N - \eta \, dL_P(\hat{\mathbf{y}}^{(i)}, \mathbf{y})/d\theta_N$, we first calculate the derivative of $dL_P(\hat{\mathbf{y}}^{(i)}, \mathbf{y})/d\theta_N$. Then we replace $\hat{\mathbf{y}}^{(i)}$ with $\tilde{\mathbf{y}}^{(i)}$ in the calculated gradient equations in line 8 and 9. We denote the procedure as $\theta_N \leftarrow \theta_N - \eta \, \tilde{d}L_P(\tilde{\mathbf{y}}^{(i)}, \mathbf{y})/\tilde{d}\theta_N$. Thus, ISI would not be limited by non-differential shallow models.

5 Experiments

We conduct comprehensive experiments to evaluate the performance of ISI on the three interpretation aspects and accuracy. In particular, we aim to answer the following questions: (1) Can ISI provide reliable interpretations for its predictions, in terms of giving proper feature contributions and unveiling feature influences? (2) Can ISI provide reasonable interpretations for feature adaptability in its learning process? (3) Does ISI at the same time have a good precision compared to the state-of-art methods?

5.1 Data and Setup

We use three datasets including one synthetic data (SD) and two real-world datasets, i.e., MNIST and the default of credit card clients (D_CCC) [17] for classification tasks. Parametric distributions of different classes in SD are known as the basis to assess the faithfulness of the three interpretation results from ISI. The two real-world datasets are used to evaluate ISI interpretation utility and accuracy. Specifically, MNIST [16] is for handwritten digit classification, and D_CCC is to explore features that have an influence on the occurrence of default payment (DP). D_CCC is randomly partitioned into 80% for training and validation, and 20% for testing. We use widely used and relatively robust interpretable shallow models [6]: Logistic regression (LR), Decision Trees (DT), linear SVM, the state-of-art interpretable neural network model mimic learning [6], and also neural networks ANN and CNN as baselines. Specifically, for the neural network module in ISI and mimic learning, we use the same structure of ANN with three layers where $tanh$ and $sigmoid$ are used as activation functions with considering the trade-off between performance and computation complexity as well as for fair comparison. Cross-entropy is used as the loss function. The CNN with two convolution layers, a pooling layer and a densely-connected layer are used. Hyperparameters for all methods are tuned by five-fold cross validation. Prediction accuracy is measured by AUPRC (Area Under Precision-Recall Curve) and AUROC (Area Under receiver operating Characteristic Curve) [6]. Results are reported by averaging over 100 random trails.

5.2 Interpretation Evaluation

We first test the interpretation ability of ISI in SD since ground truth is known. The task is a binary classification where data samples are generated from a mixture of multivariate Gaussian distributions $\{\mathcal{N}(\mu_1, \Sigma_1), \mathcal{N}(\mu_2, \Sigma_2)\}$ of two classes. For each sample $\mathbf{x}_i \in \mathbb{R}^{(d_1+d_2)}$, d_1 and d_2 are dimensions for informative and noise features respectively. Informative features are used to separate the two classes. Noise features are appended to evaluate interpretations, as those features are not expected to affect classification. $N_1 = 1200$ and $N_2 = 1500$ denote the number of samples in each class. $d_1 = 6$, $d_2 = 6 \times 20 = 120$ and $\Sigma_1 = \Sigma_2$ are identity matrices. Each noise feature is generated from independent standard normal distribution $\mathcal{N}(0, 1)$. To distinguish contributions among different features, we set $\mu_1 = [6, 5, 4, 3, 2, 1]^T$, $\mu_2 = [-1, -1, -1, -1, -1, -1]^T$, so the contributions of features are already sorted in a descending order according to their importance. Figure 2(a) shows a 3-D visualization of SD.

Table 2. ISI performance of different interpretable models.

ISI	AUPRC	AUROC
ANN + LR	0.8567 ± 0.0000	0.8850 ± 0.0000
ANN + DT	0.7200 ± 0.0438	0.7357 ± 0.0438
ANN + SVM	0.8731 ± 0.0016	0.9018 ± 0.0004
ANN + LASSO	$\mathbf{0.8802 \pm 0.0096}$	$\mathbf{0.9082 \pm 0.0067}$

Table 3. Feature selection performances of different methods.

	Selected features indices	NM	NP
LASSO	1, 2, 3, 4, 5, (10, 15, 31, 30)	18%	0.20
MIMIC	1, 2, 3, 4, 5, (50, 68, 99, 103, 122)	22%	0.26
ISI	1, 2, 3, 4, 5, (71)	3%	0.03

Table 2 shows the prediction accuracy of ISI embedded with different shallow models. When the mimic shallow model part ξ uses a linear model such as LR, linear SVM and LASSO, ISI has higher AUPRC and AUROC than that of tree-based models. This indicates that linear classifiers are preferred, which matches the features associations in synthetic data. The best accuracy performance is achieved by using LASSO in ISI, so we use it for subsequent interpretation analysis. For the first interpretation aspect "importance" of each feature, we first test the percentage of selected noise features for different models in Table 3 where parameters are tuned based on their best accuracy in Table 4. Indices in the parameter represent noise features. "NM" in the table denotes the possibility that the corresponding model contains noise features out of 100 trails. "NP" is the average ratio of noise features in each model. Specifically, noise features appear in 22% and 18% models over 100 random trails for LASSO and MIMIC respectively, while noise features appear in only 3% of the models for ISI. Moreover, we calculate the contributions of each feature by normalized coefficients of each linear model and Gini importances of DT. The results of each interpretable method is shown in Fig. 2(b). We notice feature importance calculated by ISI are close to the true value. For second aspect "impact", since LASSO is selected in ISI shallow model part (shown in Table 2), if feature F_i changes $\triangle F_i$, the probability that it belongs to a certain group changes $\alpha_i \triangle F_i$, where α_i is the coefficient of F_i in LASSO. Those results indicate ISI can provide more reliable interpretations.

For the third aspect "adaptability", the NN f can be intuitively explained using the embedded shallow models in ISI. The approximated variation of each feature contribution is shown in Fig. 3(a)(b). They are calculated by using features weights (coefficients) of embedded shallow models in each iteration, since LASSO is selected as shallow models. The variation rates in Fig. 3(b) are

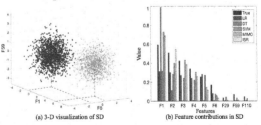

(a) 3-D visualization of SD   (b) Feature contributions in SD

Fig. 2. Fi is the ith input feature of SD. (a) uses three features to give a 3-D visualization of SD. Different colors mean different groups. (b) calculates feature contributions of different interpretable methods based on the accuracy in Table 4.

the weight differences of two adjacent iterations. As the learning process proceeds, contributions to the final results of each informative feature becomes more clear, at a fast rate especially in the early stages of training. The weight of noise features approaches 0. It also matches the converge process in NN learning iterations. Such information may help people understand the final parameters of neural network.

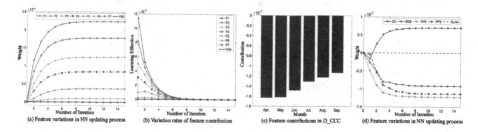

Fig. 3. (a)(b) show the variations (adaptability) and variation rates of features contributions during the learning process in SD. (c) illustrates contributions for some features in D_CCC. (d) depicts some parameter variations (adaptability) for D_CCC.

For real-word dataset, ISI also shows extraordinary and reliable interpretability in terms of the three interpretable aspects that we targets. For MNIST, we select LASSO as the shallow model part of ISI to interpret classification results according to best accuracy. Figure 4(a) shows input pixels contributions measured by corresponding coefficients in LASSO. The darker area means that the corresponding pixels have higher negative relations to the class, while the lighter area means the weights have more positive relations. Specifically, gray area means that the coefficients of corresponding pixels in the shallow part are approximate to zero. For example, for pixels in an image with high values, if they are in the lighter area, there is a higher probability that the image would be classified to the corresponding digit. While if those pixels are in the darker area, the image is less likely to be classified to the corresponding digit. Gray area means the corresponding pixels have little contribution to detecting digit. Here we can observe that the white areas sketch the outline of each digit and dark areas are near them. Gray areas are far from the outline of digits. Figure 4(b) shows five examples of feature variations interpretation results from ISI in the first 100 iterations. The interval between two columns is 10 iterations. The results show that there are no specific patterns at the beginning regarding how to classify a digit. But after more iterations, we can see that the sketch of each digit highlighted by white areas becomes more obvious. For D_CCC, LR is selected as the shallow model ISI to explain the three interpretation aspects based on the accuracy performance. Figure 3(c) shows the contribution of the amount of previous payment in each month. It indicates that the amount of previous payments in April and May strongly influence DP. Since linear model LR is selected, each feature influence is the product of the corresponding coefficient of LR and the changes of the feature. Figure 3(d) shows the feature adaptability. It also gives reasonable explanation of each features to the final DP [23].

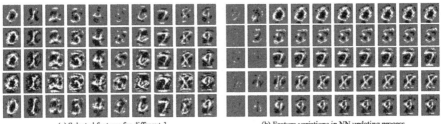

(a) Selected features for different λ   (b) Feature variations in NN updating process

Fig. 4. (a) Selected features by ISI with LASSO in MNIST dataset. The first four rows are calculated with $\mathcal{L}_1$ regularization parameter $\lambda = 0.5, 0.1, 0.05, 0.01$ and 100 hidden units. The results in the last row are calculated with $\lambda = 0.01$ and 500 hidden units. (b) Variations of feature weights in NN learning process of five examples with $\lambda = 0.1$ and 100 hidden units.

5.3   Prediction Accuracy Evaluation

In this section, we evaluate the prediction capability of ISI in AUPRC and AUROC, compared with other classification models as baselines cross the three different datasets. Here MIMIC learning uses the same NN structure as ISI for fair comparison. For the shallow part in MIMIC and ISI methods, we try difference shallow models (LR, DT, SVM, LASSO) in each dataset and reports the best performed ones. Table 4 shows the accuracy of ISI compared with baseline methods. "NA" here means the corresponding method takes more than 10 times longer than the other methods.

Overall, we see the full-blown ISI improves upon all the other interpretable models cross the three different datasets. Moreover, the performance of ISI is comparable to that of ANN and CNN while ISI is also easier to interpret. From the first three rows of LR, DT and SVM in Table 4, ISI improves versus the next-best alternative an average of 3.24% in AUPRC and 1.06% in AUROC. It may

Table 4. Accuracy performance on the three datasets. MIMIC learning uses SVM, LASSO, LASSO respectively for the three datasets. ISI is embedded with LASSO, LR and LASSO for MNIST, D_CCC and SD, respectively. Here different shallow models are used for different datasets because we choose the best NN-shallow models combination for each case. ISI outperforms all the interpretable models. The performance of ISI is comparable to that of ANN and CNN, and sometimes is even better.

Method	MNIST		D_CCC		SD	
	AUPRC	AUROC	AUPRC	AUROC	AUPRC	AUROC
LR	0.8159 ± 0.0113	0.9589 ± 0.0021	0.5954 ± 0.0017	0.6482 ± 0.0030	0.8721 ± 0.0109	0.9007 ± 0.0075
DT	0.7570 ± 0.0140	0.9189 ± 0.0019	0.5177 ± 0.0451	0.5321 ± 0.0091	0.8595 ± 0.0079	0.8128 ± 0.0109
SVM	NA	NA	0.5349 ± 0.0400	0.5661 ± 0.0048	0.8626 ± 0.0072	0.8944 ± 0.0072
ANN	0.9726 ± 0.0023	0.9946 ± 0.0006	0.6792 ± 0.0189	0.6133 ± 0.0049	0.8891 ± 0.0139	0.9119 ± 0.0093
CNN	0.9894 ± 0.0007	0.9982 ± 0.0001	0.6002 ± 0.0597	0.5010 ± 0.0018	0.8706 ± 0.0104	0.8987 ± 0.0104
MIMIC	0.7219 ± 0.0086	0.9261 ± 0.0029	0.5446 ± 0.0028	0.5790 ± 0.0028	0.8789 ± 0.0151	0.9062 ± 0.0123
ISI	0.8722 ± 0.0033	0.9710 ± 0.0003	0.6066 ± 0.0101	0.6553 ± 0.0083	0.8802 ± 0.0096	0.9082 ± 0.0067

contributes to ISI neural network structure. Comparing with traditional NN, the AUROC of ISI is significantly higher than the AUROC of ANN in D_CCC dataset. Based on the row of MIMIC method, ISI outperforms the state-of-the-art interpretable model MIMIC 10.78% on average in AUPRC and 6.08% in AUROC. It shows by jointly training shallow models and neural network, ISI can gain a higher accuracy. Moreover, based on the standard deviation of each experiment, ISI is also more robust than MIMIC learning in terms of stability. This further shows that ISI has desirable discriminative power after being incorporated into the shallow model to enable interpretability.

6 Conclusions and Future Work

We have proposed a novel interpretable neural network framework ISI which embeds shallow interpretable models in NN learning process, and they are jointly trained to gain both accuracy and interpretability. Through experiments over different datasets, ISI not only outperforms the state-of-the-art methods, but also can be reasonably explained in three aspects: feature importance, feature impact and the adaptability of feature weights in NN learning process. Notice here ISI is mainly applied in areas where interpretability is necessary and traditional models are still widely used [6,22], such as political and economics area. For the future work, how to choose proper interpretable shallow models and applying ISI to more complex data and other neural network architectures are promising directions for future explorations.

References

1. Balan, A.K., Rathod, V., Murphy, K.P., Welling, M.: Bayesian dark knowledge. In: NIPS (2015)
2. Bau, D., Zhou, B., Khosla, A., Oliva, A., Torralba, A.: Network dissection: quantifying interpretability of deep visual representations. arXiv preprint arXiv:1704.05796 (2017)
3. Cano, J.R., Herrera, F., Lozano, M.: Evolutionary stratified training set selection for extracting classification rules with trade off precision-interpretability. Data Knowl. Eng. 60(1), 90–108 (2007)
4. Che, Z., Liu, Y.: Deep learning solutions to computational phenotyping in health care. In: ICDMW. IEEE (2017)
5. Che, Z., Purushotham, S., Khemani, R., Liu, Y.: Distilling knowledge from deep networks with applications to healthcare domain. arXiv preprint arXiv:1512.03542 (2015)
6. Che, Z., Purushotham, S., Khemani, R., Liu, Y.: Interpretable deep models for ICU outcome prediction. In: AMIA Annual Symposium Proceedings, vol. 2016. American Medical Informatics Association (2016)
7. Choi, E., Bahadori, M.T., Sun, J., Kulas, J., Schuetz, A., Stewart, W.: RETAIN: an interpretable predictive model for healthcare using reverse time attention mechanism. In: NIPS (2016)
8. Du, M., Liu, N., Hu, X.: Techniques for interpretable machine learning. arXiv preprint arXiv:1808.00033 (2018)

9. Du, M., Liu, N., Song, Q., Hu, X.: Towards explanation of DNN-based prediction with guided feature inversion. In: KDD (2018)
10. Freitas, A.A.: Comprehensible classification models: a position paper. ACM SIGKDD Explor. Newslett. **15**(1), 1–10 (2014)
11. Frosst, N., Hinton, G.: Distilling a neural network into a soft decision tree. arXiv preprint arXiv:1711.09784 (2017)
12. He, X., Liao, L., Zhang, H., Nie, L., Hu, X., Chua, T.S.: Neural collaborative filtering. In: International World Wide Web Conferences Steering Committee, WWW 2017 (2017)
13. Hinton, G., Vinyals, O., Dean, J.: Distilling the knowledge in a neural network. arXiv preprint arXiv:1503.02531 (2015)
14. Jennings, D., Amabile, T.M., Ross, L.: Informal covariation assessment: data-based vs. theory-based judgments. In: Judgment Under Uncertainty: Heuristics and Biases (1982)
15. Karpathy, A., Johnson, J., Fei-Fei, L.: Visualizing and understanding recurrent networks. arXiv preprint arXiv:1506.02078 (2015)
16. LeCun, Y., Bottou, L., Bengio, Y., Haffner, P.: Gradient-based learning applied to document recognition. Proc. IEEE **86**(11), 2278–2324 (1998)
17. Merz, C.J., Murphy, P.M.: {UCI} repository of machine learning databases (1998)
18. Ribeiro, M.T., Singh, S., Guestrin, C.: "Why should i trust you?" explaining the predictions of any classifier. In: SIGKDD. ACM (2016)
19. Schmidhuber, J.: Deep learning in neural networks: an overview. Neural Netw. **61**, 85–117 (2015)
20. Sundararajan, M., Taly, A., Yan, Q.: Axiomatic attribution for deep networks. arXiv preprint arXiv:1703.01365 (2017)
21. Ustun, B., Rudin, C.: Methods and models for interpretable linear classification. arXiv preprint arXiv:1405.4047 (2014)
22. Wang, J., Fujimaki, R., Motohashi, Y.: Trading interpretability for accuracy: oblique treed sparse additive models. In: SIGKDD (2015)
23. Yeh, I.C., Lien, C.H.: The comparisons of data mining techniques for the predictive accuracy of probability of default of credit card clients. Expert Syst. Appl. **36**(2), 2473–2480 (2009)
24. Zhang, Q., Wu, Y.N., Zhu, S.C.: Interpretable convolutional neural networks. In: CVPR, pp. 8827–8836 (2018)

Multivariate Time Series Early Classification with Interpretability Using Deep Learning and Attention Mechanism

En-Yu Hsu[1] , Chien-Liang Liu[2] , and Vincent S. Tseng[1(✉)]

[1] Department of Computer Science, National Chiao Tung University,
1001 University Road, Hsinchu 300, Taiwan, ROC
b203287580@gmail.com , vtseng@cs.nctu.edu.tw
[2] Department of Industrial Engineering and Management,
National Chiao Tung University, 1001 University Road, Hsinchu 300, Taiwan, ROC
clliu@mail.nctu.edu.tw

Abstract. Multivariate time-series early classification is an emerging topic in data mining fields with wide applications like biomedicine, finance, manufacturing, etc. Despite of some recent studies on this topic that delivered promising developments, few relevant works can provide good interpretability. In this work, we consider simultaneously the important issues of model performance, earliness, and interpretability to propose a deep-learning framework based on the attention mechanism for multivariate time-series early classification. In the proposed model, we used a deep-learning method to extract the features among multiple variables and capture the temporal relation that exists in multivariate time-series data. Additionally, the proposed method uses the attention mechanism to identify the critical segments related to model performance, providing a base to facilitate the better understanding of the model for further decision making. We conducted experiments on three real datasets and compared with several alternatives. While the proposed method can achieve comparable performance results and earliness compared to other alternatives, more importantly, it can provide interpretability by highlighting the important parts of the original data, rendering it easier for users to understand how the prediction is induced from the data.

Keywords: Early classification on time-series · Deep neural network · Attention

1 Introduction

Multivariate time-series early classification has received much attention in data mining, in which the goal is to predict the class label of time-series data using only the starting subsequence of the time series. In real-life scenarios, time-series data often have multiple variables, where the variables exist at each time stamp.

© Springer Nature Switzerland AG 2019
Q. Yang et al. (Eds.): PAKDD 2019, LNAI 11441, pp. 541–553, 2019.
https://doi.org/10.1007/978-3-030-16142-2_42

Hence, the relation between different variables should also be considered. Thus, it is a challenging task for the traditional machine-learning and data-mining methods to handle the multivariate time-series early classification.

Several methods have been proposed to handle the early classification on time-series (ECTS) problem, including the one-nearest neighbor (1NN) [3,6,17, 18], shapelets [5,19,20] and deep learning approaches [2,15,21]. Data mining method such as shapelet could provide interpretable results, but feature extraction using shapelets is a time-consuming and high-complexity task. In contrast, deep learning could learn discriminative feature representations from data, but the deep architecture with nonlinear transformation renders it difficult for the practitioners to understand how the prediction results are induced from the data. The concerns about whether the user should trust the results of machine-learning models especially those of deep-learning approaches have arose in recent years [11,14], as the predictions must be trusted for further decision making. Thus, this work focuses on devising a model which could provide accurate and explainable results.

The attention mechanism provides a means for deep-learning methods to mimic the visual attention mechanism found in humans. Chen et al. [1] focused on the image classification problem and proposed a visual-attention-based convolutional neural network (CNN) to simulate the process of recognizing objects and determining the area of interest, which is related with the task. The attention mechanism provides a means for deep learning methods to focus on a specific parts that are crucial to the prediction. We herein propose using the attention mechanism to provide accurate and interpretable results. Although the proposed algorithm is a framework, we focused on the setting when the learning algorithm is of the deep neural network (DNN) method.

The contributions of this work are listed as follows: First, we propose a framework that handles the multivariate time-series early prediction problem. Next, the proposed attention mechanism enables the important segments of the time-series data to be identified. We believe that the proposed approach could be applied to other problems. As compared with the previous studies, this work applies the attention mechanism to offer effective multivariate time-series early classification with interpretability. Subsequently, we conducted experiments on three datasets, and compared with several alternatives. The proposed method is comparable to other methods in terms of prediction performance; meanwhile, the proposed work can provide interpretable results. The visualization results of the proposed work are presented and investigated.

2   Related Work

The literature survey involves the recent studies on early classification on time series data and time series classification using deep learning.

2.1   Early Classification on Time Series Data

Xing et al. [16] studied the early classification of sequences. They managed to minimize the length of the sequence prefix while maintaining comparable accuracy to the state-of-the-art methods simultaneously. Xing et al. [18] further introduced the idea of minimum prediction length (MPL) to the early classification on time series (ECTS), combining with the 1NN classification method. The concept of MPL is to determine the earliest time stamp for each time series, where the correct nearest neighbor can still be determined.

The 1NN is one of the most powerful tools for time-series classification, and the simplicity and effectiveness of the 1NN have inspired researchers to focus on this approach [3,6,17,18]. However, this method does not summarize or extract features from the data itself, therefore Ye et al. [20] proposed the concept of shapelets.

Shapelets are subsequences of the time-series data obtained by calculating the entropy and used as a distinctiveness to classify time-series instances. The nature of the shapelet approaches causes the prediction results to contain an explanation, which are the shapelets themselves. Based on shapelets, Xing et al. [19] proposed a method to obtain the local shapelets using a measurement called the best matching distance (BMD) that considers the earliness. As the shapelet approaches are designed for univariate time-series data, they could not be directly applied to multivariate time-series problems. Ghalwash et al. [4] devised a method called multivariate shapelet detection (MSD) to extract multivariate shapelets from all dimension of the time series and uses them as the pattern to match and classify the target class.

2.2   Time Series Classification Using Deep Learning

Zheng et al. [21] proposed an algorithm called the multichannels deep convolution neural networks (MC-DCNN) to learn features by corresponding each variable of the multivariate time series to its channels, and subsequently feed those features into a multilayer perceptron (MLP) to make the final prediction. Cui et al. [2] proposed a multiscale convolutional neural network (MCNN) to extract features at different scales and frequencies, and the experimental results on the benchmark datasets show that they have outperformed the state-of-the-art approaches. Subsequently, Wang et al. [15] proposed the earliness aware deep convolutional networks (EA-ConvNets) to train the stochastic truncated training data, where they predicted the target label at any given time. Liu et al. [12] proposed an attention-based approach for identification of misinformation on social media, in which the attention mechanism comprises content attention and dynamic attention. The content attention focuses on textual features of microblogs, while dynamic attention is related to the time information of the microblogs. Qin et al. [13] proposed a dual-stage attention based RNN for time series prediction, which was inspired by the theory that behavioral results are best modeled by a two-stage attention mechanism [8]. They used an input attention mechanism in the first stage to extract features at each time step, and a temporal attention mechanism in the second stage to select relevant candidates across all time steps. Notably, the above two approaches proposed to use attention mechanism to improve prediction performance, whereas our work focuses on using attention mechanism to provide interpretable results.

3 Proposed Method

This section introduces our proposed framework and the training algorithm. Besides, we briefly describe a model called the multi-domain deep neural network (MDDNN), as the MDDNN achieved remarkable results in early classification on multivariate time series. The proposed work uses MDDNN as a pre-trained model, but other deep-learning algorithms on time-series data could be used in the proposed framework.

3.1 Problem Definition

The goal of MTS early prediction is to correctly predict the class label c of a multivariate time series $MT = \{T^1, T^2, \ldots, T^N\}$, where T^j is a time series that represents the jth variable of MT, and all T^js are with the same length L. We use the early subsequences $es = \{s^1, s^2, \ldots, s^N\}$ of MT where s^j is a subsequence of T^j that has the same starting point with T^j, namely $s^j = \{t_1^j, t_2^j, \ldots, t_\ell^j\}$ for $j = 1, 2, \ldots, N$ and ℓ is the length of es with $\ell < L$ to obtain earliness $E = \frac{\ell}{L} < \tau$, where τ is the user-defined threshold of earliness.

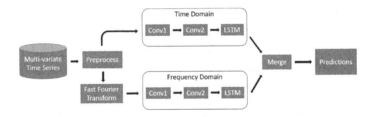

Fig. 1. The structure of MDDNN model

3.2 MDDNN Model

The MDDNN is a neural network model that contains the time-series data in the time domain and frequency domain as the two inputs of the model. The time domain input is the original time series, while the frequency domain uses the fast Fourier transform (FFT) to transform the time-series input into a frequency representation. The MDDNN structure is presented in Fig. 1, in which the two domains are processed by DNNs with two convolutional layers followed by one long-short term memory (LSTM) layer. The design of the MDDNN aims to capture the discriminative features with CNN and extract the temporal characteristic of the features obtained from CNN with LSTM. We chose MDDNN as the core model of our architecture because it performs well on a multivariate time-series early predicting task.

3.3 Attention Architecture

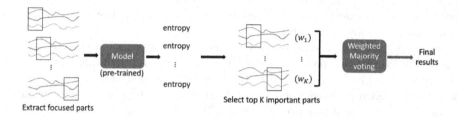

Fig. 2. The explainable time series classification model

The proposed explainable time series classification model (ETSCM) uses attention mechanism to discover important segments as presented in Fig. 2, and includes four steps. The first step is to use a sliding window to obtain the subsequences of the input multivariate time series input, each of which is considered a candidate segment. Once the first step is completed, for a N-dimensional multivariate time series $MT = \{T^1, T^2, \ldots, T^N\}$, the segments obtained from MT are defined in (1).

$$f_i = \{\{t_i^1, t_{i+1}^1, \ldots, t_{i+l-1}^1\}, \ldots, \{t_i^N, t_{i+1}^N, \ldots, t_{i+l-1}^N\}\}, \tag{1}$$

where i is the starting timestamp and l is the sliding window length, which can be specified by the user.

The second step is to feed f_i obtained in (1) into a pre-trained MDDNN model. It is noteworthy that the MDDNN requires the time-series data of the time domain and frequency domain as the inputs; therefore, f_i should be transformed into the representations of these two domains before feeding into the MDDNN. Additionally, the MDDNN model is a pre-trained model on time-series data of full length, therefore we used the padding technique to concatenate the segments with zeroes until the length is equal to the full length of the time-series data.

Next, each segment is predicted by the pre-trained model, and the outcome is a probability vector $[p_1, \ldots, p_C]$ for the classification of C classes. We uses entropy as listed in (2) to calculate the information embedded in each sliding window, because entropy is a measure of the state unpredictability.

$$E = -\sum_{c=1}^{|C|} p_c \log p_c, \tag{2}$$

where p_c is the probability for class c the MDDNN model predicted, and $|C|$ is the number of total classes.

The information entropy E is used to determine whether the candidate sliding window is discriminative. If the prediction probabilities tend to follow a uniform distribution, subsequently the information entropy is large, indicating that this candidate fails to provide discriminative information for the model to make a prediction. Consequently, we selected the candidates with the least values of information entropy in this step.

We select the top K candidates with the lowest information entropy as the selected discriminative parts, where K is a parameter representing the number of focused parts to be considered in the proposed method. The next step is to use these K entropy to perform a weighted majority vote to decide the final prediction. The weights w_k are defined in (3).

$$D_k = E_{max} - E_k \tag{3}$$
$$w_k = \frac{D_k}{\sum_{i=1}^{K} D_i}, k = 1, 2, \ldots, K$$

where E_{max} is the maximum entropy when the probabilistic result is a uniform distribution, and D_k could be viewed as the discriminative capability of the kth selected part because the lower the entropy, the more important it is, and w_k is the weight coefficient for the focused part. Once the weights are available, the prediction for this iteration is determined by majority vote as listed in (4).

$$y^* = \max_k \sum_{i=1}^{K} w_k I(y^i = y^k), k = 1, 2, \ldots, K$$
$$I = \begin{cases} 1, & y^i = y^k \\ 0, & y^i \neq y^k \end{cases} \tag{4}$$

where y^i, y^k are the MDDNN prediction results of the focused parts and y^* is the final prediction which is calculated by obtaining the maximum of the sum of weights where $y^i = y^k$.

Finally, the top K segments will be selected as the interpretation of the prediction results, because the final results are decided by the majority voting using them. We used such weighted ensemble approach to ensure that the final prediction would not be decided by only the number of votes; instead, their importance would also be considered. Using these segments, which are subsequences of the original input time series, as important features is intuitive and reasonable. This is because they are directly obtained from the data without any form of transformation into another feature space, rendering it easy for end users to comprehend and apply [19].

The time complexity of the proposed method is $O(NL)$, where N is the number of time series instances and L is the maximum length along the time series inputs, since the number of segments generated by the sliding window is L minus the sliding window size. The space complexity is $O(BsL)$, where B is the batch size, s is the sliding window size, and L is the maximum time series length.

Algorithm 1. Training Process

1: **procedure** TRAIN($input, swl, criteria$)
2:    $f_i \leftarrow SlidingWindow(input, swl)$ $\triangleright$ f_i from (1)
3:    **while** $EarlyStop(criteria) \neq true$ **do**
4:        $Prob \leftarrow Model(f_i)$
5:        $E, L \leftarrow EntropyAndLoss(Prob)$
6:        obtain E_K by selecting the top K from E
7:        $W_K \leftarrow CalculateWeight(E_K)$
8:        $L_{total} \leftarrow \sum LW_K$
9:        $AdamOptimizer(L_{total})$ to update model
10:    **end while**
11: **end procedure**

3.4   Training Process

As mentioned above, the final prediction of the proposed method is based on the K sliding windows, each of which uses cross-entropy loss as listed in (5) to measure the prediction performance.

$$l(f_k, y) = -\log p_y, \tag{5}$$

where f_k is the selected k^{th} segment and y is the ground truth label for the input multivariate time series.

The loss L for the model is the summation of all K parts, and the definition of L is presented in (6).

$$L(MT, y, w) = \sum_{k=1}^{K} l(f_k, y)w_k, \tag{6}$$

where MT denotes the input multivariate time series, and w_k are the weights from (3).

We minimize L to fine tune the model parameters in the MDDNN model. It is noteworthy that this is an iterative process. For each iteration, we input all input segments into the model and select the top K focused parts for this iteration, which can be used to calculate the weights. Subsequently, the total loss L for this iteration can be obtained, and the network can use the backpropagation algorithm to update the model parameters. The training process will continue until convergence is reached. This work uses a number of epochs or the early stopping criteria to determine the convergence of the proposed methods.

The key idea behind the proposed method is to separate the input time-series data into sliding windows, each of which can make a correct prediction. The final result is determined by the majority vote of these sliding windows, indicating that the sliding windows with large weights are discriminative ones. Consequently, the proposed method could not only help a DNN model, which is MDDNN in this work, improve the prediction performance, but also identify the essential segments that can contribute to the model prediction.

4   Experiments

4.1   Datasets

We used three datasets to evaluate our proposed method, including ECG, Wafer, and AusLan10. The ECG and the Wafer datasets are available on Bobski's website[1], and the AusLan dataset could be downloaded from [9]. These datasets are typically used in multivariate time series early prediction research, and are thus used in the experiments. The detailed information about these datasets could refer to the websites of data providers owing to the limit of page length.

4.2   Experimental Settings

For the input time-series data, they were zero-padded at the end to be of the same length. For the MDDNN model, we set the number of filters to 64 and 32 for the first convolutional layer and the second convolutional layer, respectively, on both domains. The filter size was set to 10% of the maximum input length. We also added batch normalization layers and a dropout layer with 0.5 dropout rate. For the attention architecture, we used Adam optimizer to minimize the total loss in (6), and we set $K = 5$ to obtain five focused parts for each instance. Additionally, we set the size of sliding window to be 20% of the sequence length. For the early stopping mechanism, we set two parameters: *patience* and *criteria* for each dataset. Once the change in total loss L from (6) does not bypass *criteria* for over *patience* epochs, the training process stops.

4.3   Comparison Methods and Evaluation Metrics

For the methods we choose for comparison, we selected the classical methods in early classification on time series problem, including 1 Nearest Neighbor (1NN-Full) [3], Multivariate Shapelet Detection (MSD) [4], Reliable Early Classification (REACT) [10] and Multi-Domain Deep Neural Network (MDDNN) [7]. These are the state-of-the-art multivariate time series early predicting methods.

The metrics in the experiments are the F1-score and earliness. Here, we used the non-weighted average of F1-score among all classes. We chose the F1-score

---

[1] Bobski's World: http://www.cs.cmu.edu/~bobski/.

as our metric because accuracy may be inappropriate to evaluate the datasets we used, as some of them are imbalanced. Regarding earliness, it was used to assess our model's ability to make correct predictions as early as possible while maintaining a decent F1-score.

Table 1. Earliness and F-score results of each method.

Dataset	ECG		Wafer		AusLan	
	Earliness	F-score	Earliness	F-score	Earliness	F-score
MSD	0.08	0.59	–	–	0.06	0.62
REACT	**0.06**	0.77	**0.23**	0.92	0.08	0.86
INN-full	1.0	0.79	1.0	0.87	1.0	0.97
MDDNN	**0.06**	0.81	**0.23**	0.91	**0.05**	**0.99**
ETSCM	0.13	**0.89**	0.5	**0.93**	0.16	**0.99**

4.4   Experimental Results

We compared our proposed model with four other methods in terms of the F1-score and earliness, and the results are listed in Table 1, in which the hyphens indicate that the model failed to make predictions on the given dataset.

The experimental results indicate that the proposed ETSCM can produce comparable or even better results than other alternatives in terms of F-score. However, our method does not outperform all other methods when considering earliness. Regarding earliness, REACT achieved the best earliness, but it could not provide interpretable results for the end users. In contrast, the proposed method can balance the F1-score and earliness, while providing interpretable results to the users. The shapelet-based method MSD can provide interpretable results as well, but its performance is poor. It is well known that the trade-off between prediction performance and earliness is an important consideration in designing an early classification algorithm. In this experiment, we conjecture that it is difficult to consider interpretability and earliness simultaneously in the proposed method, because the most important segments may be missing if the model emphasizes on earliness.

4.5   Interpretation

The interpretations captured by the proposed ETSCM are shown in Fig. 3. We used the top one interpretation among the top K ones for convenience. Further, we assessed the interpretation of the full-length prediction (i.e., $earliness = 1.0$), as we believe that using a full-length time-series input would provide the model with the maximum amount of information to make accurate interpretations.

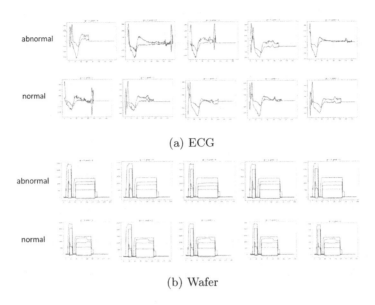

(a) ECG

(b) Wafer

Fig. 3. Interpretation (Color figure online)

Further, we only showed the correctly predicted ones. We visualized the multivariate time series by drawing all the variables on one image such that we have one image for one instance. We drew the original data in blue lines, and the focused parts highlighted by our model are drawn in red lines.

4.6 ECG Interpretation from Doctor's Perspective

Since the ECG dataset is related to the medical field, we invited two doctors to help us understand the interpretations produced by our proposed method. However, from their perspective, the highlighted parts do not seem to have medical meanings because they're too short to include a whole interval or complex. They also mentioned that it's hard for them to look at the raw data and tell whether the ECG time series is normal or abnormal, since the timespan of the collected data is also too short to make the waves complete.

We use another ECG dataset provided by Taipei Veterans General Hospital (VGHTPE) to examine the robustness of our proposed method. This is a 12-lead ECG dataset with 2 classes, namely atrial fibrillation (abnormal) and sinus rhythm (normal), and there are 50 instances for 2 classes each. We can reach 84.21% F-score on full length data, and near half of the interpretations of the abnormal class are correct according to the doctor's judgement. Figure 4 shows two of the the correctly predicted and correctly interpreted figures, the doctor said that the main trait of an atrial fibrillation ECG is that the P-waves are tattered and do not form a complete P-wave. The red highlighted parts in Fig. 4 are the place where a P-wave should occur, however they turned out to be incomplete shapes. Hence we believe our method can capture correct patterns

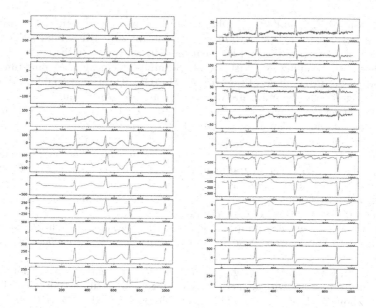

Fig. 4. Correctly captured fragments on atrial fibrillation data (Color figure online)

that lead to atrial fibrillation, and with further improvements, our method can correctly interpret more atrial fibrillation time series.

5   Conclusion

This work proposes a framework to provide accurate and interpretable results on multivariate time-series data. Central to the proposed method is using the attention mechanism to identify the essential segments, while retaining model performance. The future work involves extending the proposed work to other application domains.

Acknowledgment. This research was partially supported by Ministry of Science and Technology, Taiwan, under grant no. 107-2218-E-009-005 and 107-2218-E-009-050.

References

1. Chen, Y., Zhao, D., Lv, L., Li, C.: A visual attention based convolutional neural network for image classification. In: 2016 12th World Congress on Intelligent Control and Automation (WCICA), pp. 764–769. IEEE (2016)
2. Cui, Z., Chen, W., Chen, Y.: Multi-scale convolutional neural networks for time series classification. arXiv preprint arXiv:1603.06995 (2016)
3. Ding, H., Trajcevski, G., Scheuermann, P., Wang, X., Keogh, E.: Querying and mining of time series data: experimental comparison of representations and distance measures. Proc. VLDB Endow. **1**(2), 1542–1552 (2008)

4. Ghalwash, M.F., Ramljak, D., Obradović, Z.: Early classification of multivariate time series using a hybrid HMM/SVM model. In: 2012 IEEE International Conference on Bioinformatics and Biomedicine (BIBM), pp. 1–6. IEEE (2012)
5. Grabocka, J., Schilling, N., Wistuba, M., Schmidt-Thieme, L.: Learning time-series shapelets. In: Proceedings of the 20th ACM SIGKDD International Conference on Knowledge Discovery and Data Mining, pp. 392–401. ACM (2014)
6. He, Q., Dong, Z., Zhuang, F., Shang, T., Shi, Z.: Fast time series classification based on infrequent shapelets. In: 2012 11th International Conference on Machine Learning and Applications (ICMLA), vol. 1, pp. 215–219. IEEE (2012)
7. Huang, H.S., Liu, C.L., Tseng, V.S.: Multivariate time series early classification using multi-domain deep neural network. In: 2018 IEEE International Conference on Data Science and Advanced Analytics (DSAA) (2018)
8. Hübner, R., Steinhauser, M., Lehle, C.: A dual-stage two-phase model of selective attention. Psychol. Rev. **117**(3), 759 (2010)
9. Kadous, M.W., et al.: Temporal Classification: Extending The Classification Paradigm to Multivariate Time Series. University of New South Wales, Kensington (2002)
10. Lin, Y.-F., Chen, H.-H., Tseng, V.S., Pei, J.: Reliable early classification on multivariate time series with numerical and categorical attributes. In: Cao, T., Lim, E.-P., Zhou, Z.-H., Ho, T.-B., Cheung, D., Motoda, H. (eds.) PAKDD 2015. LNCS (LNAI), vol. 9077, pp. 199–211. Springer, Cham (2015). https://doi.org/10.1007/978-3-319-18038-0_16
11. Lipton, Z.C.: The mythos of model interpretability. arXiv preprint arXiv:1606.03490 (2016)
12. Liu, Q., Yu, F., Wu, S., Wang, L.: Mining significant microblogs for misinformation identification: an attention-based approach. ACM Trans. Intell. Syst. Technol. **9**(5), 50:1–50:20 (2018). https://doi.org/10.1145/3173458
13. Qin, Y., Song, D., Chen, H., Cheng, W., Jiang, G., Cottrell, G.: A dual-stage attention-based recurrent neural network for time series prediction. arXiv preprint arXiv:1704.02971 (2017)
14. Ribeiro, M.T., Singh, S., Guestrin, C.: Why should i trust you? Explaining the predictions of any classifier. In: Proceedings of the 22nd ACM SIGKDD International Conference on Knowledge Discovery and Data Mining, pp. 1135–1144. ACM (2016)
15. Wang, W., Chen, C., Wang, W., Rai, P., Carin, L.: Earliness-aware deep convolutional networks for early time series classification. arXiv preprint arXiv:1611.04578 (2016)
16. Xing, Z., Pei, J., Dong, G., Yu, P.S.: Mining sequence classifiers for early prediction. In: Proceedings of the 2008 SIAM International Conference on Data Mining, pp. 644–655. SIAM (2008)
17. Xing, Z., Pei, J., Philip, S.Y.: Early prediction on time series: a nearest neighbor approach. In: IJCAI, pp. 1297–1302. Morgan Kaufmann (2009)
18. Xing, Z., Pei, J., Philip, S.Y.: Early classification on time series. Knowl. Inf. Syst. **31**(1), 105–127 (2012)
19. Xing, Z., Pei, J., Yu, P.S., Wang, K.: Extracting interpretable features for early classification on time series. In: Proceedings of the 2011 SIAM International Conference on Data Mining, pp. 247–258. SIAM (2011)

20. Ye, L., Keogh, E.: Time series shapelets: a new primitive for data mining. In: Proceedings of the 15th ACM SIGKDD International Conference on Knowledge Discovery and Data Mining, pp. 947–956. ACM (2009)
21. Zheng, Y., Liu, Q., Chen, E., Ge, Y., Zhao, J.L.: Time series classification using multi-channels deep convolutional neural networks. In: Li, F., Li, G., Hwang, S., Yao, B., Zhang, Z. (eds.) WAIM 2014. LNCS, vol. 8485, pp. 298–310. Springer, Cham (2014). https://doi.org/10.1007/978-3-319-08010-9_33

Author Index

Printed in the United States
By Bookmasters